Horsedrawn Tillage Tools

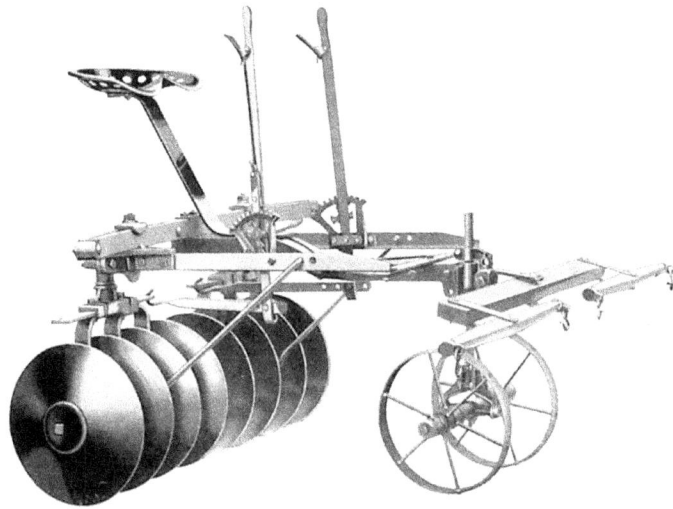

Horsedrawn Tillage Tools

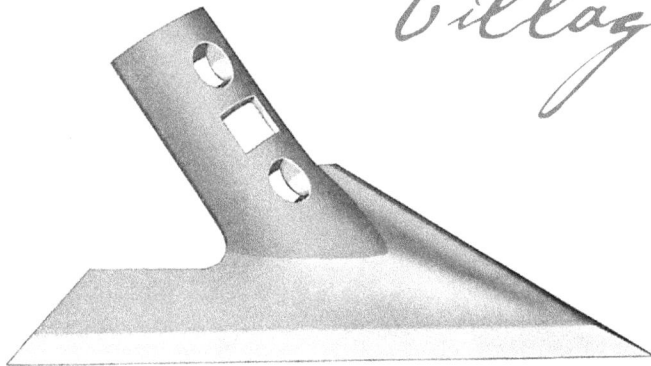

Lynn R. Miller

Horsedrawn Tillage Tools
Copyright © 2020 Lynn R. Miller

Publisher
Davila Art & Books, LLC
PO Box 1627
215 N. Cedar St.
Sisters, Oregon 97759
541-549-2064
www.davilabooks.com
email: info@davilabooks.com

distributed through:
www.davilabooks.com
www.smallfarmersjournal.com
author info:
www.lynnrmiller.com

Printed in the United States of America
First printing first edition 2001
Second printing 2020

ISBN 978-1-885210-34-0 Soft Cover

Front Cover • Top Left: Rock Island No. 35 Bonanza Disc Harrow • Top Right: Moline See Saw Riding Cultivator • Center: Four Abreast Belgians on a Pioneer 8' Cultimulcher • Bottom Left: Moline Four Row Beet Cultivator

Back Cover (Left to Right, Top to Bottom) • Rock Island No. 86 Seat Shift Cultivator • Moline Reversible Disc Harrow • Le Roy No. 5 Adjustable Potato Coverer or Hiller • Balance Frame Dandy Riding Cultivator • Moline Easy Rider Stalk Cutter • Le Roy Land Roller • Dutch Tango Seat Guide Cultivator • Moline Single Row Rotary Harrow • Moline Captain Kidd Disc Cultivator • Strong Boy Walking Cultivator • Percherons pull a Pioneer harrow at Horse Progress Days

Dedication

For any farmer who has ever embraced craftsmanship,
or grumbled over urgency, or found a private moment
to shed a tear of gratitude for the life.

"But let the stirring be carefully
repeated, and all is life again."

Edmund Morris
Ten Acres Enough
1864

other books by Lynn R. Miller

nonfiction
  Work Horse Handbook
  Training Workhorses / Training Teamsters
  The Horsedrawn Mower Book
  Horsedrawn Plows & plowing
  Haying with Horses
  Art of Working Horses

novels
  The Glass Horse
  Brown Dwarf
  Talking Man

sundry prose
  elastic signature: notes on painting
  Old Man Farming: Essays of a Rewarded Life
  Why Farm
  Farmer Pirates & Dancing Cows
  thought small: poems & drawings

out of print
  Setting Up Your Farm
  Horses at Work
  Small Farm Bookkeeping
  Farm Animal Coloring Book
  The Small Farm Dream is Possible (in concert with Ten Acres Enough)

Horsedrawn Tillage Tools

It Cultivates — It Packs

Table of Contents

PREFACE

T his book is primarily a compilation of materials from the exhaustive archival library of Small Farmer's Journal, from SFJ friends and from a handful of present-day manufacturers of new implements.

I write these words, in the year 2000, as an opening statement for this modest book effort and find myself reflecting on today's date in history and what it suggests about the importance of this work. By 'this work' I refer not just to the book you hold but to the larger effort of which it is a part, the *Work Horse Library*. That title is not

Mike Jensen of Oregon cultivates sweet corn with two beautiful grey Percheron mules and a straddle row riding cultivator.

so important, it may not even hold up. But I do believe that the collected works it refers to are important and most especially today. To this date the 'library' includes:

Work Horse Handbook
Training Workhorses / Training Teamsters
Art of Working Horses
Horsedrawn Plows & plowing
Haying with Horses
Horsedrawn Tillage Tools
The Horsedrawn Mower Book

Within each of these last titles, an effort has been made to reach three goals:

A walking straddle row cultivator and two mules from 1915.

A.) The collection of a definitive body of archival information on the targeted identified group of implements from the near past and present (late 19th century forward) in a form that farmers would appreciate. To this end we are less interested in history and historical anecdote than we are in the understanding of practical applications. We have a passion for the subject which we trust comes through in spite of the demanding reality to consolidate. We want this book, these books, to show and tell in large and useful ways.

B.) An overview of modern innovations in the horsedrawn implement world, from adaptations and updates of older implements through convergence systems permitting use of animal power for tractor implements and concluding with complete new implement designs.

C.) A review of the possible uses of these implements, then and now, with a holistic view towards alternative farming systems and procedures. "Would I use them? When? Why? and How?"

This has been a challenging project for many reasons. The one reason with perhaps the most interest to the reader encompasses scope. There are such a large number of successful implement designs, literally thousands, from dozens of manufacturers for over 125 years. For this volume alone, had we been able to offer illustrations of every disc, harrow, cultivator, roller, etc. (with all the cutaways and engineers diagrams) it would easily have run to in excess of 1000 pages and proved unwieldly and less useful.

Because, after all, 'useful' is the name of the game. Just as with the preservation of endangered life forms, endangered technologies (and the methods they are predicated upon) need to be kept alive to be saved, to be preserved. To "keep alive" a so-called 'relic' technology you must encourage its actual use, its application. This

Charlie Jensen (friend but no relation to Mike Jensen) with 4 abreast on two row cultivator.

Bulldog Fraser of Montana with a four abreast of trainees hitched to forecart and springtooth harrow.

book, these books, are meant to do this. To help keep a way of working, a culture of working, from sliding into oblivion.

Yes, what I'm saying with these books is that this is a way of working which must be preserved, not just for historical significance but, I repeatedly argue, for its practical living value as a viable option for our farming's future.

Those of you who believe that farming is and must be an industrial process will find ample cause to scoff at my contention. But our position comes from a deep conviction that farming is best performed as a craft within a vast community of small independent outfits. Within this arena a regard for appropriate technologies rules the day. And within that regard sensible animal-powered systems are a viable option.

In order for animal power to be a viable option the peculiarities and intricacies of the 'craft' must be information in a form that is readily available. Books such as this one are important but in this form the information they contain is less accessible than what direct experience and one-on-one tutelage provide. A craft is best learned by immersion, by apprenticeship, by inheritance. The books should be easy-to-use repositories of the calibrations, levered relationships, and the wishful thinking of engineers, all coupled with ideas in flux. In this way they may become extensions and tools for the hands-on learning. We hope our books work in this way.

Times have certainly changed. I speak of the last forty years as much as of the past 150 years.

When I first became interested in animal power I was an eighteen year old in 1965. At that time it was generally considered that animal power existed solely as obscure antiquity, something from the far distant past, a historical bridge to the dark ages. Now, thirty six years later, I realize that a cultural myopia coupled with blind acceptance of industrial propaganda was, to a large extent, the reason that it *felt* like it was a thing

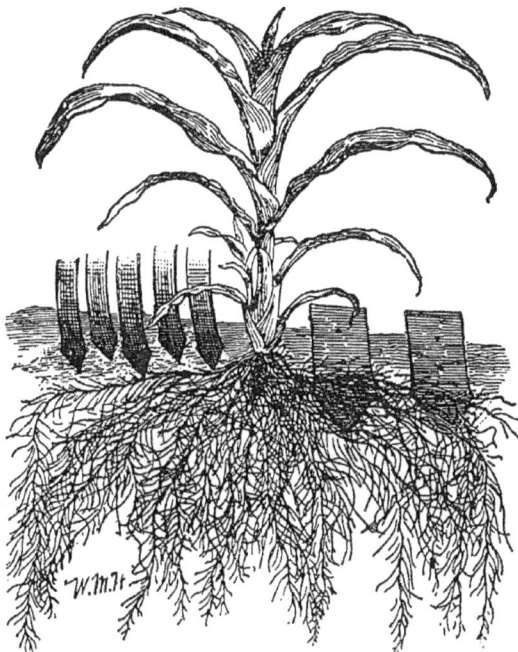

Demonstrating how shallow cultivation of corn plants will save the roots.

*The Ohio 1998 Horse Progress Days witnessed this hitch of six Percherons,
4 and 2, drawing six sections of Pioneer-built spike tooth harrow.*

from so very long ago. Is thirty years a long time? For it was just thirty years before my first thirst took hold that there were still over a million North American farms choosing to rely on animal power. 1936. World War II coming, the depression was still an issue for most rural folks. During those difficult times the inherited thrift of their farming was made eminently possible because of the understood availability of animal power. With WWII, and the industrial age required to 'tool up' for the war effort, our economy grew. Even before the war ended politicos and industrialists knew that something had to happen to avoid the doldrums that would follow the all-out produc-tivity during the war. Farming was targeted and massive sales campaigns ensued to get tractors and farm chemistry into the hands of every farmer. In ten short years millions of draft horses and mules were slaughtered by tractor dealerships which took the animals in trade for shiny new machines. During the first half of the 20th century, engineers and industrialists supplanted finesse and balance and leverage and natural syncopation for the brute force that seemingly unlimited fossil fuels and internal combustion provided. Now, as we begin the 21st century tangled in what seems like gravity-drawn rolling balls of genetic manipulation, electronic intelligence and random information access, the engineers and corporatists of the west are supplanting brute industrial force for the intoxi-cant of a vision of a new aristocracy (or meritocracy?) where whosoever controls information, (and the sciences which pretend to manipulate the natural world) rule supreme. This without regard for the powerful urges and needs of the individual and of the small communities.

Those urges and needs push us to find ways of working and living which make us feel alive and connected to the natural world. One path many people feel drawn to is craft-based, small-scale agriculture. And within that world the notion of animal power is a strong attractant. It touches on our collective working memories in undeni-ably interesting and useful ways. That draw continues to feed a growth in the number of people who choose to use horses and mules in harness as motive power.

I was recently interviewed by a journalist friend who asked about the size of this community of animal-power practitioners. I told him that in North America there are in excess of 400,000 people dependent by choice on horses, mules (and to a lesser degree oxen) for power. He asked where that number came from. Was there a single authoritative source? To which I had to answer no. The number is an extrapolation from a wide group of sources which include the sales figures of cottage industries who supply this community, event participation and attendance figures, work animal sales records, and the circulation base of publications. We know there are in excess of 170,000 Amish and Mennonite horsefarmers. We know that spread all across the continent there are tens of thousands of far flung individual horsefarmers, horseloggers and carriage service operators. And that there are concentrated pockets of prevalent specific use (i.e. oxen in New England, mules in the south, Belgians in Iowa, etc.). We know from

Montana veterinarian Dr. Doug Hammill discing with two of his Clydesdales in the 1970s.

association records that there are over 30,000 non-Amish outfits raising and breeding purebred and cross-bred workstock from draft horses to mules, from carriage horses to draft ponies. We can add up the available partial attendance figures from 1998 draft horse and draft horse equipment auctions in Pennsylvania (20k), New York (3k), Ohio (20k), Indiana (10k), Iowa (20k), Wisconsin (3k), Colorado (2k), California (5k), Idaho (4k), and Oregon (8k) and come up with 95,000 people, and that does not include any Canadian sales. We know from our own auction management experiences, measuring mailing lists against actual attendance, that approx. 15% of the work horse community in our region attends any given sales event. And we know that we have the names and addresses of sixty thousand people who purchase animal-power related products from our little business, *Small Farmer's Journal*. We look at all this evidence forward and backward and come away convinced that our community is substantial and growing. We want to see it grow safely and sensibly and with a foundation of inheritable technologies, methods and systems. To that end we offer this book and the others before it and after it.

Should you, the reader, come to these pages without having seen the volumes which come before you might be disappointed to find that there is very little here about the actual workings of an equine in harness (the harnessing and hitching). That information is contained in the ***Work Horse Handbook***, and ***Training Workhorses / Training Teamsters.***

This text is heavy on illustration and light on text. The books on plowing and haying contain far more procedural information because the complexities of setting up and operating mowers and plows require this. Tillage tools are simpler in their application with the complexities being primarily simple procedural choices (i.e., when to disc?, whether to harrow after rolling? etc.) and attachment choices (i.e., which shovel or point). A large part of the text in this book comes, with slight to heavy editing, from a variety of books listed in the bibliography.

To some, this book will seem inappropriately titled. As much because of what is missing as what it contains. We decided to create an umbrella for at least two categories of implements; seedbed preparation tools and weed cultivation tools. "Tillage" oft times includes plows and plowing. Not as any argument of fact, but for the simple sake of organizing the collected information of this book series, this book on Horsedrawn Tillage Tools does not include plows. They were covered in ***Horsedrawn Plows & plowing***. The book in hand includes those other implements and procedures which cut, stir, prepare, crumble, compact, loosen, mulch and weed the soil. Work which might be done ahead of the plowing, certainly after plowing, and often after crops are growing. The preparation of land to plow, the preparation of a seedbed, the maintenance of a soil mulch and the cultivating or mechanical removal of weeds are the procedures this book tries to cover.

Reaching back to the late 1900s, and looking across North America, there were and are hundreds and even thousands of variables in implement design. Most of the variables are insignificant, affecting shape more than mechanical and procedural engineering. A seat or seat stem or bracket might be different from one disc or cultivator to another but blades and shovels attach and function much the same. With field implements such as spike tooth, spring tooth and drag harrows there is very little difference from one manufacturer to another and indeed very little change over these last hundred years. The same is true of disc harrows and rollers. There is far more variety found in row implements as will be shown. In this arena the simple innovations of the last twenty years have been remarkable.

I want to think that there are good simple implement ideas from the not so distant past which may see a rebirth; ideas such as the mumbler, the rotary harrow, and the acme harrows.

The culture of our agriculture was artificially pushed forward, stripped of craft, and converted to a positract techno-culture without benefit of parentage. Instead of gradual evolution we have been and continue to be pushed away from heritage and tradition and into the lap of goofy high-priced new systems and gadgets. By reaching back just a little and trusting our instincts we should be able to get on track with an honorable and fulfilling way of working worthy of passing on to our kids.

Aknowledgements:
Thanks go to lots people for helping with this book. Kathy Blann, Amy Evers,
Suzanna Clarke, Lisa Booher, Lynn Woodward and Brook Wills, worked hard on
shuffling, typing and proofing. Kristi Gilman-Miller worked hard with getting images
ready. Stuff was borrowed and stolen from Eric and Anne Nordell, John Nordell, John and
Heather Erskine, Doug Hammill, Bulldog Fraser, and about 20,000 SFJ subscribers.

Twelve teams and cultivators were at work in this field in 1912.

Good, hard, rewarding work with a straddle row cultivator, 2 nine hundred pound mules, fine soil, and a good young crop. Could be spring of 1900 or 2000.

Chapter One

Horsedrawn Tillage

Tillage

The simple view might be; we plow up the ground, then we break that plowed ground up into a fine seedbed for planting, perhaps later we stir the ground to kill or set back weeds, period.

To many our farming seems to have progressed, out of necessity - out of some suffocating synthetic logic, to the highly visible western industrial model. The modern industrial view says that tillage must now be performed by large tractor-drawn implements and that animal power does not represent adequate power and speed and scale to get such heavy-duty work properly done. This is, of course, not true. It replaces the art and craft of farming with illegitimate industrio-corporate necessity and the dictums of false (say paper) economies. In an art and craft-based farming the farmer's proximity to the work is essential. He or she can see it, feel it, touch it, taste it. She or he learns of its resistance to the implements, its feel under working foot, its moisture content, in part from the partners who pull the tools, and by necessity because they must design procedures which are sensitive to the soil and plant life conditions of that very next moment.

Horses do a better job. Well treated horses and mules enjoy the work, even miss it. And the soil where horses are used has less subsoil compaction.

A Canadian-built 4 row cultivator with 4 belgians demonstrating at Ohio Horse Progress Days.

Definitions

tillage: art of tilling land, the improving of land for agricultural purposes, 'in tillage' under cultivation.

tillen: to strive for, obtain, work, cultivate.

till: to turn or stir as by plowing, harrowing (or hoeing), and prepare for seed; sow, dress and raise crops from; cultivate.

tilth: the state of being tilled, the state of aggregation of a soil.

cultivate: to prepare for the raising of crops to loosen or break up the soil about (growing crops or plants) for the purpose of killing weeds and modifying moisture retention of the soil esp. with cultivator. - intertillage to destroy weeds and loosen soil (field brought under).

horsedrawn: refers to the use of horses or mules as motive power to draw or pull the tilling and cultivating implements across and through the soil. As versus tractor-drawn. May also say horse tillage as versus tractor tillage, may also say human-scale as versus industrial-scale, may also say craft as versus industry.

It's a question of scale and diversity. Imagine that a conductor of an orchestra were limited to only trombones, forty thousand of them, and they stretched out in front for a mile, so that the conductor had to sit perched fifty feet in the air in order to see most of them, the music might have an undeniable power but lack grace, it might be a presence that was impossible to ignore but would not spawn the fertile intricacies which become the elemental juices of tonal poetry and of self perpetuating creativity. If the farming is enormous and monocultural it will also lack grace and fertility and true productivity.

An absorbent and engaged human, travelling over tens of acres of soil with horses or mules and suitable tillage implements, fits the picture of farming as craft. The human in drudgery, or complacency-induced absentia, travelling over hundreds of acres with behemoth tractors and their implements fits in the picture of agribusiness as industrial process.

Many farmers and academics think of tillage as any stirring and or turning implement-driven action of the top soil definitely including the work of the moldboard plow. All these positions stem from a view of agriculture as industrial process with the top layer of sterile dirt as a three dimensional bed or table playing host to the growth of crops much like a test tube might play host to the growth of an isolated(?) fungal culture in a scientist's laboratory. This is of course destructive and limiting.

Such industrial models deny the magnificent mysterious microscopic interrelationships at play in the soil and in the atmosphere. And that these relationships beg of us to approach them with the posture and attitude of a humble craftsman rather than a lab technician or an industrial manager. They beg this of us because such a posture on our part will and does add to the productive mysteries, to the fertility, to the very productivity, to the sustainability of the soil and life on this planet.

The use of animal tractive power for tillage brings the farmer/craftsman closest to the work with the soil. To see, hear and

16

A closeup of a modern I & J Mfg. horse drawn cultivator at work.

smell the soil turn, move, stir, vibrate, crumble, and rise is to know that piece of dirt. Using horses, mules or even oxen to do the field work allows a speed, a quiet, a proximity, which in turn allows the attentive farmer a muscianship with his or her farming. Marrying implement, animals, timing, procedure and tillage goals affords the farmer a piece of the interdependence and a glorious sense of right livelihood.

We like to think of tillage as a form of grooming for health. We massage, stir, comb, interrupt, brush, shape and placate our top soil with a goal of greater fertility, greater productivity, greater juiciness, greater biological life.

As a student of farmer inquiries into tillage - from across more than one hundred years of modern history - it was sometimes easy to feel frustrated with how right-sounding observations and conclusions seem to regularly collide in contradiction. Good farmers are frequently disagreeing about what constitutes the "right" way to till the soil - what implement to use and when. And good farmers have solid comparable results with those conflicting systems and approaches. I have come to believe, after years of my own experiences in the field and as an editor, that these contradictions are healthy and may even be helpful for the open-minded, curious and dedicated farmer craftsman. So to honor that diversity

Above and right, a single horse rod weeder. The farmer in both photos on this page is in close proximity to his work and able to judge, measure and value the crop and the soil.

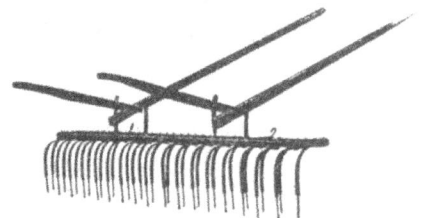

of opinon we follow this with many differing views about timing and what should follow what. You will ultimately make the right choices for youself. When you see "rules" and /or "absolutes" here take them as your own if they feel right to you.

Tillage vs. Cultivation?

"In a very important sense, tillage is manure."

For the sake of this discussion we make a fine but important distinction between what is loosely referred to as primary (open) and secondary (inter-) tillage. (First and second levels of tillage not as a rating of importance but rather as a common progression - from plowing to harrowing to weed cultivation).

"Divide tillage into two general kinds,- tillage which covers the entire ground, and tillage which covers only that part of the ground which lies between the plants. The former we may call open or general tillage, and the latter inter-tillage or cultivation. We practice open tillage before the seed is sown: it therefore prepares the land for the crop. We practice inter-tillage in fruit plantations and between the rows of crops: it therefore maintains the condition of the soil.

In a general way, tillage is deep when it extends more than six inches into the ground. We also speak of surface tillage, when the stirring is confined to the one, two or three uppermost inches of the soil."*

We are concerned here with seedbed preparation and cropbed maintenance.

Plowing does not make a seedbed. Seldom can a seed bed be properly prepared through plowing only. The notion that the only way to arrive at a seedbed is through first inverting or flipping topsoil over (or on edge) is limiting at least and possibly, in some specific locales and conditions, completely

*The Principles of Agriculture 1912 L.H. Bailey
** Refers to Vol. 17 Issue 2 SFJ

counter-productive. That is not to say that in all instances moldboard plowing is a bad or negative action. The moldboard plow is a tool, a most ingenious and valuable tool, which may be employed regularly to outstanding advantage by good farmers whose primary concern is for soil health. And it may be employed in different ways and at different times; from Nordellian** skim plowing to the harvest of root crops - from the incorporation of green manure to the disruptive front porch action for the fallowing of lands - and more.

What to do to complete the preparation of the seed bed depends on how clearly the farmer understands what constitutes a good seed bed, or good tilth. And what value the farmer places on sustainable or increasing levels of fertility. And also what tradeoffs the

Six Percheron horses, the equivalent of 90 hp (in tractor lingo) and one manifestation of the quintessential beauty of farming as craft.

farmer agrees to when opting for the long view of fertility and soil health over immediate maximum crop production levels.

As Liberty Hyde Bailey noted in 1912:

"Tillage improves the physical condition of the soil: by fining the soil and extending the feeding area for roots; by increasing the depth of the soil, or loosening it, so that plants obtain a deeper root-hold; by causing the soil to dry out and warm up in spring; by making the conditions of moisture and temperature more uniform throughout the growing season.

It aids in the saving of moisture: by increasing the water-holding capacity of the soil, or deepening the reservoir; by checking the evaporation (or conserving, or saving moisture) by means of the surface-mulch. The former is the result of deep tillage, as deep plowing, and the latter of surface tillage.

It hastens and augments chemical action in the soil: by

aiding to set free plant-food; by promoting nitrification; by admitting air to the soil; by lessening extremes of temperature; by hastening the decomposition of organic matter, as of green-crops or stable manures which are plowed under; by extending all these benefits to greater depths in the soil. In a very important sense, tillage is manure.

Tillage by means of surface-working tools--as hoes, rakes, cultivators, harrows, clod-crushers--has the following objects: (a) to make a bed in which seeds can be sown or plants set, (b) to cover the seeds, (c) to pulverize the ground, (d) to establish and maintain an earth-mulch, (e) to destroy weeds. Aside from these specific benefits, surface

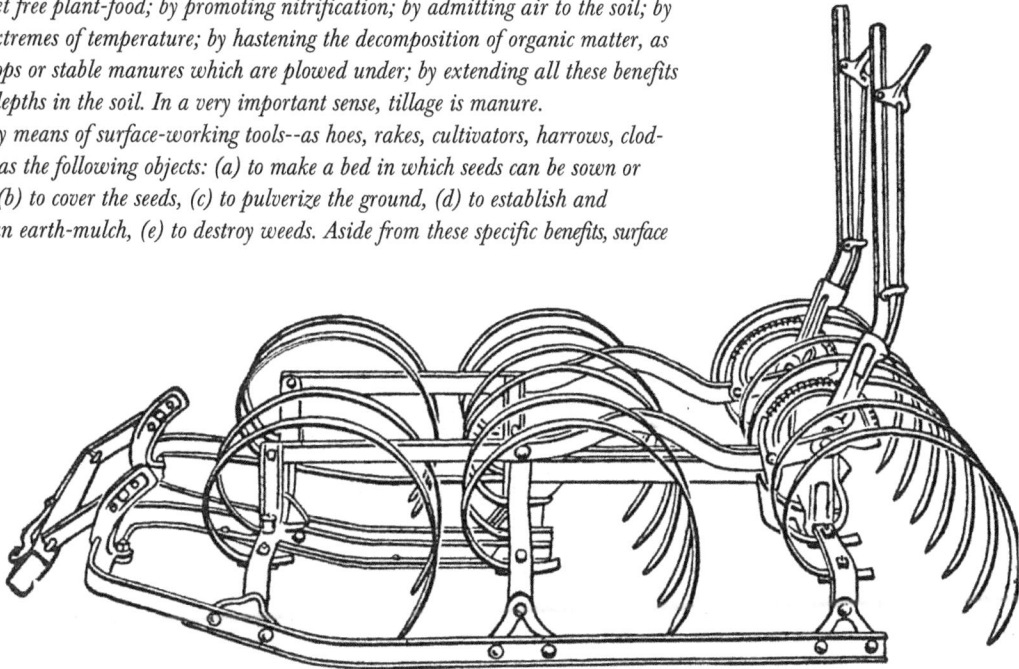

Spring tooth harrow.

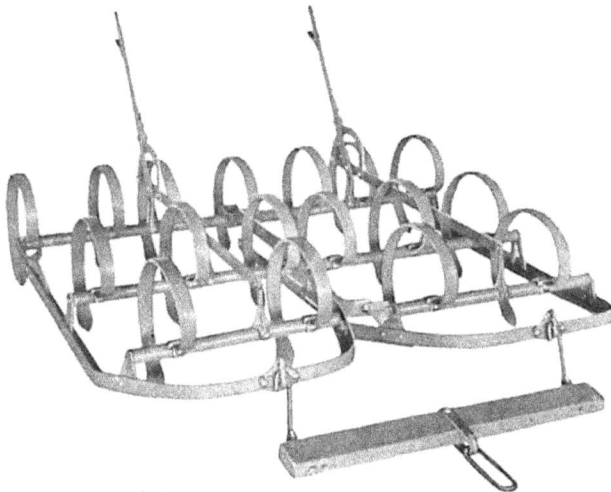

| Regular Tooth for Spring-Tooth Harrow | Tooth for Destroying Quack Grass | Special Spring-Tooth for Cultivating Alfalfa | Spring-Tooth with Separate Reversible Points |

tillage contributes to the general betterment of soil conditions.

In making the earth-mulch the other objects of surface tillage are also secured; therefore we may confine our attention to the earth-mulch for the present. The mulch is made by shallow tillage--about three inches deep, in field conditions--before the seeds are sown. The first tillage after plowing is usually with a heavy and coarse tool,--as a clod-crusher, cutaway harrow, or spring-tooth harrow,--and its object is pulverization of the ground. The finishing is done with a small-toothed and lighter harrow; and this finishing provides the seed-bed and the soil-mulch.

The earth-mulch is destroyed by rains: the ground becomes baked. But even in dry times it becomes compact, and capillarity is restored between the under-soil and the air. Therefore, the mulch must be maintained or repaired. That is, the harrow or cultivator must be used as often as the ground becomes hard, particularly after every rain. In dry times, this surface tillage should usually be repeated every ten days,--oftener or less often as the judgment of the farmer may dictate. The drier the time and the country, the greater the necessity for maintaining the soil-mulch; but the mulch is of comparatively little effect in a dry time if the soil moisture was allowed to evaporate earlier in the season.

Surface tillage is usually looked upon only as a means of killing weeds, but we now see that we should till for tillage's sake,--to make the land more productive. If tillage is frequent and thorough--if the soil-mulch is maintained--weeds cannot obtain a start; and this is the ideal and profitable condition, to which, however, there may be exceptions.

The compacting tools are rollers, and the implements known as plankers or floats. The objects of rolling are: (a) to

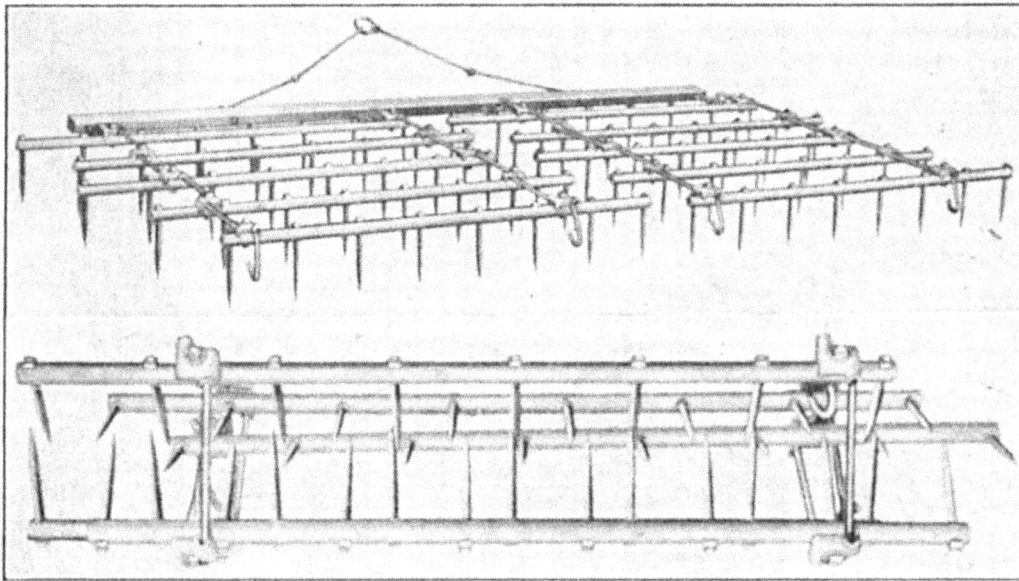

Flexible spike-tooth harrow, flat and field ready or rolled for storage.

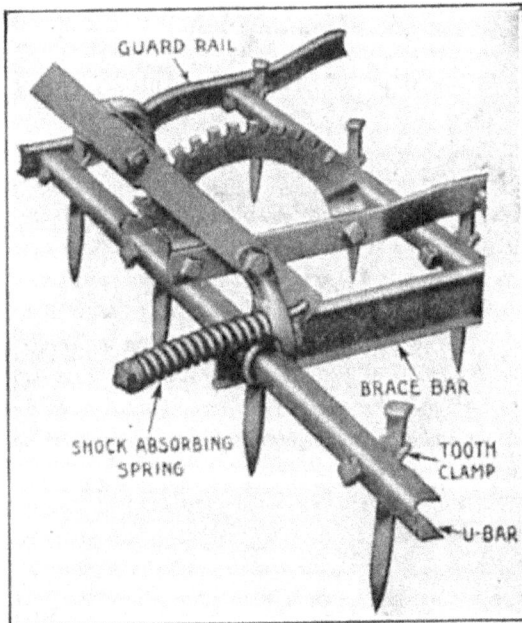

Part of a section of one brand of spike tooth harrow showing guard rails and shock absorbing spring.

Types of teeth used on spring-tooth harrows: left to right, regular, quack grass, alfalfa, detachable point.

crush clods, (b) to smoothen the ground for the seed-bed, (c) to hasten germination of seeds, (d) to compact and solidify soils which are otherwise too loose and open, (e) to put the land in such condition that other tools can act efficiently, (f) to facilitate the marking-out of land.

By compacting the surface soil, the roller re-establishes the capillary connection between the under-soil and the air: that is, it destroys the earth-mulch. In its passage upwards, the soil moisture supplies the seeds with water; and the particles of the soil are in intimate contact with the seeds, and, therefore, with the soil moisture. If the surface of rolled lands is moister than loose-tilled lands, therefore, it is because the moisture is passing off into the air and is being lost.

The rolling of lands, then, sacrifices soil moisture. The rolled or compacted surface should not be allowed to remain, but the earth-mulch should be quickly restored, to prevent evaporation, particularly in dry weather. When the object of rolling is to hasten germination, however, the surface cannot be tilled at once; but if the seed is in rows or hills, as maize or garden vegetables, tillage should begin as soon as the plants have appeared.

It should be observed that surface tillage saves moisture by preventing evaporation, not, as commonly supposed, by causing the soil to absorb moisture from the atmosphere. When moisture is most needed, is the season in which the air is dryer than the soil.

COUPLING POINT FOR REAR SECTION
PRESSURE LEVER

SCRAPER LEVERS

SEAT SPRING BRACE
AND SPRING GUIDE

GREASE CUPS

ANGLING LEVERS

TONGUE

CLEVIS

SCRAPER
GANG BOLT
HEAT-TREATED
STEEL DISKS

PIVOT GANG
CONNECTION

TONGUE TRUCK
ANGLE STEEL FRAME
PRESSURE SPRING
PIVOTED PRESSURE YOKE

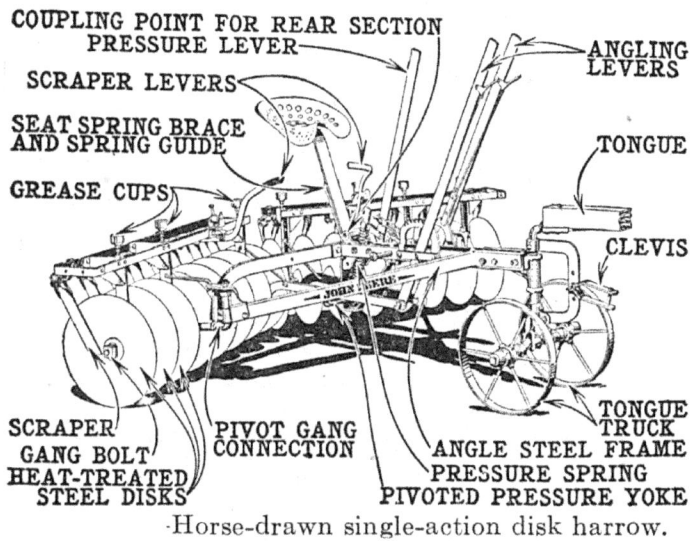

-Horse-drawn single-action disk harrow.

To illustrate the importance of air, select a thrifty plant, other than aquatic plant, growing in a florist's pot, and exclude all the air by keeping the soil saturated with water, or even by keeping the bottom of the plant standing deep in water, and note the checking of growth, and, in time, the decline of the plant. The remarks on draining show how undrained soils are often saturated with water; and no matter how much raw material for plant-food may exist in such a soil, it is unavailable to the plant. The reader can now guess why crops are poor and yellow on flat lands in wet seasons.

By hard-pan is meant very hard and more or less impervious subsoil. Some subsoils are loose; others are so hard as to prevent the downward movement of water and roots.

Observe how moist the soil is in forests, even in dry times. This condition is due partly to the forest shade, but perhaps chiefly to the mulch of leaves on the ground.

Some farmers are always asking how to kill weeds, as if this were the chief end of farming. But good farmers seldom worry about weeds, because that management of the farm which makes land the most productive is also the one which prevents weeds from gaining a foothold. But there are some cases, as in which weeds may be allowed to grow with profit.

To determine when and how much to roll land, is one of the most difficult of agricultural operations. This is because the good effects are so often followed by the ill effects of loss of moisture and of puddling of hard lands when heavy rains follow. Whenever the object of rolling is to compact loose lands or merely to crush the clods, the work should be quickly followed by the harrow or cultivator."

Good Seed Bed Favors Planting.

The advantages of a firm seed bed thus far discussed have been in relation to the germinating seed and the plant. Another advantage in having a firm seed bed is that the depth of planting can be easily controlled. Too often, grain, for example, is planted too deeply because little or no consideration is given to the looseness of the soil. If a drill is set to sow at a depth of one and one-half inches and the wheels sink down three inches in the loose soil, then the seeds are

Young C.J. Shopbell drives an Erskine team of Shires on a disc. Photo by Heather Erskine.

When Disking is Better Than Plowing

When oats are to be sown on potato or corn land in a high state of fertility, disking proves better than plowing. Many farmers have found that they can also grow better barley when the land is disked instead of plowed. This is especially true on black, crummy prairie soils, and crummy silt loams on many dairy farms. It is the experience of many farmers that oats and even barley stand up better when such lands are not plowed.

Double Concavity   Conventional   Inside Bevel

Soil-Throwing Action of Disc-Blade Shapes.

dropped at a depth of four and one-half inches. Grain seeds planted so deep may die for lack of sufficient air, the stems may meet with too much resistance and never get through, or the food stored in the seed may become exhausted before the shoots reach the surface.

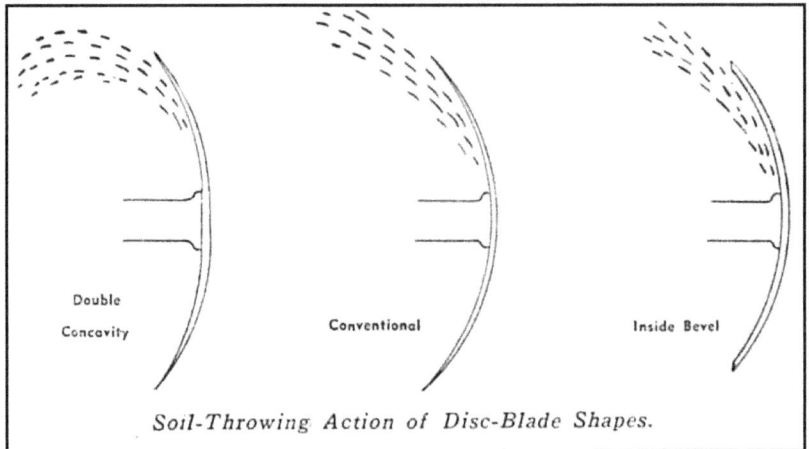

Spading and Smooth Disc Blades

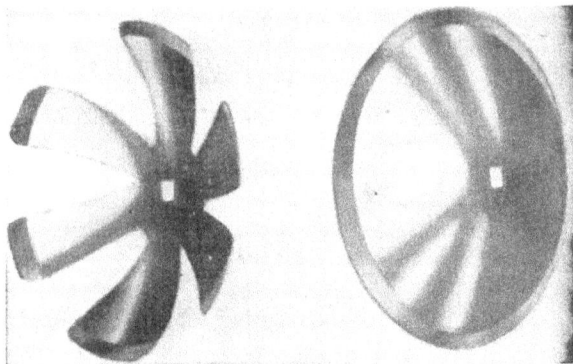

Disc Type Spading Harrow Gang

Knife Blade Spading Harrow

why disc first?

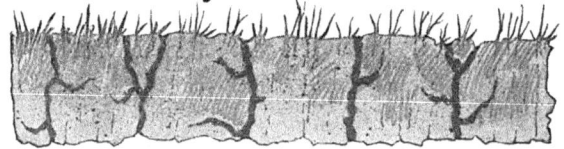

Showing Cracks in the Land Through Which Moisture Escapes

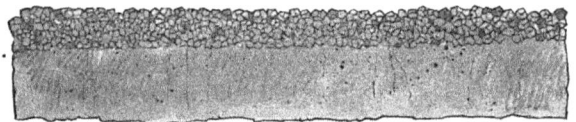

Showing How Effectively a Surface Mulch Prevents the Escape of Moisture

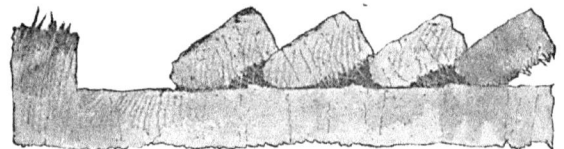

Was Not Disced Before or After Plowing

Disced Before Plowing, Making the Contact Between the Bottom of the Furrow and the Furrow Slice

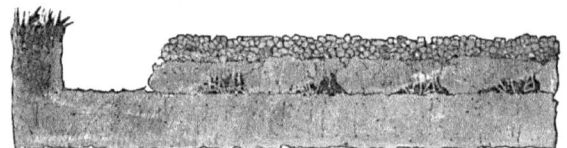

A Poorly-Made Seed Bed—Disced After Plowing, But Not Before—The Large Air Spaces Prevent Capillary Action

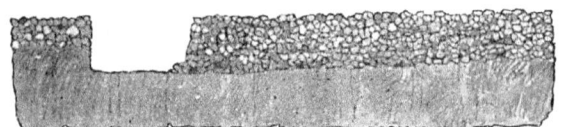

A Perfect Seed-Bed—Disced Before and After Plowing—Seed-Bed to the Left Is Too Shallow. Note—Soil Particles Magnified 1,000 Times

Doug Hammill and his two Clydes on small disc harrow.

Short Notes on the Conventional

As with all aspects of a craft based farming, tillage may provide an almost endless supply of interesting variables to the fully engaged farmer. Variables in implements, combinations of implements, procedures, timing and perhaps most interesting, variables in goals and motives. For the farmer who chooses to employ draft animal power this plate of choices expands rather than contracts.

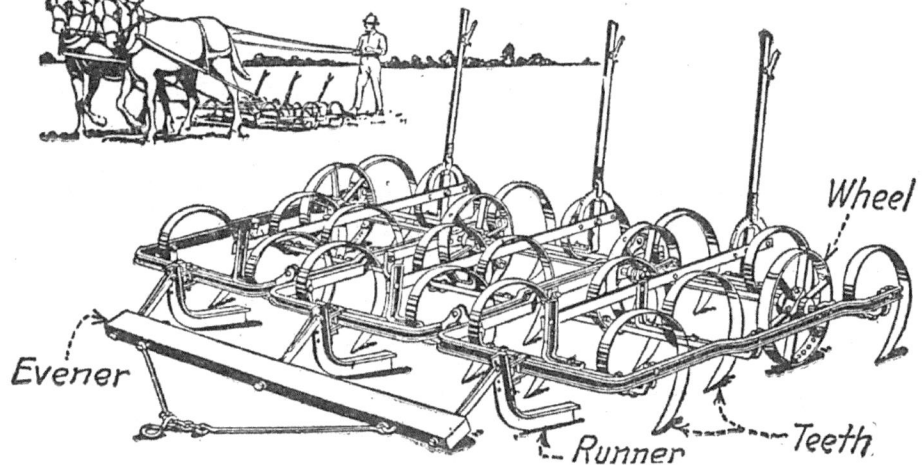

Spring-tooth harrow.

Tilling the Soil: THE NORMS.

(Sometimes the ground is disced before plowing.) After the soil is properly plowed*, and just as soon as the surface begins to look slightly dry, the field should be thoroughly harrowed. With many soil types it is not good practice to delay the harrowing until the whole field is plowed; the harrow should follow the plowing before the clods have become dry. (A notable exception would be sandy soils which might afford the attentive farmer additional time as they are less likely to form hardened clods.) If the soil is harrowed soon after being plowed these clods will be easily broken

Please see Horsedrawn Plows & plowing by L.R. Miller

and pulverized, but if the plowed chunks are allowed to become thoroughly dry, it will be hard or impossible to completely pulverize the soil by any ordinary means. Waiting to harrow means a lot of additional labor is then required to get the field in approximate good condition.

As a rule, in good weather, a harrow should follow not more than one day behind the plow.

The harrowing should be just as deep as it is possible to set a good spring-tooth harrow, and so thorough that every clod, whether on the surface or beneath the surface, will be thoroughly pulverized. If clods of considerable size and number remain after the harrow, then roll the land, breaking down the clods in that way, and harrow again.

Many soils which, in past years, were plowed when too wet and were badly puddled will continue to furnish clods for years after unless they are broken down before drying.

As different soils under different conditions and at different seasons of the year will require longer or shorter periods of time between the plowing and the harrowing in order to be in proper condition to secure the best results, it is important that there should be some test which may be applied to each individual field to determine the proper time to harrow. Such a test may be made by taking a lump of earth in the hand, and if it crumbles at the slightest touch and doesn't form a paste between the thumb and fingers, it is time to start the harrow.

Conserving soil moisture

Besides the benefits just mentioned which result from starting the harrow at the proper time, another and quite as important a consideration is that of conserving the moisture in the soil. One of the chief factors in successful tillage is the conservation of soil moisture, and the very first step taken in that direction is starting the harrow at the proper time.

Conventional Tillage

To set the stage for some of the variety which has been suggested we offer a simplistic conventional outline of tillage implements and procedure.

After Plowing, Harrowing. - The implements commonly used after the plow are the disc and smoothing harrows. Some will say no other tillage tool can pulverize the soil so thoroughly and quickly as the disc harrows. Of these, the full disc (with round blades) is most common. When the ground is very hard, stony, or when it is tough sod, the cutaway disc (with scalloped blades) is especially good. The spading disc harrow is frequently used on fields infested with quack grass, to bring the roots to the surface. It is always best to lap the disc harrow half to get best results in pulverizing fall plowed lands, sod or lumpy soil - except when two disk harrows are used in tandem. The

Double disk for horses.

degree of pulverization depends upon the angle at which the discs are set.

Commonly, the disc harrow is all that is required to prepare an adequate seed bed for small grains or potato and corn land. The disc works well for weeding and summer fallowing.

Sometimes it is desirable to use the disc harrow on land before it is plowed; to break up the surface crust or lumps, to cut up and work trash into the seed bed, and to conserve moisture. When discing is done for the first two reasons, the furrow slice comes into more intimate contact with the subsoil.

Smoothing or Drag Harrows

When the seed bed has been made sufficiently mellow or loose through discing, the smoothing or drag harrows are used to give the finishing touches. Spike tooth harrows, chain (or pasture) harrows, plankers, and homemade floating tire drags all fit in this category of procedure.

Spring-Tooth and Acme Harrows

The spring-tooth harrow is a most efficient tool on rough and stony ground, and on new land in wooded sections. It may also be used instead of the disc harrow, for pulverizing sandy and gravelly soils; and it

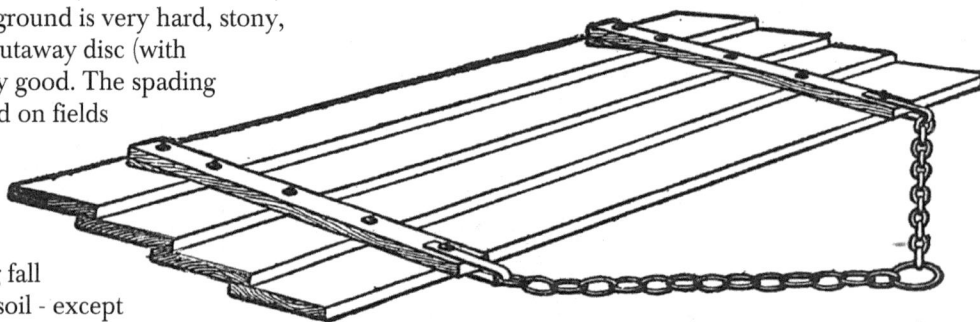

Planker.

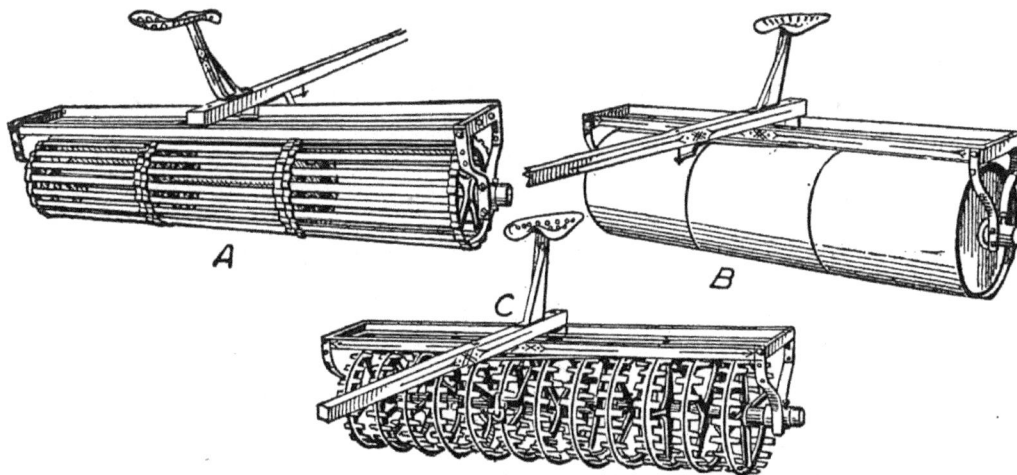

Three other types of rollers. *A*, pipe or tee bar roller; *B*, smooth or drum roller; *C*, crowfoot pulverizer (roller).

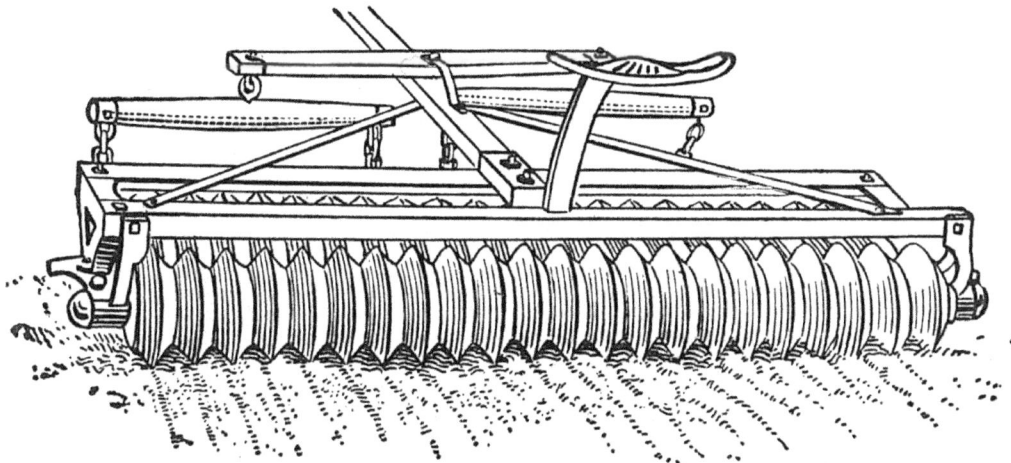

The corrugated roller or cultipacker.

may be used effectively in alfalfa fields to loosen the soil and eradicate weeds and grass. It is a more effective tool than the spike tooth harrow.

The Acme or blade harrow is used to a considerable extent in some sections for pulverizing, compacting and for killing weeds. This tool gives best results on loamy soils free from stones.

Plankers.

Sometimes, as in tobacco culture, and in market gardening, the planker is used when it is desirable to leave the surface particularly even and finely pulverized without firming it.

This tool is usually made out of three to four eight-inch or ten-inch planks bolted together with their edges overlapping. The planker is not a compacting implement.

Rollers and Clod Crushers.

Lumps are easily broken by means of rollers and clod crushers. Very often, after harrowing, the seed bed is too loose or has insufficient contact with the subsoil. If this is the case, rolling is necessary to compact the soil. Of all tools used for this purpose there is none better than the corrugated roller, or, as it is sometimes called, the cultipacker. This machine crushes lumps, compacts the soil and at the same time leaves a thin mulch in the form of a corrugated surface. Whenever a smooth or drum roller is used, it should be followed by a light spike-tooth harrow, with the spikes tilted, to form a mulch to conserve the soil moisture.

Muck, peat, sand, sandy loams and many loose silt loam soils are especially benefited by a cultipacker. When muck and peat soils are made firm they warm up quicker than when left loose.

The roller should never be used on the heavier soils when they are wet, but instead when they are in good working condition.

Soil conditions determine largely the different methods and types of machines used in seeding and planting; and the preparation of the seed bed is a most important factor in getting seeds well planted.

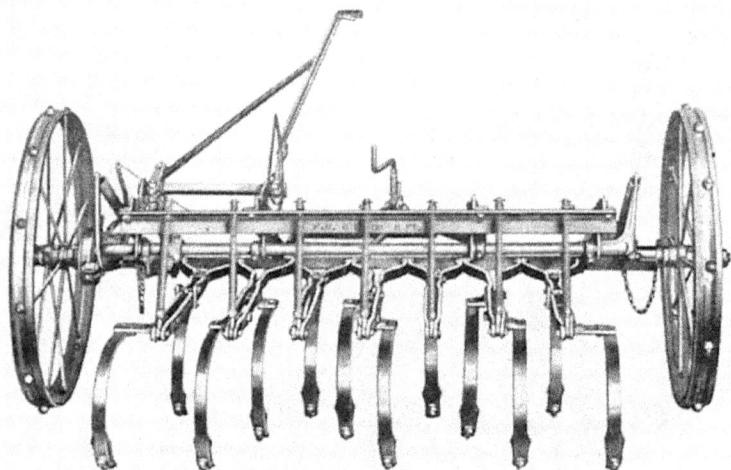

Pull-Type 6-Foot Field Cultivator with Spring Teeth.

Naming Conventions

As these implements were developed and produced, names were assigned and evolved as much by prevalent use as for any "proper" reason. For example some manufacturers referred to the corrugated roller as a "culti-packer" while others saved the name for the implement which combined a roller with a harrow.

And spellings were also up for grabs. For example the vote is tied 50/50 between 'disc' or 'disk'.

Homemade Planker

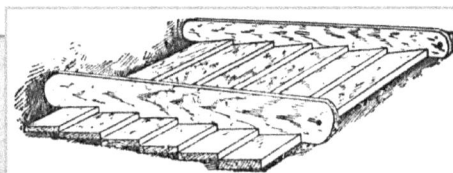

Acme harrow.

Acme harrow equipped with cart and three sections of curved knives.

Comparing Tilth

The loose mulch on forest soils.

The soil-mulch on tilled lands.

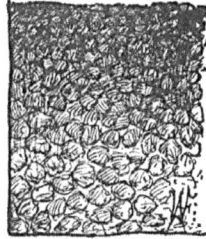

Showing the effect of the roller in compacting the surface layer.

Showing how the soil-mulch should be restored by tillage after the roller has been used.

Compaction

At some places in this text you read and see why, it is argued, that the soil must be thoroughly loosened and stirred as preparation for planting. Then good arguments are given for "packing" the top soil with a roller to establish capillary action for moisture and seed germination. Then it starts to get even more confusing when we learn that after rolling it may be necessary to harrow the top surface to loosen the soil again, creating a "mulching" effect to prevent rapid evaporation of soil moisture.

Now we must add a special concern and observation, that being the beneficial aspect of draft animals on soil compaction. Tractor wheels, especially large dual rubber tires, as they pass back and forth pulling implements, create a rolling pin action which compacts (especially moist clay-based soils) just below the reach of the drawn implement resulting in a subsoil hard pan. When the same tillage procedures are done with draft animals, the rolling pin action is replaced by the "shifting plate" impact of the hooves which creates a beneficial three dimensional tapestry of varying compact in the soil. Simply put: plant roots prefer ground which has been worked by draft animals.

Rich Hotovy with his Norwegian Fjords and a new I & J Cultivator at the PA Horse Progress Days 2000.

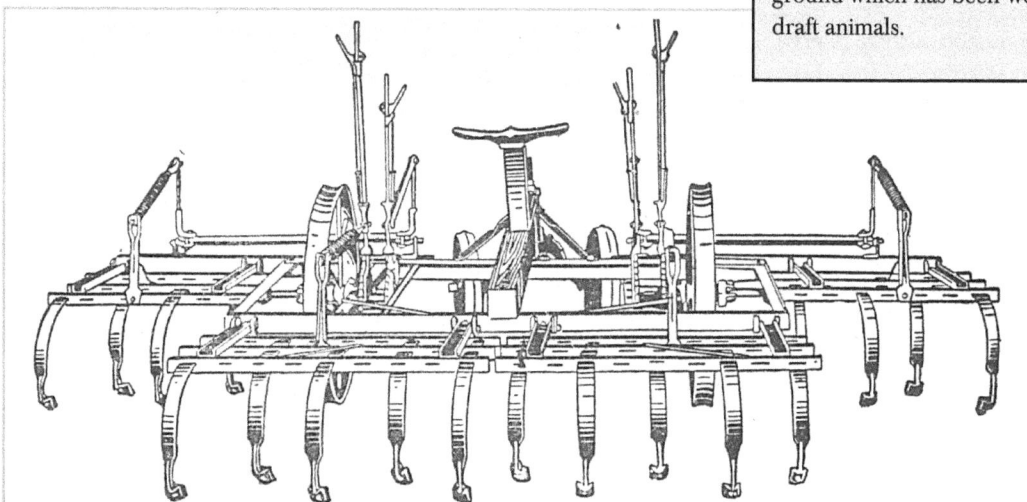

Spring-tooth orchard harrow.

27

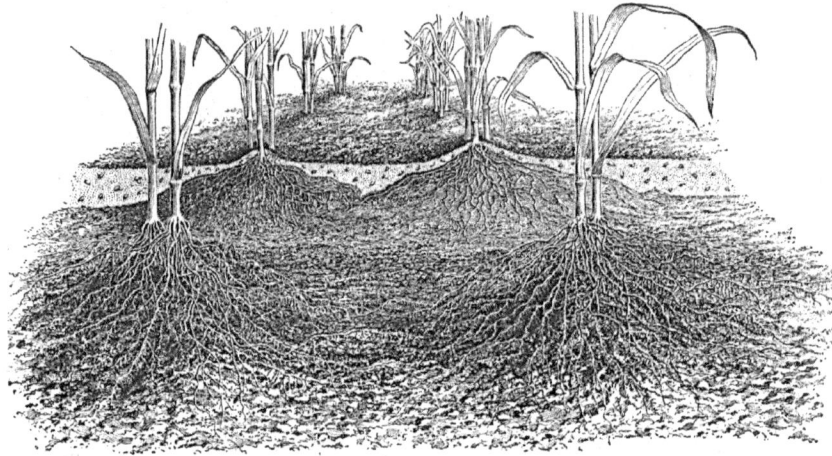

Secondary or Inter-tillage

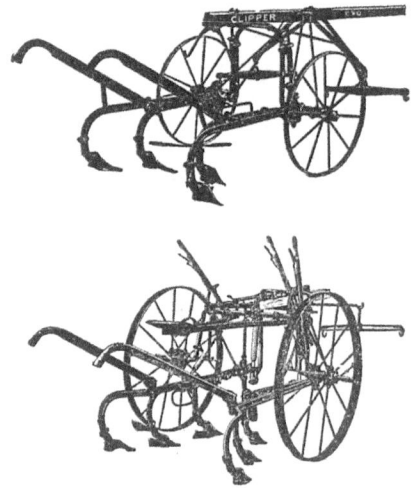

Cultivation, in its broad sense, means the act of tilling - but it is commonly understood to mean tillage done by cultivators. There are some tools designed to cultivate the land before planting, others that cultivate to cover the seed sown by them, and still others are designed for alfalfa fields. The ordinary cultivators, however, are used for intertillage.

Why Crops Are Cultivated.

The objects of intertillage are commonly given as: (1) to kill weeds, (2) to conserve moisture, and (3) to aerate the soil.

In humid-climate farming it is generally recognized that the killing of weeds is the primary importance of cultivation. This is especially true on soils in good tilth, and when frequent rains occur.

Cultivation to conserve moisture is good practice in all dry-land farming, and in sand manage-ment. On silt loams in humid sections conservation of moisture and aeration are sometimes questioned because different results have been attained under different conditions.

Interesting results from tests* done in the 1930s dramatically demonstrate the possible impact of intelligent cultivation:

In Illinois. On the common corn-belt soil of Illinois (brown silt loam) the following nine year averages in corn were obtained:

Method of cultivation	Yield per acre
(a) Land plowed, seed bed prepared, weeds allowed to grow	7.4 bushels
(b) Land plowed, seed bed prepared, no cultivation, weeds kept down by scraping with hoe	48.9 bushels
(c) Land plowed, seed bed prepared, cultivated 3 times	43.3 bushels

*from Productive Soils, Wilbert Weir, 1946, J.B. Lippincott.

In *Minnesota*, during a dry year, the following corn yields were secured on a "black loam soil containing considerable sand":

Method of cultivation	Yield per acre
(a) When all weeds were allowed to grow	0.4 bushels
(b) When weeds were cut w/ hoe w/o stirring soil	45.8 bushels
(c) When cultivated 6 times (3 times each way)	50.6 bushels

In *Wisconsin*. - On the heavy silt loam (Miami) the following results were secured during a year in which no beneficial rain fell during the period between July 3 and August 12:

Method of cultivation	Yield / acre	Rated quality of corn	Character of growth
(a) Land plowed, seed bed prepared, weeds kept down with a sharp hoe, soil not stirred in the least.	44.6 bu.	70%	Uneven
(b) First two cultivations 3.5 inches deep; subsequent cultivation shallow, and as often as was necessary to kill weeds and maintain a good mulch.	74.8 bu.	99.5%	Excellent & uniform

During a dry summer following a wet spring (1916), the following results were obtained in growing soybean hay in rows on sand at Hancock, Wisconsin. Very little rain fell between June 30 and August 15.

Method of cultivation	Yield of hay per acre
(a) No cultivation, but weeds were cut with a hoe; soil stirred the least possible	1875 pounds
(b) Frequent cultivation	3660 pounds

Most farmers know the value of cultivation if for no other reason than to kill weeds; and when this is well done, the soil is usually kept well mulched and aerated. Crops on heavy soils, in particular, should receive careful attention in regard to cultivation. Too often cultivation is done as a matter of routine. Some plan to go through their corn, or other fields, three times of four times, with no thought as to the proper time in which it should be done, and with little thought as to why.

Cultivators. Intertillage may be done through the use of several types of cultivators each type designed to do its work in some particular way to meet particular soil conditions. The shovel cultivators are the universal or most common implements. Of these, the six shoveled sulky or riding cultivator has met with greatest favor, because of its general adaptability. Many prefer the three-shoveled gang, while others the four-shoveled. Many different styles of these and other types of cultivators are made, each with various adjustments.

The spring-tooth gang cultivator is a very effective tool and it can be used under varied conditions, though in the heavier soils cultivators with rigid teeth do better work as a rule.

The surface cultivator gives good results in loamy soils and when they are comparatively dry. In soils free from stones the blades may be sharpened to cut such weeds as thistles, quack grass, etc. When soils are comparatively moist, this machine does not stir the soil sufficiently to cover and kill small weeds, because the soil simply slides over the blades and the tiny weeds are but little disturbed.

The disk cultivator is used in some localities. They don't seem popular. This type of cultivator was looked upon by some as a fad.

Lister cultivators are made especially for listed corn for first cultivation. The ordinary two-horse, shoveled cultivator is used for the later cultivating.

Many styles of walking cultivators are in use. In some sections these are generally used, while in others the common walking type is used when corn becomes too high for the sulky. Walking cultivators are especially favored by gardeners.

When to Cultivate. The best time to kill weeds is when they are small or when the seeds are germinated. In order to do this at the proper time the farmer must observe

A field rod weeder

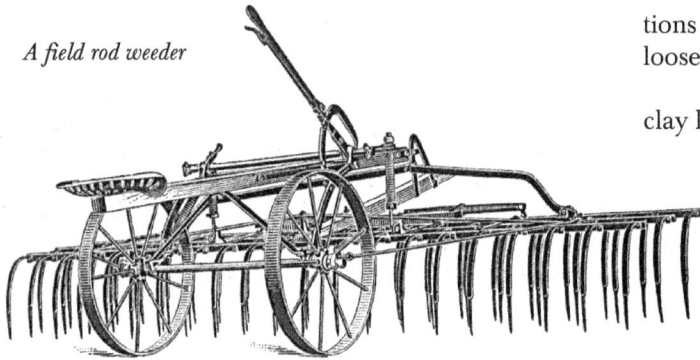

closely and often the condition of the different fields. Growing weeds may be killed through cultivation in three ways: (a) They can be loosened and exposed to the drying sun; (b) they may be covered and smothered with soil, and (c) they might be cut off or covered with soil to prevent their manufacturing any food. In the last case, diligence and close watching is required. Some prefer to kill noxious weeds by covering them with earth when summer fallowing, particularly quackgrass and Canada thistles.

Cultivation to conserve moisture should be done before the land is allowed to dry out. A good mulch should be prepared at the beginning of a dry period.

Crops on heavy soils are best cultivated, especially for the first time, when the moisture conditions are right. When this is done subsequent cultiva-

tions are made much easier because a layer of well-loosened soil prevents baking.

Some soils, particularly the black lowland silt and clay loams, shrink considerably when they dry out, causing big cracks to form. Such lands should be cultivated frequently to prevent as much as possible the formation of these cracks and to fill them when they do occur.

Shallow Cultivation Gives Best Results. In humid farming regions results are in favor of shallow cultivation. The only time when it is safe to cultivate deep at all is when the plants are very young, and before they send their feeding roots into the surface soil. Much harm results in deep cultivation (four to five inches), in cutting these feeding roots. The

Two corn plants of the same age. On the right deep cultivation has clipped the roots and stunted growth.

Cracks like these are moisture chimneys.

Cultivation of Sandy Soil

Q. I have a peach orchard on "blow sand" in Merced County, CA. I have given instruction to cultivate every two weeks, but my man objects that turning up moist sand loses moisture.

A. Sandy soil which is gotten into loose condition need not be disturbed by cultivation so long as that friable condition is not interfered with and so long as weeds do not grow. It would certainly be undesirable to use any form of disk or cultivator which turned up moist soil to the air, but on loose sandy soils there is sometimes a firm, evaporating surface formed at a little depth and not on the immediate surface, as is the case with heavy soils. When this takes place there is a loss of moisture by evaporation into the dry air which readily penetrates the granular covering to the crusting place below. To prevent this the soil should be stirred by a narrow straight-toothed harrow or a disk set upright so that there may be pulverization to a depth of five or six inches without turning moist soil up to the air. If you are sure that the cultivated layer remains loose to a satisfactory depth, it need not be stirred. But be careful that a loose layer on top does not deceive you.

only way to determine whether or not cultivation is too deep is to investigate what the cultivator teeth are actually doing. If the shovels next to the row are going too deep and cutting the roots, they may be raised; and if all the teeth are doing injury they should be set for more shallow work.

Level Cultivation Generally Best. Hilling corn does not increase the yield, hence level cultivation is more desirable. In some localities hilling the corn is a common practice because it has always been the custom. A farmer gets the hilling habit when he allows the weeds to get ahead of him. It then becomes necessary to throw much dirt on the rows in order to cover the weeds. If this must be done, and in some instances it is necessary to cover such weeds as the wild morning-glory, later cultivation should be done, if possible, at right angles to the ridges to level them. In this respect planting corn in check rows is advantageous. Because of the action of the shovels, proper cultivation leaves a slight slope between the rows.

The high hilling of potatoes has no particular advantage. When this is done more surface is exposed and hence more moisture is lost through evaporation. When potatoes are grown on the heavier, compact soils, digging is made easier when they are ridged a little; and, moreover, the throwing of some loose dirt on the hills becomes necessary as the potatoes advance in growth to protect the tubers from sunburning, since in many silt loams the growth of the

Fitting the Land after Plowing.

Fall-plowed land is usually left without other working until spring. If heavy soil is fall-plowed and too finely pulverized, it is likely to "run together."

Spring-plowed land should be dragged with a smoothing harrow or otherwise stirred before the clods become too dry to crumble readily. The drier the soil the more frequently this should be done. Under usual conditions, the harrowing should be done on the day that it is plowed. If the weather is very dry, and particularly in semi-arid regions, it may be necessary to harrow within a few hours after plowing. One may stop in the middle of each half-day for this purpose. Usually the land should be harrowed with the smoothing harrow two to four times before planting. Sometimes it may be better to use the disk harrow. On stony land or on very hard soil the spring-tooth harrow may be used. This is really a cultivator.

If corn is to be kept clean, it should be planted in a seed-bed that is free from weeds and that has been freshly stirred in order to kill any sprouting seeds. This gives the corn a chance to start even with the weeds. It is very foolish to plant on land that has germinating weeds, thinking to kill them after planting. It is better to delay the planting long enough to kill the weeds.

from *Elements of Agriculture, L. H. Bailey*

tubers causes cracks to form around the hill, which let in the light.

Weeders. The weeder is a weed-killing and mulching tool consisting of many narrow spring teeth. It is adapted for killing very small weeds in corn, potatoes, etc., either before or after the plants are up. This is not an effective tool when weeds are quite large or when the ground is at all hard or heavy.

A light, spike-tooth, smoothing harrow is often used in place of a weeder.

Emergency Tillage Operations. Sometimes it is not convenient to compact the seed bed or break lumps before planting. In this case, if the soil is still too loose or lumpy, grain and even corn land may be rolled after the crop is up. This should be done when the plants are small.

When grain is grown on heavy soils, it is best to leave the seed bed covered with a layer of small, loose lumps. This is not so favorable for the formation of crusts as in case of finely pulverized soil.

Heavy rains often pack the soil so firmly after the seed is planted that hard crusts form, which prevent the penetration of shoots and stems. A spike-tooth harrow is often used to break the crust, and sometimes a roller gives best results. Beans often break their necks in trying to get through a hard, crusty soil. In such a case the hoe or the careful use of a cultivator is best to break the crusts.

ADDITIONAL VIEWPOINTS AND SPECIFIC APPROACHES

What follows are a handful of different perspectives on tillage. Two are general in nature, one is a set of specifics targeted at certain crops, and the last is conversational in nature. Reading all of it will give the novice a much more solid starting point than any one presentation on its own. LRM

Principles of planting and cultivation

Preparation of the soil. The thoroughness with which the soil is prepared for planting determines, to a large extent, the cost of the after cultivation of the crop. It not only determines the cost of the cultivation of the crop but it also predetermines, to a very considerable extent, the stand of plants which will result from the use of good seed. Good seed upon poorly prepared soil will give an unsatisfactory and uneven stand of plants, while good seed upon thoroughly prepared land should give a perfect stand of plants. The time of germination is also governed by the thoroughness with which the seed bed is prepared. If the soil is only partially fined and compacted, the seeds will be much longer in absorbing the necessary amount of water to cause them to germinate; but with thoroughly fined and compacted soil, the process of conducting the

Three Grey Percherons on I&J Cultivator in Ohio 1998. Photo by L. R. Miller

Six Percherons on six sections Pioneer spring tooth harrow in Ohio at Horse Progess Days..

water to the seed begins immediately. These remarks tend to show the importance of properly preparing the soil in connection with the stand of the plants. The after cultivation of a thoroughly fitted area is much easier than that of one which has been only partially prepared. Soils which have been carefully handled previous to planting have a small percentage of noxious weeds than those carelessly prepared, for the reason that the operations of plowing, harrowing, compacting, and preparing the seed bed follow one another in succession, giving an opportunity for weed seeds to germinate between the successive steps, those which have germinated being killed, naturally, by the later operations. This of course destroys a considerable percentage of the weed seeds in the layer of soil which comes under the influence of implements of cultivation.

Plowing. The first and one of the most important operations in the preparation of soil is deep stirring, accomplished by either plowing or spading. In most farm operations the deep stirring of the soil is practiced but once each year. In truck-crop work, however, it becomes necessary to stir the soil deeply several times. With light sandy soils this deep stirring can be done very early in the season and little heed need be given to its moisture content; but with heavy, retentive soils it is of the utmost importance that deep cultivation be done only when the soil is in good mechanical condition. A soil is in good mechanical condition if, after being gently compacted between the fingers and the palm of the hand, it gradually falls apart when the pressure is released. If it is moist enough so that it retains its form and the soil particles seem to be broken down, it is too wet to work; and if worked in this condition, it will require several years of the most careful management to restore it to its proper physical texture. Upon clay soils or heavy clay loams it is, therefore, of the utmost importance that the work of plowing and harrowing be done when the soil is in proper mechanical condition. Not only is the physical structure of the soil destroyed, but bad treatment tends

to lock up the available plant food and render the soil unproductive.

For truck farming and market gardening, deep plowing should be practiced. If new land is to be brought under cultivation for market-garden purposes, it should be broken up to the depth of about 8 inches and a heavy coating of manure incorporated with the surface layer of the soil. After the soil has been in cultivation for one or two years, the process of deepening the seed bed should then begin. This operation should not be a radical one, but the seed bed should be deepened slightly each year; that is, by plowing 9 inches deep the second year, 10 inches deep the third year, and so on, until 12 or 14 inches of the surface soil have been brought under cultivation and reduced to the proper state of a seed bed. This will necessitate the use not only of the turning plow but of a subsoil plow as well. It does not necessarily follow that the whole 12 or 14 inches of soil which is used as a seed bed should be brought to the surface. It is sufficient if the first 8 or 10 inches of the soil be turned over. This depth will insure covering all organic matter sufficiently and will admit the air to the lower layers, if they are broken up by the use of the subsoil plow. In general, plowing should be done with a plow which has a quick turn to the moldboard, and which cuts a deep, rather narrow furrow, so that the soil as it is turned over is thrown with considerable velocity and in such a manner that the particles grind forcibly upon one another. This is a very important part of the mechanical preparation of the soil.

Harrowing. After plowing, the ground should never be allowed to lie exposed to the sun and wind more than a few hours. The harrow should follow the plow as quickly as possible. It is a good rule in cultivating the soil never to plow more during any one day than can be harrowed before night. Formerly, harrowing and compacting the soil were practically one and the same operation, but in recent years harrows have been constructed on quite different principles. The advent of the disk harrow which not only loosens and lifts the soil but pulverizes it, marks one of the great advance steps in soil cultivation. This implement, however, does not compact the soil. The acme harrow which is another important pulverizing implement, compacts the soil to a slight extent, but its action is more that of the mold board of a plow than of a compacting implement. The cutting portions of the acme harrow are so constructed that they first cut the clods, then turn them either to the right or to the left, according to the portion of the blade with which they come in contact. In the action of both these harrows the soil is pushed against itself so as to give it a grinding

It's all about being close to your work

motion, which produces the same result in the same manner as the mold board of the turning plow.

The old type of harrow was constructed with a heavy wooden frame and provided with teeth varying in size from one-half to seven-eighths inches square, sometimes even larger. The heavy frame was supposed to act as a clod crusher, and the teeth were not only to loosen but to stir the soil. This heavy implement, together with the tramping of the animals, had a tendency to leave the soil somewhat compact. It is only within recent years, however, that the importance of thoroughly compacting the lower strata of the seed bed has been appreciated; but since extensive investigations in soil physics have been carried on, and the movement of water in the soil is more fully understood, the importance of this feature in cultivation has been emphasized and is pretty generally understood by cultivators at the present time. The compacting of the

soil in the seed bed may be effected by an ordinary land roller or, better still, by a subsurface packing implement, one of the best and is constructed as follows: A series of ten or twelve independent cast-iron wheels about 2-1/2 feet in diameter are provided with a rim about 1-1/2 inches wide, and along this rim are placed offsets about 1-1/2 inches wide and 2 inches long. These are not placed opposite one another but are arranged alternately at intervals of about 2 inches along the rim of the wheel. This gives the casting a surface nearly 6 inches wide, of which only about one half is actually occupied by iron. When these wheels, which are upon a common axle, are made to pass over the soil, the implement, having sufficient weight to crush the clods, leaves the surface of the soil comparatively loose and slightly rough, but the underlying portions of the seed bed are forced compactly together. By following this implement with a spike-tooth harrow which will cut 2-1/2 to 3 inches deep, an almost perfect seed bed can be obtained. The subsurface packing closes up the capillary tubes in the soil; that is, it forces the particles of the soil close enough together so that a comparatively small amount of water is sufficient to cause an upward movement in the soil beneath. This compacting brings that portion of the soil which has been displaced by plowing and harrowing into intimate relation with the very compact substrata which haven't been disturbed by cultivation, and sets up a movement of water from a considerable depth; this is carried toward the surface until it comes in contact with the loose blanket of soil which has been produced by the spike-tooth or acme harrow, leaving a loose mulch 2 or 3 inches deep over the surface. This is the ideal way of preparing a seed bed for general farm crops. For truck crops, which are planted very shallow, it is necessary to have the harrow which produces the soil mulch work to a depth of not more than 2 inches, so as to leave the soil compact except for the last 2 inches on the surface. This will be found sufficient to bring the moisture up to within 2 inches of the surface of the ground and will provide a suitable seed bed for the more delicate seeds which are to be planted in it. The chief end to be attained in the preparation of the soil is a very fine seed bed, underlain by a thoroughly cultivated yet compact stratum at least 8 or 10 inches deep.

Water-holding capacity of the soil. The water holding capacity of the soil is increased by this method of tillage. It enables the soil not only to take up more moisture during a rain but also to retain the moisture longer after a rain. The important point is that the loose mulch of soil over the surface of the cultivated area acts as a cover to prevent direct and rapid evaporation. The opening and loosening of the soil by deep cultivation increases its power to absorb moisture quickly.

The water-holding capacity of the soil is also

increased by the fineness of the soil. While it seems a rather contradictory statement to say that the finer the soil the more water it can hold, yet this is true, because each particle of soil is enveloped by a very thin film of moisture when it is in proper condition for the growth of plants. The smaller the particles the greater the number in a given volume of soil; the greater the number of particles occupying a given space, the more surface there is exposed for this film of moisture; and the finer the soil the more tenaciously is this moisture held. This is the reason why clay soils or soils containing a considerable percentage of clay lose their moisture much more slowly than do light sandy soils. The large particles of the sandy soil allow the water to leach through as well as to evaporate more rapidly than is possible with the compact clay soils.

Implements which will be found of advantage. As before state, a turning plow, a good subsoil plow which only breaks up the subsoil without bringing it to the surface, a disk harrow, an Acme harrow, a spike-tooth harrow, and a Meeker disk-smoothing harrow will all be found of great service in the preparation of land for trucking and market gardening. There are probably no implements better suited for surface cultivation of crops in the field than one-horse and two-horse cultivators built after the Planet Jr. and Iron Age models. Those implements which do best work stir the surface only and have numerous narrow teeth rather than a few broad ones. The efficiency of surface cultivation lies in breaking up the crust as soon after each rain as the ground can be cultivated, or as soon as there is the slightest tendency to form a crust. The formation of crust indicates that direct evaporation from the soil, which is undesirable, is going on. Cultivation, then, should follow immediately upon noting these conditions. The end to be attained in cultivation is not only the destruction of weeds but the conservation of the soil moisture by the maintenance of a soil mulch. Cultivation should at all times be conducted so as to expose a minimum of soil surface to the action of sun and air; this can be secured only by level culture. Cultivators which leave the land in ridges, or methods of cultivation which tend to ridge or "bed up" the land, expose a much greater portion of it to the action of the sun and air than do the level methods. One square rod of soil, if it lies perfectly flat, has only one square rod of surface exposed to the action of the wind and sun. But suppose that the same area is raised into ridges six inches apart and six inches high; by computation it will be an easy matter to determine the exact increase of area over this plot and it will surprise one to find how much more surface of the land is actually exposed under these conditions than when it lies perfectly flat. In fact, it is possible almost to double the surface exposure of land by throwing it into such narrow

ridges. Since the object of surface cultivation is the preservation of moisture, the more nearly the area can be made to approach a perfect plane the better.

Artificial means of modifying the soil. Beside the mechanical benefits derived from culture and drainage, quite as important physical modifications can be produced by the addition of organic matter to the soil through turning under heavy crops for green manure. In those districts where cowpeas, crimson clover, vetch, and clover can be used for this purpose the fertility of the soil can be decidedly increased at the

C. J. Shopbell and Erskine Shires on disc harrow near Monroe Washington. Photo by Heather Erskine.

same time that its physical character is modified by plowing under these crops. Under conditions which permit the use of soil-improving crops only during the winter, crimson clover or rye will serve a useful purpose; while the rye adds no fertility, it will prove of great value as a means of increasing the humus content of the soil.

The effect of turning under large quantities of green manure or coarse stable manure is to modify the physical character of the soil. Heavy soils become more friable, more easily worked, and somewhat darker in color. Light sandy soils are equally benefited by such treatment. The addition of organic matter makes them more retentive of moisture, less liable to erosion, and better able to hold and withstand injurious effects of heavy applications of chemical manures. In fact, decaying vegetable matter or humus seems to be Nature's great restorative for all soil ills. In cases where continuous heavy applications of high-grade chemical fertilizers have produced injurious effects on crops, the plowing under of green crops and the use of lime quickly restore the soil to this normal cropping capacity.

SEED-BED

The seed-bed should be made deep, provided the subsoil is not a loose sand or gravel and too near the surface. Owing to the fact that the roots are inclined to grow deep, it is advisable, if the subsoil is compact, even in humid regions where the rainfall is abundant, to use a subsoil plow for the purpose of mellowing the ground, thereby facilitating deep penetration. If the seed-bed is not deep, the roots, owing to their fragile condition, will not penetrate a very compact plow sole, but will spread out, taking the course of least resistance, and in the event of drought, the plant will die or suffer for lack of moisture on account of their nearness to the surface.

If the seed-bed is deep and mellow and the substratum is permeable, delicate roots will penetrate into the soil where water is secured and where some plant food is available. The practice of drilling wheat without plowing, while it may prove successful occasionally, as a general rule means a very deficient crop. The writer had an opportunity to observe the two conditions in the west during an extremely dry season. Wheat drilled in corn ground where corn was listed, but not plowed, made a yield of from four and a half to six bushels per acre. In an adjoining locality where the ground was plowed deep, having been disced before it was plowed and subsequently disced, in spite of the protracted drought that season, made a yield of over thirty bushels per acre, showing the value of a deep, well-made seed-bed.

In a locality in South Dakota where the ground was plowed shallow, the wheat roots did not penetrate to a sufficient depth to hold the plant, and during a drought when the wheat was a few inches high, it was completely blown out of the ground by the high wind. Had the same land been plowed deep, the wheat would not have been dislodged by blowing nor would it have perished for lack of moisture, as was evidenced where the deep seed-bed was made in the same section.

It is a safeguard against the possibility of a drought to disc the ground before it is plowed in order that all trash may be worked into the seed-bed and the surface lumps pulverized so that when the furrow slice is turned, the contact is compact between the bottom of the furrow and the turned portion of dirt. The discing

Charlie Jensen in the foreground, Mike Jensen, no relation, in the background cultivating Oregon corn.

Young Pennsylvania Amishman culitvating with Haflingers.

prevents the formation of air spaces, a condition that materially interferes with the upward movement of water. Again, the seed-bed should be thoroughly disced until all of the lumps are pulverized in order to make plant food accessible to the roots. Plant food is held in solution and forms a film; or, in other words, clings to each particle of soil. The little delicate root filaments are thrown around these particles of soil and absorb, through the process of osmosis, the food and moisture. If lumps exist, the roots will not penetrate them; hence, the feeding area is restricted just in proportion to the number and density of the lumps in the seed-bed. The seed-bed should also be compact. Compactness is essential to capillary attraction, and it is also necessary in order that the plant roots can receive a firm hold in the soil.

AIR

Atmospheric oxygen is necessary to plant roots; or, in other words, to soil bacteria, which convert plant food into compounds. If the seed-bed is not deep and thoroughly pulverized, it is not well aerated. If, for any reason, the soil becomes surcharged with water, so that the air spaces between the particles of soil are filled up, the air is driven out and the growth comes to a standstill, and if the clogging continues even for a day or two, the plant will smother. Every farmer has seen this condition where water has stood for twenty-four or forty-eight hours in a wheat field.

ROTATION

It is well known to every wheat-grower that if he plants that cereal on the same land for a series of years,

the production will become less each year, until, finally, he will hardly get his seed back.

It is thought by some that plant roots throw off a deleterious excreta which is a poison to its own kind, but that the excreta is a food or stimulant to plants of a different variety; while others claim that a plant exhausts its specific requirements from the soil to such an extent that there is not enough fertility left to make a crop. Beyond question, both theories have merit to a degree, but certainly the second one is far from being correct, for we know that after wheat has been grown on soil until a crop cannot be produced, the same soil will make a remarkable crop of barley, rye, buckwheat or millet, using practically the same plant food elements, showing that fertility still exists, but for some reason cannot be utilized by the wheat.

A piece of land which produced two hundred bushels of potatoes per acre the first two years, finally failed to grow twenty bushels after it had been cropped for sixteen years, but did make the seventeenth year, seventy-five bushels of oats per acre. Many other like experiments might be given.

CROP Specific

WHEAT

Rolling and Harrowing. Soon after ground for wheat has been plowed, it should be rolled to make a firm seed-bed. This settling of the ground is particularly beneficial in wheat culture. If it is found necessary to plant the seed very soon after plowing the field, a corrugated roller is a better implement to use than a smooth-cylinder roller, as it packs the soil more firmly.

The field, after having been made firm by the roller, should be well harrowed. The number of times that it is harrowed depends largely on the soil and on the climate of the locality. In one field two or three harrowings may be sufficient, while in another four or five may be necessary. The farmer's eye and knowledge of conditions must be his guide in deciding how much harrowing the soil needs. The main consideration when harrowing ground for wheat is to get the soil fine. If it is found that one type of harrow will not do the work, others should be substituted.

In some sections farmers roll the wheat ground after it is harrowed, sometimes both just before and just after the grain is planted. A word of caution in regard to this practice is here offered. The settling of the soil tends to improve the conditions necessary for capillary attractions, and as a result there is an undue loss of water. If, however, it is desired to level the field just before planting the seed, a roller may be used, but it should be followed by a light harrow that will scuff the surface. The mulch thus formed on the surface will prevent the excessive loss of water.

Oats

The seed-bed for oats is similar to that for wheat, although it is not usually made so deep. If the grain is to be planted in corn-stubble land that is not exceedingly hard and compact, it may be sown on the unprepared land and then covered by means of a cultivator, a disk harrow, or a similar implement. However, such practice under most conditions is considered to be poor.

For spring oats, it is a good plan to plow the ground in the fall so as to have it ready for use in the early spring. In such cases, the seed-bed may be plowed to the same depth as that for wheat, but the rolling and harrowing must not be done until the seed is to be planted. If the work must be done in the spring, the soil need not be plowed to a depth greater than 4 inches, after which the seed-bed should be rolled and the oats sown broadcast and harrowed in. If a drill is to be used in sowing the oats, the rolling and harrowing must be done before drilling.

Barley

The seed-bed for a barley crop should be carefully prepared. Deep fall plowing is preferable, as it permits an earlier preparation of the seed-bed in the spring. The soil should be thoroughly disked and put into good tilth as early in the season as possible. Good tilth should be secured before planting, especially if the barley is intended for malting purposes, which requires the grain to be uniformly matured. To secure this tilth if the soil is lumpy, it should be gone over, after disking, with a roller or a drag, and if these operations do not produce the desired fineness of tilth, the Meeker harrow should be used.

Rye

For the growing of rye, the soil should receive practically the same preparation as has been recommended for winter wheat. It may be mentioned, however, that rye will often yield a fair crop if planted in a seed-bed in which wheat would be sure to fail.

Buckwheat

In preparing a seed-bed for buckwheat, it will be advisable to plow the soil long enough before seeding to allow the land to settle and to permit of two or more harrowings at different times so as to kill the weeds. If this plan is followed, any sod that is turned under will have time to begin to decay; the settling will reestablish capillarity, insuring a more rapid and uniform germination of the seed; and the field will be free from weeds when the buckwheat is sown. Such a seed-bed should produce as large a crop as can be obtained from the soil in which the buckwheat is sown.

Corn

After ground to be planted to corn has been plowed it must be made ready for planting by means of a roller, a drag, and harrows. These implements should be used in such a way as to make a finely pulverized, uniform seed-bed. All clods should be broken up and the field made smooth. If that land has been plowed in the fall no further work is done on it until early spring, when the land is disked to prevent loss of water, as before explained. Later in the season it is further prepared for the planting. When land is plowed in the spring, clods will be found; if these are not pulverized soon after the plowing they are likely to become so hard that they can be broken only with difficulty. To avoid this condition, a roller or a drag should be used on the land before the final work of finishing the field with harrows is accomplished.

CULTIVATION OF CORN DURING GROWTH

Corn is a crop that is cultivated during the growth of the plants. The chief reason that most farmers cultivate corn is to kill the weeds in the field. Incidentally, however, the working over of the soil is of much benefit to the crop, as it pulverizes, aerates, and mellows the soil, and increases its water-holding capacity.

Three draft paints stirring the soil at PA Horse Progress Days 2000.

Implements Used in Cultivation. *Harrows, weeders,* and *cultivators* are used in cultivating corn. Spike-tooth harrows are the implements best adapted for the work, but their use is confined to but a few days of the season, as explained later. The **weeder** is an implement that, like the harrow, can be used on a corn field for a few days only, but its use is very effective. The most-used implement during the corn-growing season is the **cultivator.** Of these there are both single- and double-row types. Corn cultivators with shovels are more generally used than those provided with disks although some farmers prefer the disk type. Corn fields are weeded by drawing a cultivator up and down the corn rows, allowing the shovels or disks, as the case may be, to loosen the earth between the rows. Double-row cultivators are made to straddle one row of corn at a time when being used, but with the single-row cultivator the horse, implement, and driver all pass between the rows.

Frequency of Cultivation. The number of times a field of corn is cultivated, and the methods of doing the work, vary in different localities. It is important to cultivate the land early, for the weeds are then small and easily killed. Corn plants, as a rule, will be above the ground on about the fifth day after planting, but the work of cultivating can be begun even before this time. About the third day after planting, unless the ground is wet, a weeder or a harrow can be used on the field. If a crust has formed on the surface of the soil, by reason of a heavy rain just after planting, this early working will break the crust and make it easier for the plants to get

above the ground. Some persons advocate the use of a cultivator before the corn plants are above ground, the marks left by the planter serving as a guide to show where the rows will be. The use of a harrow or a weeder follows that of the cultivator, and thus the field will be practically free from weeds when the corn plants appear above the ground.

If the ground is wet on the days before the corn plants appear, it is a good plan to postpone cultivation until about the eighth day after planting. Under such circumstances some persons use a harrow for the cultivation, while others condemn this practice, as the teeth of the implement uproot some of the corn plants. If a harrow is used in the afternoon, it is less likely to injure the corn plants than if it is used in the morning when they are brittle and easily broken. Although a harrow will uproot a few corn plants, it will rid the field of many weeds, which is an advantage. A weeder is sometimes used for cultivation under the conditions mentioned. Like the harrow it will pull up some corn plants but it will also rid the field of weeds.

If the corn is cultivated soon after it is planted there will be less weeds to contend with than if this early working is neglected, but whether there is early working or not, later cultivation is necessary. The cultivator is the implement that is generally used for all cultivating after the corn plants get to be over 2 or 3 inches high. Cultivating should be continued until the plants shade the ground. After that but few weeds will grow, and cultivating is likely to injure the corn roots.

The number of times a field is cultivated often

Single Ox pulls a walking cultivator at Horse Progress Days.

depends on the number of acres a farmer has in corn, and the help he is able to obtain. In the Central States farmers try to get over the fields at least three times, and possibly four, while in regions that are more intensively cultivated to go over a field seven or eight times is not uncommon. The more times a field is cultivated, provided the cultivation is not too deep, as explained later, the better for the corn crop.

Depth of Cultivation. Maize (corn) is a plant with fibrous, branching roots, and many of these roots are near the surface. It is not a good plan to cultivate so deep as to injure the roots. Experiments have been made showing that 50 per cent more roots will be cut off by cultivating 4 inches deep than by going to a depth of 3 inches. It is clear, therefore, that shallow cultivating is better than deep cultivating. From 1 to 2 or 2 ½ inches may be considered shallow cultivating, and whenever a farmer cultivates deeper than this he may be sure that he is injuring his corn crop.

GRASSES

Preparation of the Land for Grasses. As a rule, when a meadow is seeded, it is expected that several crops will be harvested before the land is again plowed. The preparation of the soil before sowing the seed should be such as will fit the land for several years of cropping. If land is poorly fitted for drilled crops, it is possible by a thorough system of cultivation afterwards to correct this error; this is not true in the case of meadows, for it is not possible to till the field after grass has begun to grow. For this reason it is of special

importance to prepare the ground thoroughly before sowing grass for a meadow.

Deep and thorough plowing is essential for grasses. If the best results are expected, the land should be as carefully prepared as for the most exacting garden crops. Thorough drainage, preferably by a system of tile drains, is advisable if the field is at all inclined to be wet. As it is impossible to remove weeds from a meadow economically after grass seed is sown, the weeds should be destroyed as nearly as possible before sowing. This can be accomplished by clean culture of crops grown during the 1 or 2 years preceding the seeding of the meadow. When grasses are sown in late summer with no other crop, it is sometimes the practice to summer-fallow the field, and by frequent harrowing kill weeds as fast as they appear. This, however, is a rather expensive method unless some early-maturing crop can be secured from the land before preparing it for grass.

The addition of barnyard manure to the soil is an aid to the growing of large yields of hay. From 20 to 30 tons per acre is often used when applied immediately before sowing grass seed. It is a more usual custom to manure a previous crop, such as corn, in which case it is expected that the effect of the manure will extend to the grass crop.

The average crop of hay from grasses in the eastern part of the United States is but little over one ton per acre. With proper weather conditions and the best cultural methods, more than four times this quantity has been produced. It is almost universally true that the meadows that produce the lowest yields of hay also produce hay of the poorest quality. A field that does not support a fine stand of good grasses is sure to have the vacant places filled with weeds or inferior grasses. Thus the farmer who does not take the pains to carefully prepare meadow lands loses in two ways--by securing a low yield of hay and by producing hay of an inferior grade, which always sells at a low price.

Legumes

Thorough preparation of the soil is an element of success in growing legumes. This means deep and careful plowing and afterwards thorough harrowing or rolling, or both if necessary. If legumes are to be sown for permanent meadows, the soil cannot be cultivated while hay crops are being taken from the land. In view of the fact that the land is usually left unplowed for a number of years, it is all the more important that the plowing and further tillage operations be done thoroughly. Time and effort spent before the crop is sown means much in the production of a maximum crop of hay. Often the drainage of fields is imperative for the production of clovers and alfalfa, especially the latter. There are many cases in which drainage not only will help the clover but will also increase the yields of other crops grown in the rotation.

Cowpeas

The preparation of the seed-bed for cowpeas is not necessarily an expensive operation, as is the case with alfalfa. It is a very common practice to sow the seed in corn, in which case no extra preparation of the soil is needed. If sown before the last cultivation of the corn, the corn cultivator covers the seed sufficiently.

The time of sowing will depend on latitude, purpose for which the crop is to be used, and variety. As cowpeas are an annual crop, seeding occurs in spring or early summer. The season may vary from April to August. Late sowing is practiced when the crop is being grown for seed. If hay is wanted, early seeding is the rule, as early seeding is favorable to a luxuriant growth of vines.

Peanuts

Cultivation of the peanut crop should be similar to the cultivation of corn. Frequent and shallow cultivation is better than deep and infrequent cultivation, as the soil by this practice is kept moist and loose, which is essential. After the first cultivation, the soil should be worked toward the row in order to provide a mellow bed of earth in which the pods may form. Cultivation should cease as soon as the pods begin to form.

Potato

The time for plowing will depend largely on the nature and situation of the field. If it is located on a slope that is liable to wash, spring plowing is preferable. If plowed in the fall, an extra amount of harrowing will be necessary in order to obtain the same results as are produced by spring plowing. No matter what time is selected for plowing, the soil should be harrowed and pulverized in a thorough manner. The farmer should not be content with stirring and leveling the surface only, but should use harrows that will work deep and leave the soil in the best condition to receive the potato seed.

Purposes of Cultivation. The purposes of cultivating the crop are to liberate plant-food, to conserve soil moisture, to maintain good texture and to aerate the soil, and to keep down weeds. Nothing will add more to the yield of the potato crop than frequent tillage with a shallow working implement. At the beginning of nearly every season there is sufficient moisture in the soil to carry the crop through, provided that moisture is conserved. Nothing will prevent the waste of this moisture better than frequent tillage with a cultivator that has a large number of fine teeth working at a depth of about 2 inches. This maintains a dust mulch that prevents the loss by evaporation of the capillary moisture. The dryer the season the more frequent should be the cultivation. The use of the cultivator has produced more than one excellent crop of potatoes in seasons of drought.

Methods of Cultivating. Immediately after the land has been thoroughly prepared, and the potatoes have been planted, shallow cultivation should be started. The most economical method is to run over the field about twice each week with a weeder or spike-tooth harrow with teeth set backwards, until the potatoes are out of the ground. This destroys all weeds while they are yet small and often obviates the need of hand hoeing even once.

When this method is to be practiced, the cuttings must be planted at least 3 inches deep. If only shallow intertillage is afterwards practiced so that weed seeds are not brought up to the surface, the field will usually be practically free from weeds. A two-row riding cultivator with pivot gangs, provided the gangs are fitted with many shallow working teeth, is the most economical implement to use.

Very little hand culture should be necessary in the potato field. The method just outlined in which nothing but cultivators are used, is known as **level culture.** It is without doubt the most effective method of saving moisture and in times of drought will give the best results. When this method is followed, the potatoes are somewhat harder to dig with a machine than where *ridge culture* is practiced.

By **ridge culture** is meant that method of cultivation in which the ridging begins at the time of planting. The planter most used has a plow so constructed that it makes little more than a mark on the soil, unless it is very light, instead of a furrow; then the disks at the rear of the machine cover the seed by throwing up a ridge perhaps 4 inches high so that the seed at the very start is practically on a level with the surface between the rows. Some farmers make a practice of going over the field with a weeder and somewhat flattening the ridge soon after planting. The plan most usually followed is to go between the rows with the cultivator perhaps 8 to 10 days after the potatoes are planted, and then as soon as they begin to break the ground to go over with an implement known as a *horse hoe* and bury them, also burying the weeds at the same time and thereby raising the height of the ridge. Covering the young plants also protects them from frosts in northern sections and is claimed to cause the plant to send out more tuber-bearing stems. This kind of cultivation is continued until the tops are too large to pass through without injury. By this time an **A**-shaped ridge has been formed about 12 to 15 inches high and, of course, the surface between the rows has been dropped by the continual scraping up of the dirt so that the tubers growing in the ridge are considerably above the surface between the rows.

It can readily be seen that in a dry season a field so handled must suffer considerably from lack of moisture. Of course, in localities where wet seasons are the rule, no lack of moisture is felt and the drains between the rows are an advantage rather than an injury; but in an extremely dry season or in light sandy soil the drainage is too great. As the ridges are high and narrow, they dry out very quickly; it would appear, therefore, that the crop must suffer more from lack of moisture than it would if the roots of the plants were below the level as they are when level culture is practiced.

A combination of the level and ridge methods of cultivation is often practiced with profit. The potatoes should not be cultivated deep, especially close to the rows, after the vines have attained any considerable size. If they are, the roots will be injured and the yield decreased.

Mangel Wurzels

Mangels respond as readily as any crop to good tillage, but no crop will be more discouraging to the grower who does not prepare the land well. Deep fall plowing is advisable to insure a compact subsurface and a mellow soil; from 8 to 12 inches is a common depth of plowing for this crop. Thorough tillage of the surface soil should be given in the spring; it will probably be wise to use the disk or the Acme harrow for the first workings and finish with the spike-tooth or the Meeker harrow. If the land is weedy, a cross-plowing may be given in the spring to a depth of 5 or 6 inches, previous to harrowing. Frequently six or more harrowings are necessary to make the seed-bed fine, although the use of the roller, either the plain type or the pulverizer with the corrugated disks, will reduce the number of harrowings and compact the soil at the same time. A fine, compact soil is required.

Turnip

In preparing land for turnips fall plowing is usually done, except on very light soils or when the land is occupied with some cover crop. The plowing should be to a good depth, 10 inches if possible, and when the soil has been well pulverized by freezing and thawing it is not advisable to plow again in the spring. It is better to keep at the surface the fine soil made by freezing and thawing during the winter and to keep the weeds down by harrowing and rolling until time to sow, thus securing a fine tilth and getting rid of many weed seeds in the surface soil. If the land is very weedy it may be profitable to plow it the previous August or September and keep it cultivated in the fall in order to clean it of weeds. It will not do to plant turnips or any other root crop on land that is very weedy unless it has been well prepared, because of the heavy expense in hand hoeing. Land plowed in August or September should be cultivated with a harrow several times the next spring before it is planted to turnips; it is much better to cultivate the land several times rather than to plow it, for in plowing good soil is buried and clods and soil that have not been weathered are brought to the surface. In addition, plowing results in a large loss of moisture from the top 3 or 4 inches of the soil and causes this soil to become so dry that the germination of the seed is uncertain. Since the seed cannot be put in very deep and moisture is required to germinate it, the aim is to preserve this necessary condition to aid in the germination. If the soil is at all lumpy it should be rolled and harrowed and then rolled and harrowed again until all lumps are reduced.

Eric Nordell cultivating market garden crops with straddle row. Photo by John Nordell

Tillage
One Hundred Years Ago
Interviewing Two Good Farmers

*The following interview originally appeared in **How The Farm Pays** published in 1902 and features William Crozier and Peter Henderson.*

Q. After plowing comes the harrowing. Please describe your method.

A. In my experience with help, I have found ten men competent to plow where I have been able to get one competent to harrow; not that there is any more skill required in harrowing than in plowing, but from the fact that it is not so easy for the eye of the master to detect bad work in harrowing, and consequently men indolent or careless can run over the surface so that it may appear to be well done when it is not. For this reason, it is all-important to have a full examination made of the work, for harrowing has everything to do with the welfare of the crop--to have the soil thoroughly disintegrated and pulverized. This harrowing should penetrate to a depth of five or six inches, in order that the soil may be thoroughly and deeply worked.

Q. You take pretty strong ground in regard to harrowing. Give me your ideas of what is good work and bad work in harrowing?

A. Let us take a newly plowed field; the soil is mostly in lumps, small and large. A poor workman runs a harrow over the surface and smooths it and makes it fine; it looks well, but it is bad work; it is bad because when one sows seed on such ground it works down under the fine surface and among the lumps and clods, where it may sprout, but soon dies because the soil is too loose and open and is filled with air spaces. A good workman makes his harrow teeth work down in the soil among the lumps at the bottom, and breaks these up, or brings them to the surface, and so works the fine, pulverized soil down where the seed will lie in it, and sprout and grow perfectly because the soil is fine and compact around it. This is good work. It may not look so smooth to the eye, but it is better for the crop.

Q. But this rough surface would not be suitable for seed; then I presume the use of a roller would be necessary?

A. Yes--then the roller is used, followed again by the chain harrow, so that the surface may be made level and smooth for the seed.

Q. What harrow as a pulverizer do you consider the best?

A. I have heretofore used the imported Scotch harrow, which I had found to be the best; but this season a trial of the American harrow known as the Acme leads me to believe that it will supersede the Scotch as a pulverizer or leveler, for it is the best implement I have ever used for these purposes.

(Mr. H.) I am pleased to agree with you in this matter. After a thorough trial this season with this harrow, I find it to be the best implement I have ever

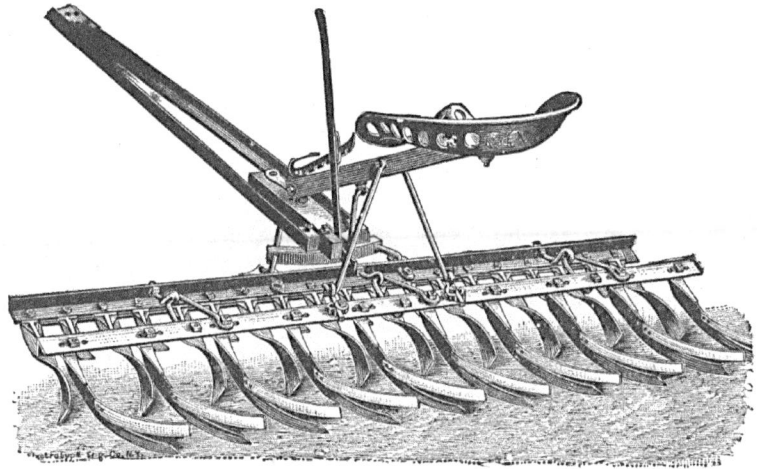

Acme Harrow

used for the purpose of pulverizing and leveling the soil. It is not only a harrow, but under certain conditions of the soil it is to all intents and purposes a gang of small plows; or, in other words, in a soft or light soil you can plow the ground just as thoroughly for six feet wide as you can do it with the ordinary plow eight inches. The great value of this implement induces us to use more space for a description of it, and its uses, than will be probably given to any other implement in this

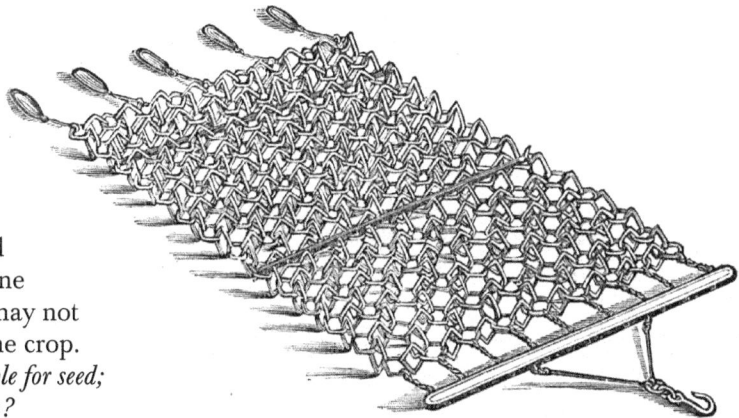

CHAIN HARROW.

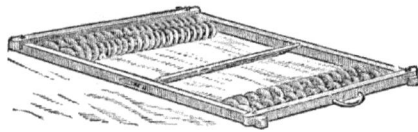

THE DISC HARROW.

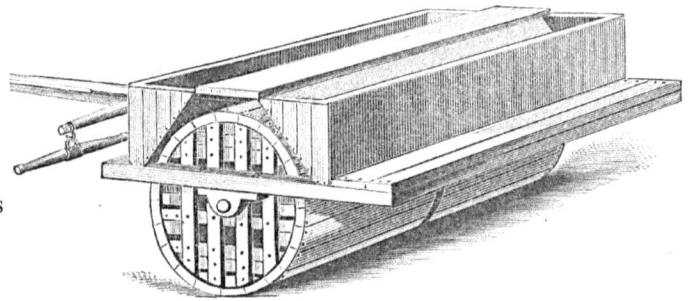

A FIELD ROLLER.

work. Upon this account I would like to give the views of a well known farmer, whose experience with this implement has been longer than mine, and who is a high authority upon such subjects. This is Henry Stewart, of Hackensack, N.J., who, after using the harrow for six or seven years, says: "After plowing, the soil is worked over with the Acme harrow and is thoroughly broken up; the furrows are leveled; the whole soil to the depth of four inches at least is disturbed as though a series of small propeller screws passed through it; it is thoroughly mingled; the upper portion, which has been exposed to the air, is turned under and buried, and the whole soil is loosened up, broken and made mellow. This is the only implement, so far as I know, that does this necessary work, and with this the best preparation for crops is easily possible. That is to say, that the full effects desired cannot be obtained by, or through, any other one implement than this; because it does all that a plow could do, and it does all that the harrow can do to pulverize the soil, but it does what no mere harrow can possibly do in the way of turning over the soil and presenting a fresh surface to the atmosphere, and it does all that a cultivator can do, without the objectionable effects of that implement; and lastly, it does all that a roller can do in the way of pulverizing cloddy soil, without the objectionable effects of that implement in packing the soil so closely that the air cannot penetrate it.

Q. You make a distinction between what you would call leveling the soil and smoothing it, do you not?

A. Yes. For instance, the Acme harrow levels and pulverizes the soil, while the Chain harrow smooths the surface.

Q. When you say that you harrow your manure after spreading it on the land (which I believe is an excellent plan, and one that was entirely new to me), what harrow do you use for that purpose?

A. I would by all means use the Acme or a similar harrow, as for that purpose we require to mix in part with the soil. The great advantage of the Acme harrow for working up the manure, would be that you can regulate the depth of the teeth at will.

(Mr. H.) In my experience among our market gardeners, where the pulverization of the soil is as perfect as we can get it for the reception of small seeds, I have used for the past two years a smoothing harrow known as the Disc harrow, which consists of some sixty sharp discs placed on revolving shafts, so as to cut the soil to a depth of three inches by one

inch in width, which fines and levels the ground as completely as can be done with a steel rake in the hands of an expert workman, but whether such an implement would answer the purpose as well for the requirements of a farm as the Chain or Acme harrow I am not able to say.

(Mr. C.) One great advantage of the Acme harrow over all others is the disposition of the teeth, which are so placed that on sod that has been plowed it cuts and pulverizes it, without dragging it to the surface. The present season I turned down a piece of sod on which I sowed mangels and planted potatoes. The thoroughness of the cultivation by the use of this implement was such, that I was enabled to work the land up in ridges--which is my usual practice with such crops--as easily as if it had been stubble land.

Q. What do you deem a proper day's work for plowing for a man and team?

A. One acre on sod land and one acre and one-fourth on stubble.

Q. What area should a man and a pair of horses harrow in a day, to do it properly, with the Acme or other harrow?

A. From four to five acres, to do it thoroughly.

Q. Of course you are aware that about twice that area is harrowed when done in the ordinary way?

A. Yes, and even more. But I consider that such labor thoroughly done is the best investment the farmer can make. My experience of thirty years has been varied and extensive, and

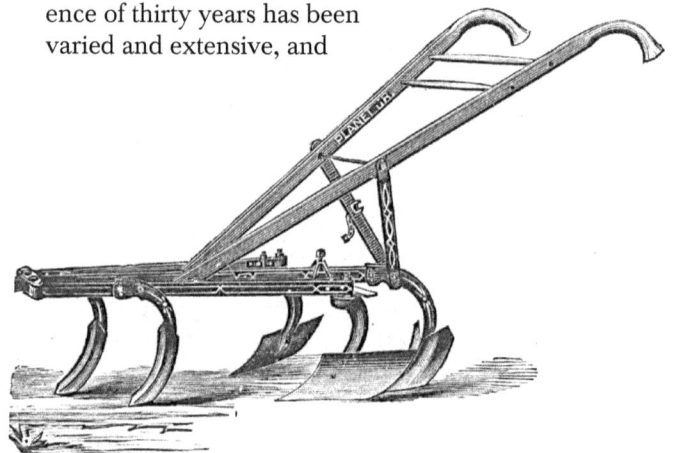

PLANET, JR., CULTIVATOR AND HORSE HOE.

every succeeding year only impresses the more strongly upon me the fact, that to get good crops you must have thorough pulverization of the soil.

Q. Of course you use the various kinds of cultivators for the various crops. What implement do you at present use for cultivating corn?

A. Cultivators are now so varied and improving every year, that it is hard to say that any particular one is the best. There are many patterns more or less valuable. My rule in all such things, when purchasing at an implement or a seed warehouse, is to ask what tool is in largest demand for a certain purpose, and I usually find that the public in the long run finds out which is the best article, and that the article most in demand is the one usually having the most merit. At present I have found that the cultivator known as the Planet Jr. Horse Hoe, does the best work in this way, and as it is mostly used in this vicinity, public opinion bears me out.

(Mr. H.) I agree with you in that entirely, and as a seedsman I can well endorse it; for whenever a customer asks for any particular tool, the answer I make to him (unless I have certain knowledge myself of the subject), is to go and ask the clerk having charge of that department to select for him the kind that is in most general demand, and as a rule it will be such as is the best. However, I may state that I have used for nearly twenty-five years a simple form of cultivator--which any blacksmith can make--known as the Harrow-tooth Cultivator. It is merely a triangular harrow having from twelve to sixteen teeth, which we use to stir up the soil almost immediately after a crop has been sown or planted, and this we continue to do once a week or so, between the rows, until it may become necessary to use (in particular crops) a cultivator to work deeper, such as the Planet Jr. but the use of this Harrow-tooth cultivator is of great importance in checking the first growth of weeds, and as it is light and easily worked, a vast amount of labor can be saved by using it often enough, so that the weeds will never be allowed to be seen.

Q. Do you make much use of the roller on your farm, Mr. Crozier?

A. I used it on all crops and particularly on my pastures early in the spring. I thoroughly believe in the practice which you so persistently advocate, of firming the soil for all seeds and plants. You, in your limited areas in market gardening, can afford to do this with the feet, which probably there answers the purpose of firming the seeds or plants better than the roller, but on a farm that, of course, would be impracticable; but, whatever method is used, the principle should never be neglected, of compacting the earth around newly sown or planted crops, especially in hot, dry weather, and particularly so on loose and porous soils.

While you, as a gardener, advocate the use of the

feet to firm the soil, in sowing and planting, I, as a farmer, advocate the use of the roller. The object in both is the same; and I am satisfied beyond any shadow of a doubt, that millions and millions of dollars are annually lost to the farming community, through a want of the knowledge of the vast importance of firming the soil over the seed. This is particularly the case with buckwheat, turnips and other crops that are sown from the month of July until September, as at such seasons we very often have long-continued droughts, and the soil is like a hot ash heap, and to expect germination from small seeds when sown in such soils, without being firmed against the entrance of the hot air, is just about as useless as if we threw them in the fire.

(Mr. H.) I consider this subject of so great importance, that I think we should take the liberty to again print here the article which I read before the National Association of Nurserymen held at Cleveland, OH., in June of 1879, entitled, "The Use of the Feet in Sowing and Planting." I have written a great deal on horticultural subjects in the last twenty years, but I think (and I say this advisedly) that the value of this article to the horticultural and agricultural community is more than the whole I have ever written, put together, and I have great satisfaction in knowing that thousands of men have thanked me for impressing so strongly the necessity for this work. This article has been reprinted in thousands of newspapers in the past four years, but if it, or some other similar advice on the necessity of firming the soil after sowing, was ever placed before the eyes of the farming community and acted upon, thousands would be saved from mourning the loss of wasted seed, manure and labor; for in a country vast as ours, a new crop of inexperienced men are annually engaging in farming and gardening. In no European work on farming or gardening that I have ever seen, has the importance of what we have so strongly argued for been referred to, probably for the reason that in the cooler and more humid atmosphere of most European countries the necessity is not so great.

I'm going farming with my mules, what's the best, cheapest short list of implements I'd need for tillage after plowing?

If your farming plans include no row crops you can get by with; a disc harrow, spring tooth harrow, spike tooth harrow, and a home-made planker. Later when funds allow, add a corrugated roller and field cultivator. With row crops add at least a single or straddle row cultivator.

How The Draft Animals are Used

This book is not about how to harness, hitch and drive draft animals. Those subjects are covered in the *Work Horse Handbook* and the *Training Work Horses* book. We must insist that you not attempt to use any implement described in this book without adequate schooling in the teamster's craft. With good well-trained calm animals and normal intelligent safety precautions operating most HD tillage tools is safer than their tractor equivalents. Without experience (or experienced hands-on help), without sensible precaution, and/or with arrogance, the use of HD Tillage tools can be dangerous and even deadly to humans as well as animals.

There are brief hitching notes (i.e. how many horses for the tool and for the procedure as well as special setup concerns) with each appropriate chapter. Here are some general dynamics you should pay close attention to for good work and for safety.

Understand Your Chosen Implement Application.

With most tillage implements, most but not all, you will have the choice of additional depth and/or pressure in the working operation. Teeth may be set deeper, weight may be added (in the case of disc harrows) varied shovels and points may be chosen for cultivating effect, etc. how your animals are hitched may have an effect on the implement in these regards. Remember this basic rule of motive power:

Hitch close for lift - Hitch long for drag

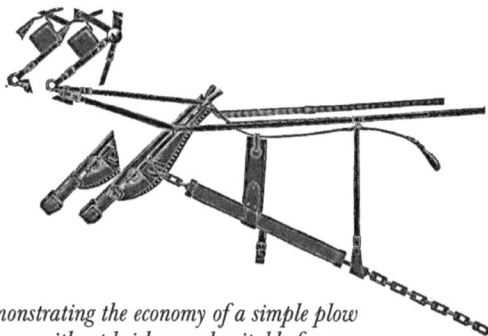

Western-style brichen harness suitable for any field work application.

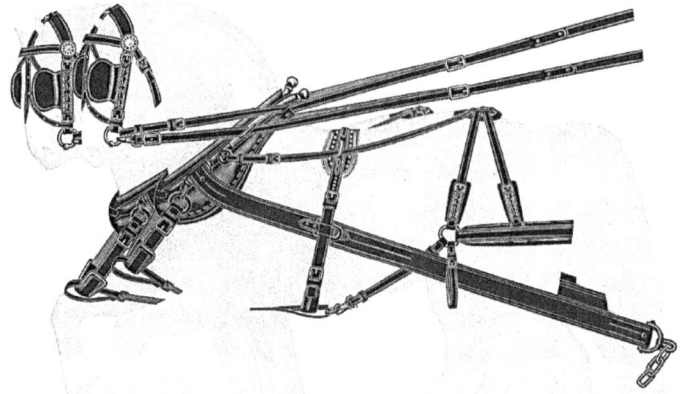

Demonstrating the economy of a simple plow harness without brichen and suitable for any field application where the backing and breaking system is not required (sans pole).

Eric Nordell and his fine crossbreed team enjoy a magnificently precise cultivating job on their Pennsylvania farm. photo by John Nordell

With spike, springtooth and disc harrows - as a rule - you will want full contact with the soil, in other words no lifting of the implement by the draft animals.

As an added note, should the teamster prefer four abreast or wider (over a strung out hitch - see WHHB diagrams) the animals will need to be hitched far enough forward to permit sharp turns without inside animals coming to contact with implement. (See diagram) A jack-knifed hitch with frightened

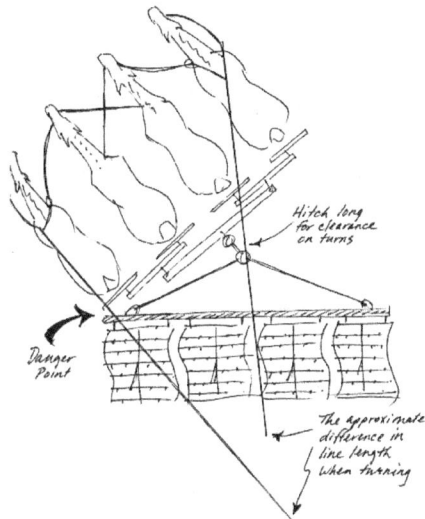

Hitch long for clearance on turns

Danger Point

The approximate difference in line length when turning

horses stepping into a harrow frame can cause an advancing wreck.

While on the subject of the widths and distances; play close attention to danger zones when you're hitching and unhitching. Never put yourself in a situation with no escape hatch. You're not much good to your animals if you're dragging behind or under an implement. And those things tend to spoil dinner for your spouse as well.

Pioneer forecart with bench seat and fender kit. This all-purpose tool allows that any draw-type tractor implement be pulled with teams of horses.

Charlie Jensen of Oregon training Belgians on his forecart.

Doug Hammill of Montana rides a harrow cart hooked gooseneck style to the front of a spike tooth harrow. This cart allows Doug to ride behind the outfit in relative safety.

Common sense should alert you to the relative dangers of particular working arrangements with these implements. For example an eight or ten foot wide harrow or cultivator drawn behind a forecart (see chapter   and photo on page) can be comfortable work for the teamster unless control is lost or for some reason the teamster should fall off the forecart and into the path of the implement. Riding directly on, as is shown page , is slightly less hazardous and riding behind is definitely preferred.

Some of these implements beg for added caution especially if young or green horses or mules are being used. Think about the relative risk when you consider taking people along for a ride. If you are busy trying to control the team or adjust or disengage a tangled implement you won't be able to help a child or slow individual get out of harm's way.

Length of Driving Lines:

Before you go to the field make certain your driving lines are long enough. Especially with abreast hitches. It is far too easy to underestimate how much line is necesssary to the outside horse on a turn and if you are behind the impement the challenge of a rapidly disappearing line length, slipping through your fingers, may be more than a little threatening at the worst time.

Training and Conditioning Value of Tillage Work.

Pulling a harrow back and forth across plowed ground is some of the best work you can do for training new horses. Precision is not so important. And anxiety seems quickly asbsorbed by the difficulties of walking in plowed ground. Young green horses learn quickly to value the rest breaks and will stand appreciatively.

My favorite training outfit features a four abreast on an eight foot disc harrow. I take three trained horses and set the trainee on the outside where it's easy to get to his or her tugs and rigging.

Seedbed preparation work is also excellent for toning up horses which have been out of the working routine. All normal precautions should be taken as to feed, over work and proper harness and fit.

Hooks and Eveners

From experience I learned to hitch all my harrows with self-closing hooks (see diagram) which I could operate with one hand should I need to maintain a grip on the lines. Bolted or pinned shackles or clevises will work but require some time to get them undone should you be anxious to seperate your animals from the implement. And open hooks will sometimes unhook at rests and on turns.

Speaking of which; be sure that should you be using rolled steel eveners the single tree hooks are bending up rather than down. If pointing down they also will sometimes come undone.

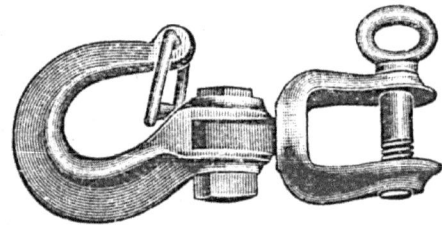

Self closing hook

Screw Pin, Ribbed

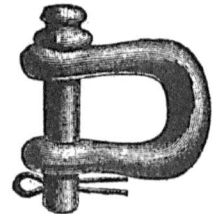

End Clevis, Oval Band

Clevis Furnished with Boss Harrow

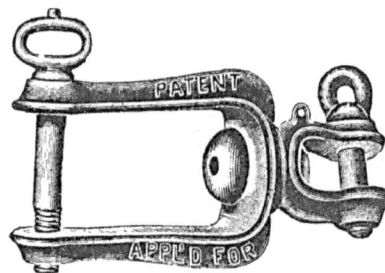

Ideal Swivel Clevis

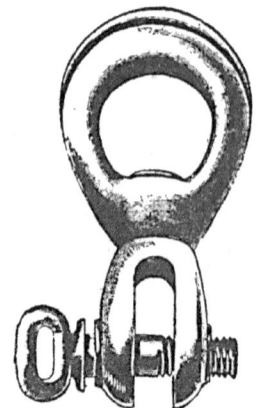

End, Swivel Clevis

Anne Nordell cultivating at Trout Run Farm in Pennsylvania. Photo by John Nordell.

The Nordell cultivator custom rigged for trash incorporation and the building of humus. Photo by John Nordell.

Chapter Two

Stalk Cutters

Strictly speaking, this chapter features an implement which doesn't till the soil, rather it sets the stage for tillage after certain crops are harvested. The emergence of corn as a large scale monoculture in North America dictated the need for a mechanical device to deal with stalk residue before good tillage practises could commence. The stalk cutter was developed to allow for a way to chop standing and loose residual corn stalks in preparation for discing and plowing. The cutter is also handy for cotton, sorghum, canary grass, sudan grass, canes, and certain coarse weed stalks. If the ground is soft and/or wet and the stalks are limp or soft, chopping may be foiled.

Parlin & Orendorff claims that its company originated the stalk cutter in 1850. The first prototype consisted of a log, with spokes driven into it to hold the knives, and upon that principle the stalk cutters of the 1900s were built.

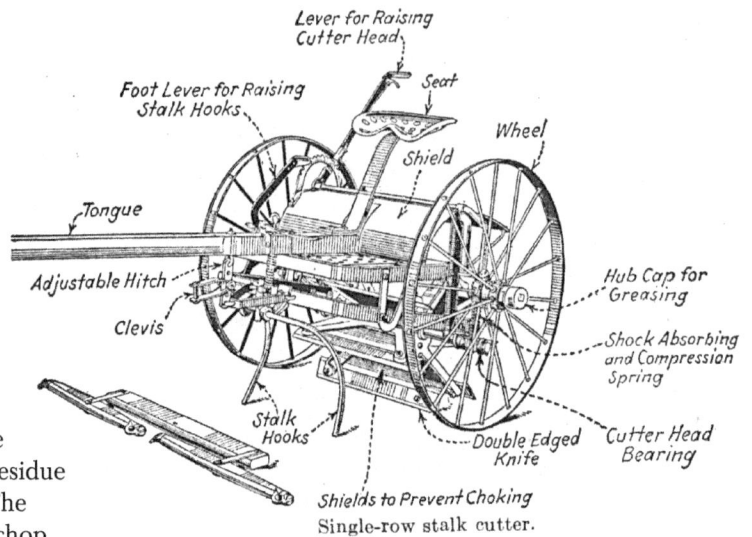

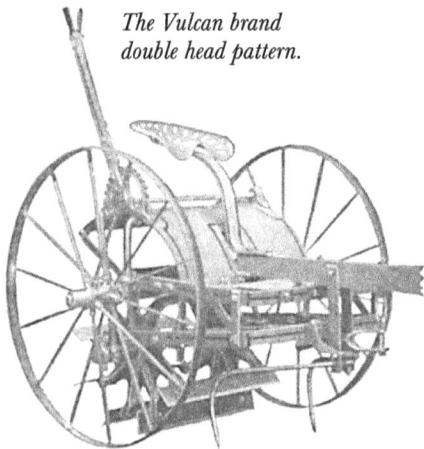

Single-row stalk cutter.

The Vulcan brand double head pattern.

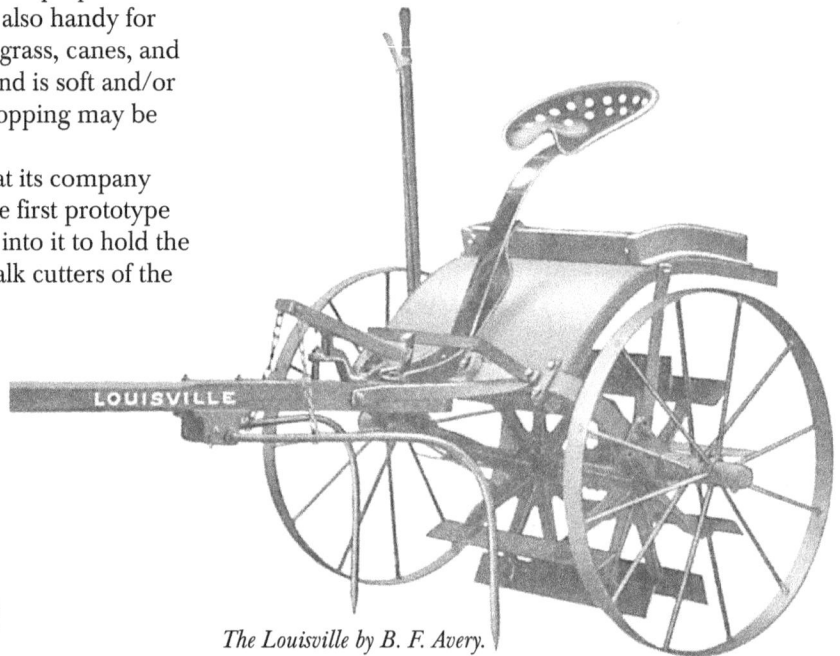

The Louisville by B. F. Avery.

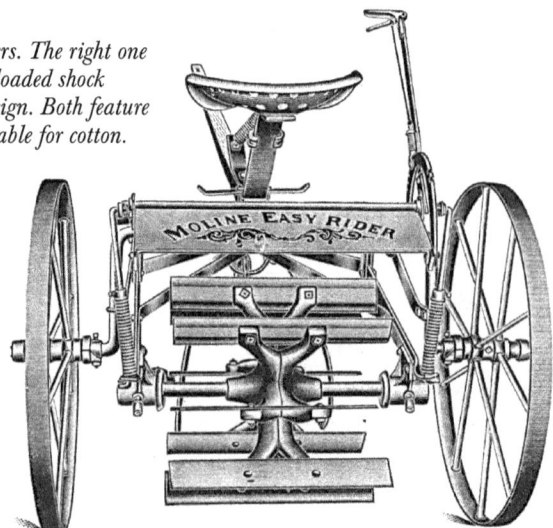

Moline stalk cutters. The right one features a spring loaded shock absorbing seat design. Both feature a single head suitable for cotton.

50

B. F. Avery's fourteen knife two row three horse machine.

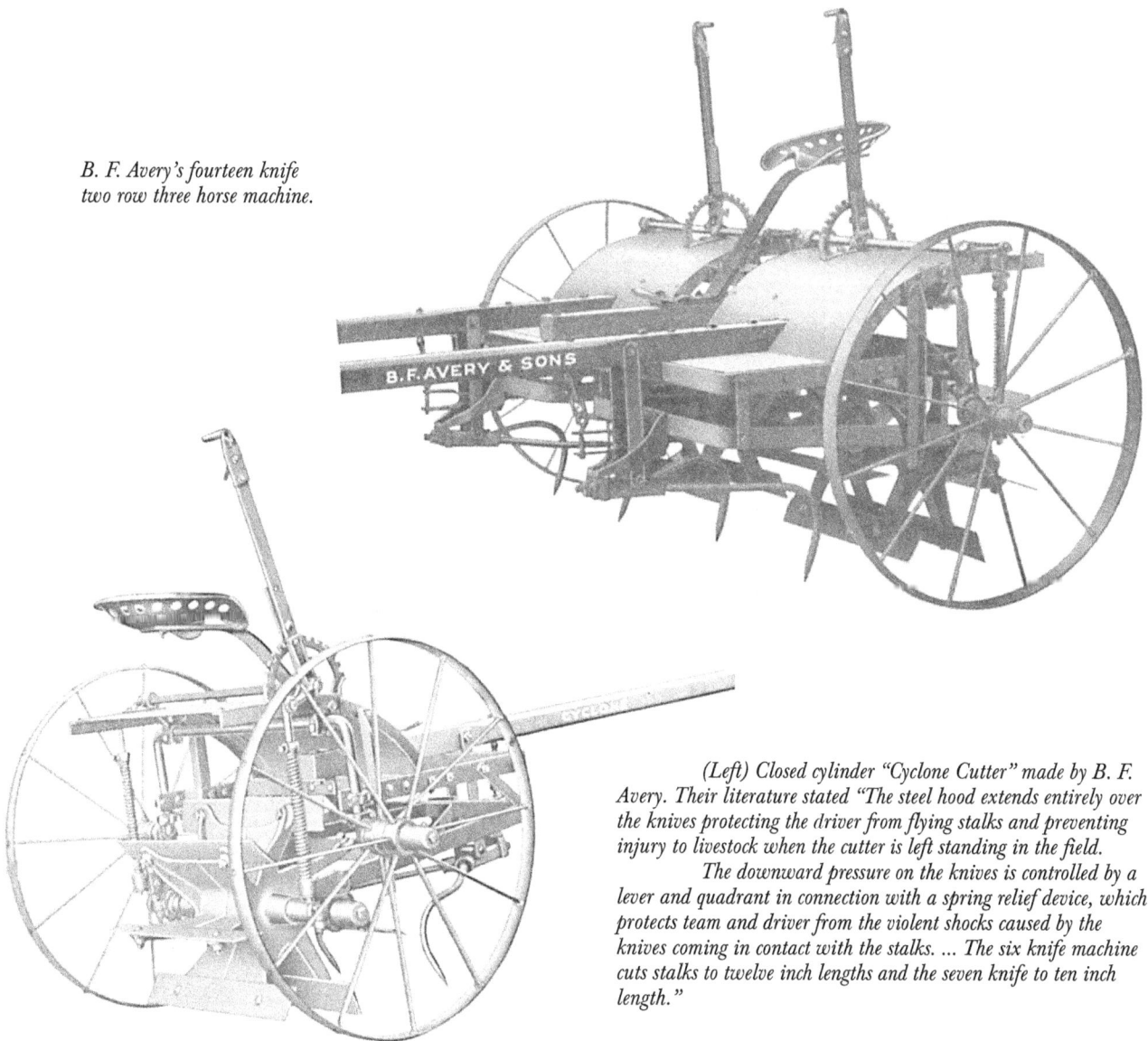

(Left) Closed cylinder "Cyclone Cutter" made by B. F. Avery. Their literature stated "The steel hood extends entirely over the knives protecting the driver from flying stalks and preventing injury to livestock when the cutter is left standing in the field.

The downward pressure on the knives is controlled by a lever and quadrant in connection with a spring relief device, which protects team and driver from the violent shocks caused by the knives coming in contact with the stalks. ... The six knife machine cuts stalks to twelve inch lengths and the seven knife to ten inch length."

Moline's two row three horse stalk cutter.

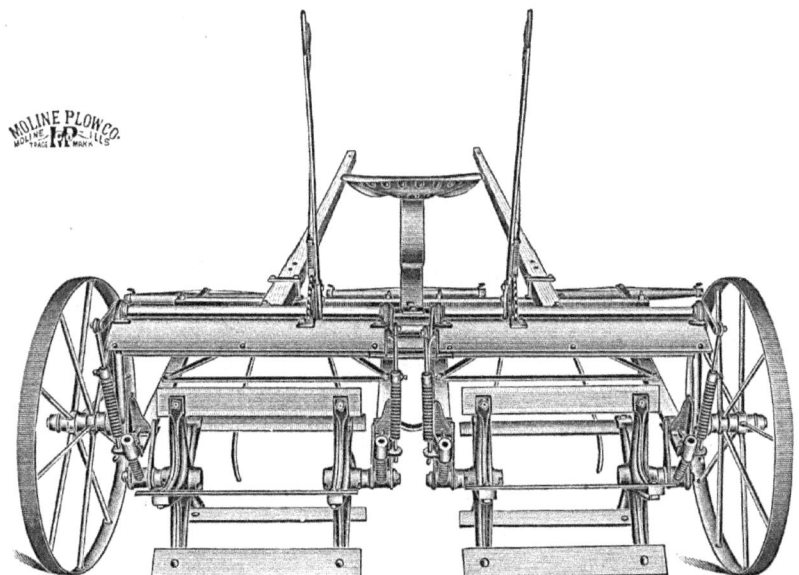

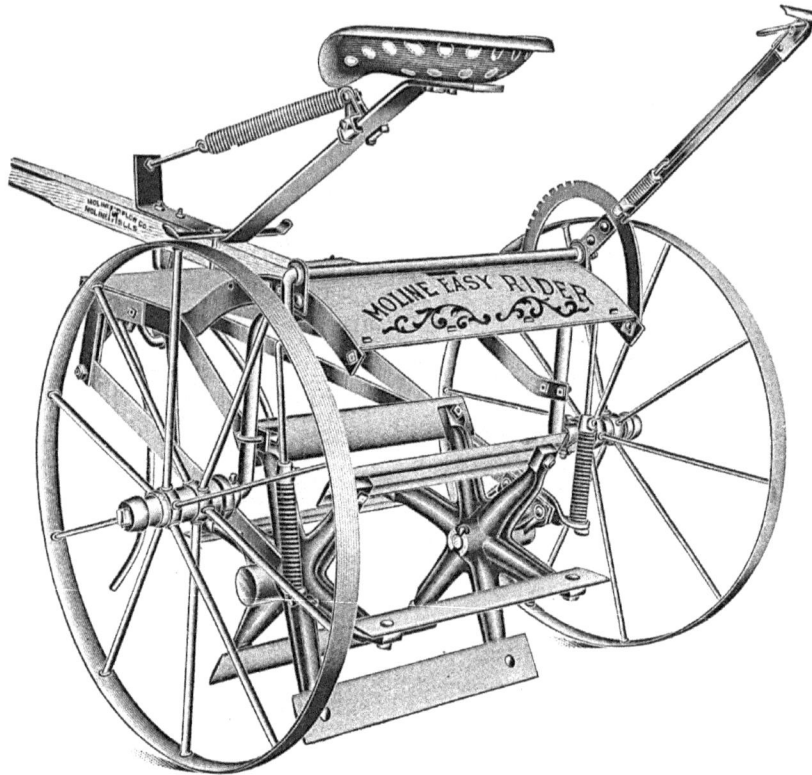

The Moline plow company built several models of stalk cutters including the Easy Rider No. 4 shown on this page. The top image is of the regular head, the bottom unit is called the 'Texas Head.' The Moline literature included this statement:

"To maintain the fertility and productiveness of the soil, it is absolutely necessary to keep up the supply of organic matter or humus. The Moline Easy Rider Stalk Cutter chops up the stalks so that they may readily be plowed under and will quickly rot, providing ample humus."

Moline and other manufacturers offered both single head and double head one row choppers. The single head was designed for chopping cotton stalks.

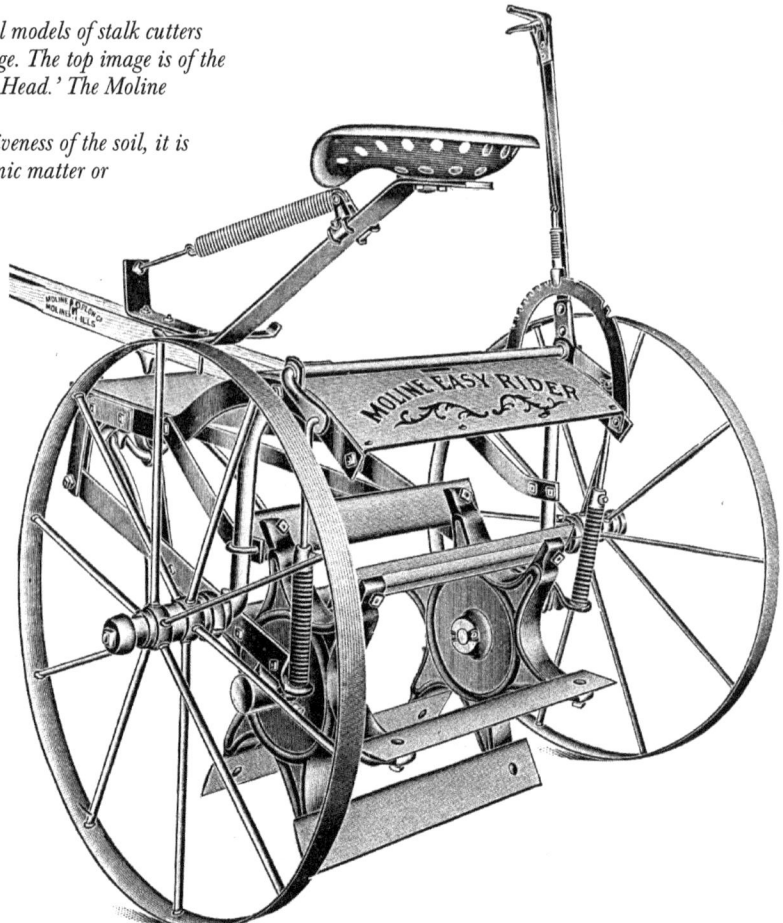

*Moline's Gladiator Steel Frame Stalk Cutter
built to be strong and light. The cutting head
was held down under double spring pressure. The
second spring action is intended to produce a
chopping motion. This spring also relieves the
frame from unnecessary vibration.*

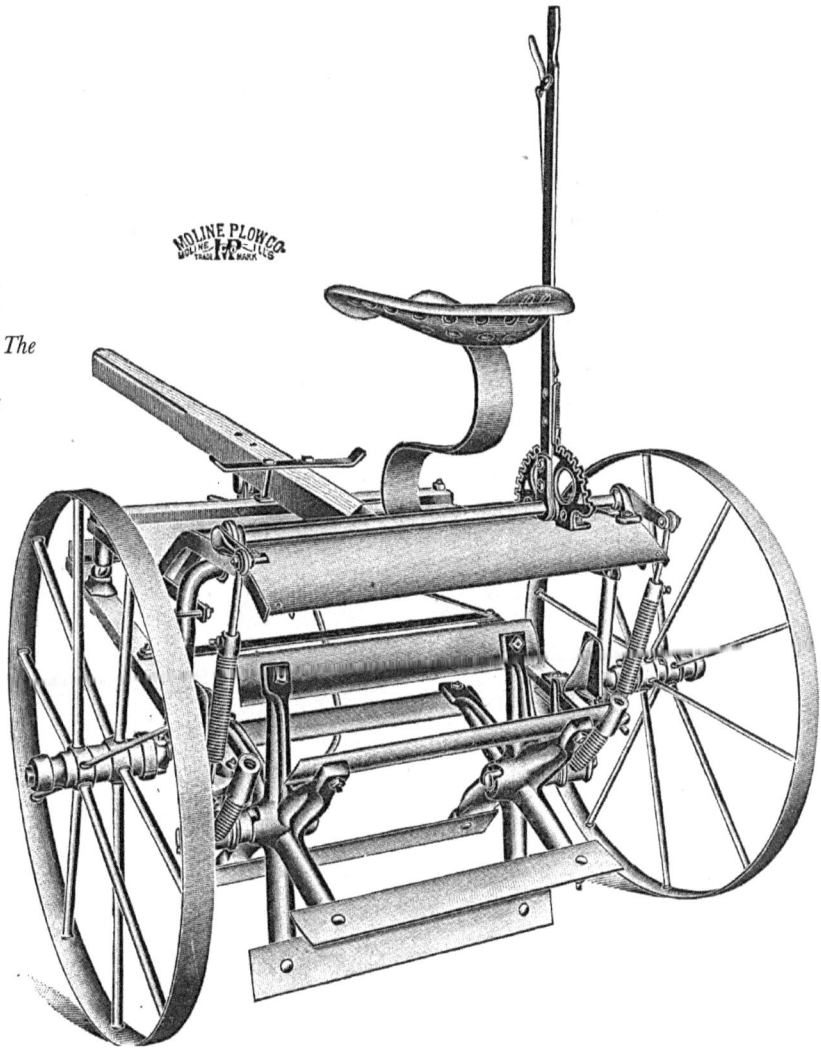

ROCK ISLAND STALK CUTTERS

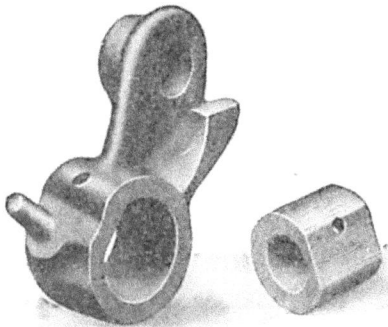

Wood (hard maple) bearing for
cutter-head

78

Rock Island No. 28 Single Row

The Vulcan Plow company had this to say about their stalk cutters:

A good Stalk Cutter is a necessity on every farm. The use of a Vulcan Stalk Cutter is by no means restricted to the cutting of corn and cotton stalks, but can also be used very effectively on larger kinds of obnoxious weeds.

The Vulcan Single Head Stalk Cutter had been designed to meet the demands for various sections and is one of the most popular tools of its kind made.

The Heads are located in the center with extension arms, which provide a strong rigid base for the blades. This style cylinder has the advantage that it clears itself and does not clog.

The work of cutting stalks is usually done at the time of the year when the ground is hard to work. This naturally makes the work severe on the machine. The Vulcan is built to withstand this kind of usage.

The Draft Equalizing Spring, upon which all of the pull comes, prevents transmission of the vibration of the cylinder to the team. This makes a steady draft and helps

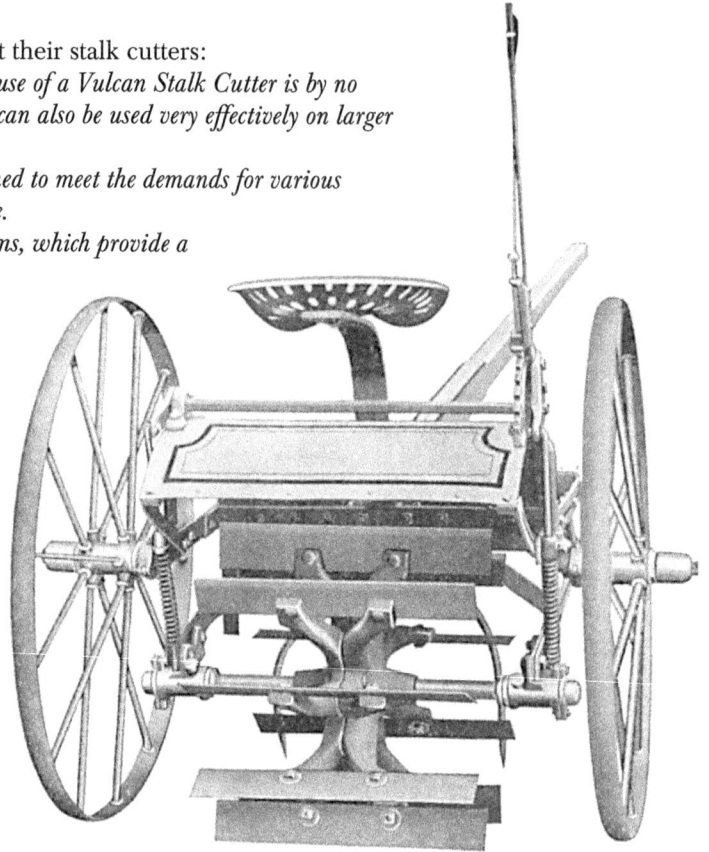

the knives to give a striking blow.

The Vulcan Stalk Cutter has a staunch, sturdy as well as graceful appearance and it is attractively painted - and above all it does highly efficient work.

10, 12, 14, 16, or 18 Blades

The number of blades corresponded to the length the stalks were cut. The larger the number, the shorter the stalks.

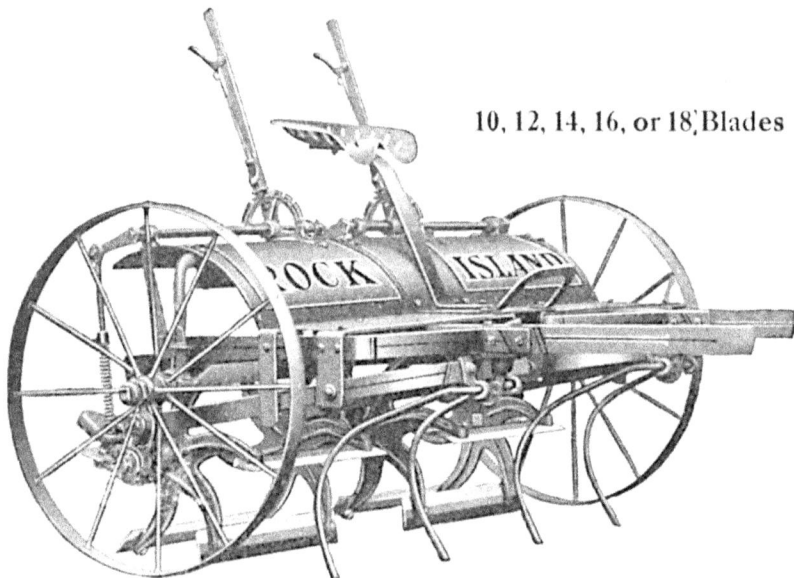

Rock Island No. 29 Double-Row

SC STALK CUTTER
14-Knife—Double-Head

This cut shows the
style of evener
recommended for the
Moline two row stalk
cutter. Note: it pins in
two places, usually on
a bar hitch whose
center is attached to a
heavy spring which
aborbs the chopping
action.

No. 23, 3-Horse Hitch for the Double-Row Stalk Cutter

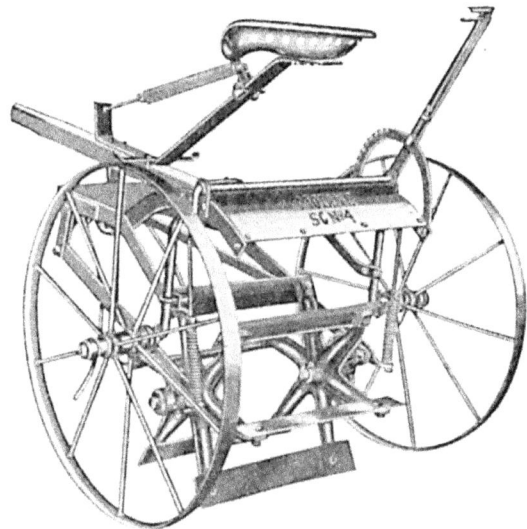

SC 7-Knife Double-Head Stalk Cutter

*The No. 1 machine. The heavy heat treated steel axle will not sag and allow the
wheels to get out of line.*

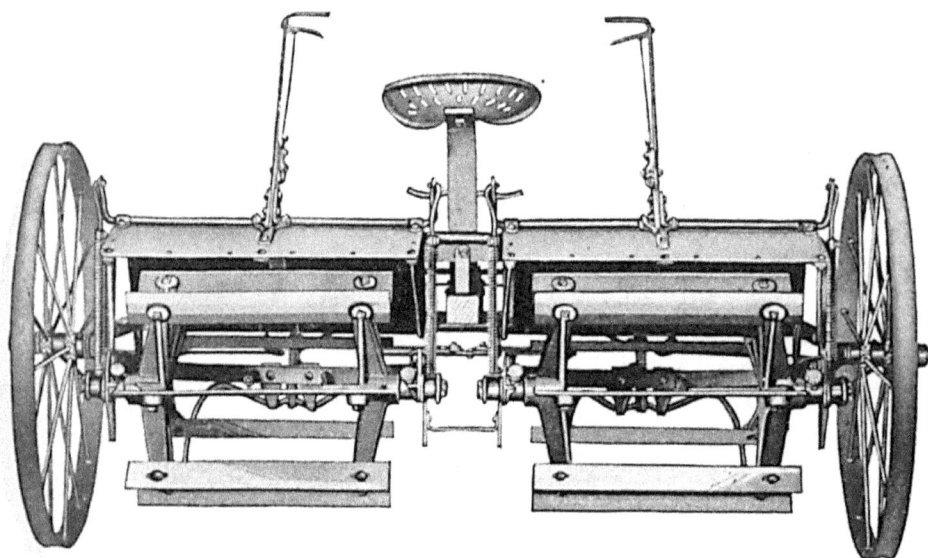

Showing how the trash guard is attached to the knives.

Oliver stalk cutters are known as easy riding machines as all jolts and jars are absorbed by springs. This rear view shows the trash guard (special equipment), the heavy springs which insure positive cutting and also absorb jars and vibration. Simple though these machines are, they do excellent work.

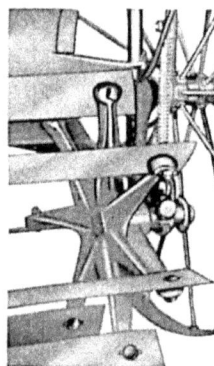

Left: Pressure is applied on the knife heads by powerful coil springs on heavy rods. The springs cannot buckle. They give a chopping motion to the knives which more readily cuts the stalks.

Stalk
Cutters

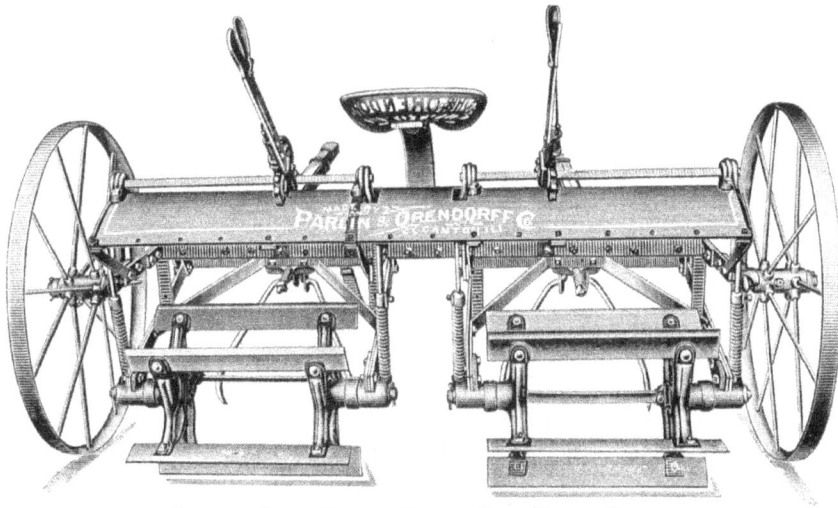

CANTON STEEL FRAME DOUBLE ROW STALK CUTTER

Our Double Row Stalk Cutter is of the same general construction as the Single Row, in fact, it is practically two single row machines combined, having two knife heads, two levers and two tongues, and is furnished with three horse evener and neckyoke, which please remember when comparing price with others. This cutter is popular on large farms, for, by the addition of one horse, a man can do twice the work that can be done with the single row.

As noted elsewhere, *Parlin & Orendorff* merged through acquisitions with *McCormick Deering* and *International* which retained P & O's many outstanding design innovations. You may have an IH or McD stalk cutter which looks exactly like these P & Os because it is, except for name, the same unit.

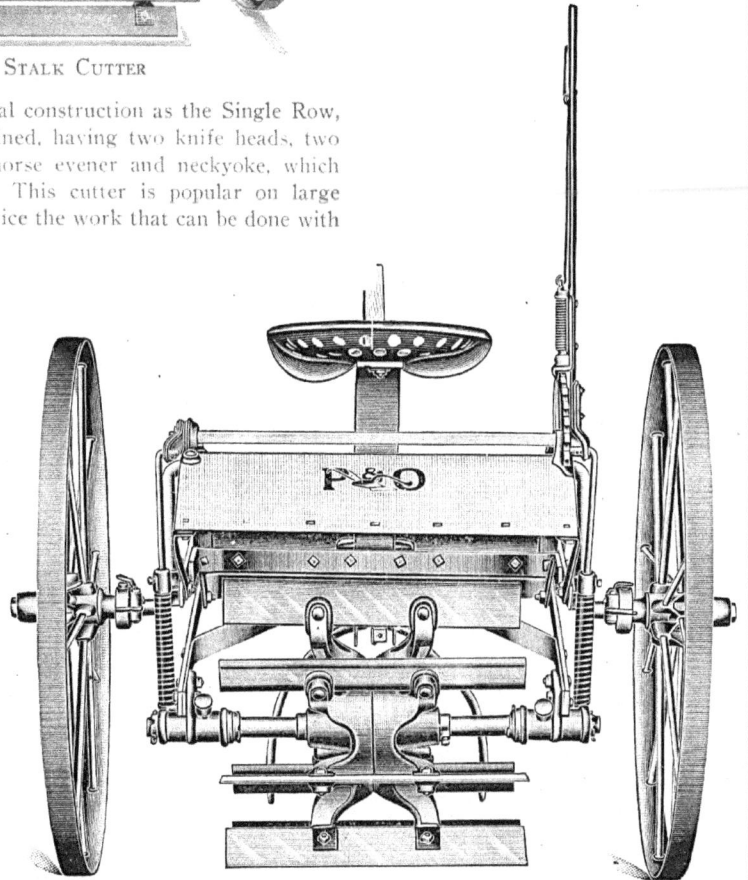

8-KNIFE SINGLE-ROW STALK CUTTER, SINGLE HEAD.

CANTON SINGLE ROW STEEL FRAME STALK CUTTER

The stalk cutter is an implement whose design should suggest many inventive spinoffs. I personally have never used, or even owned, a stalk cutter but would like to. It, although not strictly essential, could be a beneficial implement for most any mixed crop farm. Should you have an opportunity to find a cheap one in complete condition I suggest you consider adding it to your treasures.

12-KNIFE DOUBLE-ROW STALK CUTTER.

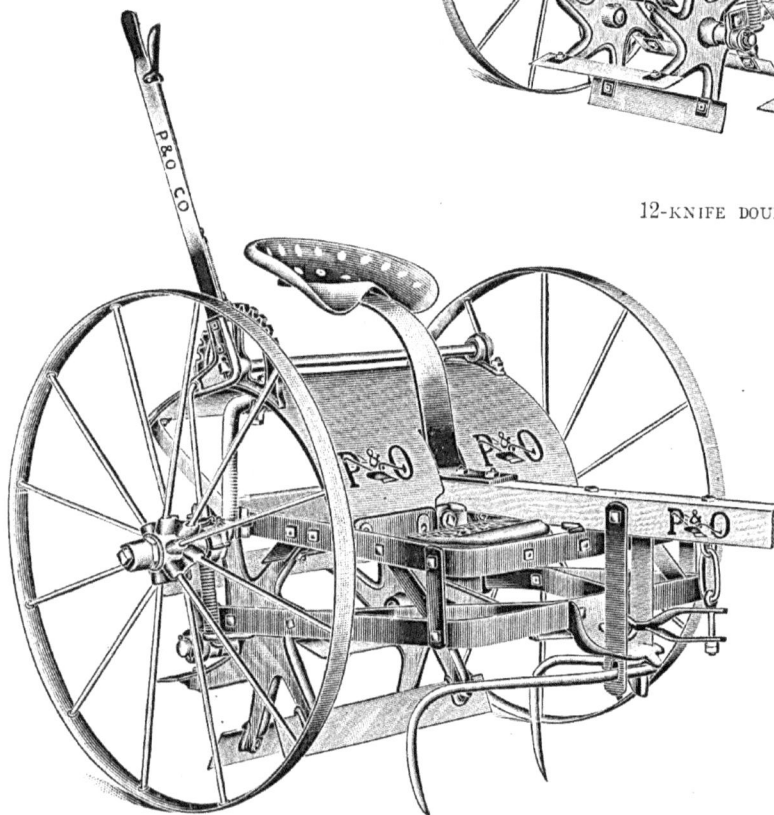

SIX-KNIFE SINGLE ROW STALK CUTTER.

Horse Notes: The stalk cutter is NOT an implement to use to break horses! A rhythmic vibration with resistant pop and release actions all occuring behind the horses can drive a nervous green broke animal to distraction. Calm well trained horses will quickly appreciate the assuring repetition and accept the process.

Always hook the neckyoke first on any tongue implement THEN the tugs. Make certain that you are NOT hooked so long that neckyoke comes off the tongue end. (Please refer to the *Work Horse Handbook* and/or *Training Workhorses*.) Care MUST be taken not to allow the driving lines to drop into the cutting reel.

Chapter Three

Discs & Discing

Throughout this chapter you will see two different spellings of 'disc' (disk). They are both correct and it is up to the user to choose. We have chosen to write our portion of the text as "disc".

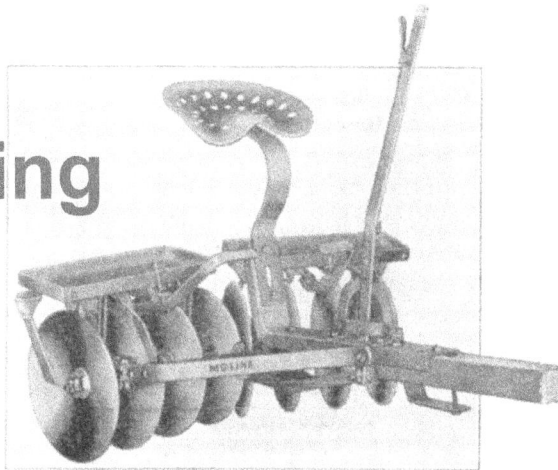

No. 10 Harrow with Straight Pole

The disc, or disc harrow, as opposed to disc cultivator*, slices plowed and/or unplowed ground into smaller particles (or pieces as in the case of crops and crop residues). The implement consists of a frame from which hang secured axles which, with spacers, hold seperated free wheeling disc blades. These blades turn from contact with the ground over which they are pulled. Each single line of standard disc harrow is usually comprised of two separate sections referred to as "gangs." The disc blades are customarily set in opposing directions and the gangs may be set at more or less of an angle. The increased angles result in more stirring action (and greater draft). Most horsedrawn disc harrows are singles (meaning they have two gangs in a single line as the implements shown on this page). A double disc is most often comprised of 2 pairs of gangs in two offsetting lines - (see photo page 61).

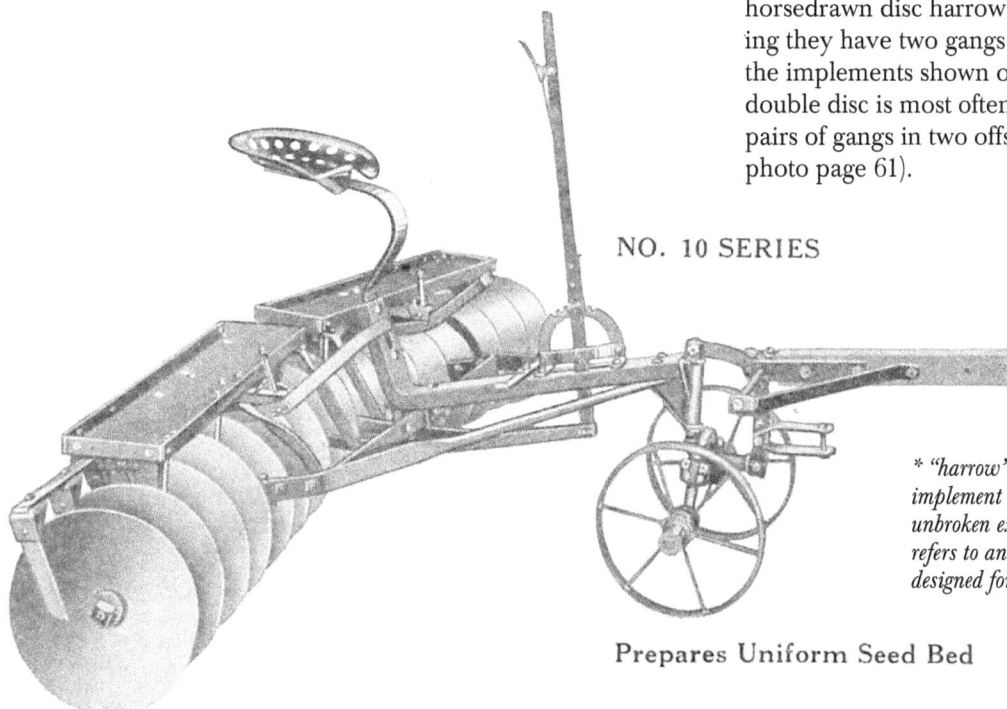

NO. 10 SERIES

* "harrow" refers by usage to an implement working across a broad unbroken expanse while, "cultivator" refers to an adjustable implement designed for use with row plantings.

Prepares Uniform Seed Bed

Disc gangs flex and may ride up at center, for this reason the better designs are equipped with adjustable apparatus which can hold the center down. This may work, at times, against the farmer because the center held down would cause the entire disc to rise above ground if it were hitting a rock or obstruction near the center. As the old photo on page 59 illustrates, with the center free to float - riding over a rock does not then cause the gang to rise out of the ground.

The discs offered, from the turn of the century to World War II, came either with or without tongues (or poles) and with or without tongue trucks (or truck wheels). Those disc harrows which came without truck wheels invariably featured a tongue attached directly to the implement frame (as seen at the top of page 59). One notable design variant is that the tongue or pole might have been attached lower or higher at the frame. Those harrows with pole attached direct to frame could be backed up slightly when the discs were set straight. Whereas those harrows with truck wheels could not be backed up without causing the implement to jackknife up or sideways.

One exciting event in modern horsefarming is the successful work developing brand new disc harrows. Most notable is the new KOTA brand shown at the end of this chapter. Given that this is a heavy implement by any design, shipping costs become the main challenge to widespread distribution. This author wishes to encourage small on-farm shops to consider designing and building horsedrawn discs for their own farming region.

Horse Notes: The single disc harrow running over plowed ground is an excellent procedure for "working down" anxious horses or mules. A double disc, by virtue of its design, can put the teamster in harm's way should something go wrong and therefore is not such a good idea. (It's darned hard to gracefully or safely leave the seat and get clear of the back run of a double disc should the implement reach 20 or 30 miles per hour.)

How many horses? A six foot single disc may be comfortably pulled by two horses. The nature of the soil, sand versus clay - wet versus dry etc., will have an effect on how hard horses will have to work. An eight foot disc will require three animals, ten foot requiring four. A double disc with countering angles might require more animals than the simple addition would suggest.

How many acres per hour? Two horses will disc ten acres per day. Four horses will disc 20 acres. Six horses will disc 30 acres. Eight horses will disc 40 acres per day.

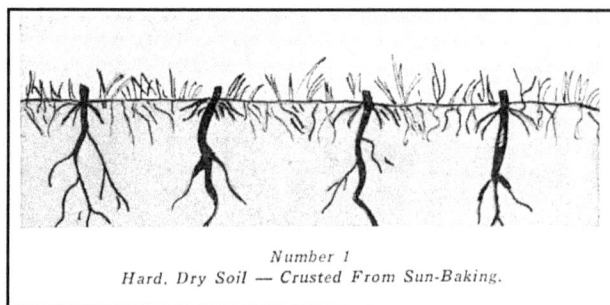

Number 1
Hard, Dry Soil — Crusted From Sun-Baking.

Why Disc Ahead of Plowing?

Number 2
This Soil Has Been Plowed But Not Disced.

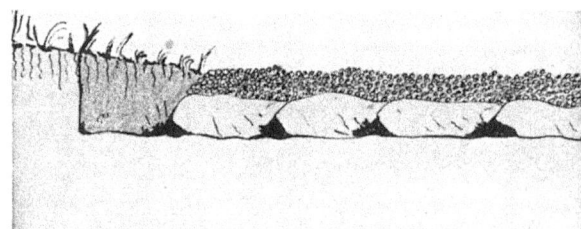

Number 3
This Soil Was Disced After Plowing. Note Open Spaces —
Called Air-Pockets

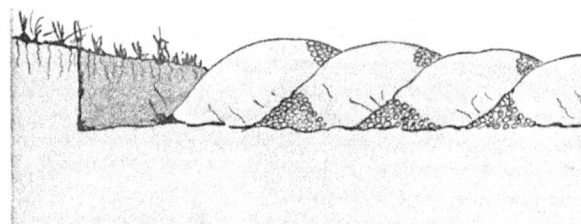

Number 4
This Soil Was Disced Before Plowing. Open Spaces Are Filled
For Good Subsoil Contact

Number 5
This Soil Was Disced Before And After Plowing.
This Makes The Ideal Seed Bed.

A double disc set extreme proves to be a hard load for these three horses.

Diss

MOLINE SINGLE-LEVER DISC HARROW
NO. 10 SERIES

Moline Features

Center Depth Control and Gang Angling
Bars, with Adjustable Roller

Flexible hinge between front frame and gangs
facilitates angling - reduces end thrust on
bearings.

Each alternate disc has been heat treated. Note the smooth edged treated discs.

Moline Transport
Protect the Discs

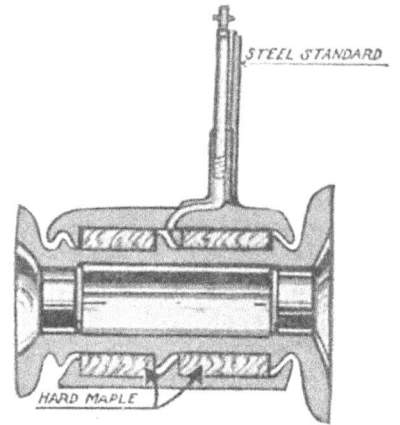

Moline Double End Thrust Bearing
with Alemite Oiler

Bearing assembly for disk harrow.

Chilled-iron bearing for disk harrow equipped with pressure oil fitting.

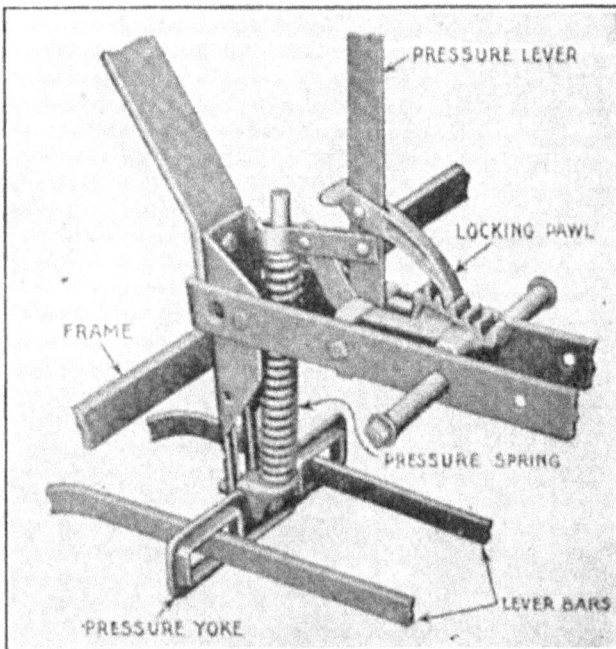

Center-depth regulator.

On the older disc harrows hardwood bushings were commonplace. Forgetting to grease the disc FREQUENTLY meant certain death or deterioration for the bushings. It's a good idea to grease them every four hours of use. And make doubly sure that grease tubes, cups and/or zerc fittings are working freely.

Sharpening disc cleaners, and in some cases even discs, will improve performance. It is a mistake to run disc blades on pavement or gravel.

Discs

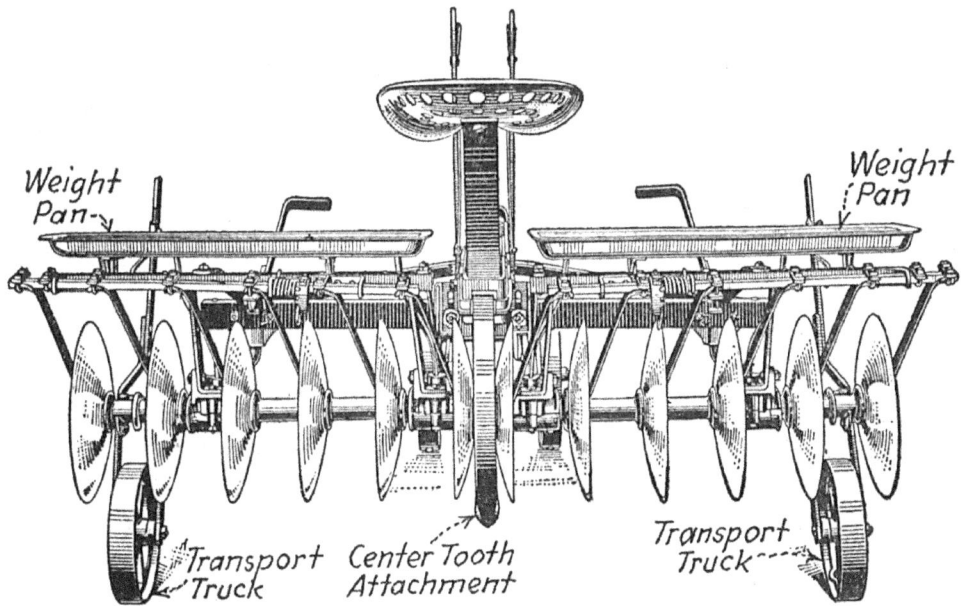

Single-disk harrow equipped with weight pans, transport trucks, and center-tooth attachment.

A few discs were offered with the option of a center spring tooth.

Reversible-disc harrows, sometimes called orchard harrows, are extremely handy and adaptable.

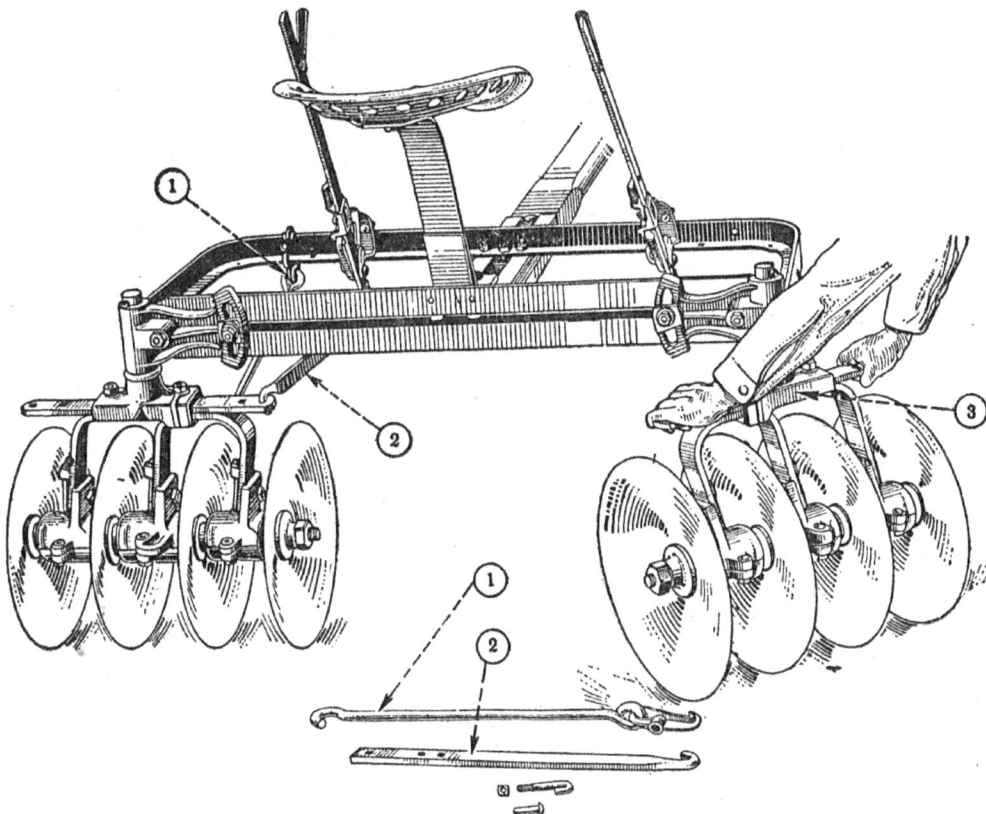

Reversible-disk harrow showing method of changing from outthrow to inthrow, or *vice versa.*

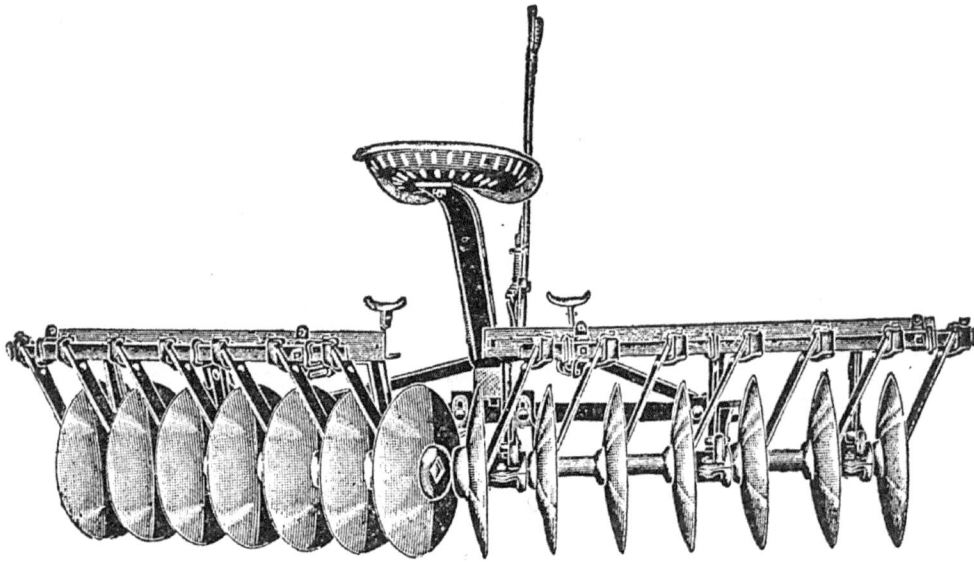

The full disk harrow.

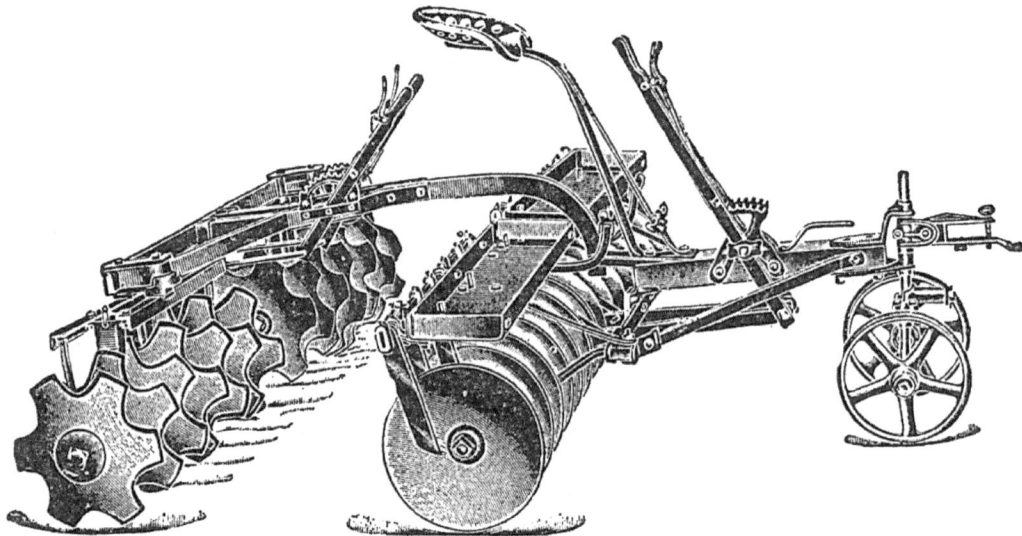

A double disk harrow   Cutaway disk in rear

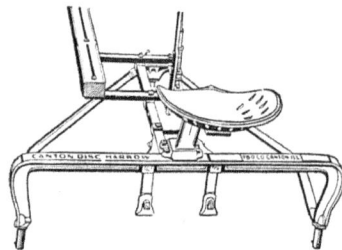

This shows the frame with the tongue set
for three horses.

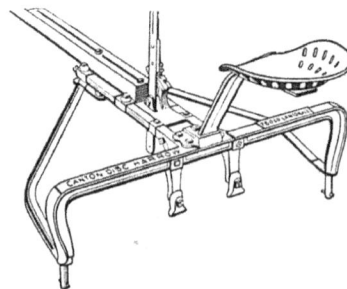

This shows the frame with the tongue set
for two or four horses

The P & O (IH) disc frame illustrating how the tongue was offset for three abreast.

The information below was taken from the Parlin & Orendorff catalog of 1912 and illustrates fairly typical descriptive language published by manufacturers of disc harrows, though P & O was certainly a cut above 'typical'.

P & O

Canton Disc Harrow, One Lever

It is simple, very strong and durable, and easily operated, and further-more, does not require weight boxes. It has the following desirable features, which make it very easy for the dealer to sell:

The arch is made of T-shaped steel, curved downward at the ends and thoroughly braced from the stub tongue. The gangs are swiveled directly on the ends of the arch. The front projection, or stub tongue of the frame, to which the tongue is attached, is constructed from angle steel, and so arranged as to place the tongue in the center for two or four horses, or at one side, to eliminate side draft when three horses are used.

Journal bearings have chilled and wood wearing surfaces, are dust-proof, and have self-closing oil holes, excluding all dirt. There is absolutely no end thrust to wear on the bearings, as the gangs come together at the center of harrow, one side offsetting the strain on the other side.

Either 16, 18 or 20 inch discs can be furnished, and are the best that can be secured.

The inner ends of the gangs are held firmly down so center of harrow can not rise out of the ground, but they can readily adapt themselves to dead furrows, or low places in the ground.

These disc harrows are made with one or two levers, as desired. The two lever is more easily handled, and the gangs can be given different angles for working on a side hill or overlapping.

Being made entirely of steel and iron (with the exception of the tongue), it can not be other than strong and durable, and experience has proven this to be the case.

Canton Disc Harrow, Two Levers

The scrapers have steel cutting edges, bearing against the blades, and each scraper is independent of the other, and held against the discs by springs.

Reversible Disc Harrow

For use in orchards and vineyards this disc harrow is unsurpassed, and meets all requirements for an easily operated, substantially constructed, Reversible Harrow. The frame is made of steel, well braced, and of sufficient width to

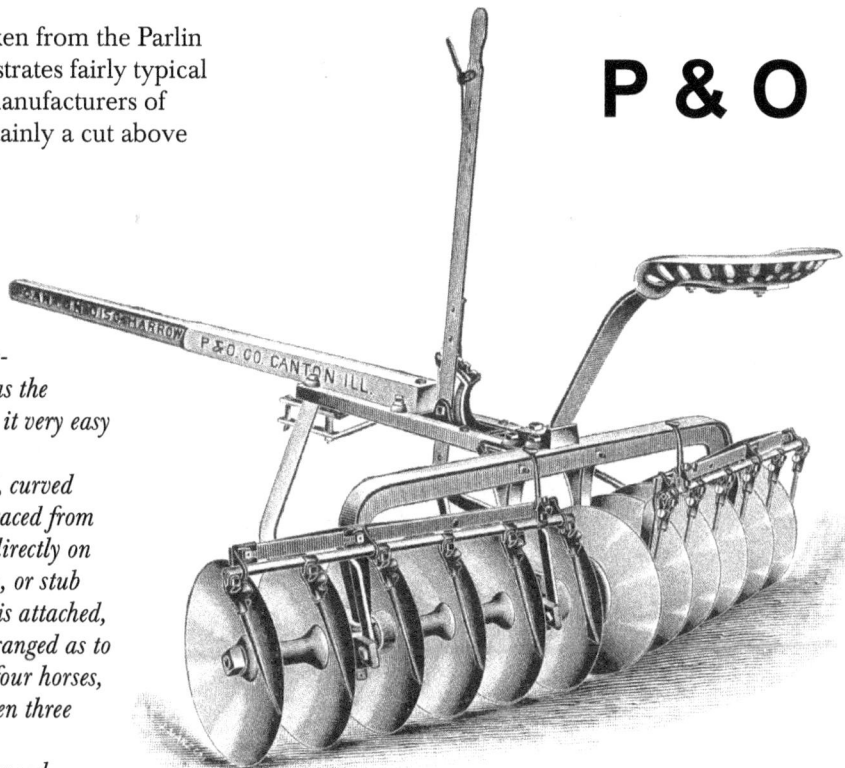

CANTON DISC HARROW, ONE LEVER

permit the gangs to be adjusted out for cultivating under the overhanging branches. Each gang is angled independently of the other, making it easily handled, and also allowing the gangs to be run at different angles to prevent crowding to one side on a side hill, or when overlapping.

Bearings are dust-proof and provided with suitable oil holes.

Tongue can be set in the center, or at the side, adapting it to the use of two, three, or four horses.

Pony Reversible Disc Harrow

This is a light, one horse Disc Harrow, much used in cultivating orchards and vineyards, as the gangs may be adjusted in or out, swiveled round to throw as much or as little dirt as desired, as well as from in-throw to out-throw, and can also be tilted either way.

Cutaway Disc Harrow, with One Lever

This Harrow is the same as the regular Canton Disc Harrow, except that the discs are notched as shown in cut on page 67, and the scrapers are different - being of a form especially adapted to the cutaway discs. The discs are ground to a sharp cutting edge throughout, giving a long cutting edge on each disc, and also a longer cutting edge on the ground. The Cutaway Disc has many friends, some claiming that it will do more thorough work than the regular plain disc, that it penetrates deeper and more thoroughly pulverizes the soil.

Canton Spading Harrow

In place of the usual round discs this harrow has a number of curved knives, securely held by iron clamps so that each knife has a long cutting edge, and will penetrate and thoroughly pulverize the ground. The knives have rounded cutting edges, and are set exactly in line with each other, as a result of which they do not gather trash and become clogged.

There are many who prefer this style of harrow to either the solid or cutaway discs, claiming that it more thoroughly pulverizes the ground and does not leave ridges; that it covers up trash and weeds, works lighter, and can be used where an ordinary harrow would clog and could not be used at all.

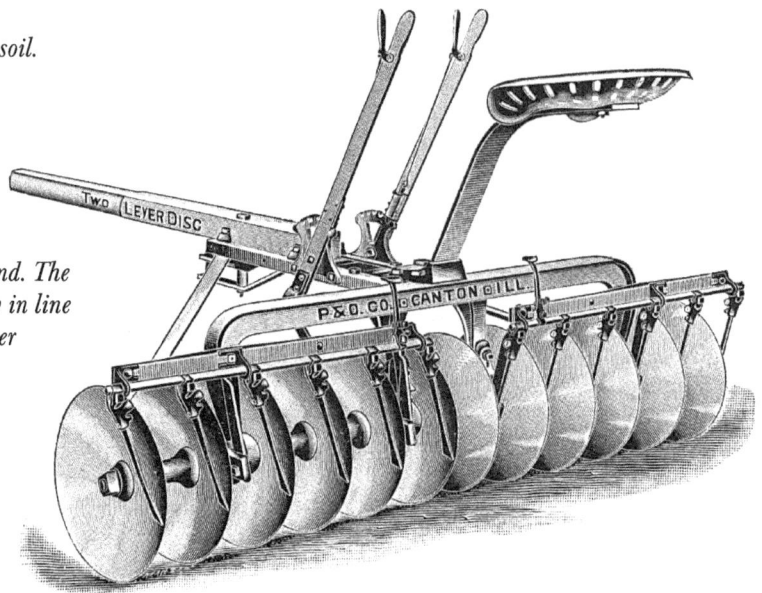

CANTON DISC HARROW, TWO LEVERS

P & O

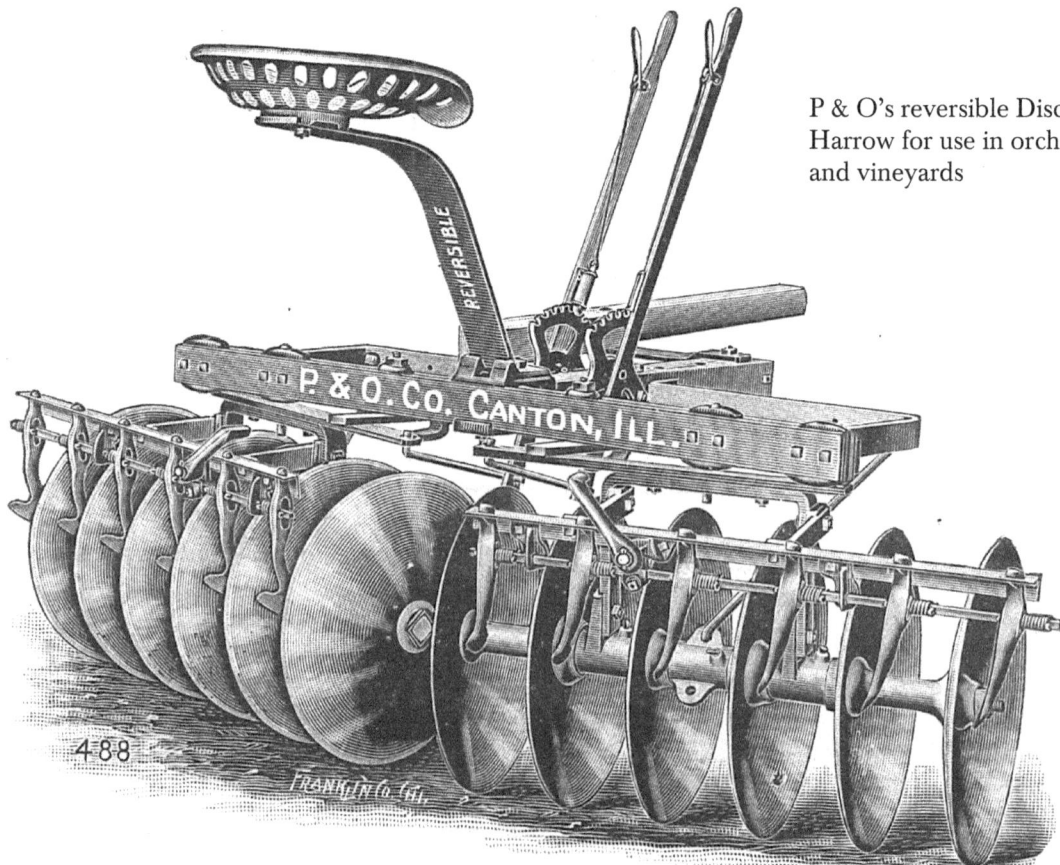

P & O's reversible Disc Harrow for use in orchards and vineyards

P & O

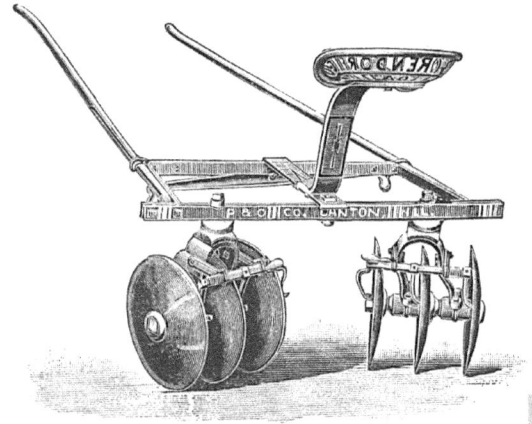

Pony Reversible Disc Harrow

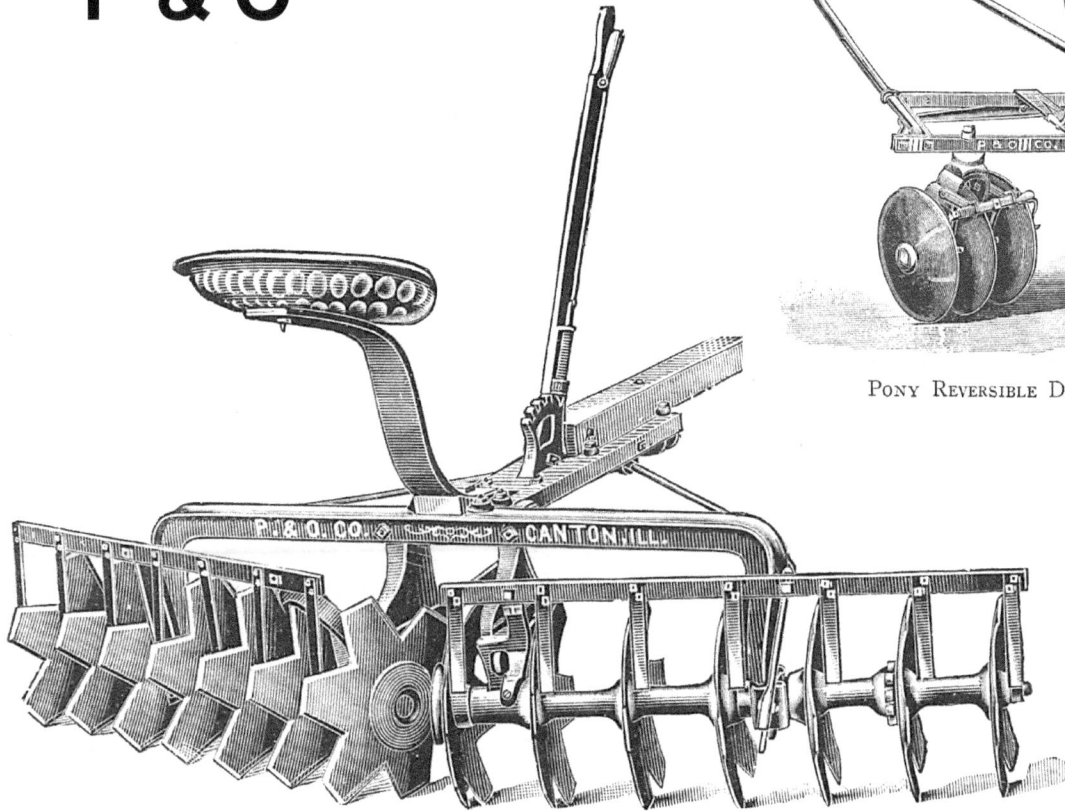

Cutaway Disc Harrow, with One Lever

> NOTE: The cutaway style of disc was not designed to be used on sod. In fact, a well established sod will frustrate the cutaway disc enough to cause a weak farmer to consider early retirement.

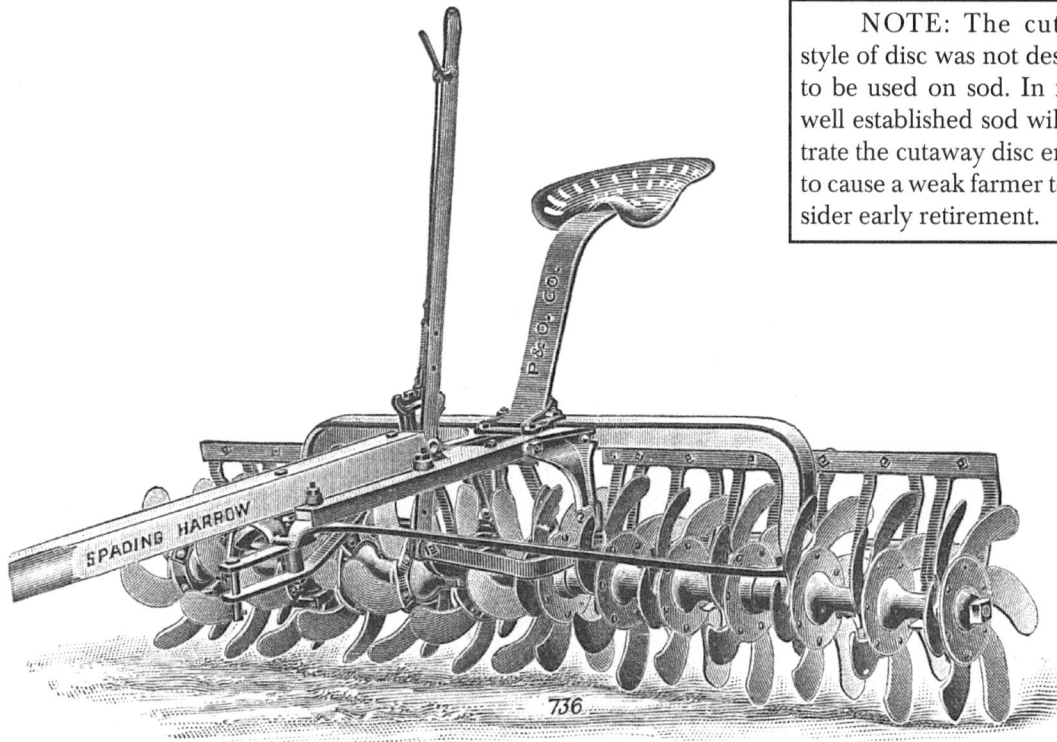

Canton Spading Harrow

Discs

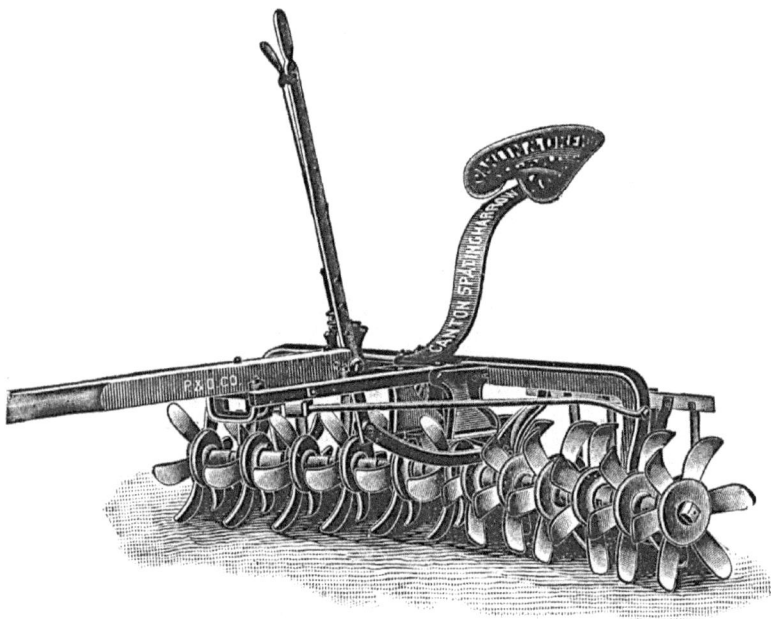

Spading Harrow

P & O

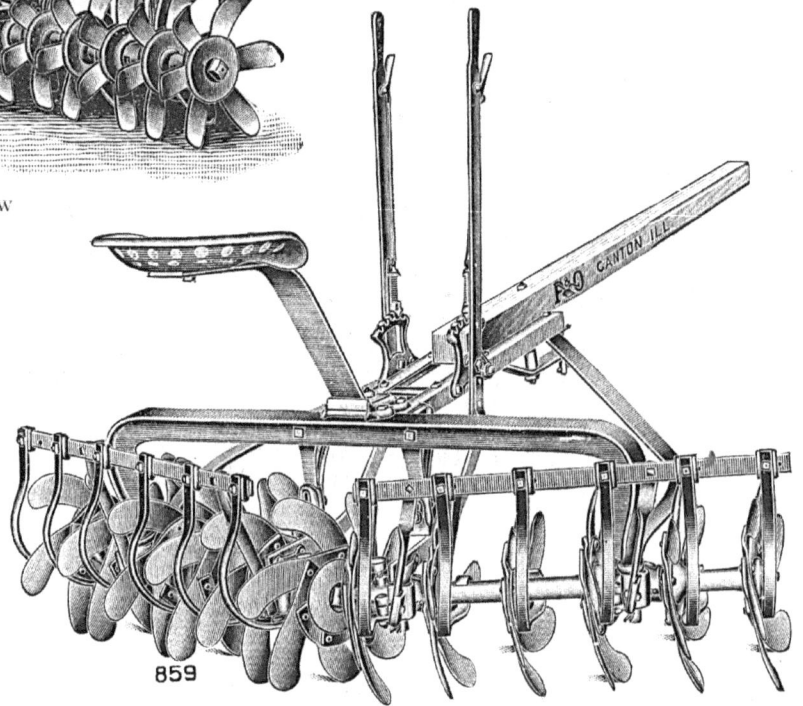

859

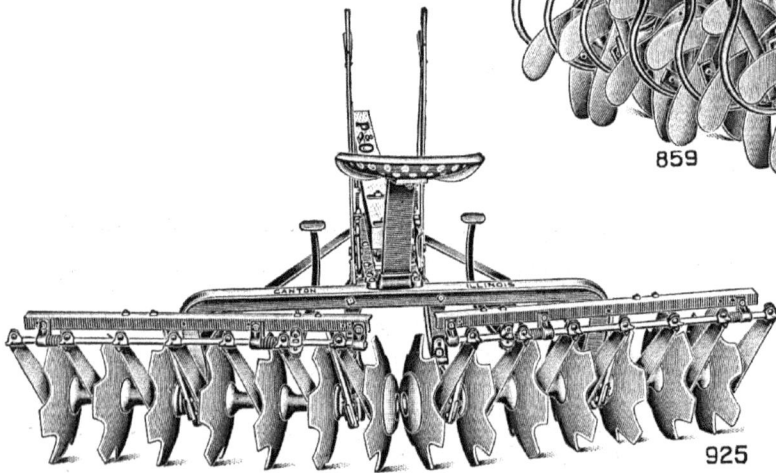

925

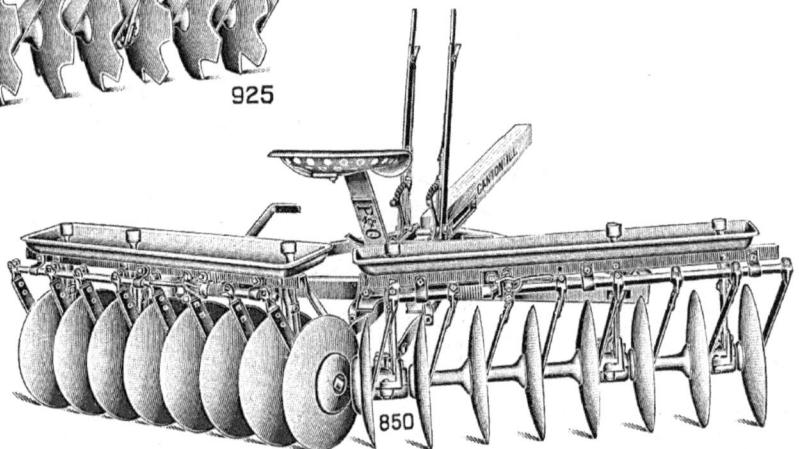

850

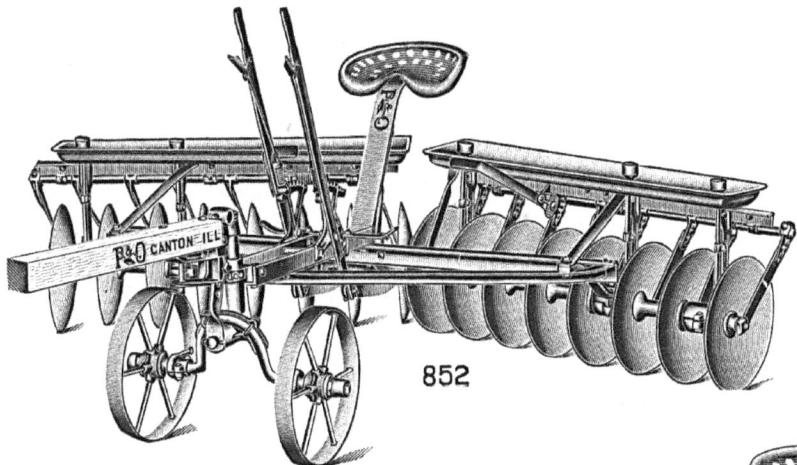

852

P & O

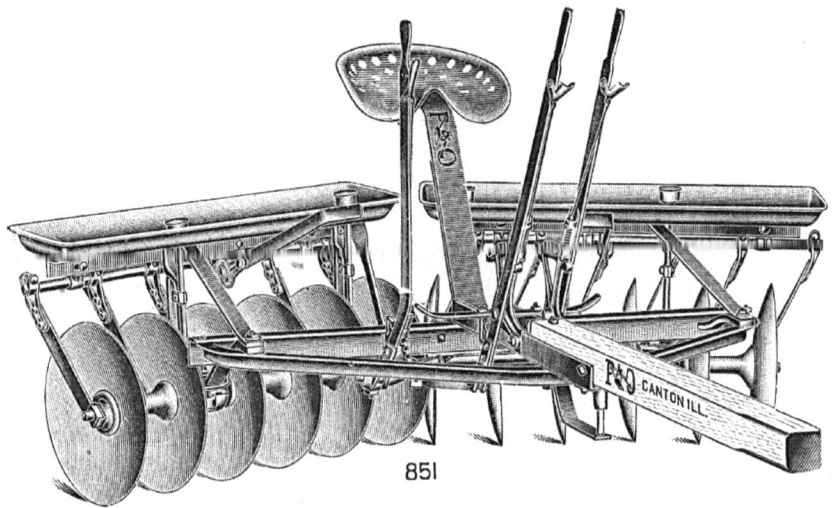

851

231

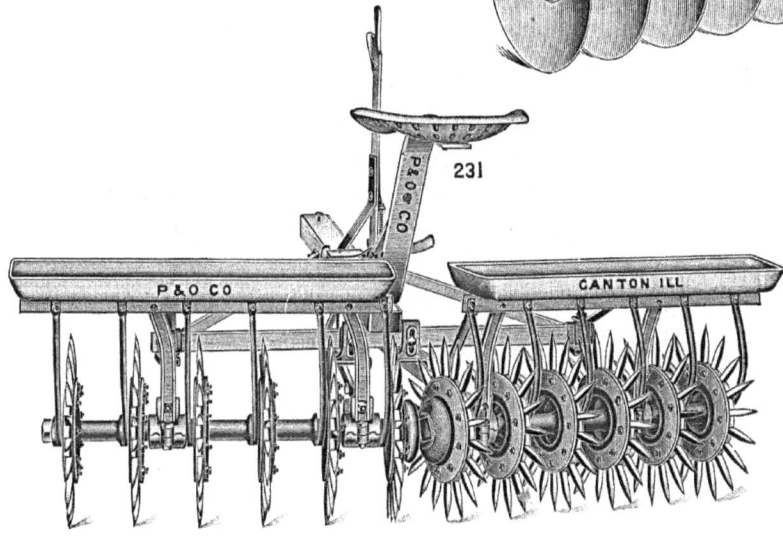

Below, P & O's Star double disc with cut-out tandem attachment and transport trucks in place. P & O advertised that this six foot wide double could be pulled by four horses. The back or tandem section was detachable.

P & O's Afalfa Harrow was designed to go over an alfalfa field after a crop had been cut in order to loosen up the hard baked soil. Also called in some circles a soil renovator, other styles may be seen in our Rotary Hoe chapter.

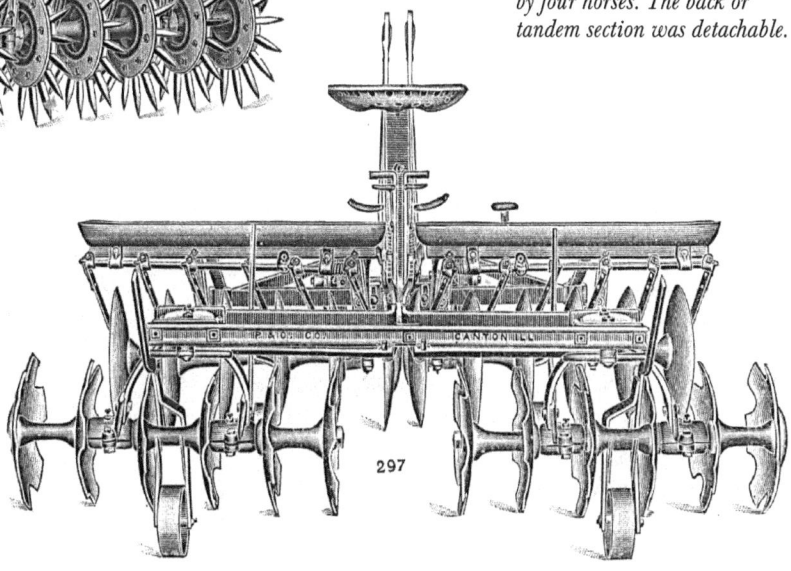

297

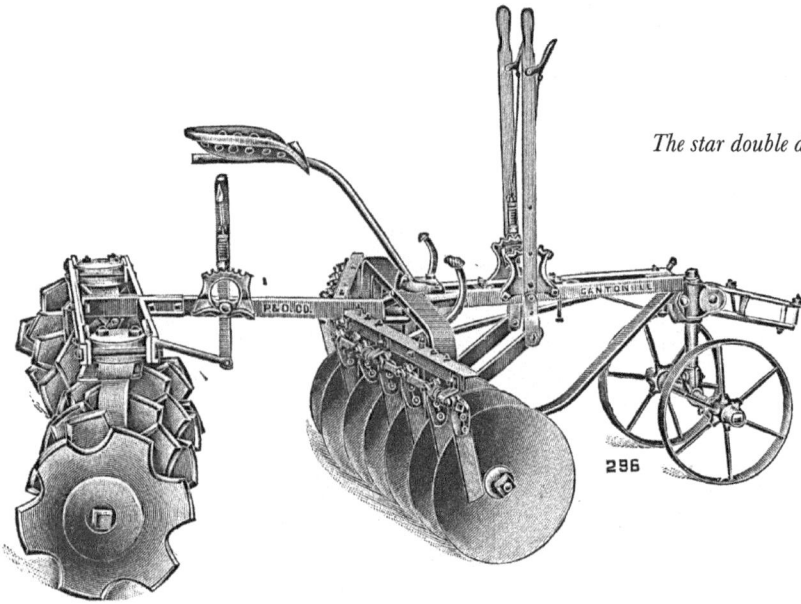

The star double disc harrow, same as on the bottom of page 69.

(Below) No. 3
Reversible Disc
Harrow designed
with truck farms in
mind.

296

P & O

523

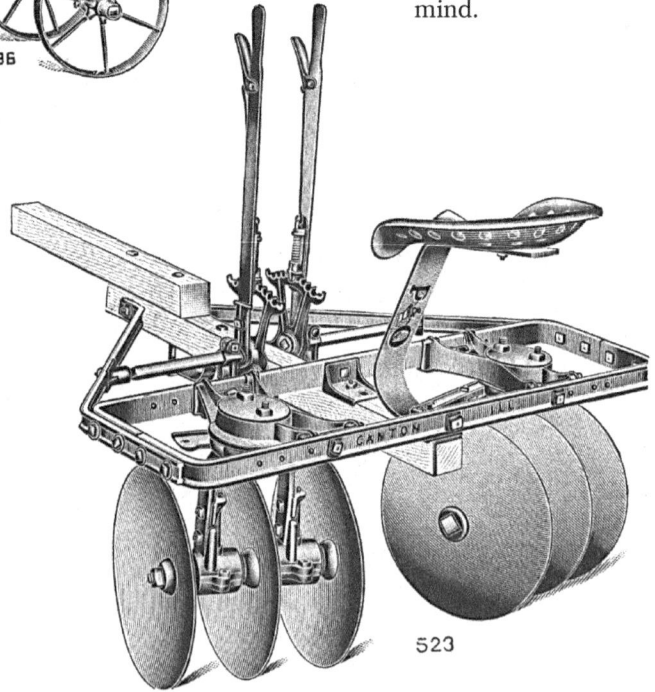

*The No. 2 Reversible Disc Harrow
with the long or wide frame.*

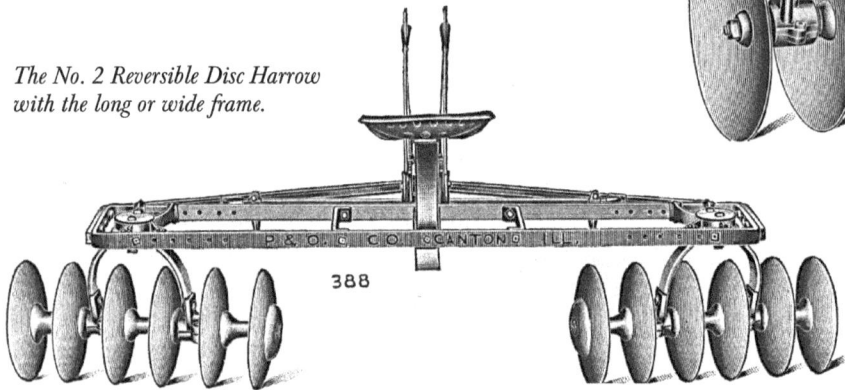

388

*P & O offered a seeder
attachment for its disc
harrows. The seeder was
thrown in or out of gear by a
foot pedal. The power was
transmitted by a chain, driven
by a sprocket attached to the
outside end of the right-hand
gang.*

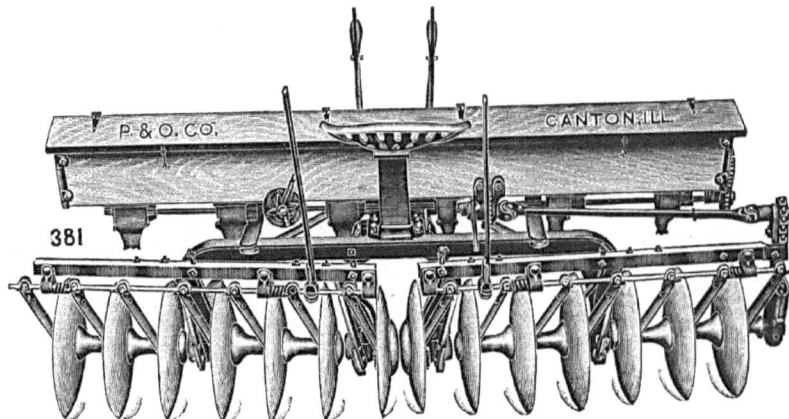

381

P & O

228

Will I Need a Disc?

Yes, If you're wanting to farm with horses you need a disc or two. Our recommendation is that you find a good serviceable reversible disc and, depending on the number of horses you have, a six, eight, or ten foot single disc. (If planning on a large acreage and significant numbers of horses you might want to consider a six or eight foot double disc.) Some mid-century pull-type tractor discs can be made to work with an add on seat (see page 93) or a forecart. In some instances a larger heavier, cover crop tractor disc might be called for and can be made to work with enough horsepower.

Discs

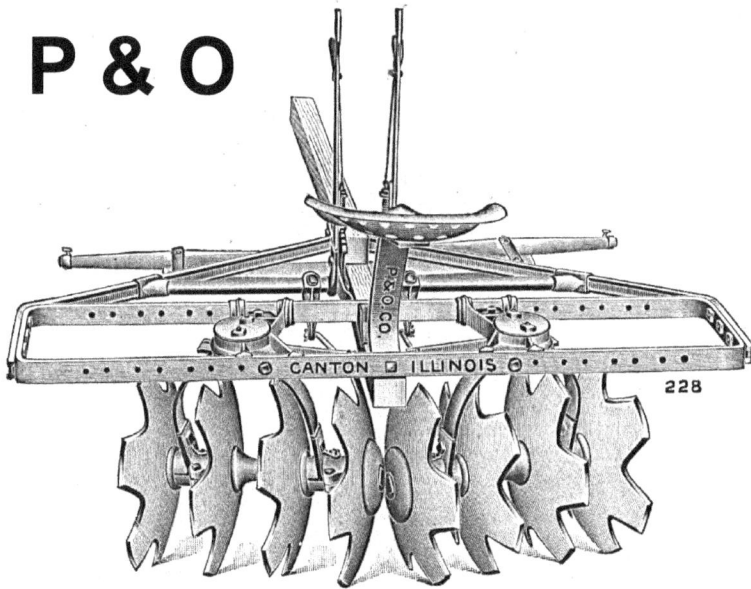

(Above and right) No. 2 Cut-out Reversible Disc Harrow - medium frame. When the gangs were set full out (below right) there was a clearance of 32 inches between the inner ends.

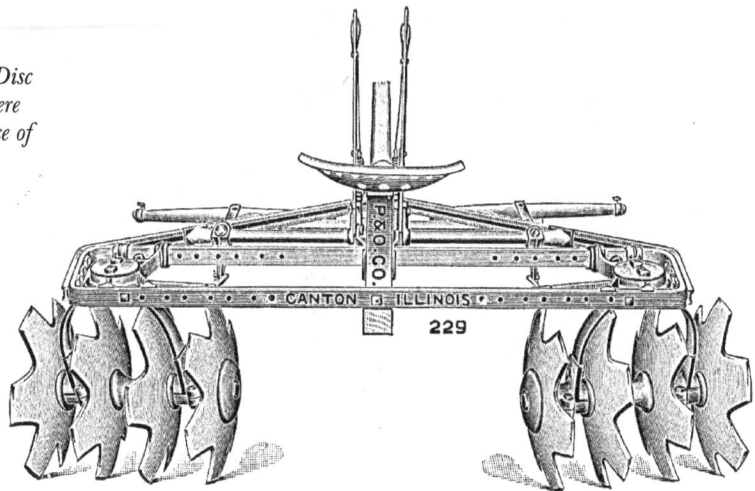

229

No. 2 Cut-out Reversible Disc Harrow - short frame.

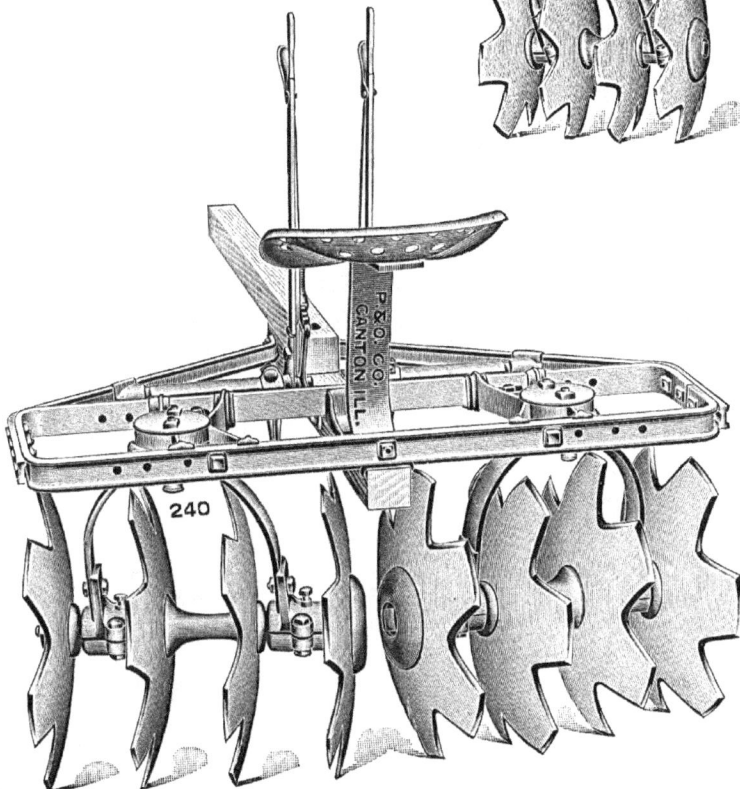

240

71

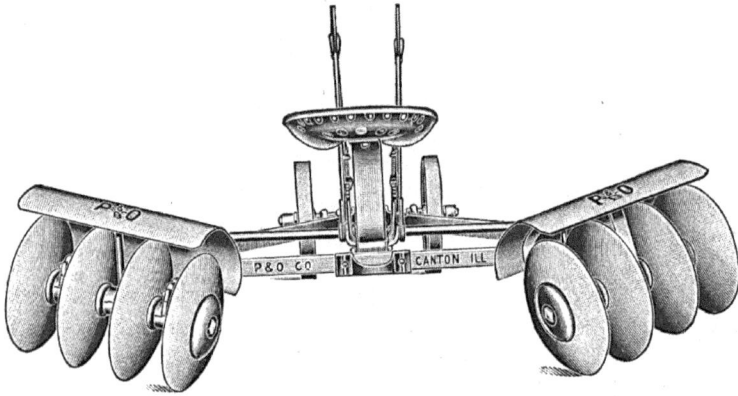

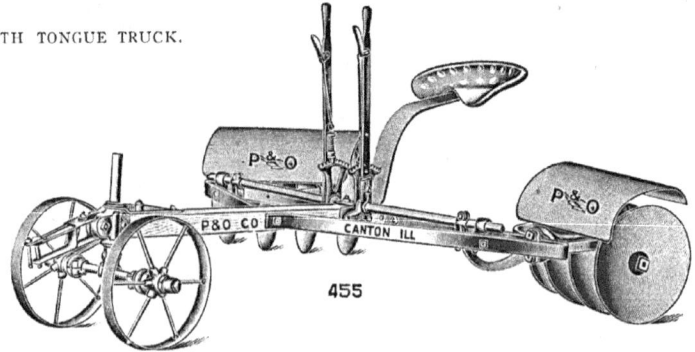

8-16 NO. 1 LOW-DOWN ORCHARD DISC HARROW WITH TONGUE TRUCK.

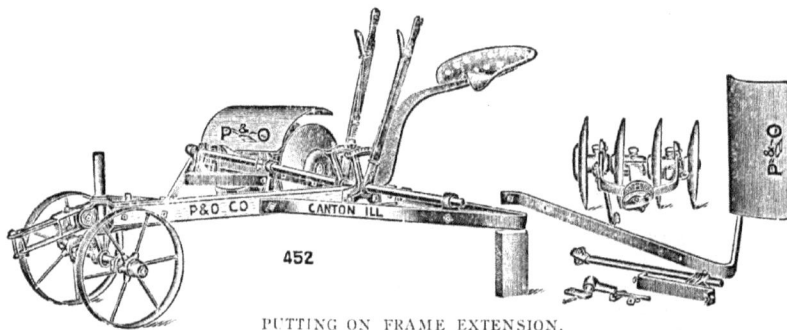

P & O

455

8-16 NO. 1 LOW-DOWN ORCHARD DISC HARROW WITH TONGUE TRUCK.

452

PUTTING ON FRAME EXTENSION.

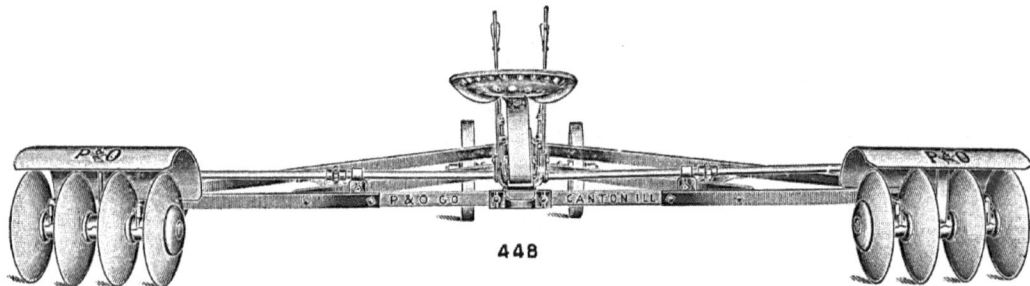

448

This Cut shows the Low-Down Orchard Disc Harrow with Gangs set at extreme ends of Extension Frame.

P & O

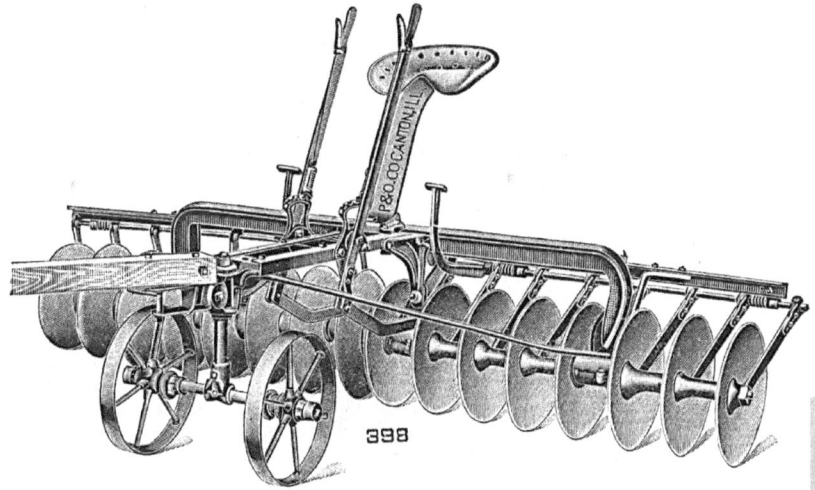

"CANTON" DISC HARROW WITH NO. 3 TONGUE TRUCK.

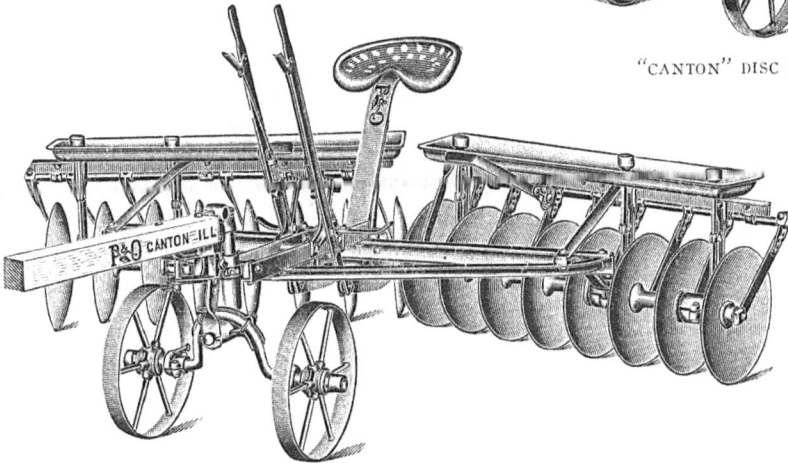

NO. 2 STAR DISC HARROW WITH NO. 2 TONGUE TRUCK.

Case

They called it the NEW J. I. CASE and it featured twelve 16 inch discs with slicer scrapers in a width cut of 6 foot.

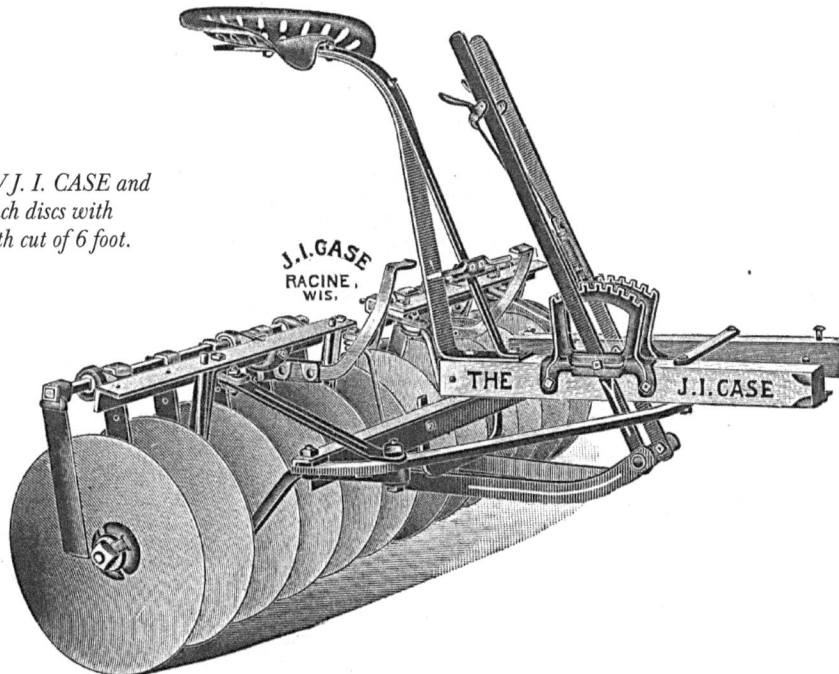

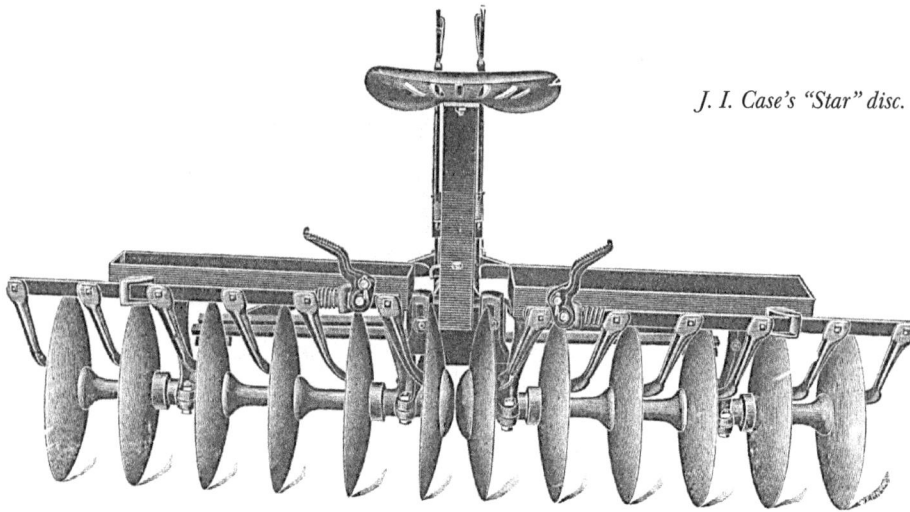

J. I. Case's "Star" disc.

Case

Ohio Reversible Disc Harrow. It will be noticed that when the gangs are moved to reverse the machine, the levers move with the gangs. With most other reversible discs time must be spent adjusting the levers.

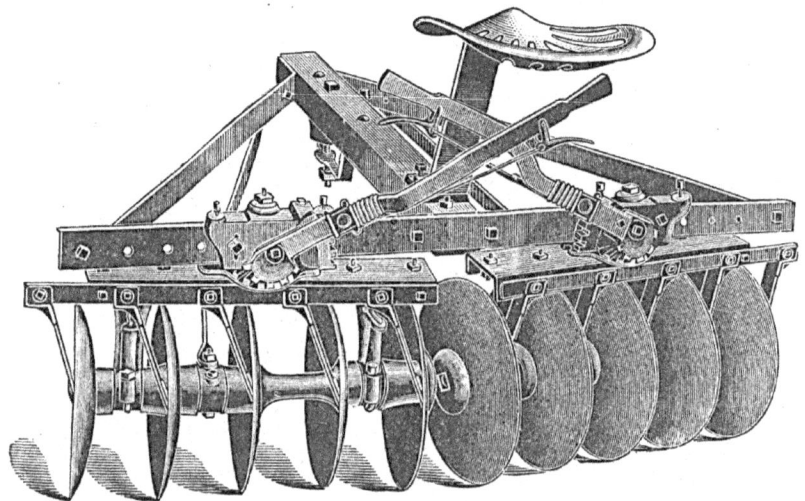

OHIO REVERSIBLE DISC HARROW.

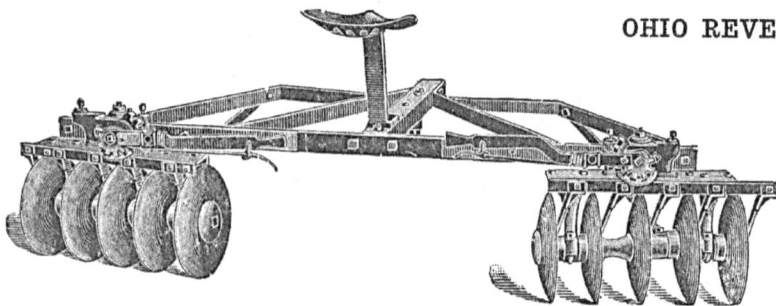

To the left is the Ohio Reversible with extension head and adjustable levers which are parallel with the frame and under perfect control of the operator while in the seat. This design puts the levers down and out of the way of any low hanging tree limbs.

Clark

A 'Clark' made Reversible Orchard Cutaway Harrow with extension head.

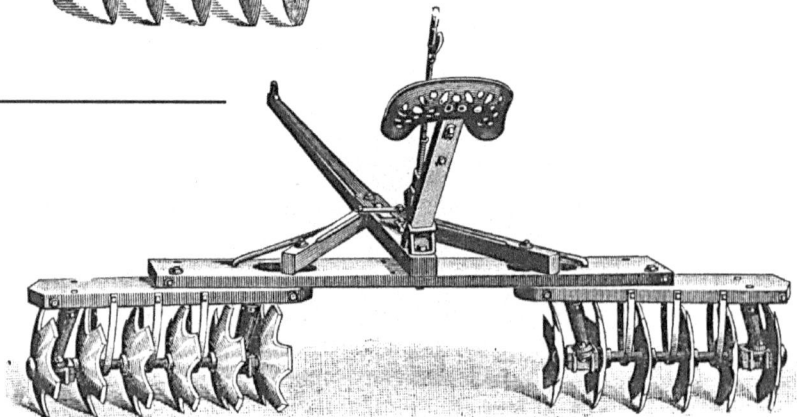

Clark

Clark Orchard Disc.

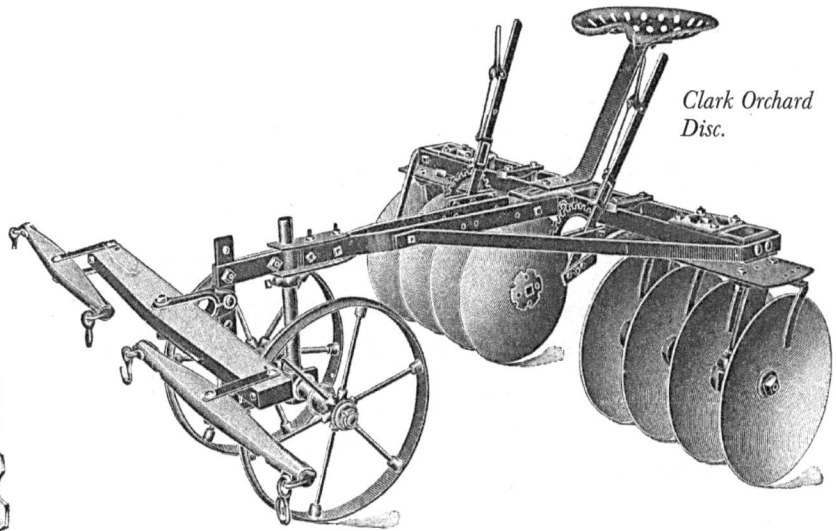

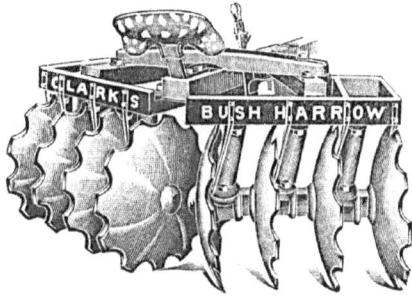

Discs

Reversible Bog and Bush Killing Plow and Harrow with 24 inch blades, cuts a 5 foot swath. Designed to attack bushes, bunch grass, witch grass, hardhack, thistles, wild rose, morning glory, milk weed, sunflower, or any wild plant.

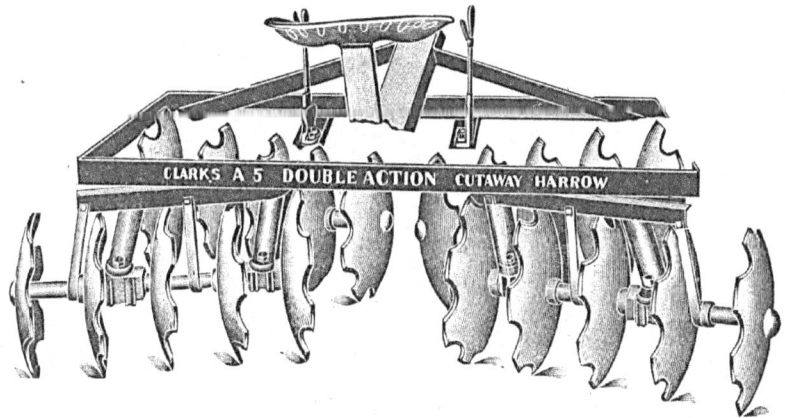

Double Action Cutaway Harrow. The manufacturer claimed: " It will cut up from 28 to 30 acres, or double cut 15 acres in a day."

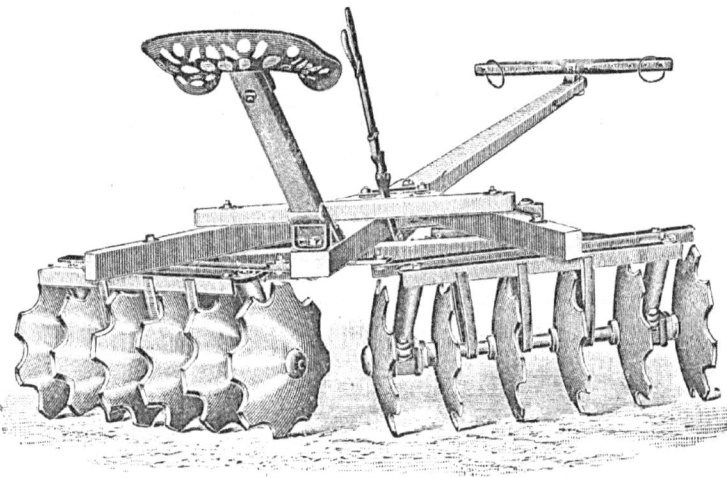

Clark's Cutaway Disc Harrow

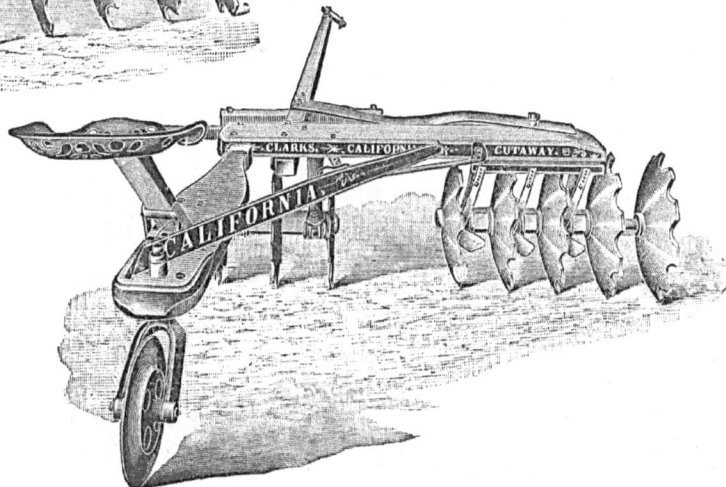

Clark's California Senior Cutaway. A very unusual setup designed to reach in under fruit trees and up against grape vines.

Roderick Lean

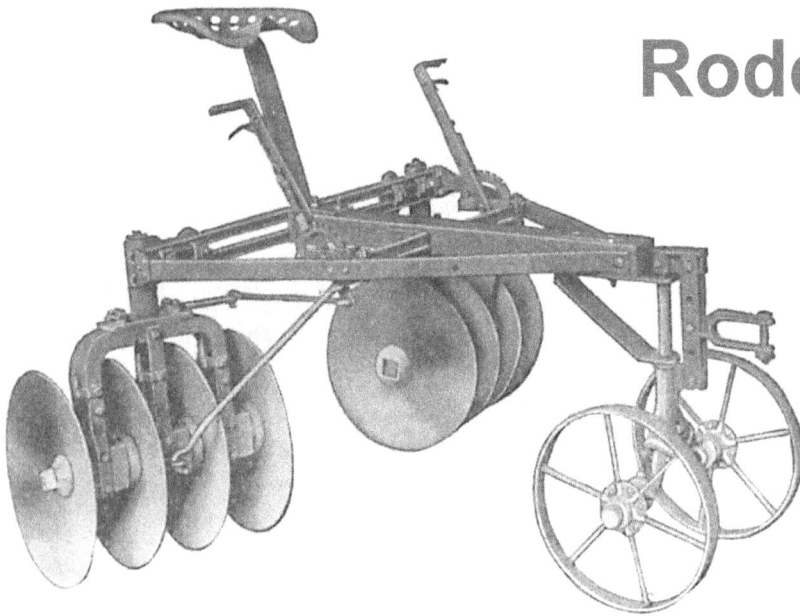

"Highlander" Reversible Disc Harrow by Roderick Lean. Left, with gangs out, below with gangs in.

Case

Showing Gangs Narrowed

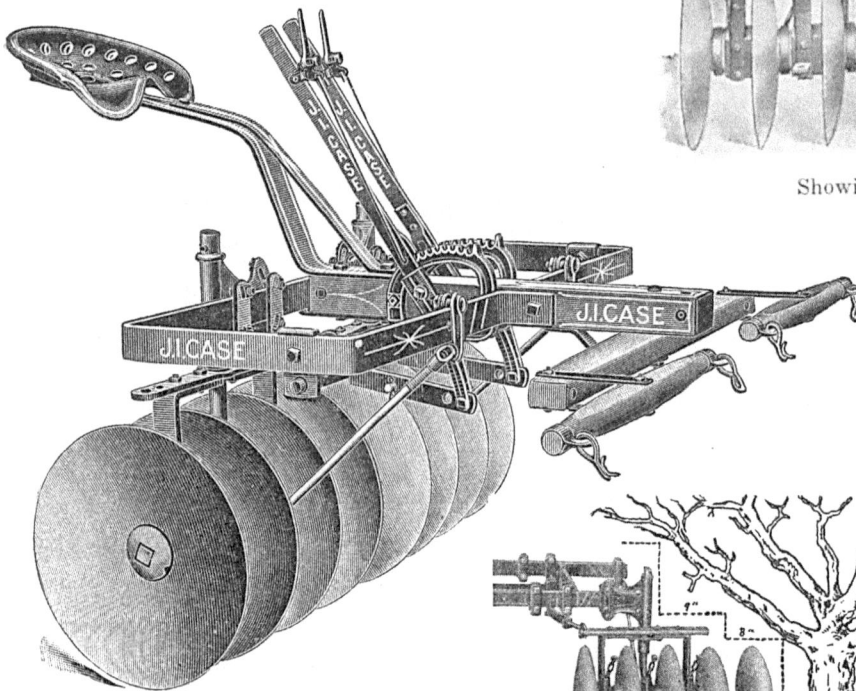

An early model of Case Reversible Orchard Disc.

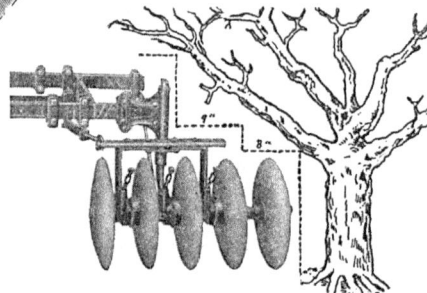

Orchard Extension.

Case

Three Horse Truck

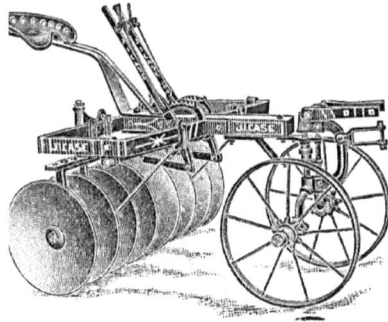

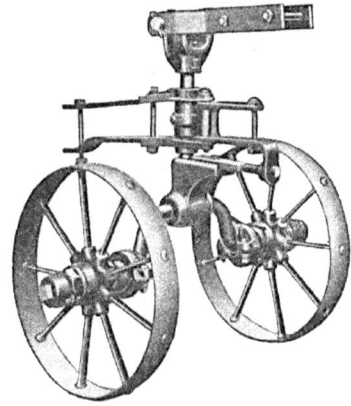

Two Horse Truck

This J. I. Case Disc Harrow featured solid enclosed boxings with chilled ring bearings

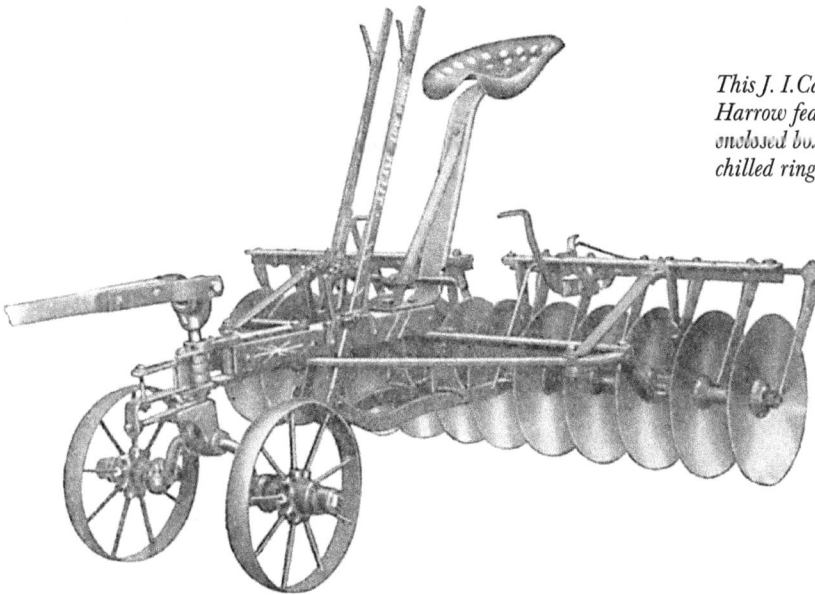

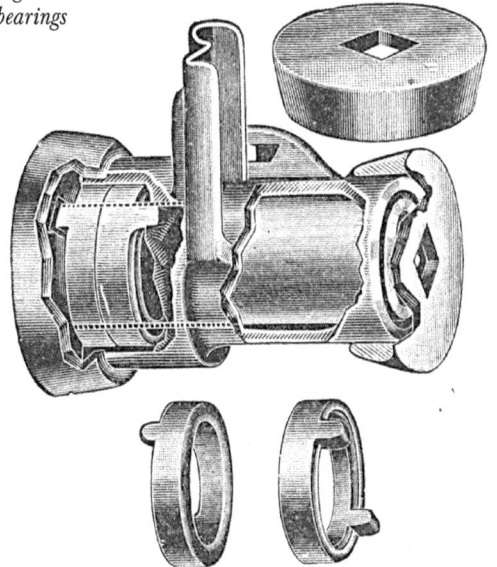

Chilled Ring Bearings

SCRAPERS RELEASED

FOOT SHIFT

SCRAPERS ON

LOOSE REVOLVING BUMPERS

Showing loose revolving bumpers on Case discs. Also illustrates the action of the scrapers.

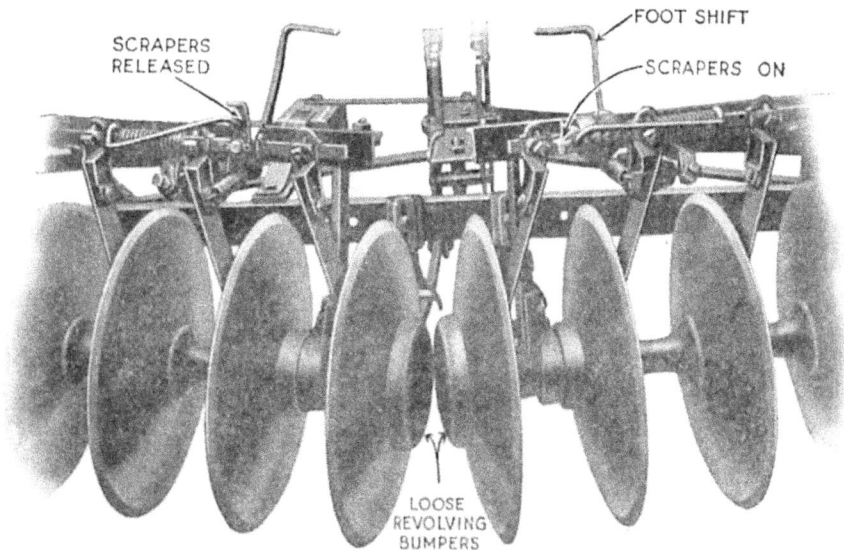

Case

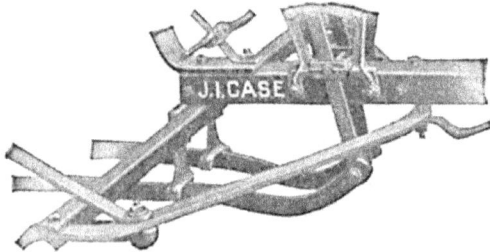

Showing Drawbars and Hold-down Castings

Clark

*One horse Reversible
Orchard Disc Harrow*

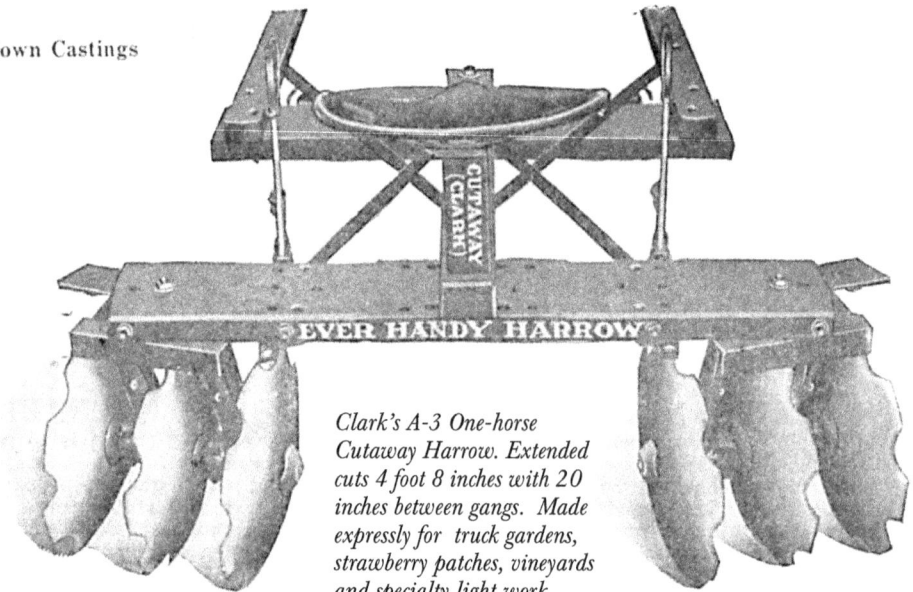

*Clark's A-3 One-horse
Cutaway Harrow. Extended
cuts 4 foot 8 inches with 20
inches between gangs. Made
expressly for truck gardens,
strawberry patches, vineyards
and specialty light work.
Closed up it cut 3 foot wide.*

*Showing a simple sliding rod setup on a truck
wheel assembly for the offset necessary for three
abreast.*

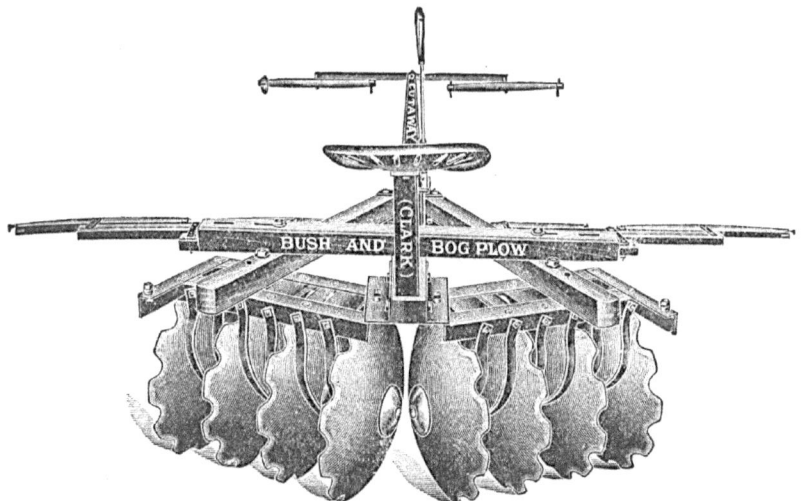

*Clark Bush and Bog Disc Harrow (also erroneously
called a disc plow) setup for four abreast*

Clark

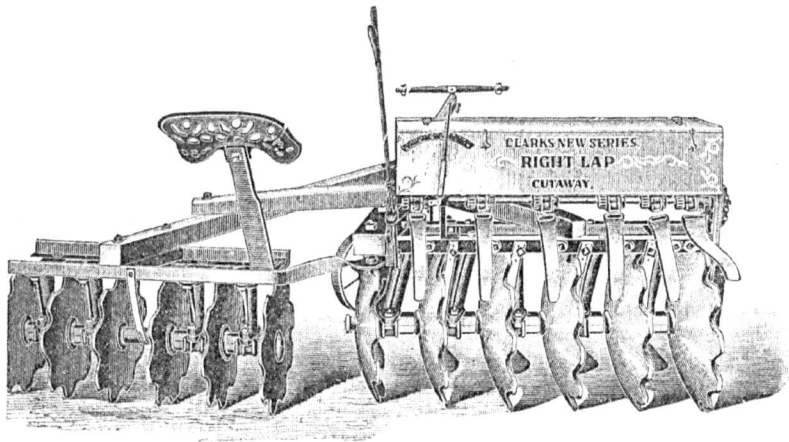

"Right-Lap Cutaway". A highly unusual implement. The left hand gang has straight notched discs, 18 inches in diameter; the right hand gang has curved 'cutaway' discs, 24 inches in diameter, and plows a furrow from five to eight inches deep. Atop the right-hand gang is a ground drive seeder. Designed with trashy ground in mind.

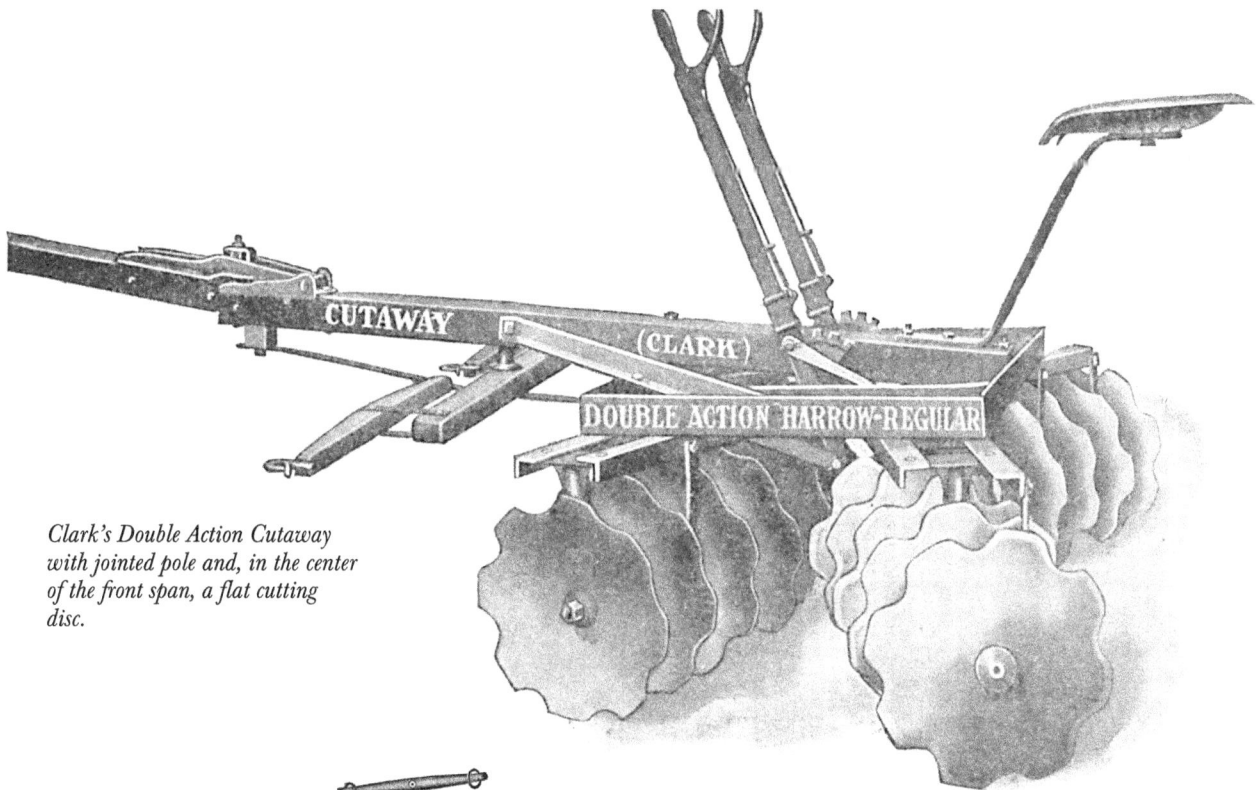

Discs

Clark's Double Action Cutaway with jointed pole and, in the center of the front span, a flat cutting disc.

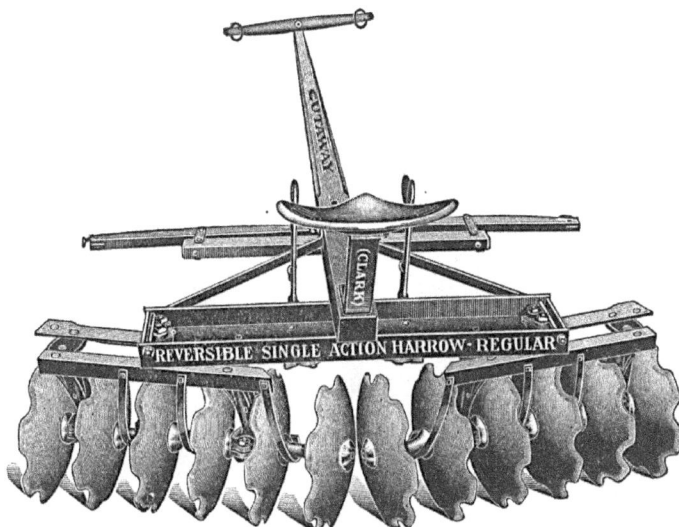

Clark's Reversible Single Action Cutaway harrow.

Avery

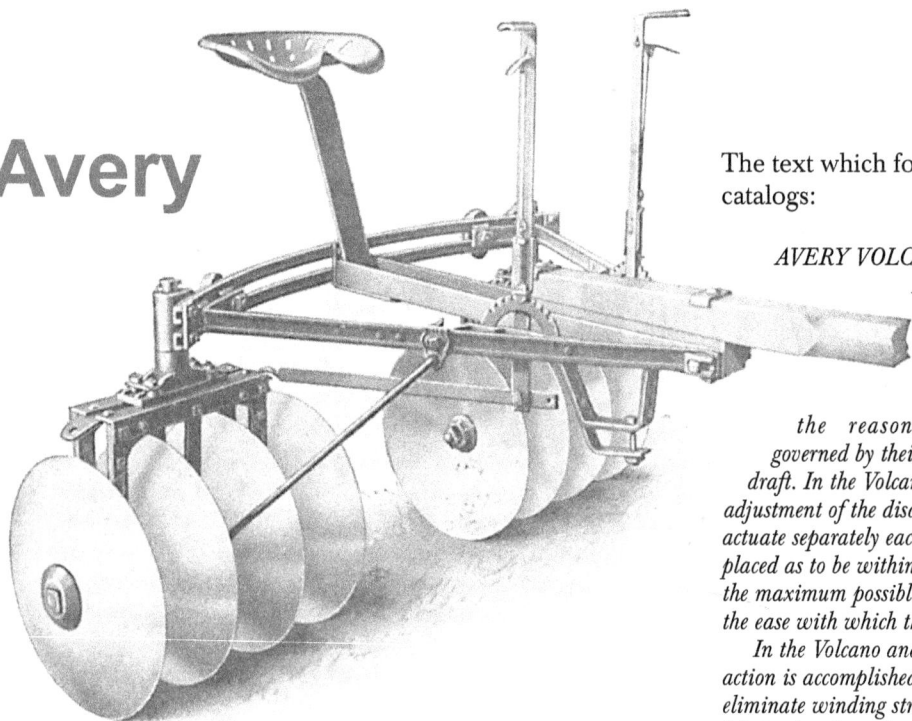

Volcano

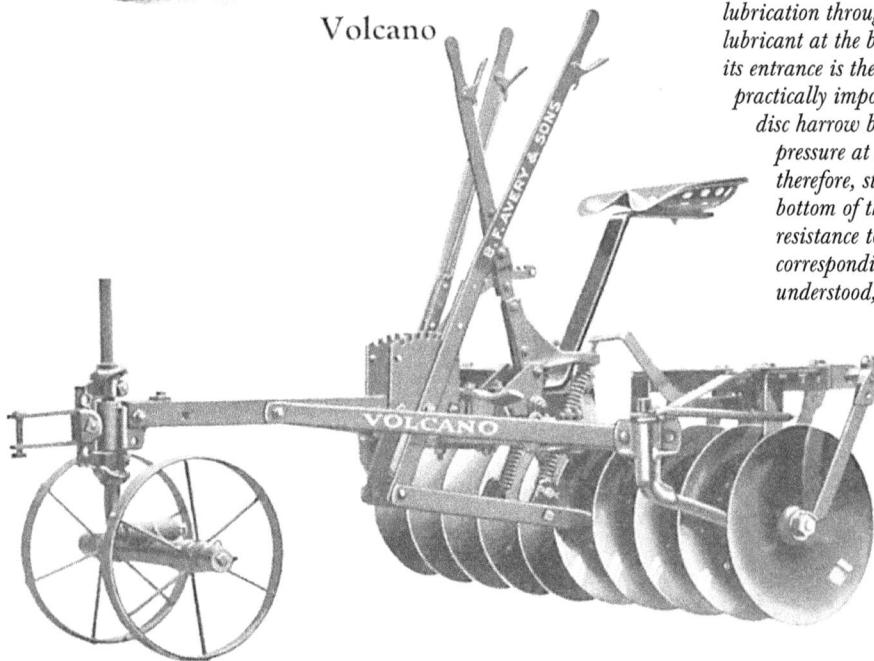

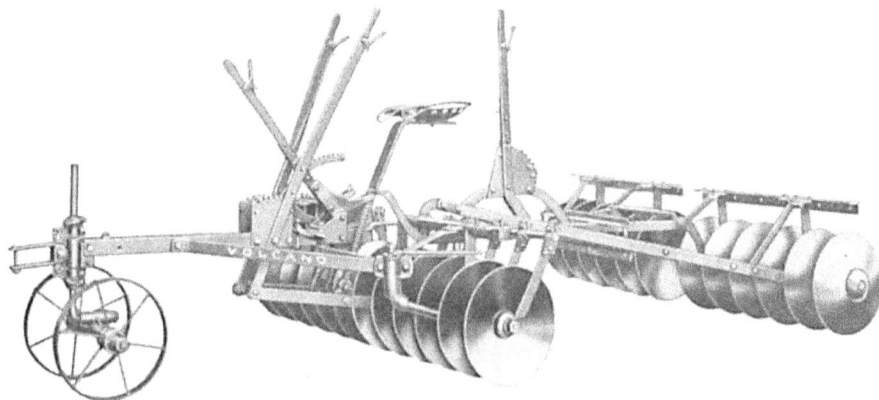

The text which follows appeared in the Avery catalogs:

AVERY VOLCANO AND VOLCANO JR. HARROWS

One of the first requirements of a disc harrow is that the angle of the individual discs to the line of draft must be easily adjustable, this for the reason that the penetration of the discs is governed by their position with respect to the line of draft. In the Volcano and Volcano Jr. Series, the angular adjustment of the discs is controlled by two long levers which actuate separately each of the disc gangs. These levers are so placed as to be within easy reach of the operator and to give the maximum possible leverage thereby materially increasing the ease with which the gangs may be manipulated.

In the Volcano and Volcano Jr. Series, freedom of disc action is accomplished, first by the full floating gangs which eliminate winding strain on the bearings, and second, by lubrication through a spiral oil tube which introduces the lubricant at the bottom of the bearing where the resistance to its entrance is the least. It has been proven that it is practically impossible to force hard grease into the top of a disc harrow bearing by virtue of the fact that the pressure at this point is of necessity the greatest. It, therefore, stands to reason that the pressure at the bottom of the bearing will be the least and that the resistance to the introduction of the lubricant will be correspondingly lessened at that point. It will be easily understood, there, when considering the mechanical means which are employed for this purpose just why the Volcano and Volcano Jr. Harrows are celebrated for their light running qualities.

Avery

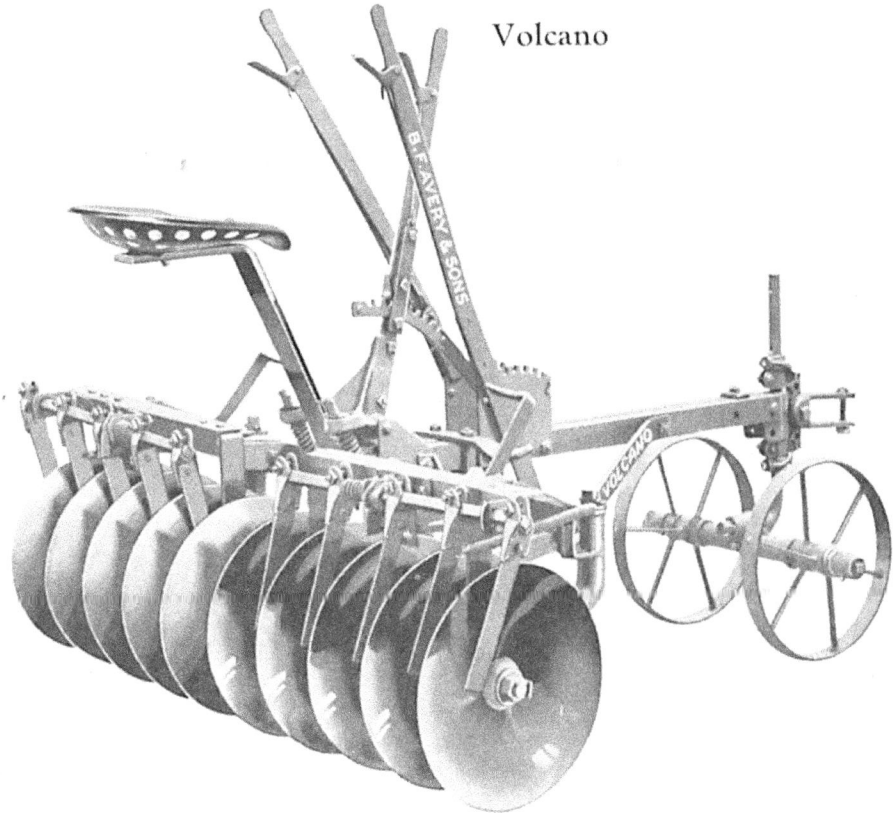

Volcano

The loosening of the discs on the spindles in the ordinary disc harrow is due to the fact that it is impossible to tighten the retaining nut on the end of the gang bolt permanently unless some spring relief is provided between the discs and the ends of the spindles.

In the Volcano and Volcano Jr. Series, loosening of the discs is definitely overcome by the construction of the spindles themselves. These are made with concave ends, one smaller than the other so that when they are strung on the gang bolt and tightened up with the discs inserted between them, there is set up a bending strain in each individual disc which serves the same purpose as a cut spring washer and allows, along the entire gang a spring relief, which provides for two complete turns of the retaining nut before the pressure on the discs is relieved. By reason of this construction, the discs on the Volcano and Volcano Jr. Harrows stay tight indefinitely and as a result give much more satisfactory and long service than is usual in harrows of this general type.

It is a matter of common knowledge among the users of disc harrows that the tendency of the gangs is to raise at the inner end as a result of the pressure of the soil against the discs. If this tendency were allowed to fully exert itself, the result would be ragged, uneven work. If, on the other hand, rigid adjustment is provided, the general result would be quite as unsatisfactory as with no adjustment whatever. In the Avery Volcano and Volcano Jr. Series the inside ends of the gangs are held down by independent spring pressure on the inner ends of the gangs which can be adjusted by means of a lever set between the two adjusting levers. This construction not only takes care of the variations in the upward pressure at the inner ends of the gangs caused by changes in the angular position of the discs, but further allows each gang to move separately without reference to the other, thus providing for irregularities in the surface of the ground and insuring an even, well distributed job of harrowing at all times.

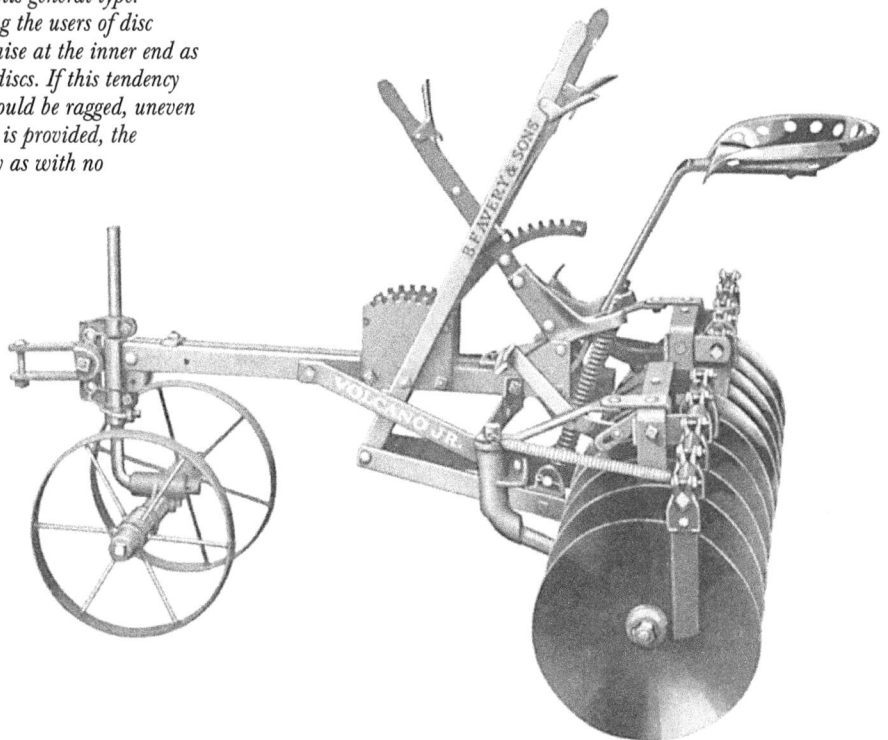

Avery

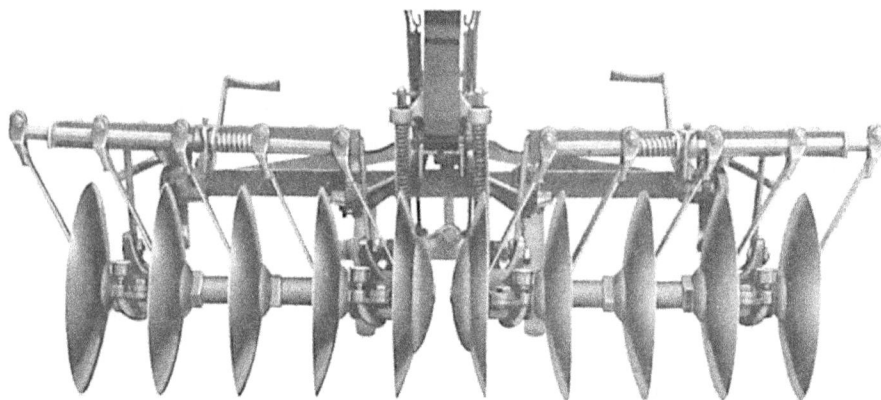

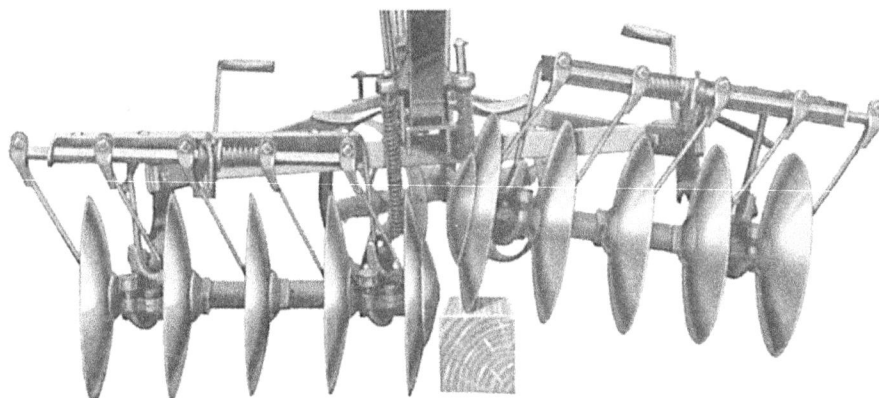

The photograph above is used to illustrate the gang construction of both Volcano Harrows. The manner in which the individual spring pressure is applied to the inner ends of the gangs is also shown. The scrapers are actuated by the two foot-levers shown in the cut which, when pressed down, move the point of the scraper from the center of the disc to its outer edge, thereby passing over its entire surface as it revolves and preventing clogging in the stickiest of soils.

The photograph above shows the action of the Avery Volcano and Volcano Jr. gangs in rough ground. The edge of the rear end inside disc is elevated about eight inches off the ground while the left hand gang remains stationary. It should be noted that the compression spring on the left hand gang is fully extended while that on the right hand is compressed to the limit. The flexibility thus illustrated, is invaluable in all uneven land and assures a clean uniform job under practically all working conditions.

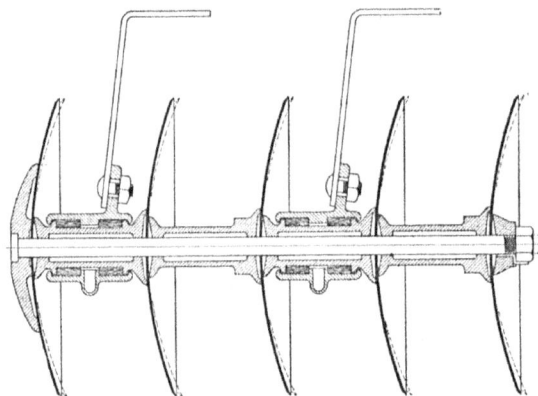

The Avery Volcano Self Tightening Disc Construction as illustrated in the diagram shows the principle on which this important feature is based. The discs are shown in two positions, before and after tightening the retaining nut. It will be seen that during the process of tightening this nut the discs actually bend, creating precisely the same effect as is gained by the pressure of a cut steel spring washer in a similar position. This feature definitely assures the operator of tight, true running discs which are essential to economical and effective operation.

The oiling feature (left) which contributes so largely to the freedom of action of the Volcano and Volcano Jr. Harrows represents a radical departure from the usual construction and accomplishes a real saving not only in unexpended energy but in wear and tear on the harrow's disc spindles and bushings. The oil, as before stated, is forced by a threaded cap grease cup through a spiral tube into the bottom of the bearing at which point there is the least resistance to its introduction. This construction provides efficient lubrication at all times and adds much to the general efficiency of the harrow.

Avery

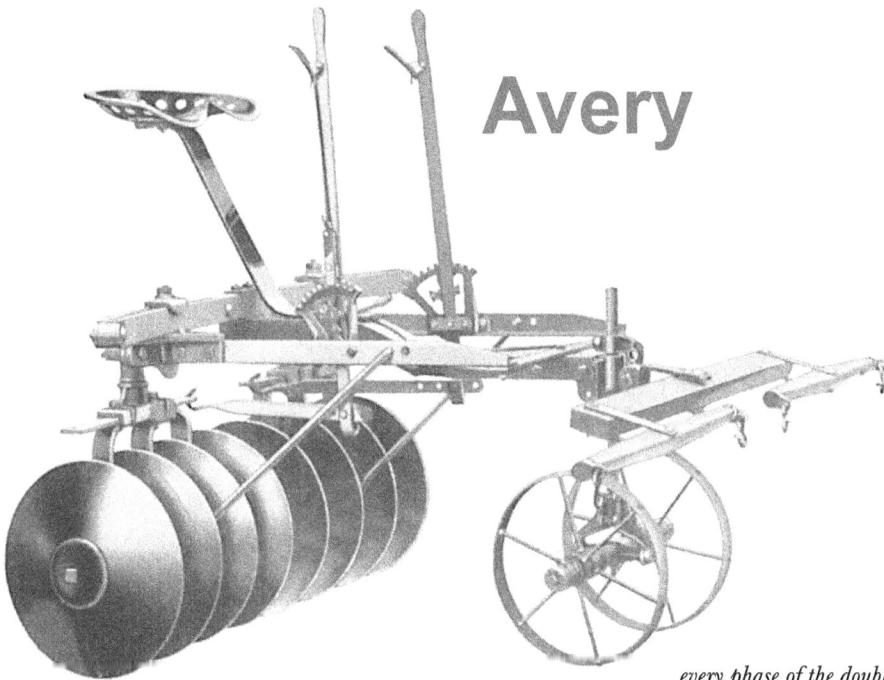

Detail View Angular Disc Adjustment

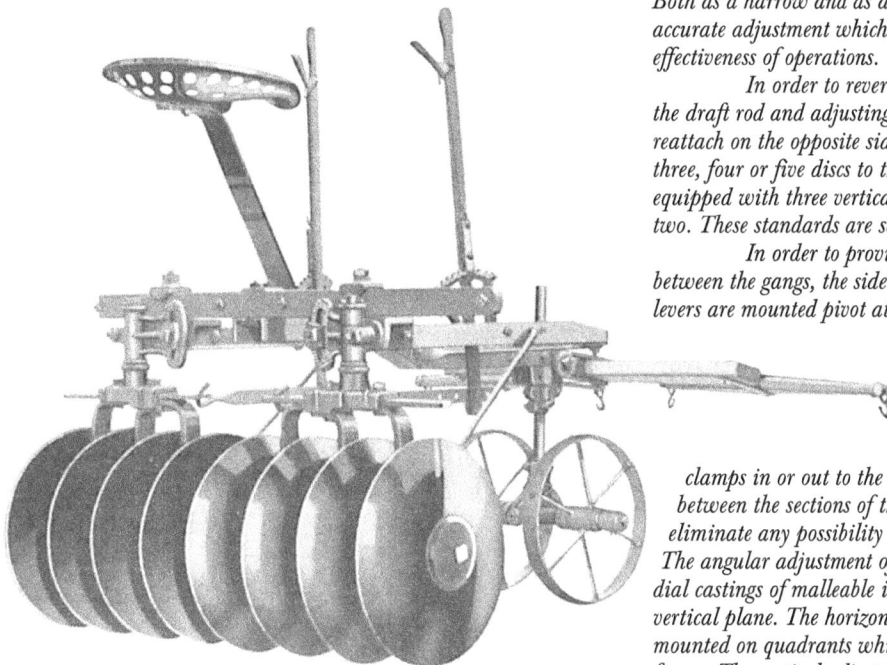

every phase of the double duty for which it was designed and built. Both as a harrow and as a cultivator it is subject to simple and accurate adjustment which made for ease of manipulation and effectiveness of operations.

In order to reverse the gangs it is only necessary to remove the draft rod and adjusting link, reverse the individual gangs and reattach on the opposite side. The Crescent is furnished with either three, four or five discs to the gang, the four and five disc gangs being equipped with three vertical standards and the three disc gangs with two. These standards are set edgeways in order to prevent clogging.

In order to provide for the adjustment of the distance between the gangs, the side pieces of the frame on which the adjusting levers are mounted pivot at the end of the front cross bar. This allows for a lateral movement of the rear end of these side members along the built-up cross bar at the rear of the machine. This adjustment is accomplished by releasing the two nuts and sliding the retaining clamps in or out to the desired width. These clamps fit snugly between the sections of the rear cross bar and when tightened eliminate any possibility of lost motion in a lateral direction. The angular adjustment of the disc is accomplished by means of two dial castings of malleable iron which rotate both in a horizontal and vertical plane. The horizontal adjustment is controlled by the levers mounted on quadrants which are attached to the side member on the frame. The vertical adjustment is controlled by a serrated washer in contact with the slotted end of the universal casting on which the gang is mounted.

CRESCENT REVERSIBLE DISC HARROW

The Avery Crescent Reversible Disc Harrow is primarily notable for its ability to handle efficiently the double duty of harrow and cultivator. In one position the Crescent is a high-grade standard Disc Harrow simply and sturdily constructed and equal in every way to the efficient handling of the heavy black land in which it is ordinarily used. By simple adjustment which reverses the disc gangs it becomes a disc cultivator with sufficient overhead clearance to work the crop well along toward maturity. As harrow and cultivator, the Crescent possesses in every particular all the refinements of design and construction which make for economical and efficient operation. It is furnished either with pole or tongue truck and is adaptable to

Avery

Crescent Reversible Disc Harrow set with gangs back and into low position.

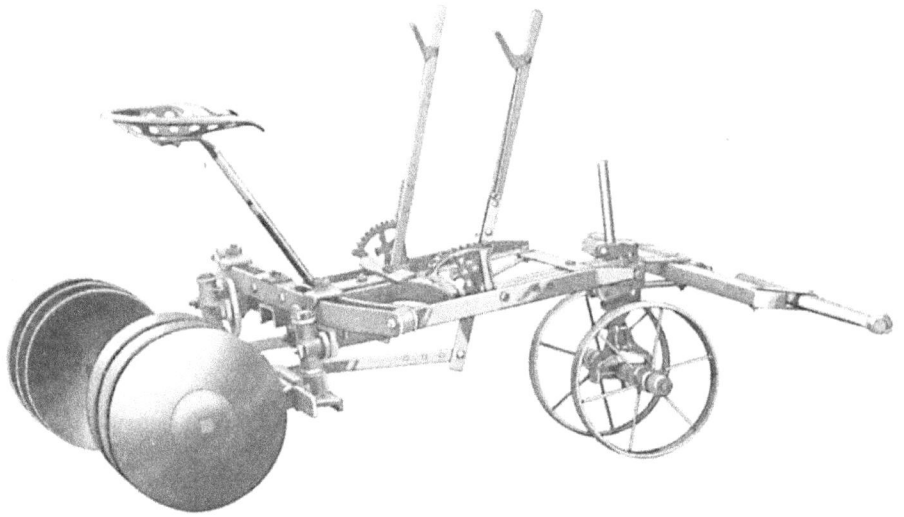

Moline

No. 10 Series Disc Harrow

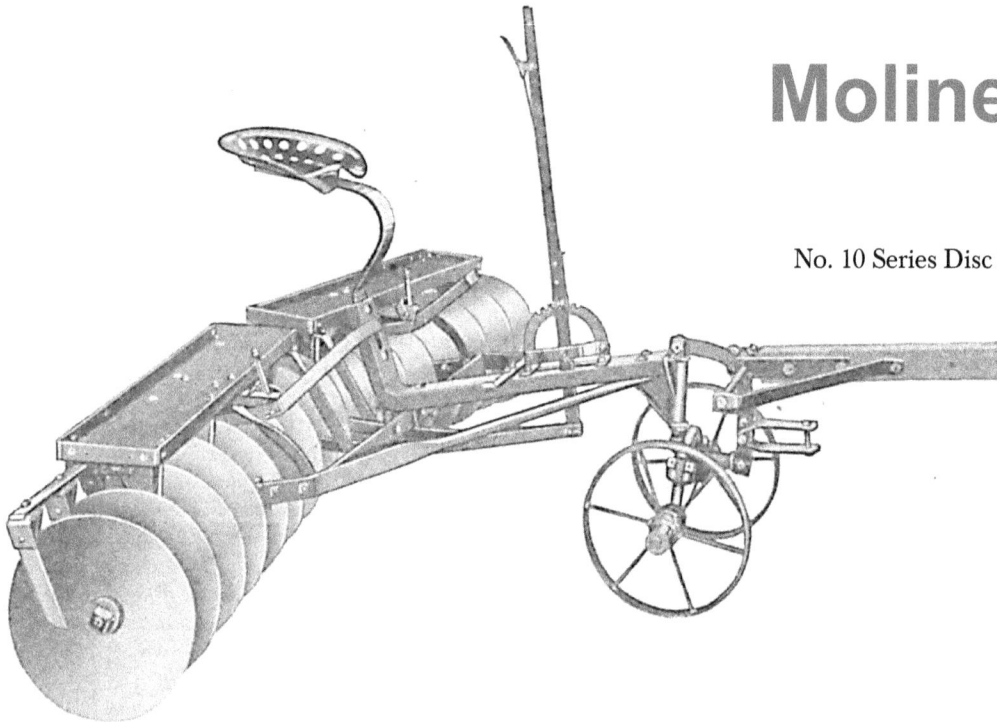

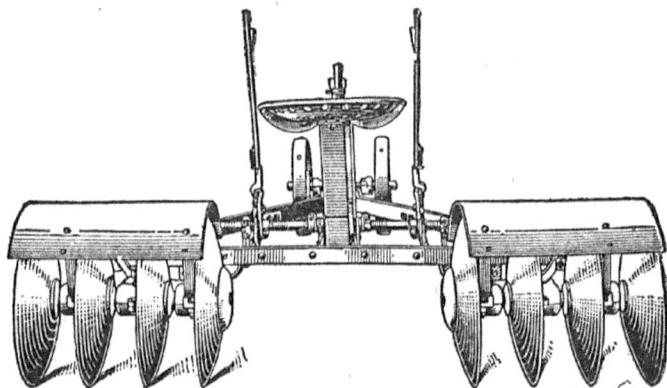

Disk harrow for orchards.

Moline

*Economy Double Lever Disc
Harrow with tongue truck*

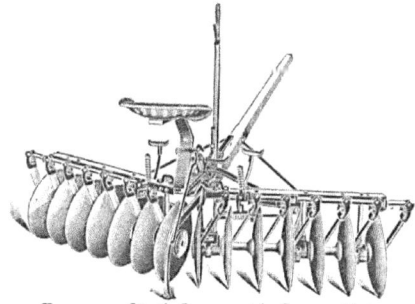

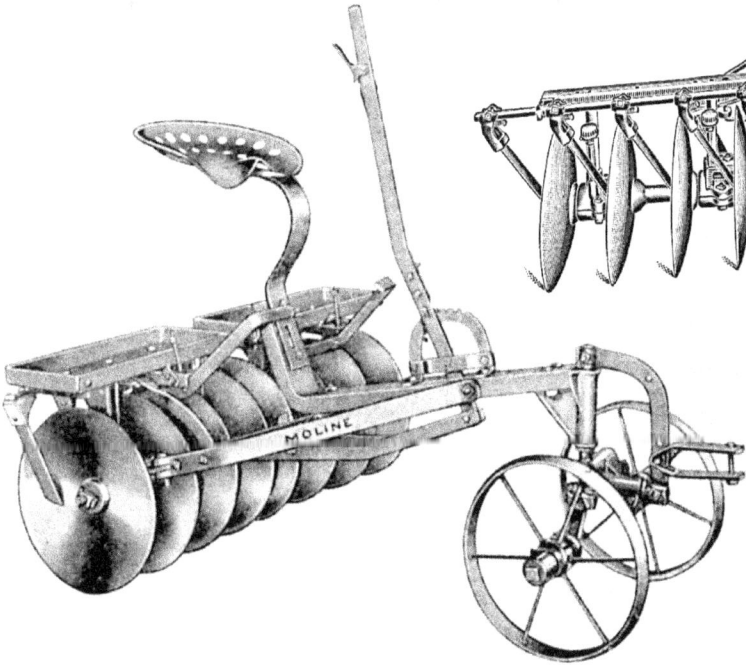

No. 10 Series single lever with tongue truck

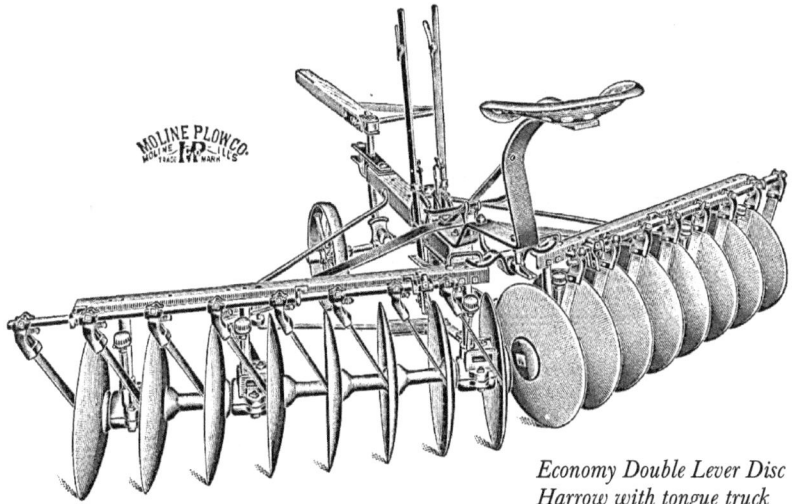

Economy Single Lever with Center-Cut.

No. 10, 8-FOOT TANDEM HARROW

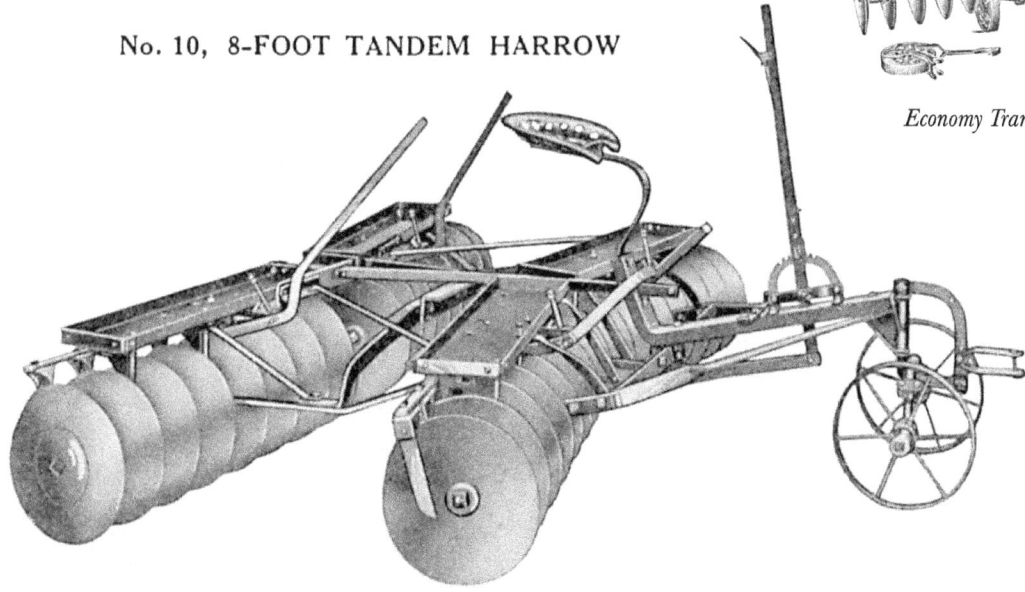

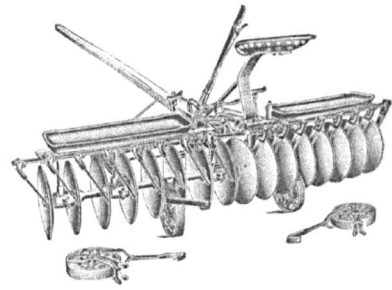

Economy Transport Attachment

Moline

Moline Double Acting
Reversible with cut-out discs
and offset pole.

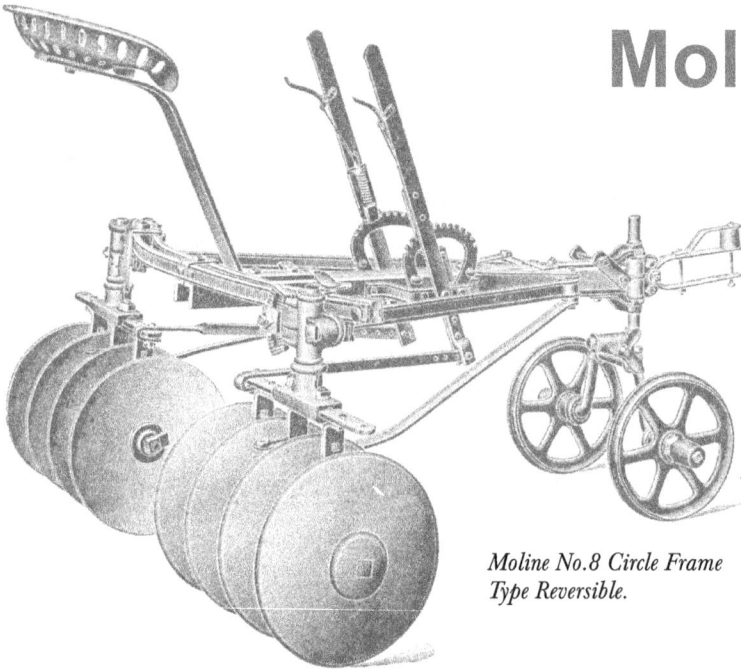

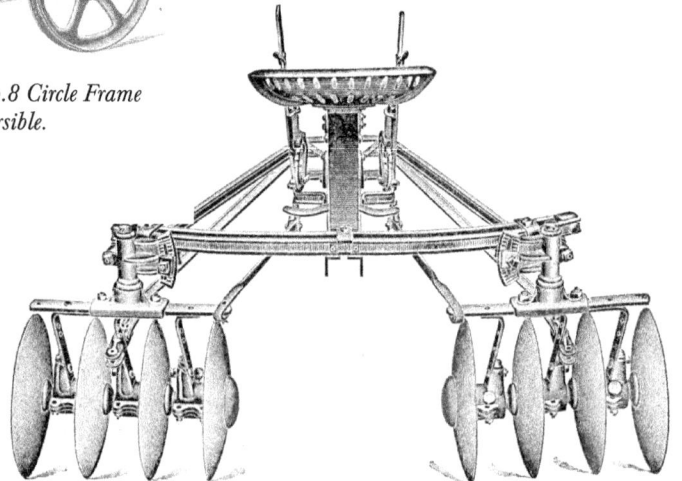

*Moline No.8 Circle Frame
Type Reversible.*

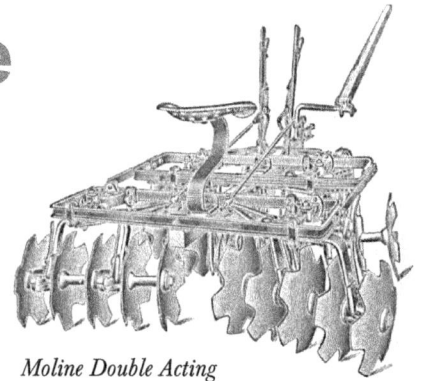

No. 8 Reversible Harrow, Rear View

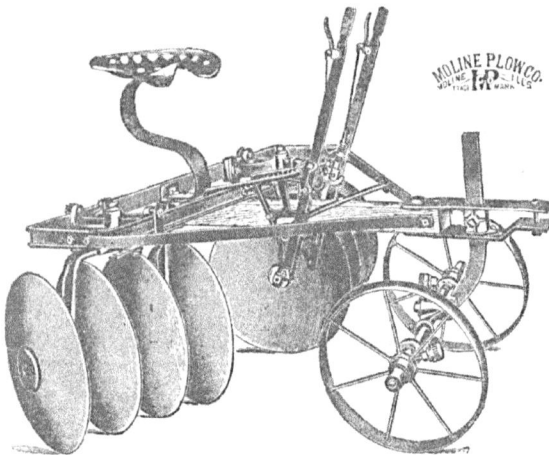

Moline Reversible Disc Harrow with No. 3 Tongue Truck.

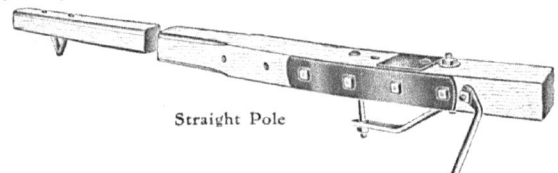

Straight Pole

Moline's No. 4 Economy Tongue Truck was said to remove all neck weight, prevent whipping or side lashing of the pole and be convenient and easy on the team when turning. The axle was longer than ordinarily used, setting the wheels farther apart for straddling corn ridges. The stub pole may be carried at any height by means of the adjustment on the upright stem. The cuts above show tongueless, center-fire, and offset.

Moline

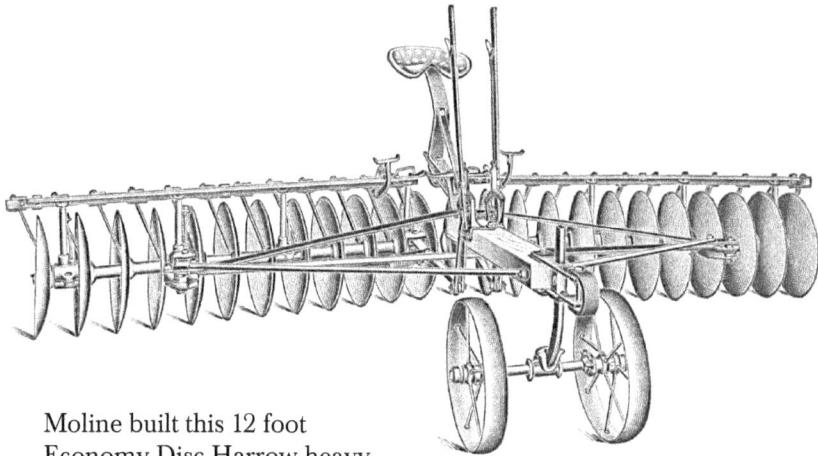

Moline built this 12 foot
Economy Disc Harrow heavy
enough to be pulled behind a
tractor

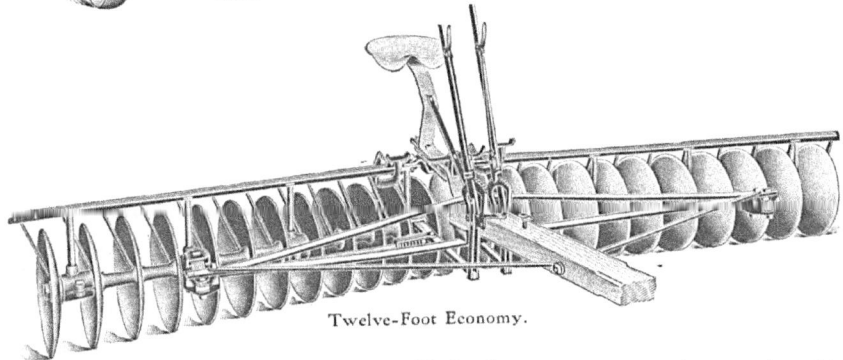

Twelve foot economy setup with tongue (or pole)

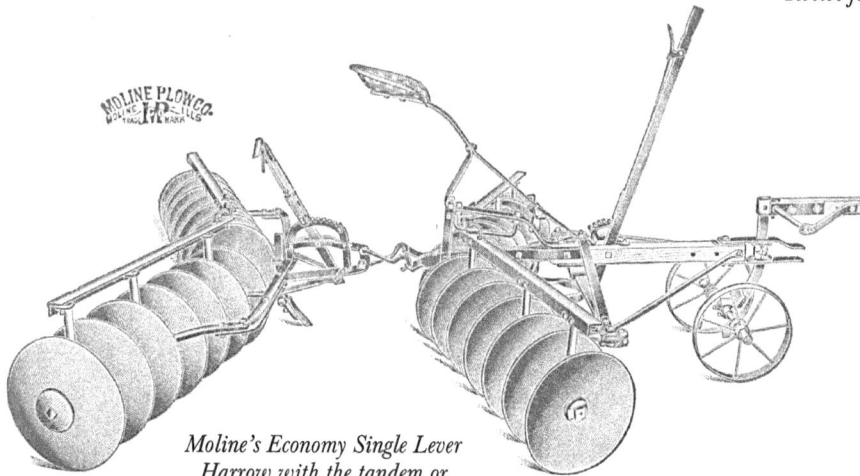

*Moline's Economy Single Lever
Harrow with the tandem or
Double Cut Attachment and
No. 4 Tongue Truck.*

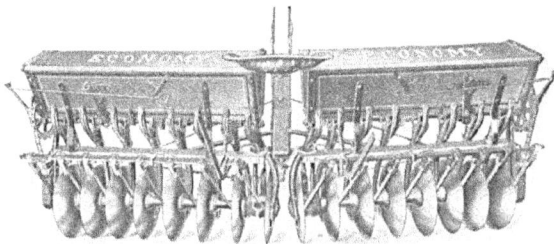

A Monitor disc harrow fitted with a Monitor seed drill.

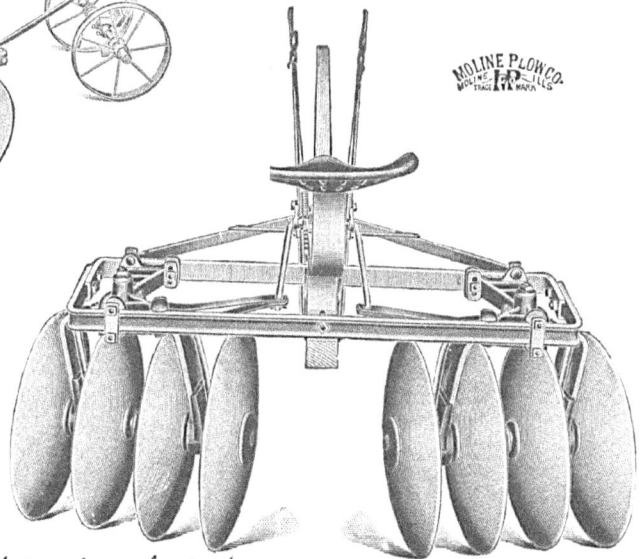

*Moline Reversible Disc Harrow showing
the detachable scraper assembly.*

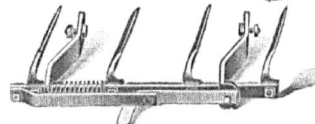

Moline

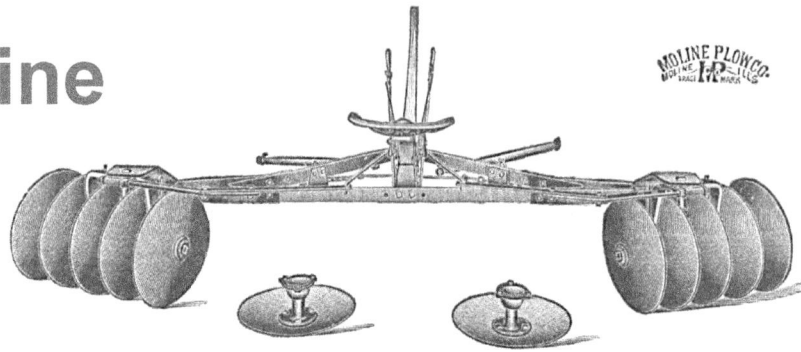

Low Down Reversible will set out as wide as ten foot with extenders.

Double Acting Reversible Disc Harrows

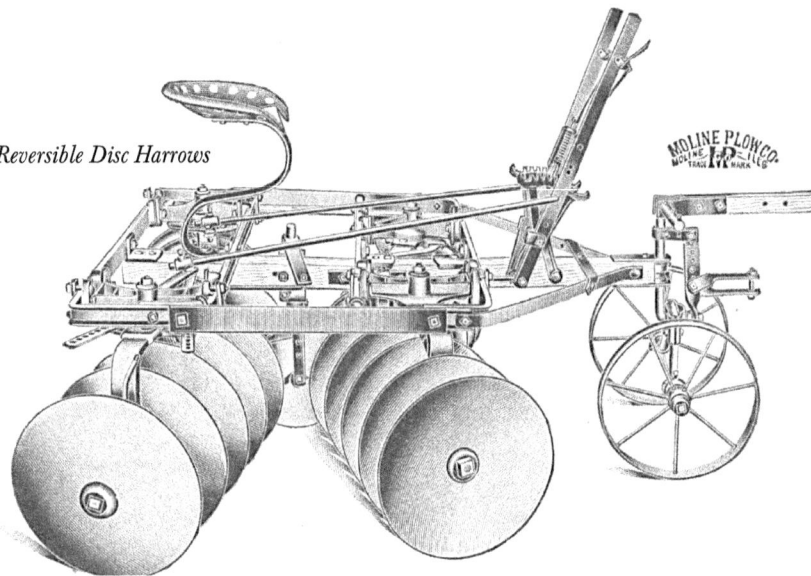

Moline's Oblong Frame style disc.

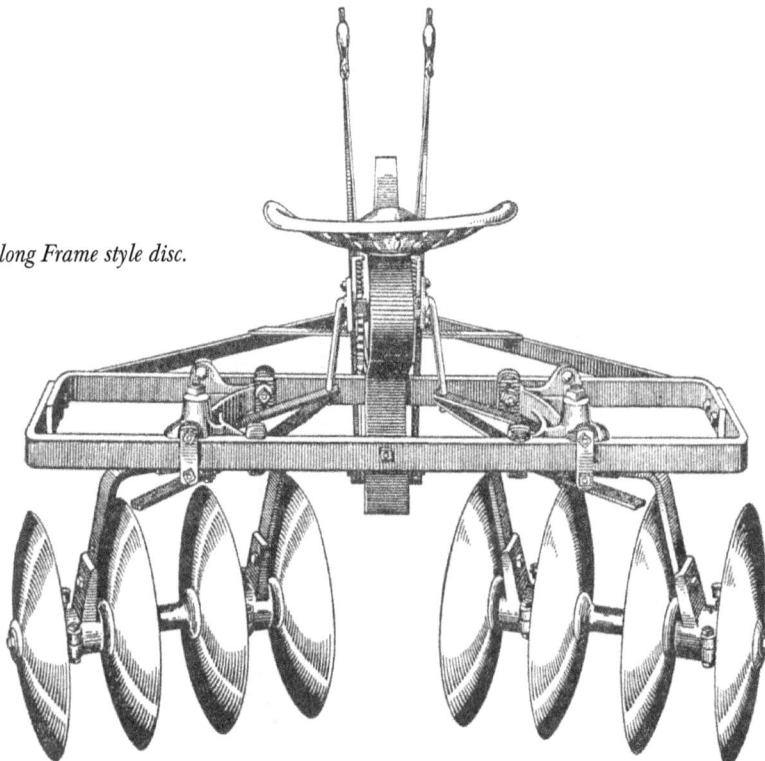

Rock Island

Gangs Angle from Outer Instead of Inner Ends

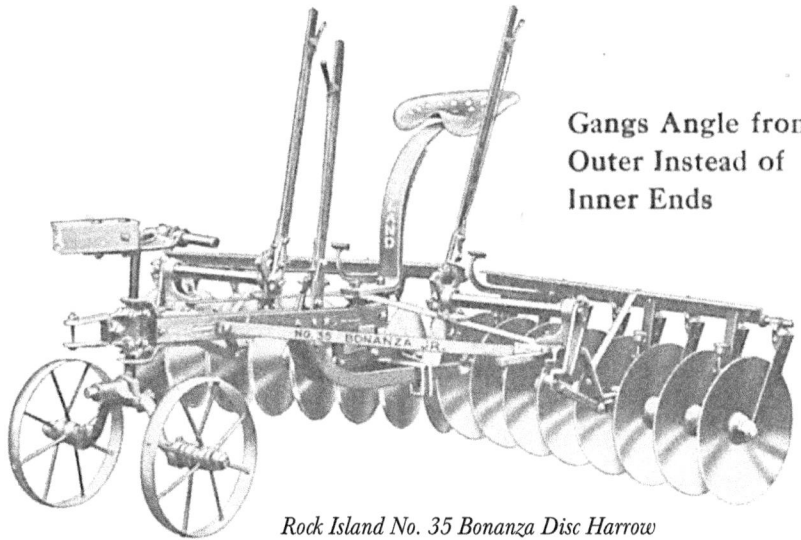

Rock Island No. 35 Bonanza Disc Harrow

Bumpers Always Bump

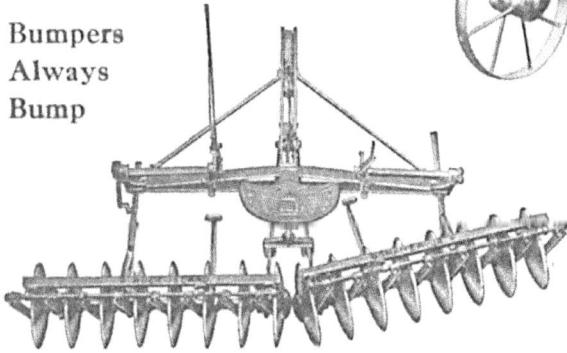

Top View of Bonanza

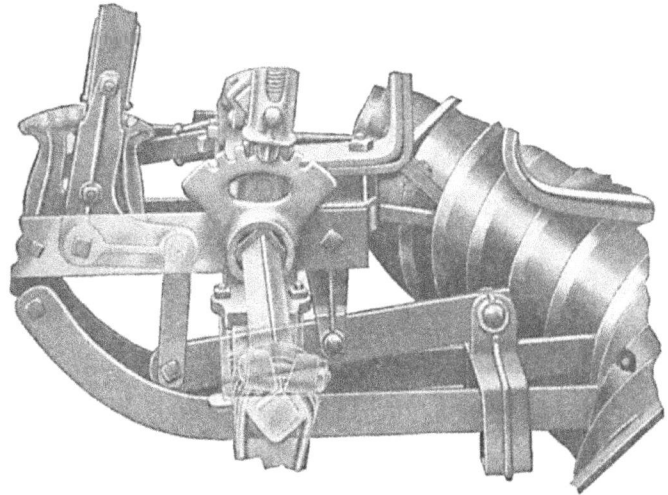

Detail view of Instantaneous Pressure Lever

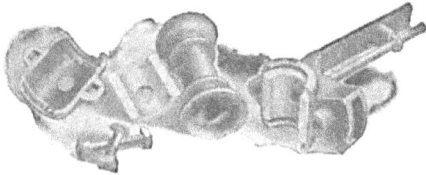

Bearing spools and boxes are interchangeable, neither right nor left.

Twenty five Rock Island No. 35 Bonanza Harrows in one field.

Rock Island Tandem Attachment

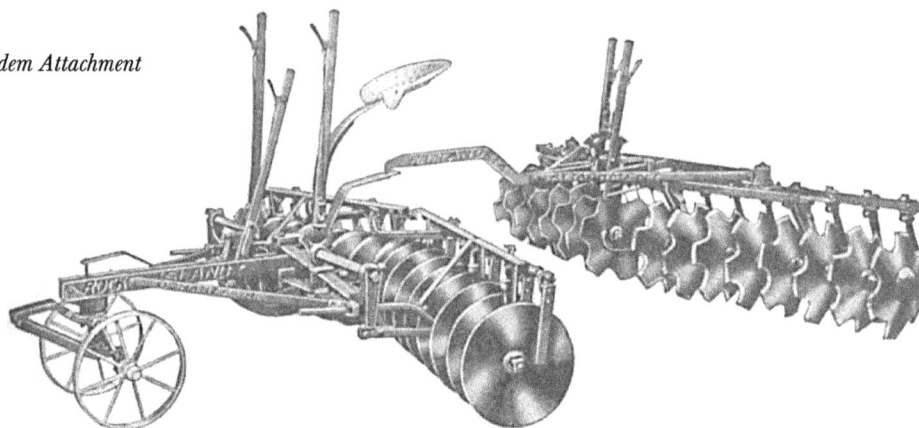

Rock Island

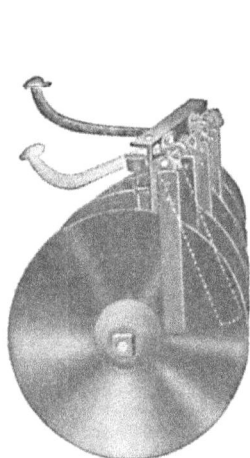

Rock Island Oscillating Scraper

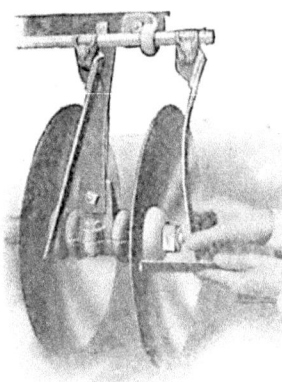

Note the flexibility of Rock Island Scrapers.

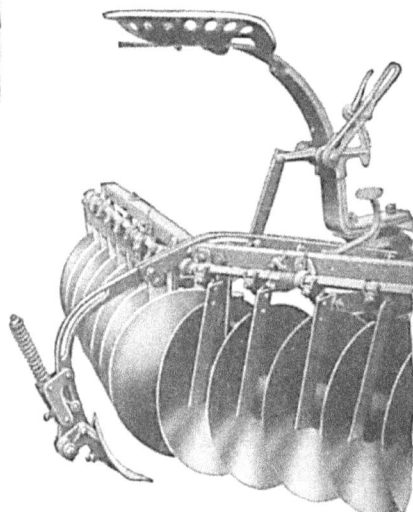

Center Tooth Attachment

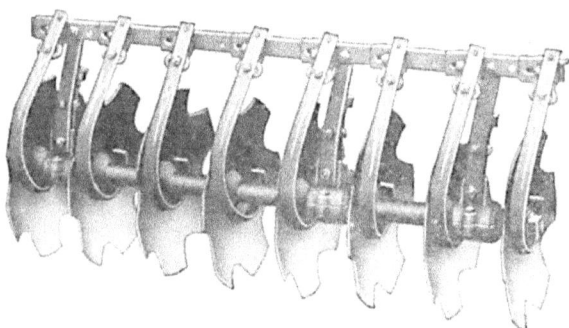

(Below) Rock Island Transport Truck

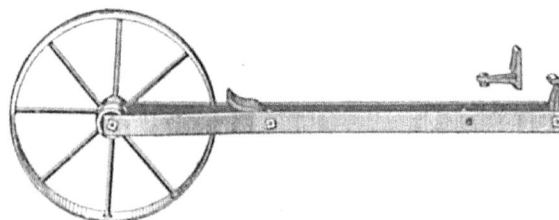

Some companies called them cutaways, some cut-outs, Rock Island called them shear cut or plow-cut discs. This gang above features an unusual scraper designed to work with the scalloped discs.

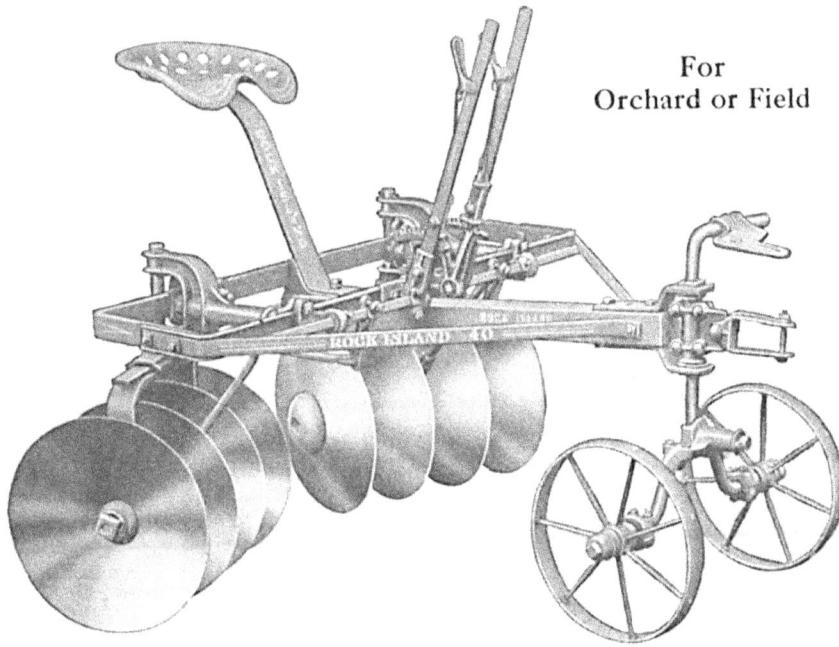

For
Orchard or Field

Rock Island

Two views of the Rock Island No. 40 Reversible Orchard Disc Harrow

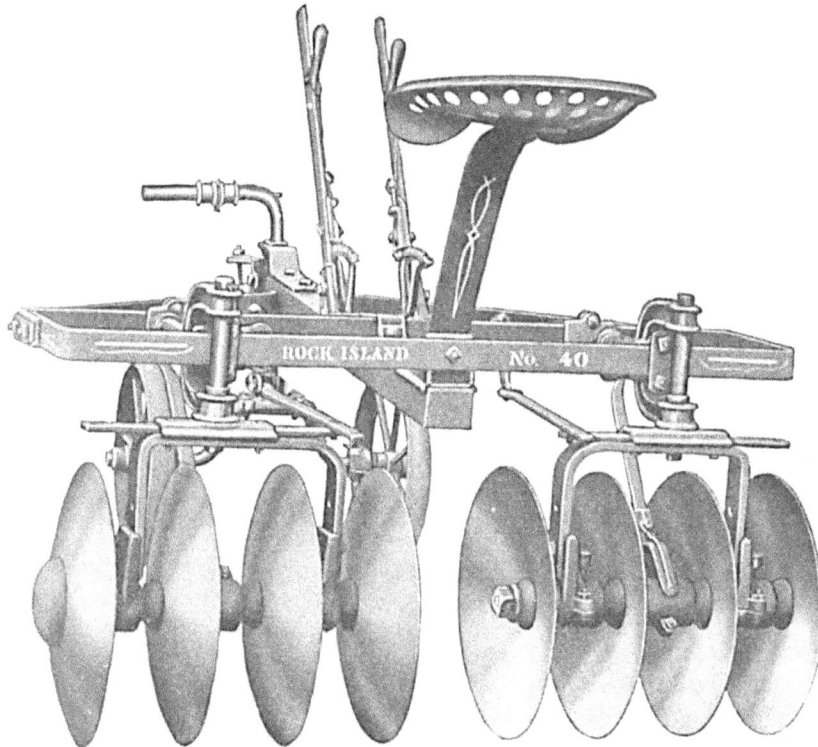

Rock Island

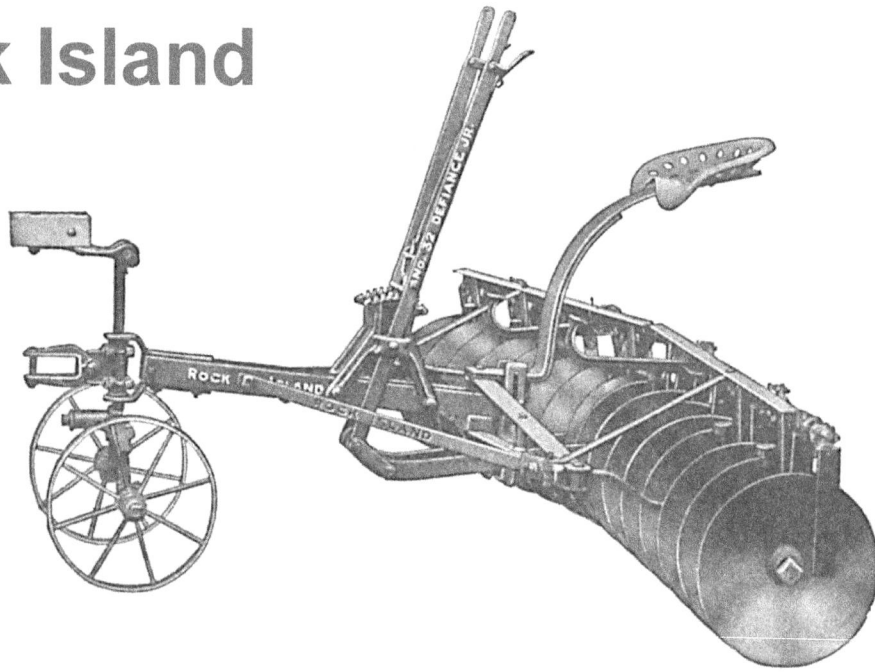

Rock Island Fore-Carriage, High
Adjustment, Tow-Horse Hitch

The Rock Island Fore-Carriage Readily
Passes Over All Obstructions—
3-Horse-Hitch

Flexible Disc Harrow Trailer

Vulcan

"M" Series

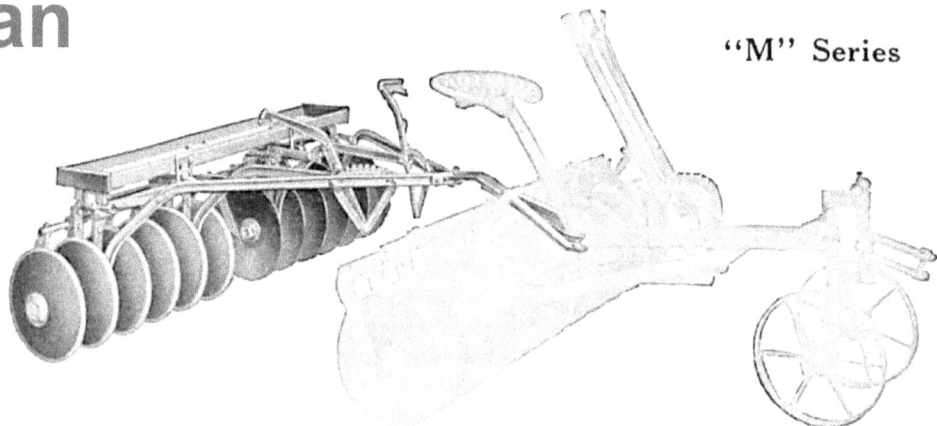

Cuts Deep, Pulverizes
Fine and is Easy on
Your Horses

"H" SERIES
DISC HARROW

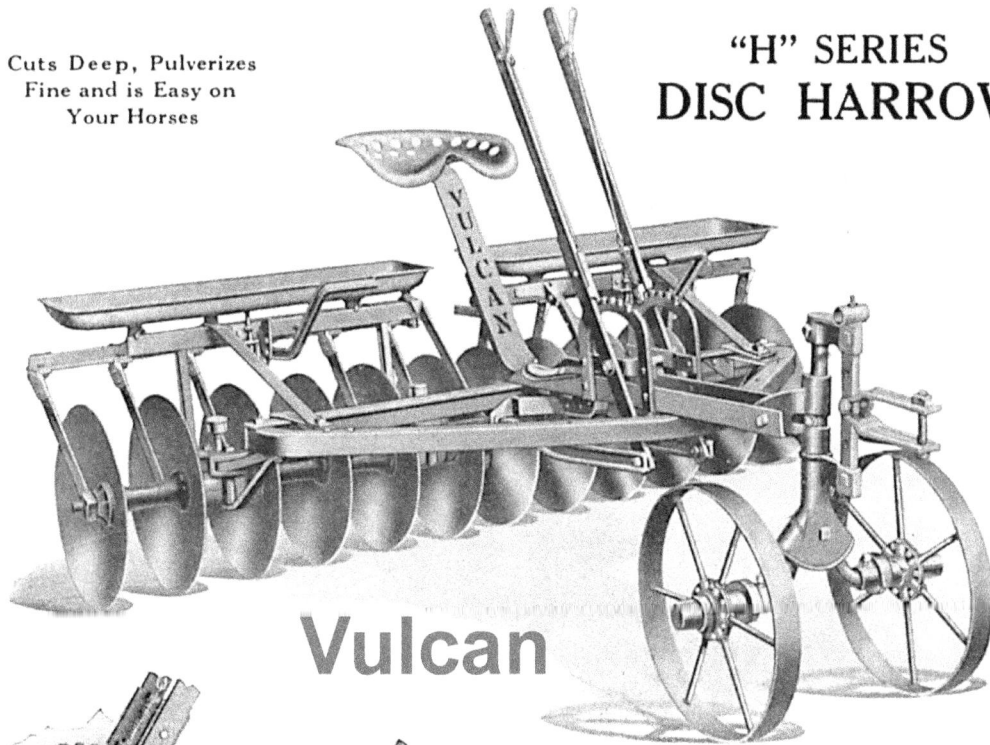

Vulcan

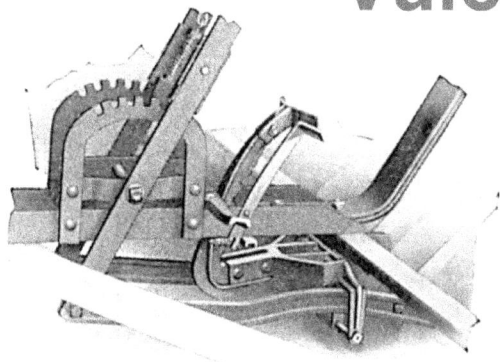

Fig. 1 shows Center Foot Lever Adjustment and Release.
Note the Lever Draft Bars are Curved to Keep Gangs Level
no Matter at What Angle They are Placed.

Vulcan Double Wheel Foretruck

John Deere

*A modified tractor John Deere
double gang setup for horses.*

Vulcan

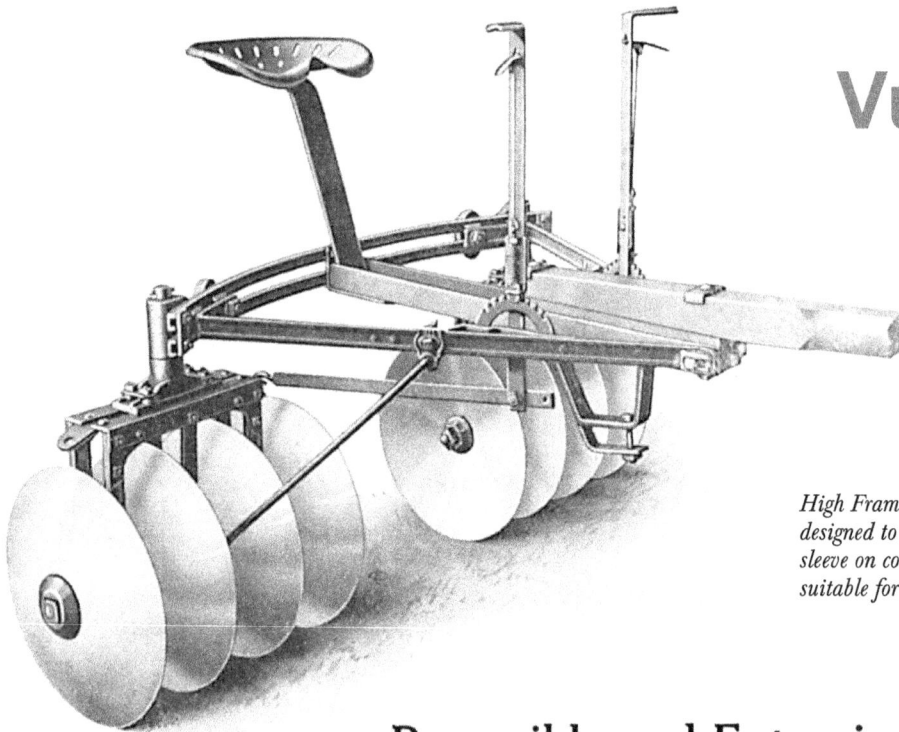

High Frame "J" Series. These gangs were designed to be easily tilted by an adjustable sleeve on coupling. This made the implement suitable for hilling purposes.

Reversible and Extension Disc Harrow

Low Frame "D" Series

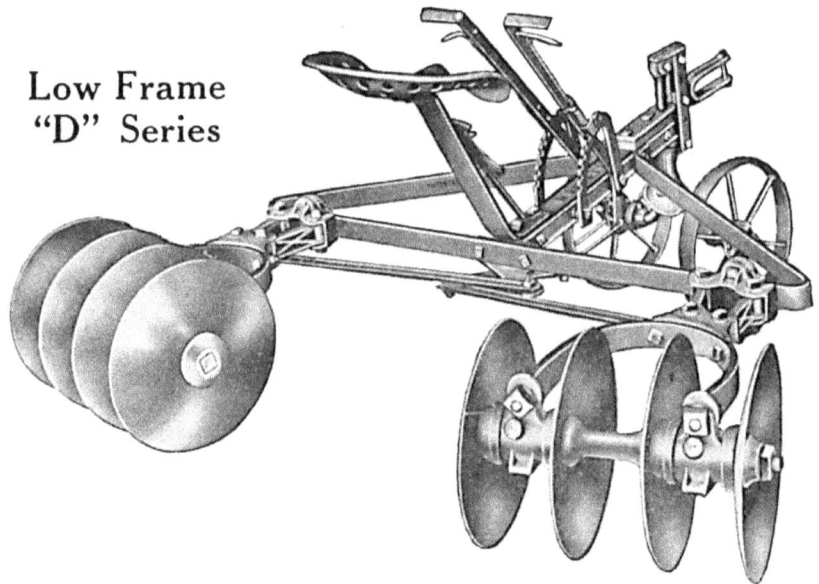

Rear View, Gangs Set For Out Throw

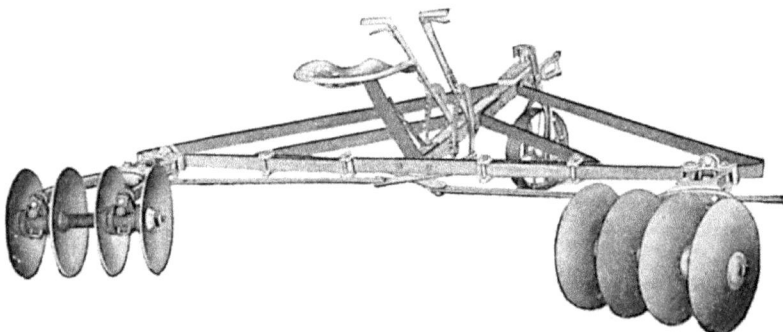

Left - shows Low Frame with extension to 10 foot wide.

The Vulcan "H" series disc at work., circa 1915.

Discs

Pole or No Pole? Cautionary notes.

As has been shown, disc harrows were available with and without tongues or poles. The pole is central to a backing and braking system with harnessed and hitched mules or horses. When you hook without a pole there is no "backbone" to the hitch to prevent the implement from rolling up on the horses heels. With disc harrows in soil, and even sod, there is enough resistance or drag, especially when an angle is set, to act as a brake and keep the tool from rolling up on the horses. When working steep ground with a poleless disc harrow, great caution needs to be exercised on the downhill. With green or anxious horses driven by a green and/or anxious teamster we recommend the use of a pole.

Oliver Double Disc with four well conditioned horses circa 1920.

Blount Disc Harrow with foretruck and weight boxes.

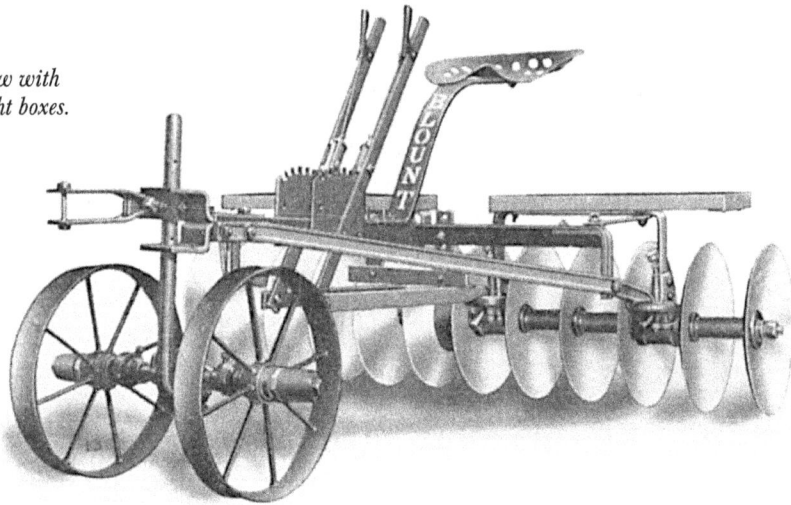

Blount

Circle Frame Reversible Disc Harrow

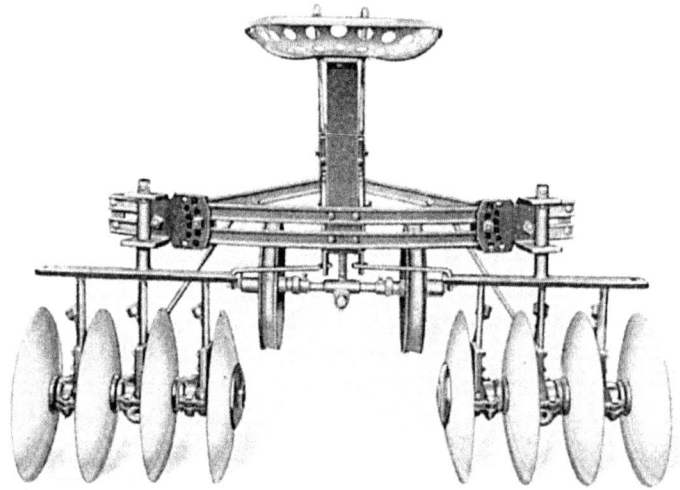

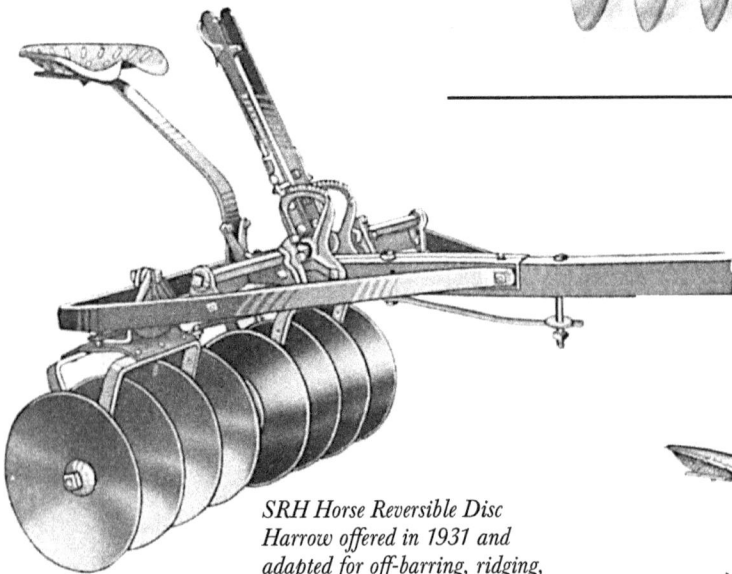

Oliver

SRH Horse Reversible Disc Harrow offered in 1931 and adapted for off-barring, ridging, throwing out middles, flat cultivation, and leveling ridges in cotton fields. Of interest is that Oliver never mentioned vineyard or orchard applications.

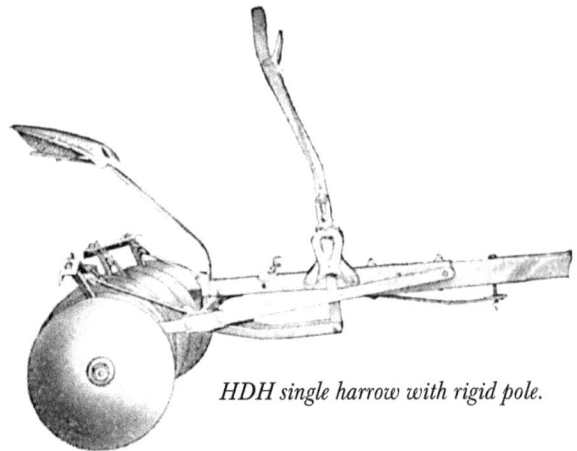

HDH single harrow with rigid pole.

Oliver

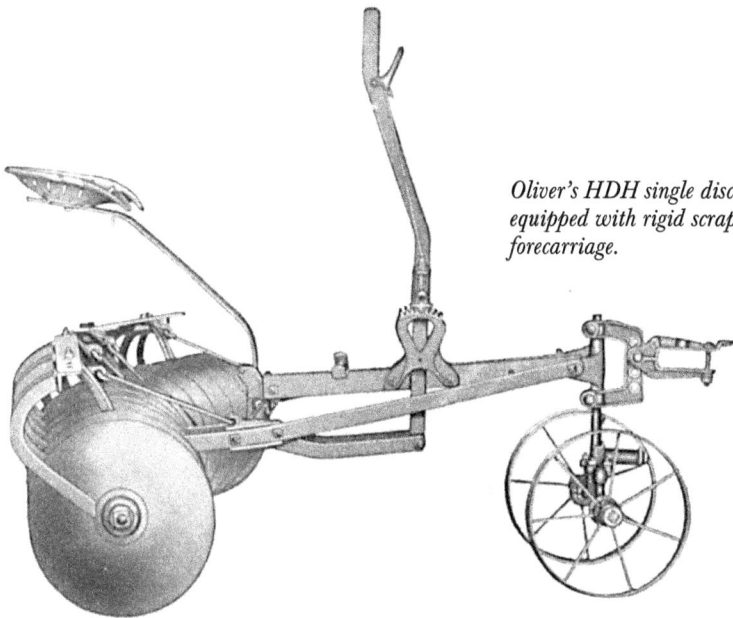

*Oliver's HDH single disc harrow
equipped with rigid scrapers and a
forecarriage.*

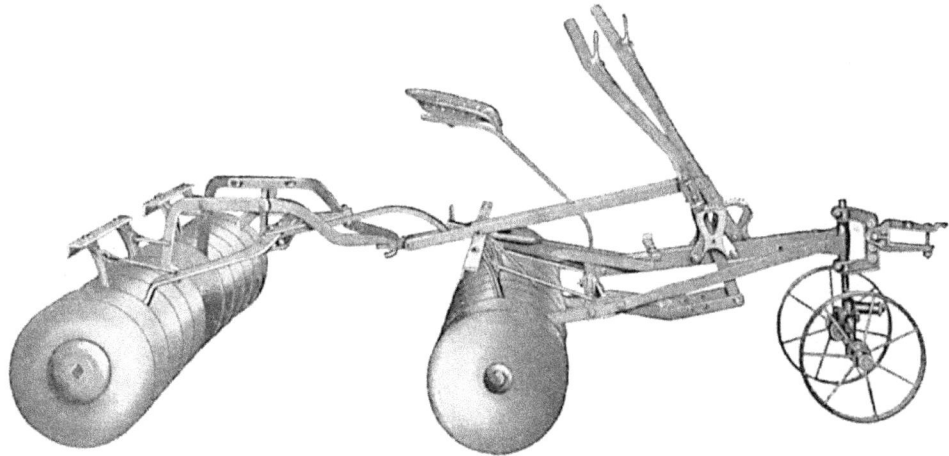

HDH double disc harrow with no scrapers and forecarriage. The tandem section is detachable.

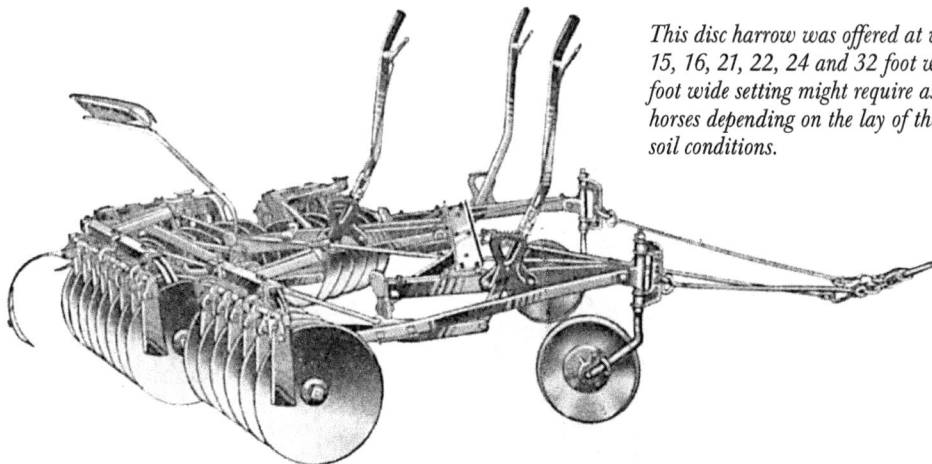

*This disc harrow was offered at widths of 14,
15, 16, 21, 22, 24 and 32 foot wide. The 32
foot wide setting might require as many as 12
horses depending on the lay of the land and the
soil conditions.*

Side view of the 14-foot WDH wide disc harrow with oscillating scrapers ("D" equipment).

Kota

KOTA Manufacturing of Lewistown, Pennsylvania, is an affliliate of the long standing horsedrawn equipment and parts supply company B. W. MacNair. They have recently come out with this handsome and rugged horsedrawn disc in six foot and five foot widths. The company address is at the rear of this book.

Chapter Four

Harrows
Spike Tooth
Chain
Spring Tooth
Acme/Crusher

Harrows with handles.

circa 1850

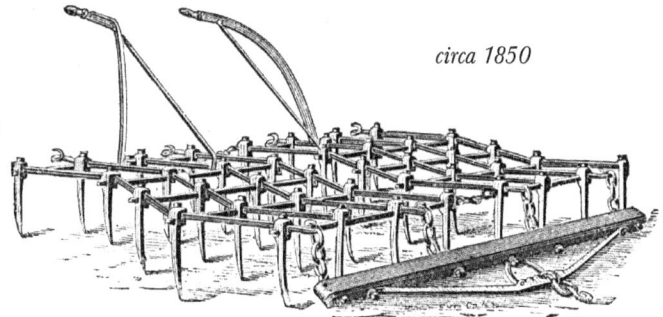

Harrow with handles. Picksley, Sims, & Co., Leigh, England.

Spike Tooth Harrows. The spike tooth harrow is the most basic of implements. A few thousand years ago, when limby logs were drug across dirt to stir and smooth in preparation for planting, someone came up with the idea that a framework of wood could be drilled to receive hardwood pegs. By the mid nineteenth century blacksmith's were constructing spike tooth harrows of elegance, beauty and perfect functionality. Since that design hiatus the tool has been simplified but remains the basic top soil comb or brush it has always been.

Chain Harrows. As shown on page 100, since mid-1800s smoothing harrows have been used which feaure a net of spiked or unspiked chain. While this style of harrow is

At Indiana Horse Progress Days an Amishman rides a Pioneer spike tooth harrow pulled by two Percherons.

The late great Addie Funk of Montana rides his saddle horse and drives two mules, center, and two Doug Hammill Clydesdales on the outside while drawing a deep-set spring tooth harrow over plowed ground.

Olde Harrows of Yore
(circa 1850)

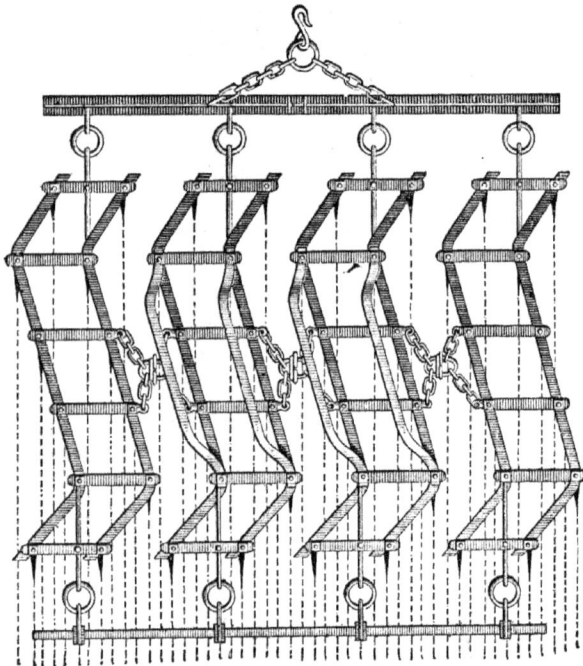

Flexible harrow. É. Puzenat, Bourbon-Lancy, France.

Picksley, Sims, & Co.

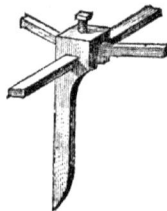

Harrow tine and frame. Picksley, Sims, & Co., Leigh, England.

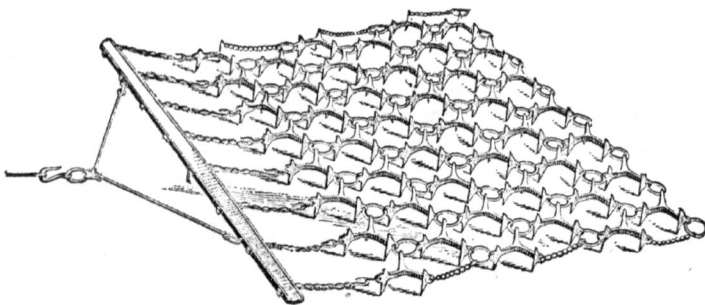

FIG. 106.—*Flexible chain-harrow. James & Frederick Howard, Bedford, England.*

most popular today for dragging pastures, it also works exceedingly well to level and finish seedbeds.

Spring Tooth Harrows. This implement was designed to reach deeper into plowed ground and do a more thorough job of tillage. The spring steel teeth, when set down, push in where allowed and crush and stir to a greater depth than possible with spike teeth. Most spring tooth harrows are built to receive replaceable points designed for specific purposes (see page 19). This harrow requires more horsepower.

French flexible harrow. Puzenat.

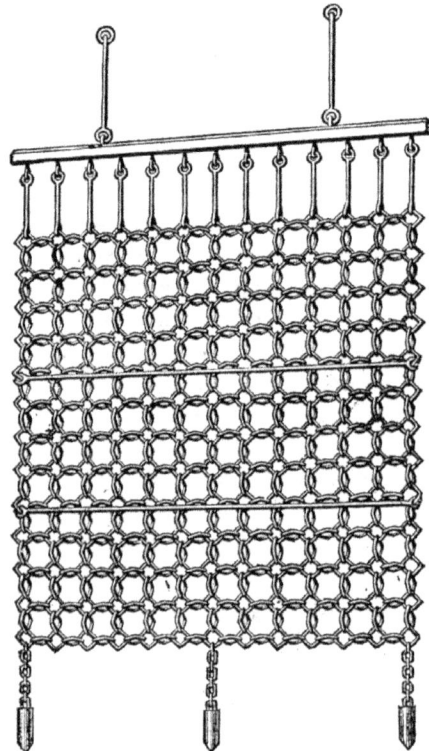

Chain-harrow. Picksley, Sims, & Co., Leigh, England.

Acme Harrows/Crushers. Though the curved blade harrow was developed in the first half of the 19th century it has the mistaken reputation of being a more recent innovation. The blades arranged in comb fashion are drug, by the framework, across the tilled soil. (see pages 120, 121) The soil is twisted and flipped as if little plow furrows. Implement companies all came up with their own names for this implement. This has been a specialty tool for farmers who take their craft seriously. Though it is not strictly necessary it does provide for an outstanding finish to prepared soil.

AP-30, 30-TOOTH SECTION
PIPE-BAR HARROW

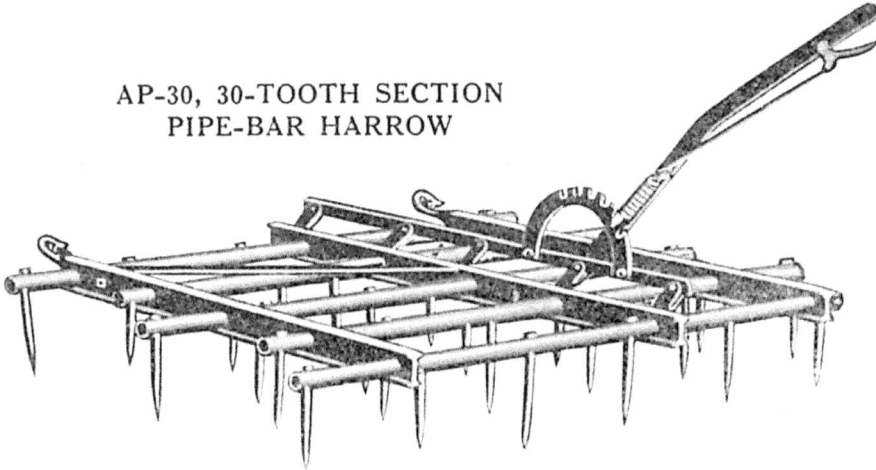

Spike Tooth
Harrows

Moline

AG-30, 30-TOOTH SECTION
END-GUARD HARROW

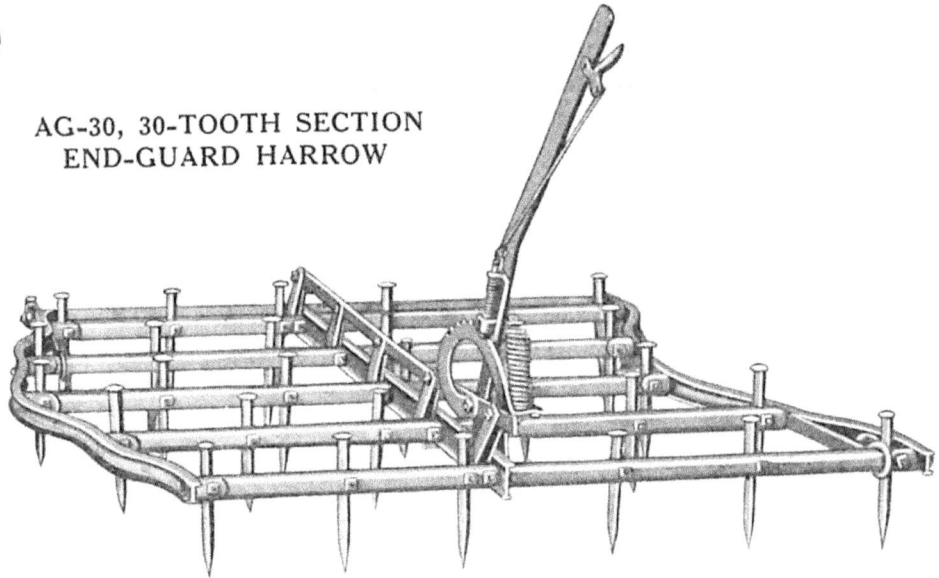

AU-30, 30-´
U-BAR HARROW

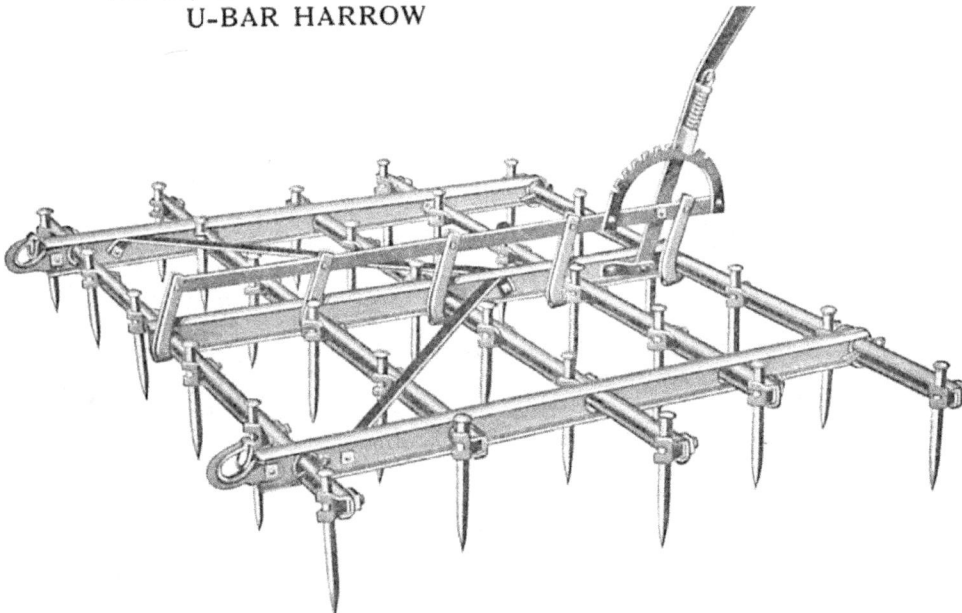

Avery Staytite Harrow

Information taken directly from company literature.

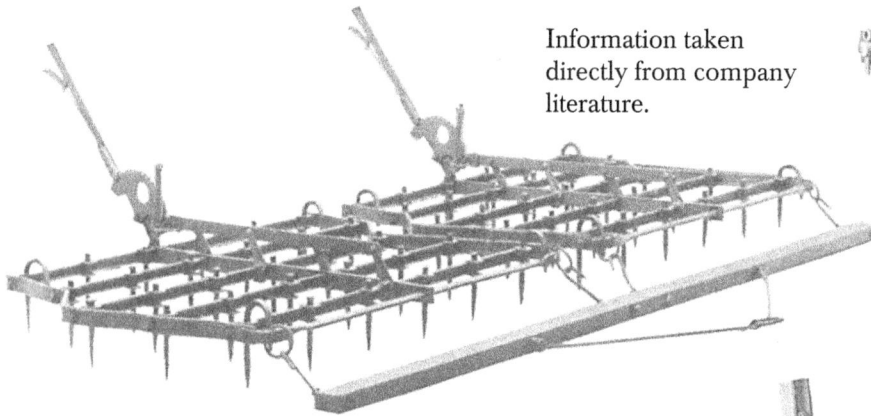

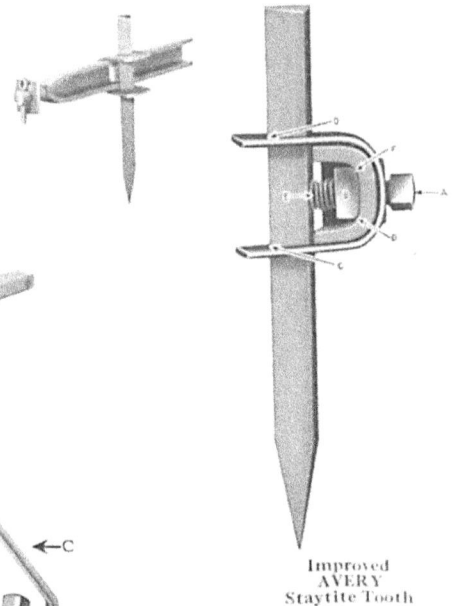

Improved
AVERY
Staytite Tooth

Avery

B→
A→

The Improved AVERY Hitch

AVERY STAYTITE HARROW

Every farmer who has ever used a harrow knows what a lot of trouble can be caused by a harrow with teeth that do not stay put. To be dependable and capable of doing a good job day after day, the teeth must be so set in the frame that they will neither loosen nor drop out when subjected to long hard use. They must stand the strain.

The Avery Staytite Harrow, with Jack Screw Clamp - with teeth fastened more securely than ever - that become tighter as the strain on them becomes greater again proves that Avery engineers work on the theory that "letting well enough alone" was not doing

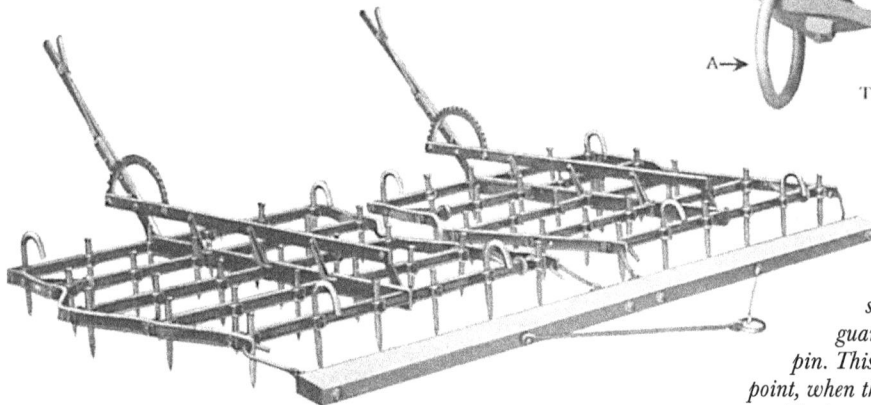

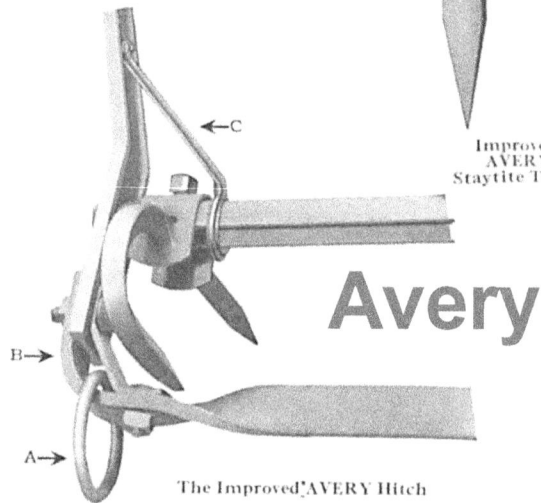

well enough.

The Jack Screw arrangement produces a pressing strain on the set screw instead of the pulling strain which in the usual bolted construction causes the bolt to stretch and the teeth to loosen. See illustration of Improved Avery Staytite Tooth.

The tooth is held in place by a cup-point set screw (set screw A) which penetrates the surface of the tooth at point "E." This pressure forces the sharp edge of the tooth into the ends of the triangular openings in the clamp at points "C" and "D" and makes any movement of the tooth impossible as long as the pressure remains constant. Keeping constant pressure against the tooth is accomplished by means of nut "B" which is set inside of the "U" section tooth bar. Nut "B" cannot turn inside the tooth bar at points "F" and "G." The pressure that is applied against the tooth is applied equally against the threads of nut "B" and any movement of the nut on the set screw is impossible without the use of a wrench.

It will also be seen that whereas the teeth cannot be disengaged

accidentally they can be removed easily and readily for sharpening and replacement at any time, by simply using a wrench.

The end of the U-bar is forged into a sturdy round stem. It is inserted in the end or guard rail and held in place by a washer and cotter pin. This eliminates the breakage so common at this point, when the ordinary casting is used.

The levers which control angle of the teeth are mounted upon spring relief quadrants. These serve as cushions against violent impacts and shocks and add greatly to the strength and durability of the harrow.

The Improved Avery Hitch construction does away with the cast or forged piece which is usually bolted to the end of the U-section crossbar to complete the connection at the guard rail. It will be seen that by the elimination of this piece, there is also eliminated the necessity for two bolts on each cross member and the certainty of loose parts and lost motion which results therefrom.
Note three things:

First - A strong forged steel ring encircles the harrow frame at "A." The usual eyebolts and hook bolts screwed or bolted to the frame are eliminated. Second - Note at "B" the auxiliary member which serves to hold the frame rigidly in place. This also keeps the steel hitch ring in place. It makes a permanent, wear proof hitch and a rigid frame. Third - Corner brace "C" provides additional strength and durability to the frame. No warping or twisting.

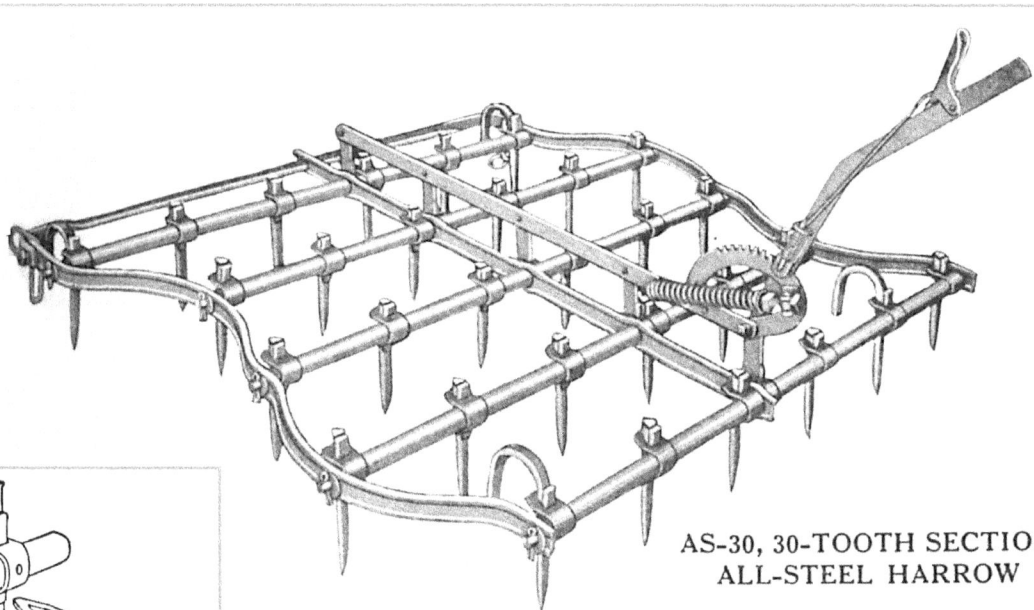

**AS-30, 30-TOOTH SECTION
ALL-STEEL HARROW**

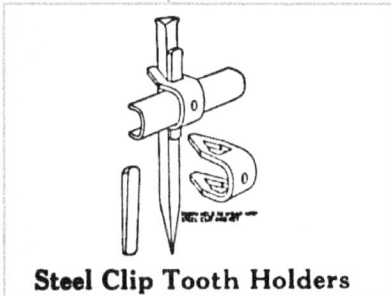

Steel Clip Tooth Holders

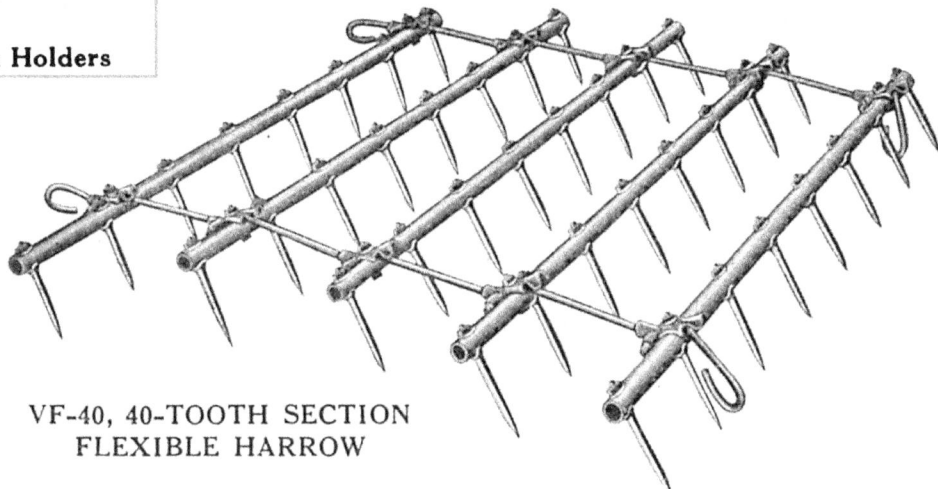

**VF-40, 40-TOOTH SECTION
FLEXIBLE HARROW**

Moline

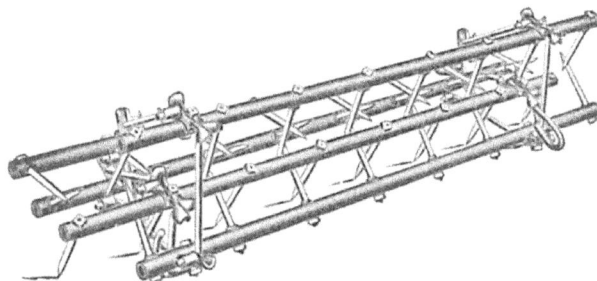

Each Section Rolls Up, Making It Convenient to Handle and Transport;
Little Space Is Required for Storage

Drawbars for Moline U-Bar and Pipe-Bar Harrows

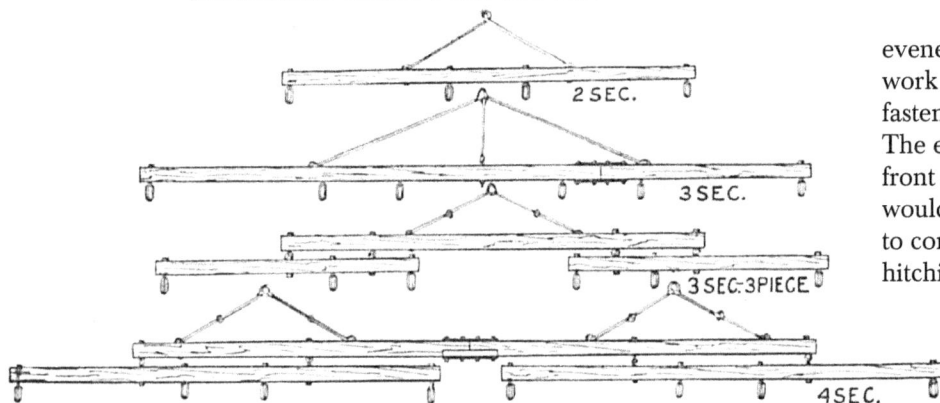

2 SEC.

3 SEC.

3 SEC. 3 PIECE

4 SEC.

Not to be mistaken for eveners which hook directly to work animals. These bars are fastened to the harrow sections. The evener is then hooked to the front ring. The bottom draw bar would require an additional stick to conclude with a single ring for hitching.

RW-42, 42-TOOTH SECTION WOOD-BAR HARROW

Moline

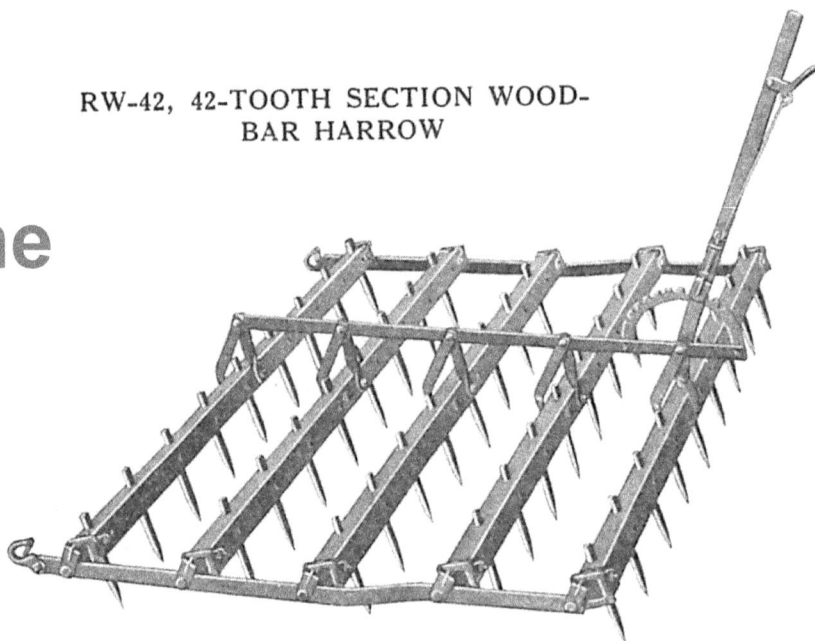

(Below, would require 8 head of horses.)

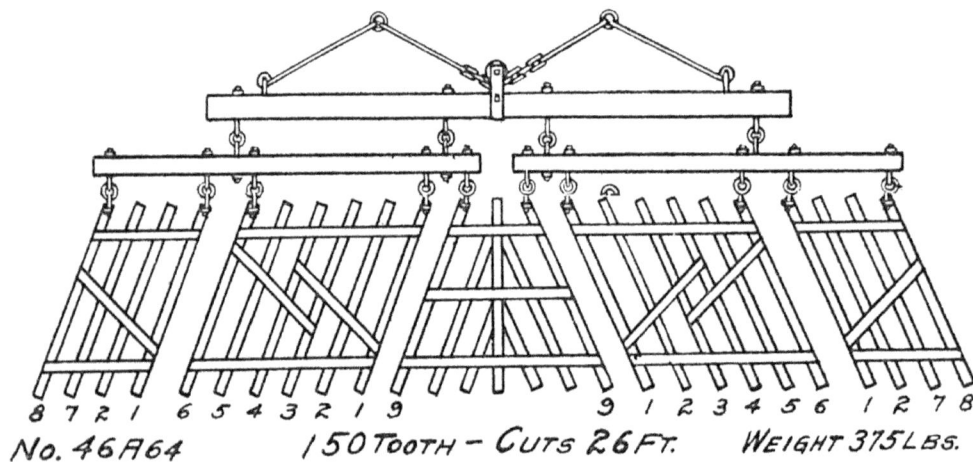

8 7 2 1   6 5 4 3 2 1 9   9 1 2 3 4 5 6   1 2 7 8

No. 46A64   *150 TOOTH – CUTS 26 FT.*   *WEIGHT 375 LBS.*

Case

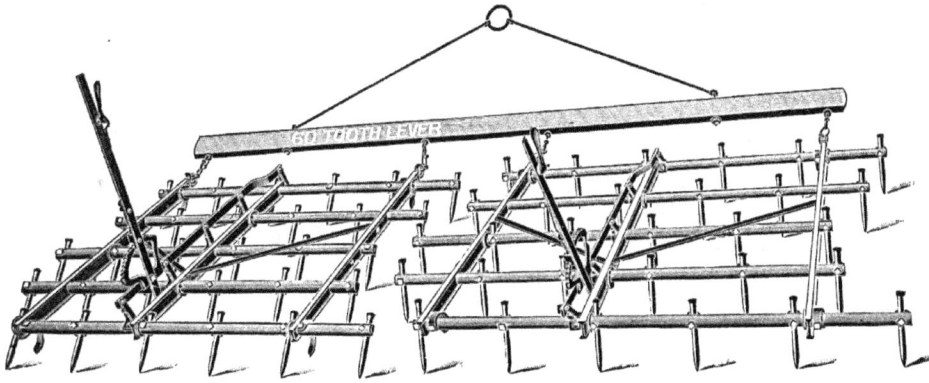

Above and below: U-bar Steel Lever Harrows

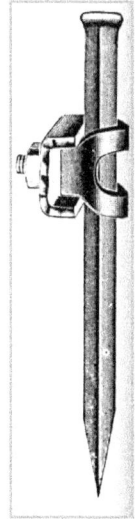

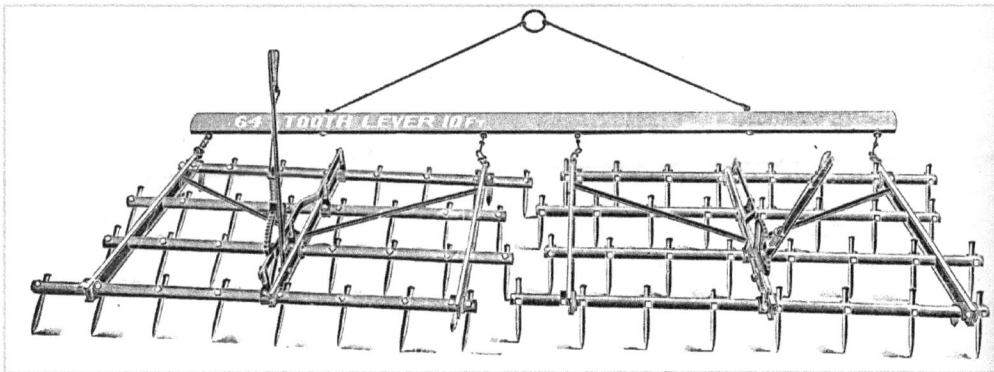

Rock Island

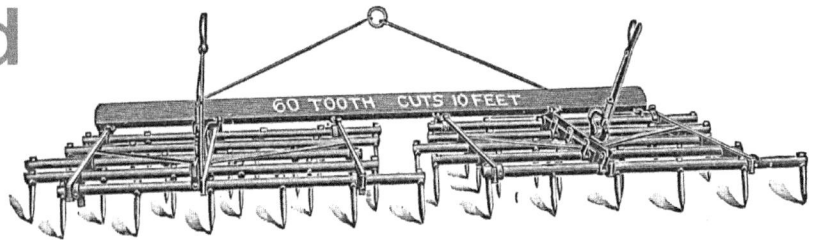

Pipe Bar Steel Lever Harrow

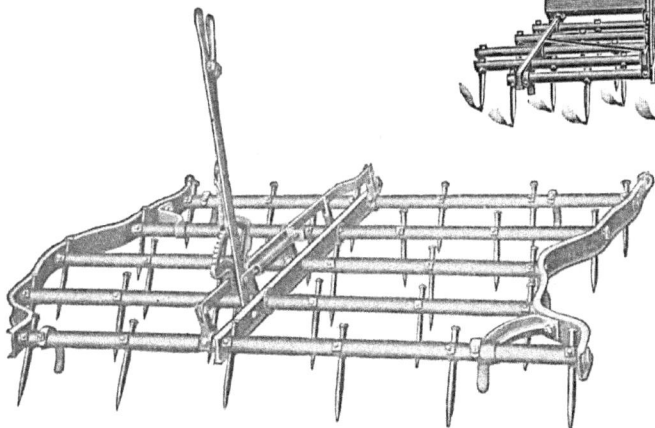

Guard End Steel Lever Harrows

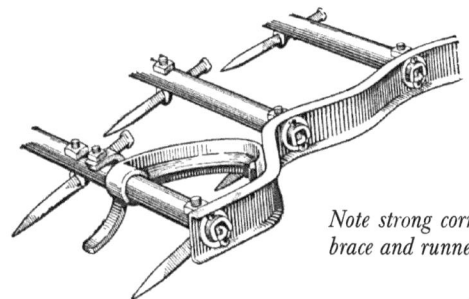

Note strong corner brace and runner.

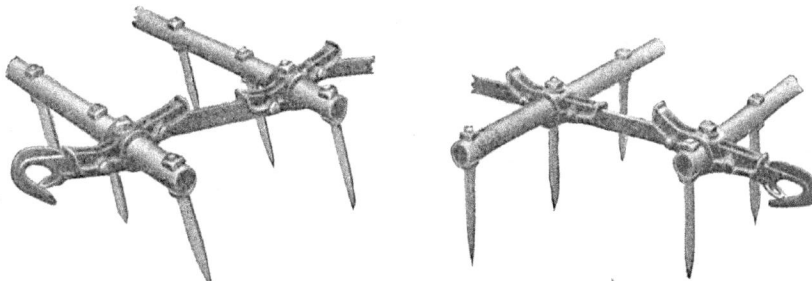

Closeup of the Fexible Bar harrow showing tooth positions.

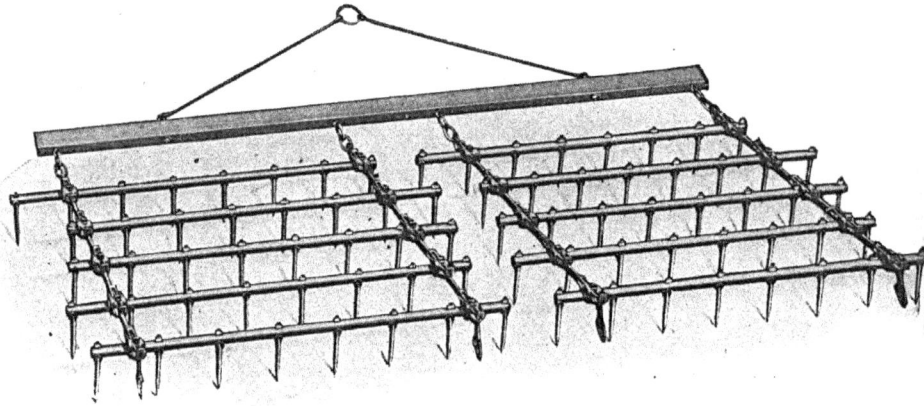

Flexible Steel Harrow

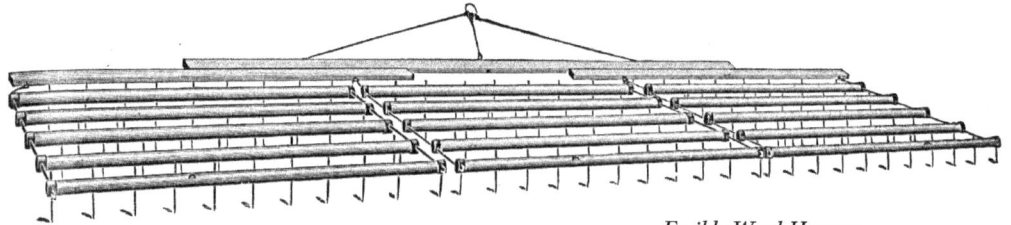

Fexible Wood Harrow

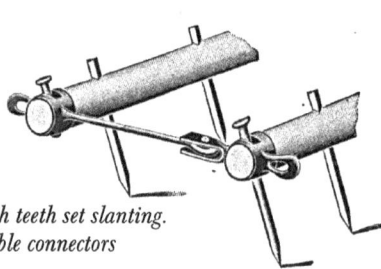

*Two wood bars with teeth set slanting.
Note the flexible cable connectors*

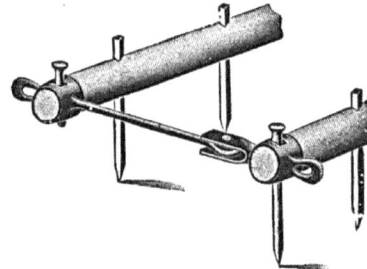

Teeth set perpendicular

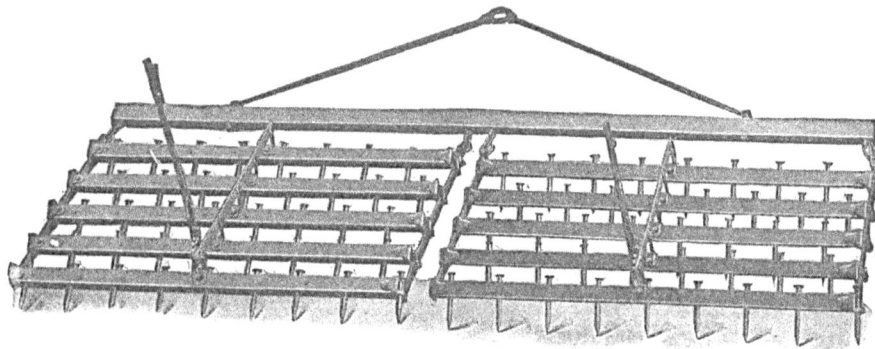

*Wood Frame
Lever Harrow*

Rock Island

Number of Horses Required for Spike Tooth Harrows: One work horse or mule in good condition 'should' be able to handle one section of spike harrow. However, the number of spikes, condition of the soil and the field's topography can affect power requirements. This author prefers to use more horses than absolutely necessary to keep the animals fresh. I would use 3 horses for 2 sections.

Number of Horses for Spring Tooth Harrows: One and a half to two animals are required per spring tooth section. Again extenuating circumstances can affect this number.

Number of Horses for Acme Harrows: Two animals can handle a six foot section rather easily.

Case

Critic Steel Frame Lever Harrow with runner tooth

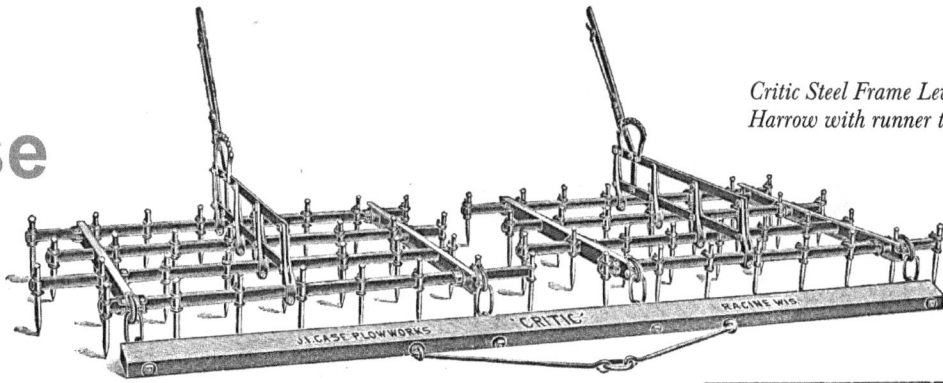

Front of bar - tilted

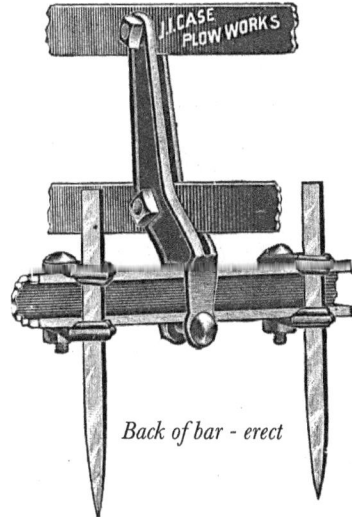

Back of bar - erect

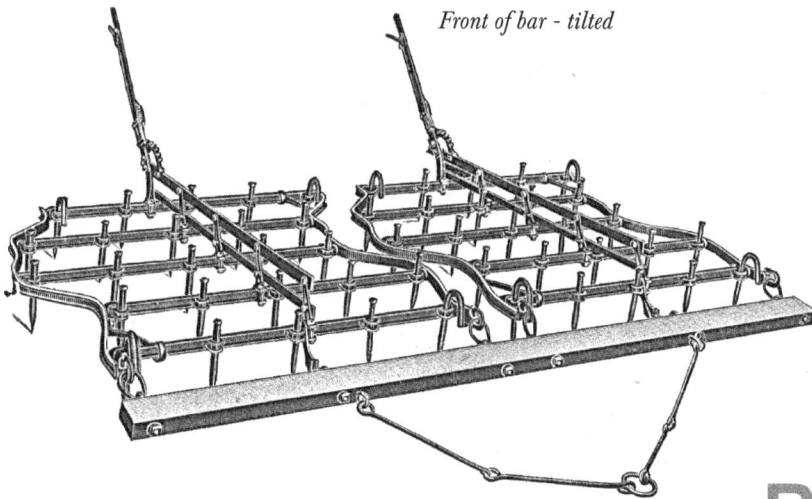

Case's Guard Rail Harrow
Designed for use in orchards or anywhere that debris or obstacles might catch on an open sided harrow frame.

Roderick

Roderick Lean Lever "A" series

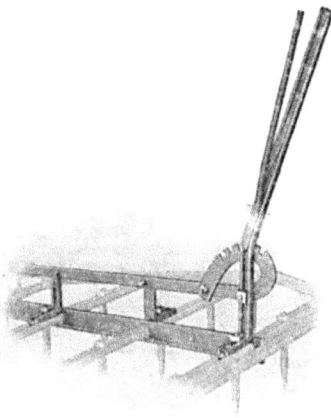

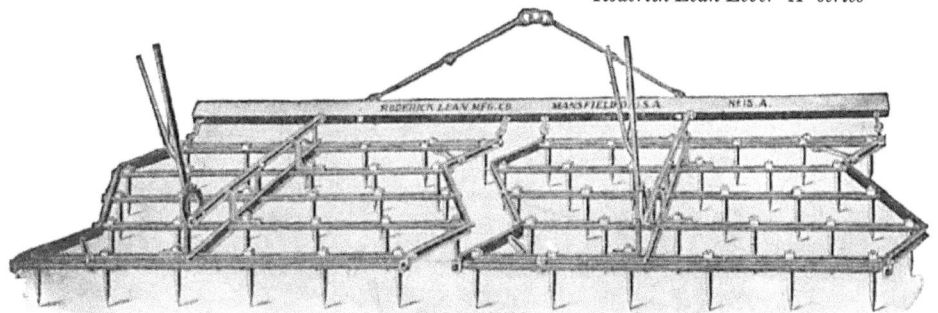

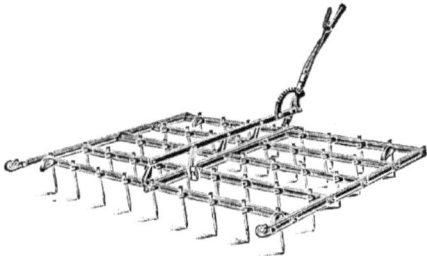

Deere Steel Square End Harrow

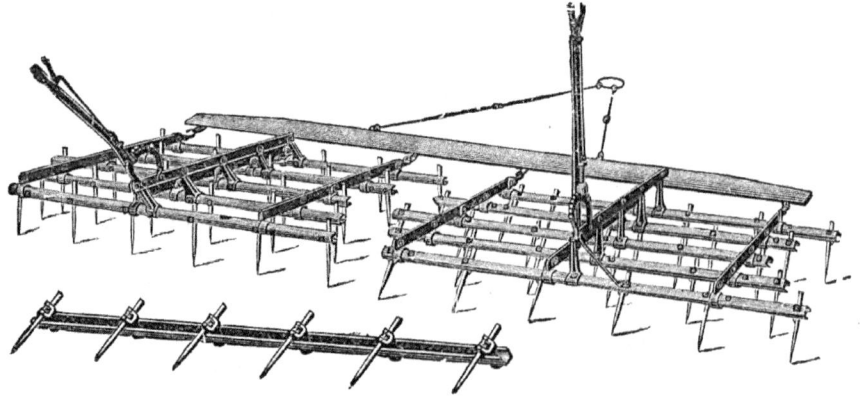

Zig Zag Harrow

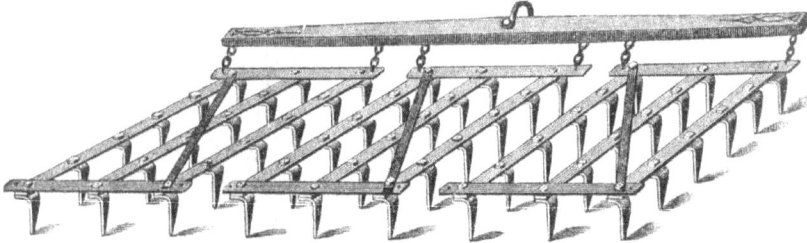

Glidden Harrow

John Deere

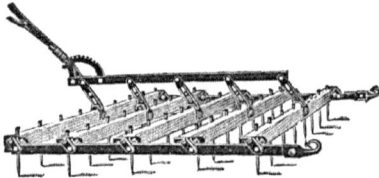

Deere Wood Square End Harrow

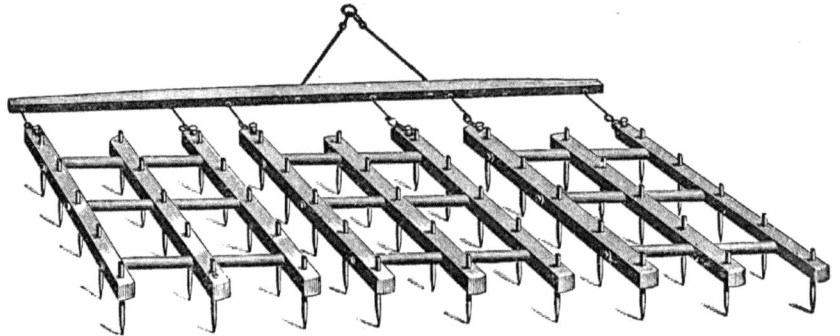

Deere Triple Harrow

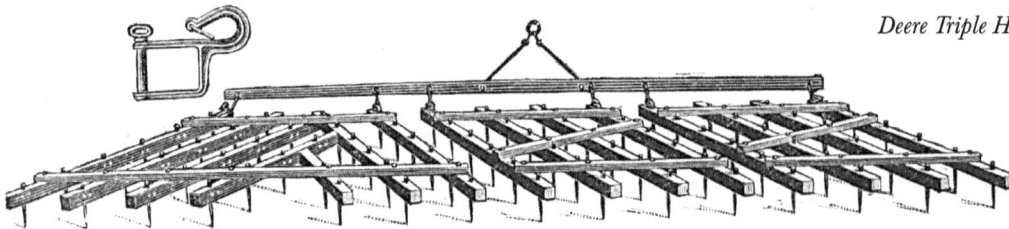

Deere King Harrow

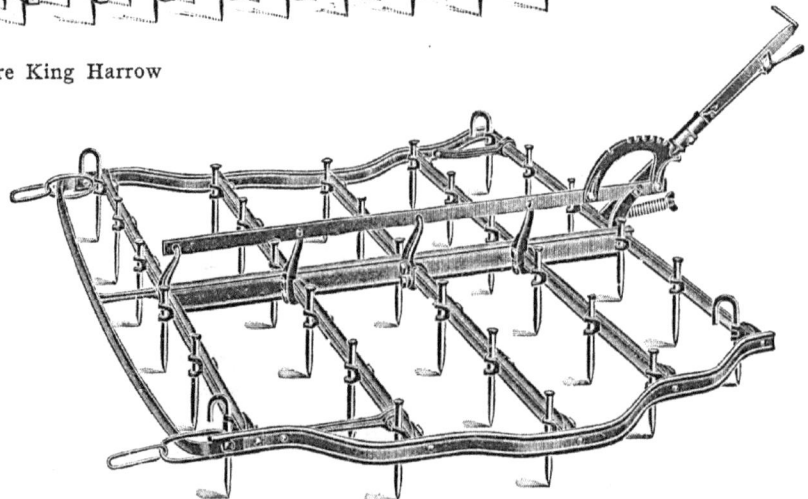

Deere Universal Harrow

John Deere

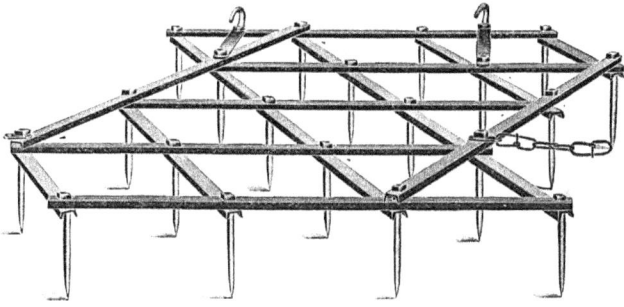

Diamond Harrow

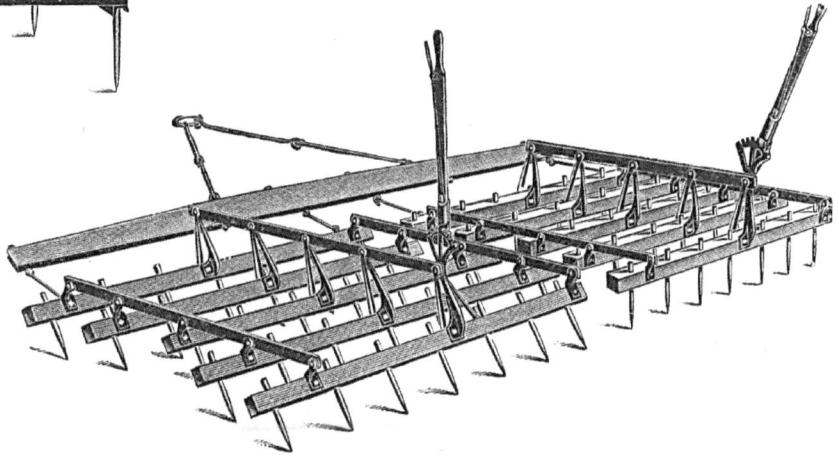

Deere Wood Lever Harrow

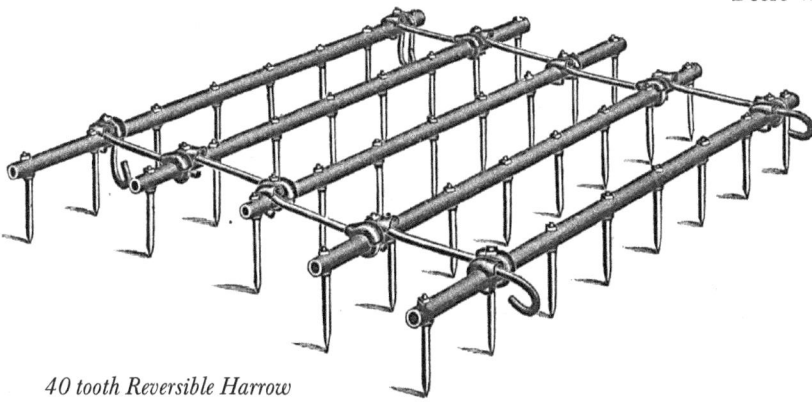

40 tooth Reversible Harrow

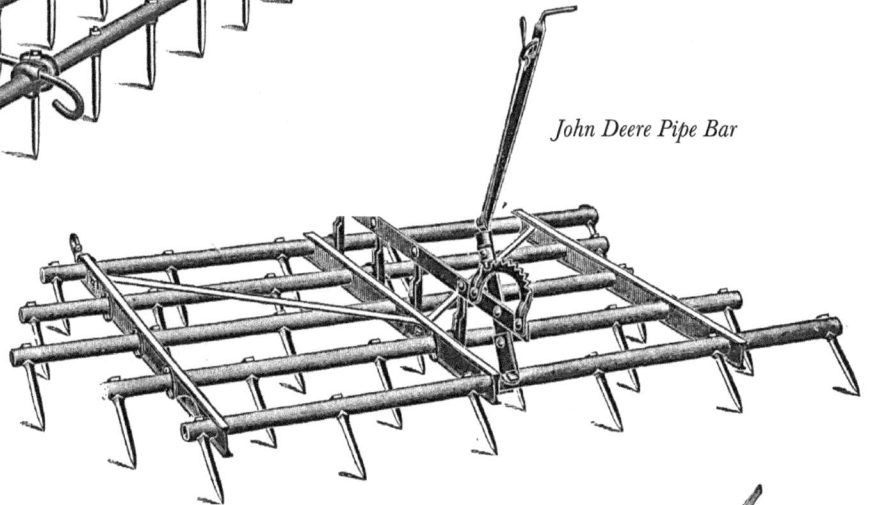

John Deere Pipe Bar

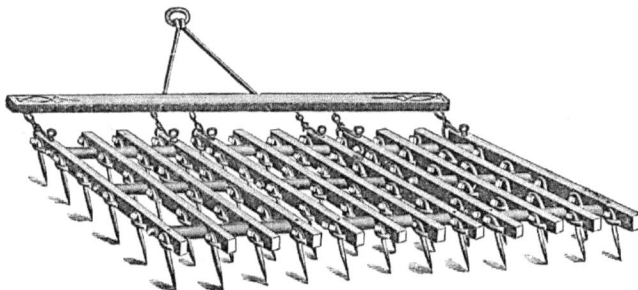

Deere Smoothing Harrow

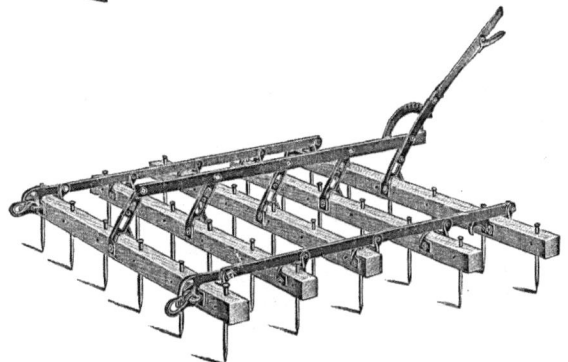

Deere Wood Lever Harrow

John Deere

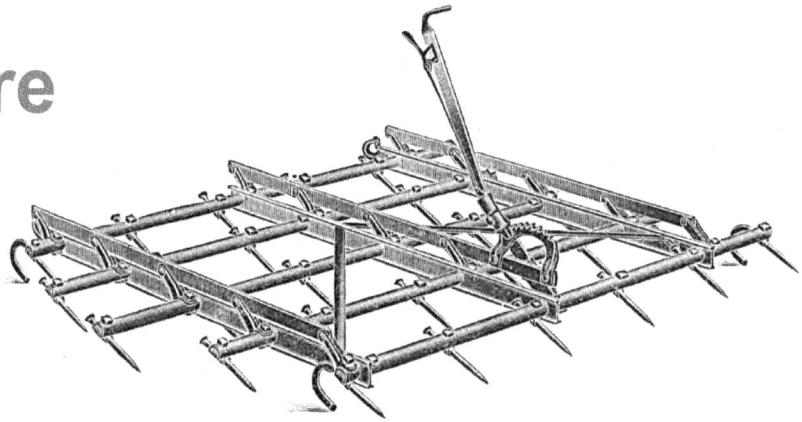

Ajax Five-bar Harrow with Top Rocker Arms and Runner Teeth

P & O

Featuring comments from the P & O catalog.

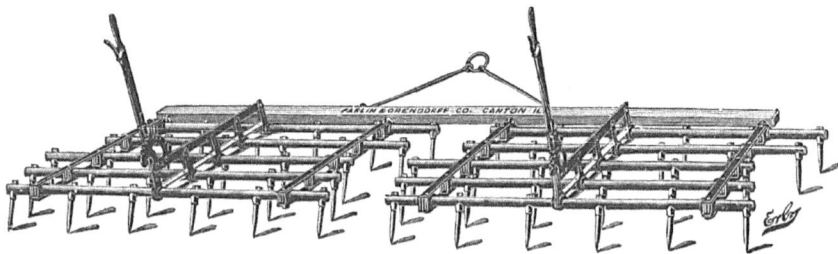

PIPE FRAME HARROW, TWO SECTIONS.

PIPE FRAME HARROW, TWO SECTIONS

These harrows are made of one inch steel pipe, with 1 1/4 by 3-16 steel cross-brace trussed. The teeth are threaded, and have a shoulder which, when bolted, holds them firmly in place. The teeth are also reversible, and can be turned around, giving a fresh cutting edge when they become worn. It can be used as a straight tooth pulverizing harrow, or a slanting tooth smoothing harrow. If trash accumulates, the position of the teeth can be changed in a moment, while in operation, by using the lever, to a slanting position to clear trash. As a smoothing harrow it never clogs, the rubbish either being cut, or it passes down and off at the ends of the teeth, and is completely buried. For cultivating corn in its early stages it is unsurpassed.   The soil is thoroughly pulverized, all weeds are destroyed, while

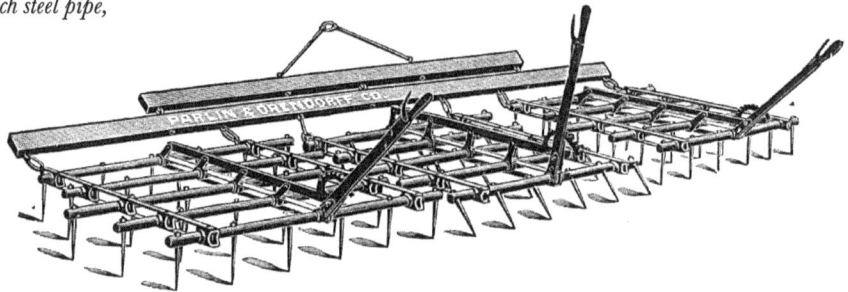

PIPE FRAME HARROW, THREE SECTIONS

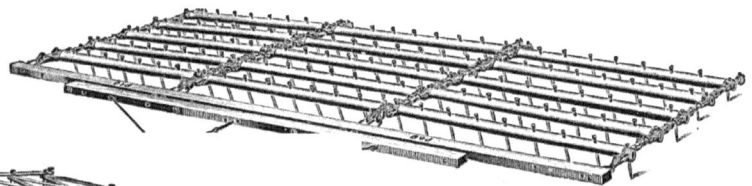

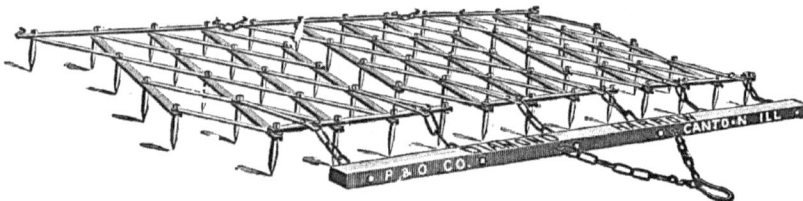

the corn remains uninjured.

This harrow is also equipped with runners, making a perfect sled for transportation. The Pipe Frame Harrows are excellently designed and constructed for high-class implements, and have long been one of our best harrows. 1/2 or 5/8 teeth.

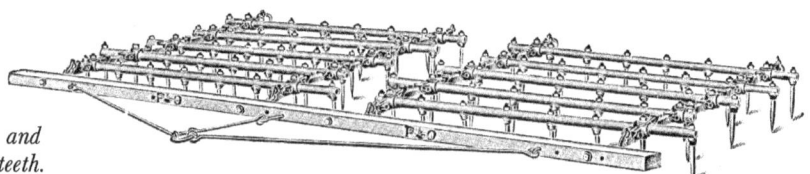

P & O

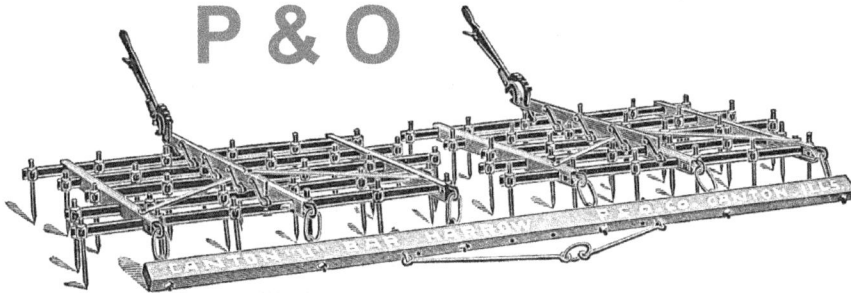

"U" Bar Harrow, Two Sections

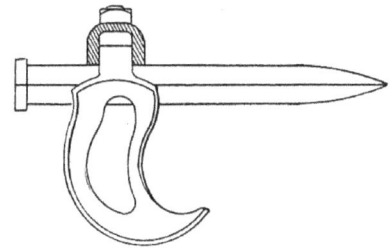

Runner for " U " Bar Harrow.

"U" BAR HARROW, TWO SECTIONS

So called because of the shape of the beams, which are made of "U" bar steel. The cross braces are made of channel steel, and the bars are held in position by flanged malleable clamps.

The teeth are eight inches long, and are adjustable up or down, besides being reversible, so that a uniformly sharp edge can be maintained. The teeth are firmly held in place by malleable iron clamps and can not work loose.

The frame is very rigid, and the center cross brace and the diagonal straps prevent any tendency to strain or twist.

The levers allow a number of adjustments of the slant of the teeth, and can be easily tripped to clear the harrow of trash.

The brackets for the teeth allow the interchanging of either 1/2 or 5/8 teeth.

A new feature on these harrows is the runner. A runner is placed on the four corners of each section, which are out of the way when the harrow is in use, but which act as perfect sled runners when the teeth are thrown back. This allows the teeth to clear the ground by about four inches in transporting.

The "U" Bar Lever Harrows have long held their position as one of the very best and strongest harrows we make, and no lever harrow was ever designed which excels it as a reliable and most serviceable pulverizer. 1/2 or 5/8 teeth.

WOOD FRAME HARROW, TWO SECTIONS

This is a very strong harrow, the beam being made of 2 1/4 inch square oak. The diamond teeth are rigidly held in place by being driven into round holes slightly smaller than the teeth. The draw bars are made of 1 1/2 by 3/8 inch steel, and clamped with heavy malleable iron brackets.

The levers will throw the beam to any angle from a vertical to a horizontal position. When transporting, the harrow is dragged on the runners, which are placed on the four corners of each section.

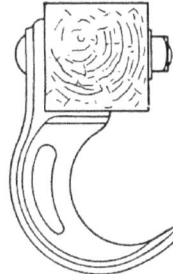

Runner for Wood Frame Lever Harrow.

FAVORITE HARROW, TWO SECTIONS

A harrow which combines lightness, simplicity and strength, made entirely of steel and malleable iron.

The bars are made of one inch pipe, with beaded steel cross beam, and an additional center brace to prevent the pipes from sagging.

The teeth are threaded and made with shoulder, and held in place by malleable iron washers and nuts.

Either ½ or 5/8 inch teeth can be used, as the threaded ends of each are of the same size.

The lever adjustments are such that the tooth bars can be set at any angle, and when transporting, the harrow rides on the runners.

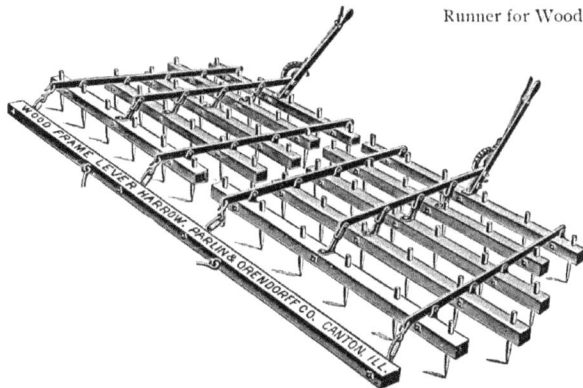

Wood Frame Harrow, Two Sections

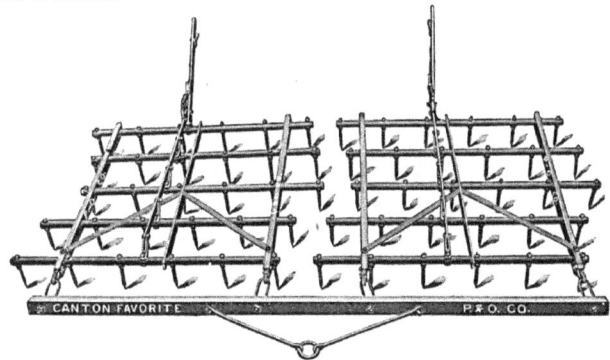

Favorite Harrow, Two Sections

DIAMOND HARROW, THREE SECTIONS

The Diamond Harrow is simple and compact, and is an extremely light implement. The frame is perfectly rigid, and can not be telescoped, each bar and brace being a factor in adding to its strength and rigidity.

The sections are chained together in the rear, and thus any tendency to undue side swing is overcome.

This harrow is equipped with 9/16 inch diamond teeth, 7 inches long, giving good clearance, and are set on the frame so they will not track.

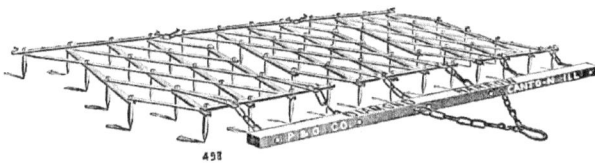

Diamond Harrow, Three Sections

Harrows

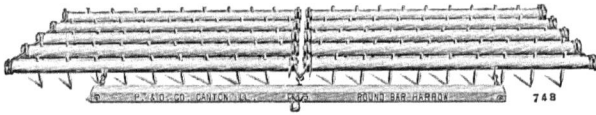

ROUND BAR HARROW, TWO SECTIONS

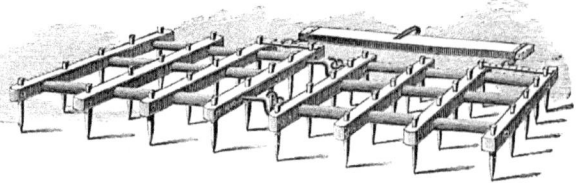

SCOTCH HARROW, TWO SECTIONS

ROUND BAR HARROW, TWO SECTIONS

This is a new combination smoothing and pulverizing harrow, that possesses a number of features not embodied in drag harrows.

It is a combination pulverizing and smoothing harrow, as it can be pulled from either the front or rear. The coupling irons are swiveled and self-adjusting, and when dragged from the front the teeth assume a perpendicular position. When dragged from the rear the teeth run in a slanting position.

Another point of this harrow is the manner in which each bar acts independently of the other in rough or uneven ground. When a low piece of ground is encountered each bar will drop down as it passes, and every portion of the ground is pulverized.

When not in use, or when transporting, the harrow can be rolled up with the teeth pointing to the inside.

The beams are made of 2 3/4 oak, chamfered, and the teeth are firmly held in place by being driven into round holes slightly smaller than the teeth. ½ inch teeth only.

SCOTCH HARROW, TWO SECTIONS

A very light harrow, and therefore a very serviceable one in rough, stumpy or uneven ground, as it can be easily handled. It is one of our old harrows, made to stand the wear and tear for years. The rails are made of 2 by 2 3/4 oak, connected by spools and rods, and the rods can be tightened up at any time.

The hitch is made on the side and the teeth will not track. There are no projections of any kind underneath to gather trash.

The harrows are equipped with two sets of spools and rods, and a third set can be added if ordered. Past experience with these harrows has demonstrated that they are equal to any kind of work.

If a new section is wanted it can be added by boring two holes for the connecting loops. The sections are hooked together, presenting a uniform width between the bars. These connecting loops are so large that plenty of freedom is allowed each section.

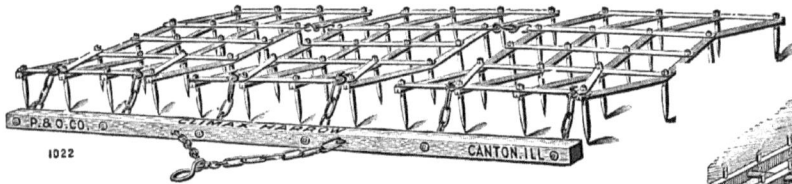

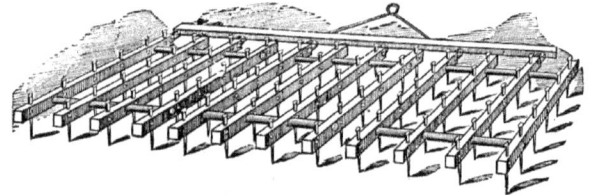

CLIPPER HARROW, THREE SECTIONS

ROUGH RIDER HARROW, THREE SECTIONS

The name of this harrow is not a misnomer, as it partially describes this, one of our latest drag harrows.

The bars are made of 1 3/4 by 3/8 inch steel, reinforced by a 1 by 5-16 inch steel brace. The teeth are firmly bolted to the bars, and for a small section harrow it is quite heavy. The shape of the sections is such that a direct pull will allow the teeth to run diagonally, and the implement is sufficiently heavy to pulverize any kind of soil.

The teeth are cut with a straight edge in front and a wedge edge in the rear. The hitch can be made on either side, the front giving a pulverizing and the rear a smoothing harrow.

CLIPPER HARROWS, THREE SECTIONS

The rails are made of 2 by 2 3/4 oak, connected with spools and rods. The spools are cut at a slight angle, and the sections are diamond shaped.

Owing to the number of teeth used, and the manner in which they are placed, they are excellent pulverizers in rough ground, as every portion of the soil is thoroughly harrowed.

The teeth are diamond shaped, with points on the front edge, and are very strong, cutting into any kind of lumpy ground.

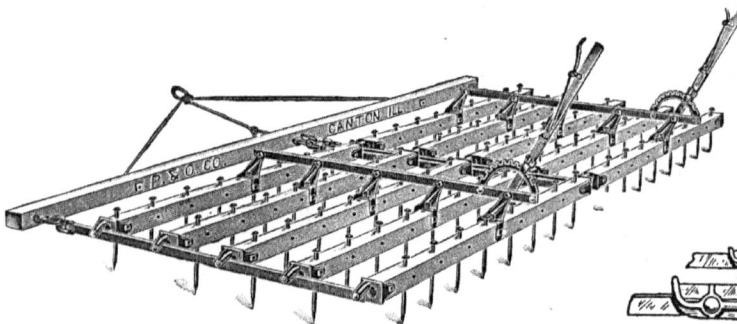

P & O

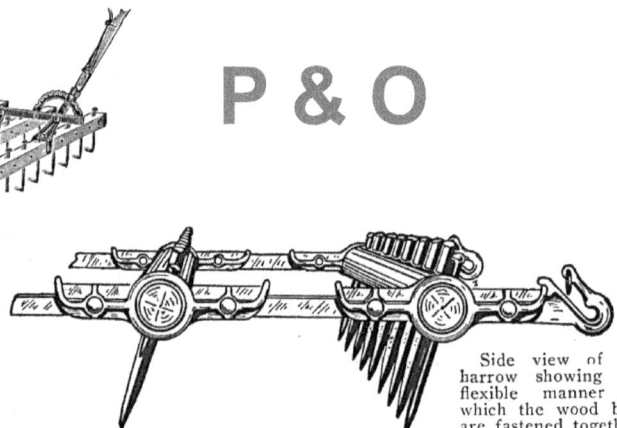

Side view of the harrow showing the flexible manner in which the wood bars are fastened together.

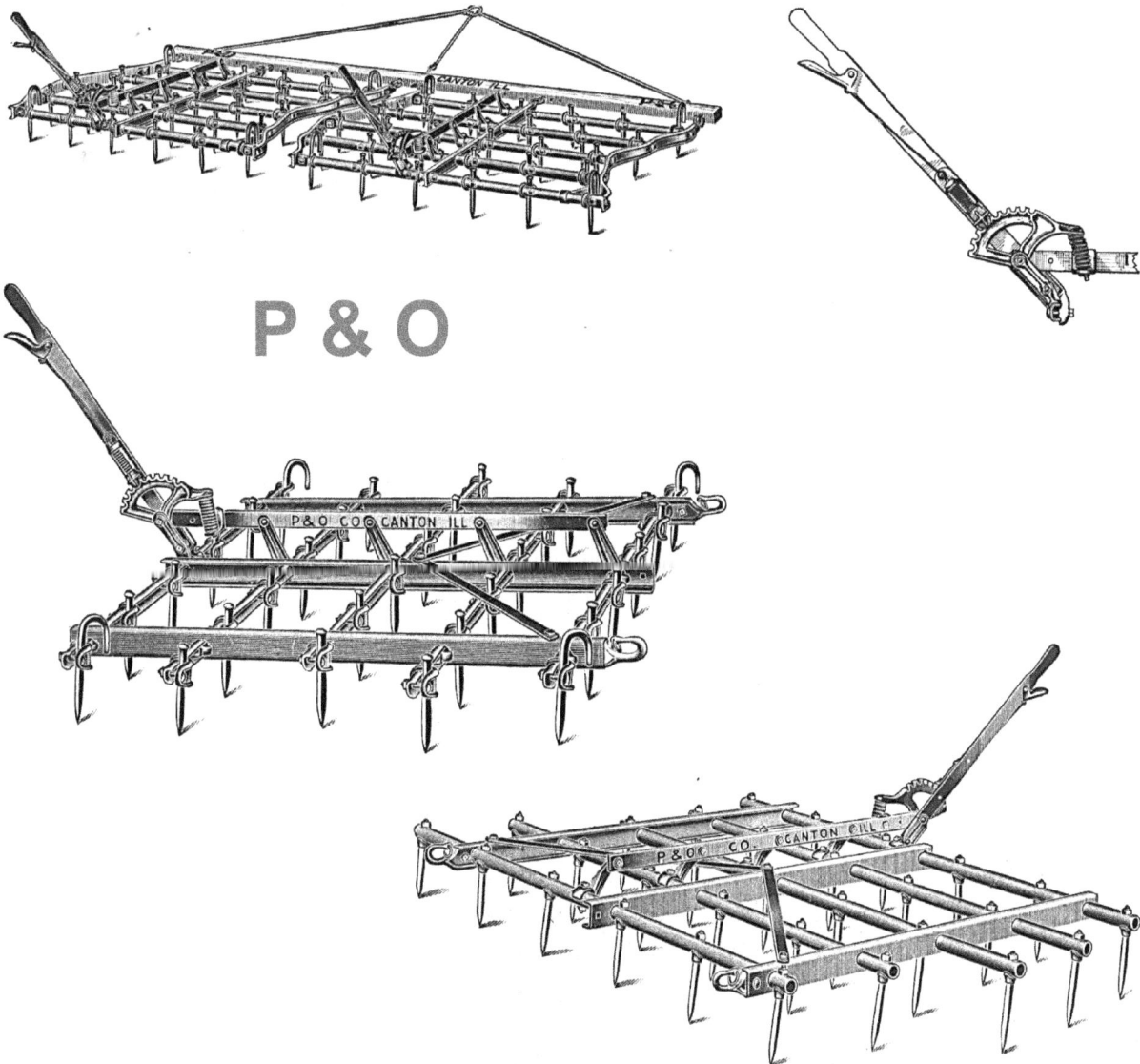

P & O

Horse Notes:

Hooking Close. When hooking to the harrow drawbar keep in mind that if you hook close and tight two things result. First, as the animals step ahead the front of the harrow is lifted. And second, when you turn sharp you are at greater (or earlier) risk of having the animals step into or onto the harrow itself.

Hooking Long. When you hook further away from the drawbar (or long) there is no lifting action and the entire surface of the harrow is making maximum contact with the soil causing full draft. Hooking long will require slightly more line length.

Length of Lines. Most standard team lines are twenty feet long. In this author's experience four additional feet are handy if not necessary. Lines can be custom ordered longer OR additional length can be added either by buckling (not recommended because of the hazard posed by the hardware catching on clothing) or tying and employing the plowman's knot (See *HD Plows & plowing*).

Riding the harrow. Some teamsters will tie a board across the harrows and stand on that to ride while working. I wish to discourage this as the need to maintain your balance on that board will have you occasionally or constantly leaning back against the lines and putting too much pressure on the horses' mouths. It is true that a good teamster with good horses may accomplish this riding position with little or no pressure necessary on the lines. But for most beginners this is not the case.

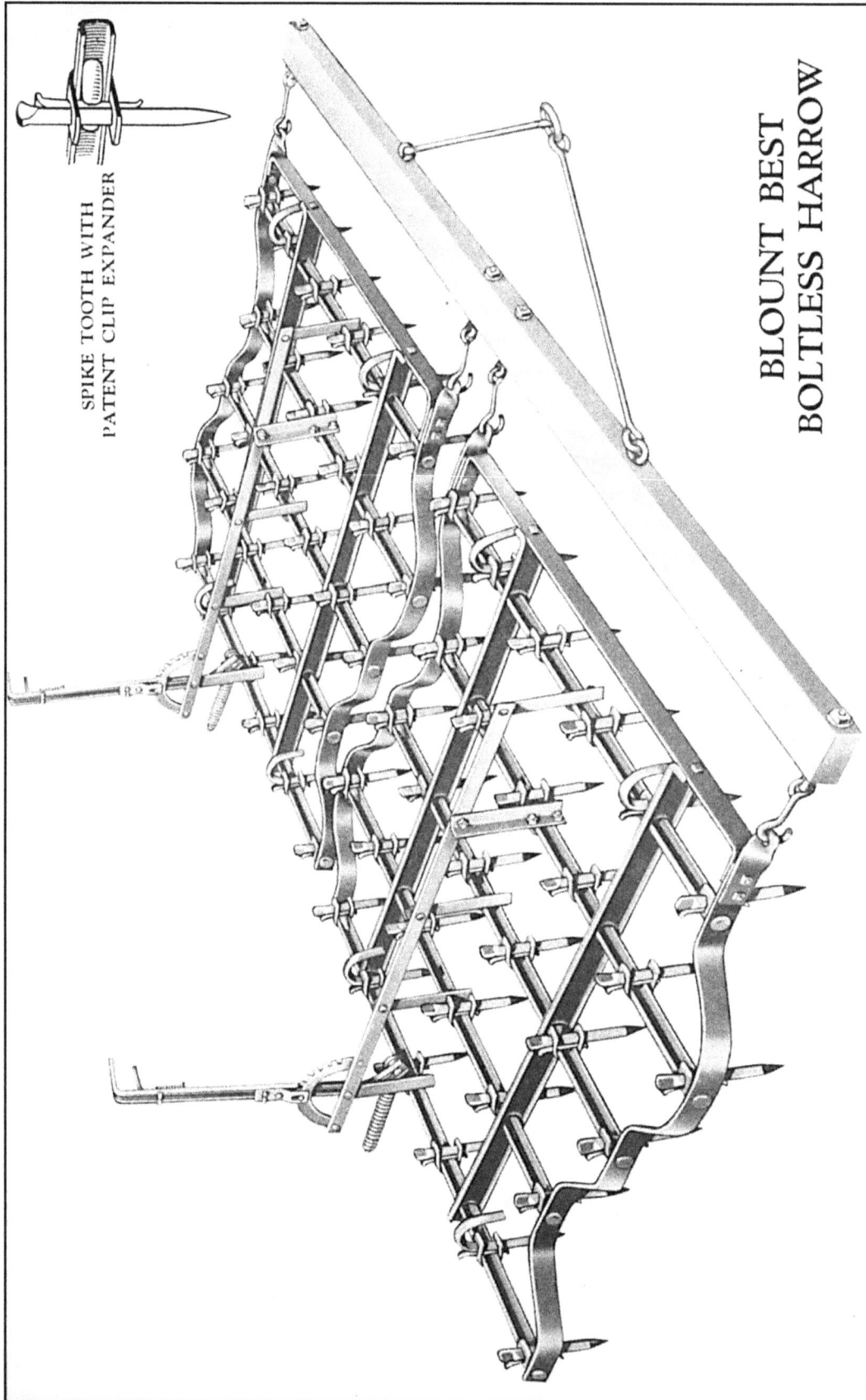

SPIKE TOOTH WITH
PATENT CLIP EXPANDER

BLOUNT BEST
BOLTLESS HARROW

Channel Bar
"TCB" Series
One Lever

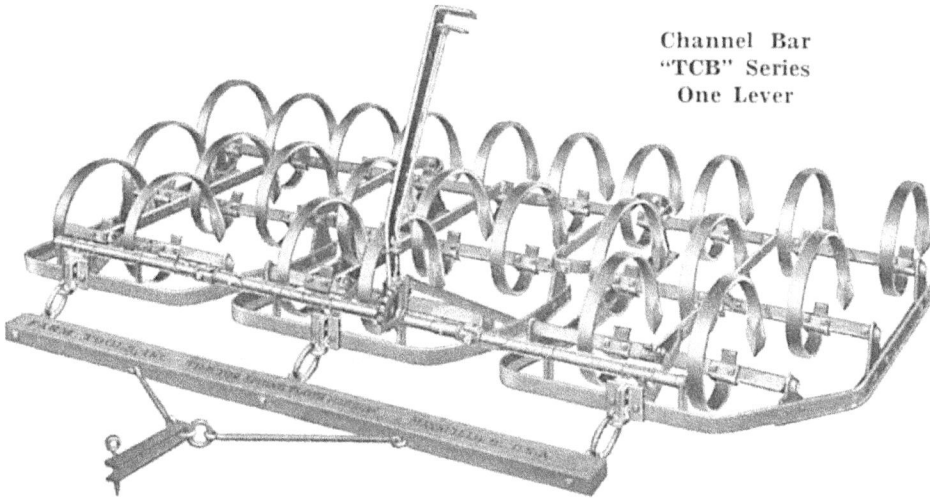

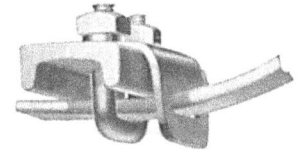

Roderick
Lean

Adjustable
Tooth Clamp

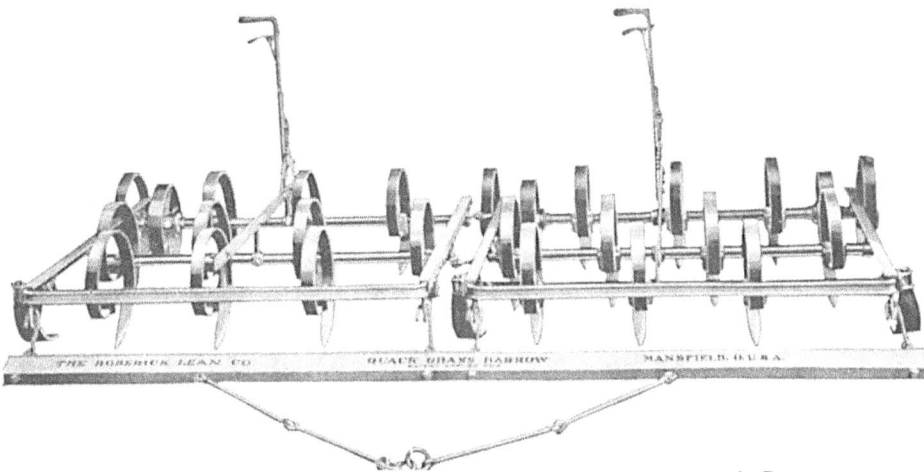

"T" Series with Low Frame

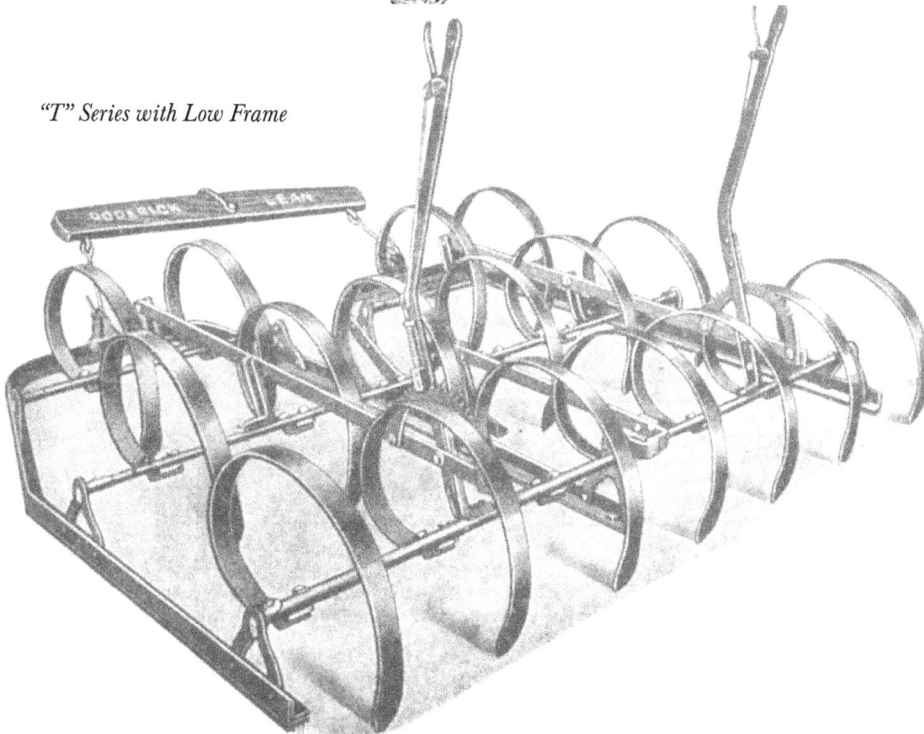

Moline

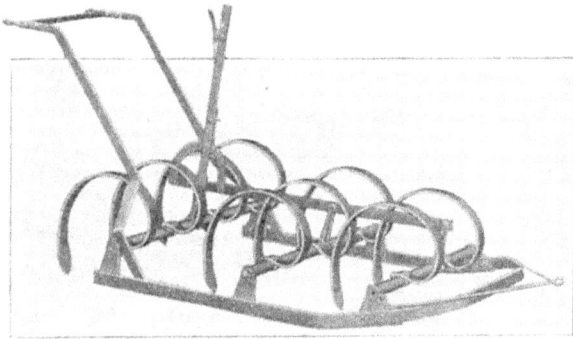

RS-9, Harrow with Handles

Alfalfa Tooth

Regular Tooth

Reversible Point Tooth

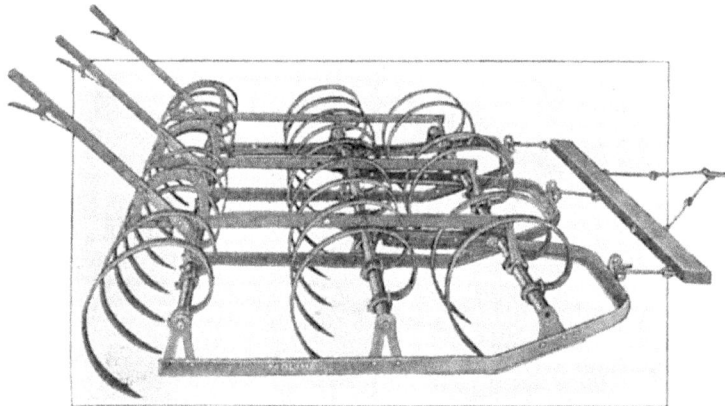

RS-25, 3-Section Harrow Equipped with Quack Grass Teeth

RS-15, 2-SECTION HARROW
Equipped with Riding Attachment

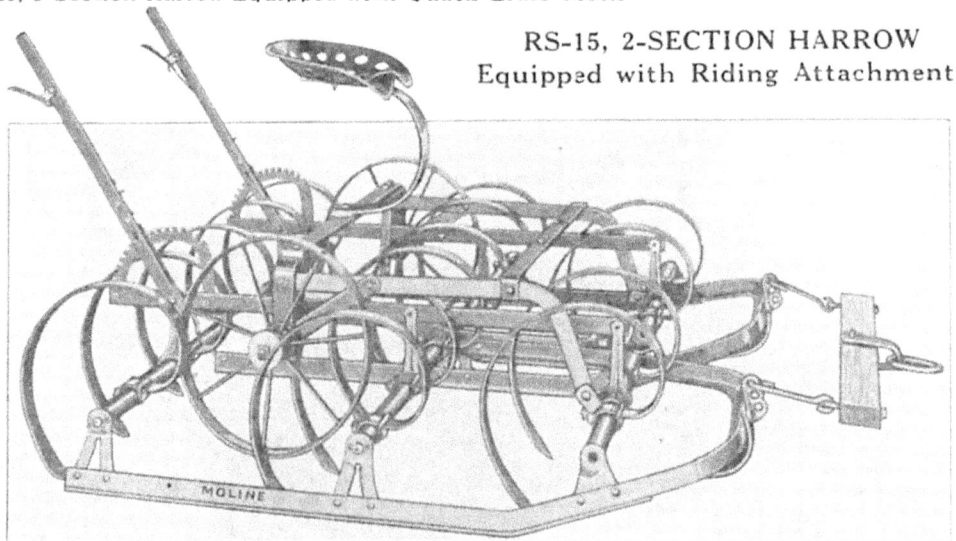

Quack Grass Tooth

Bucher & Gibbs

The Imperial Low-down Harrow. The frame serves as the runners.

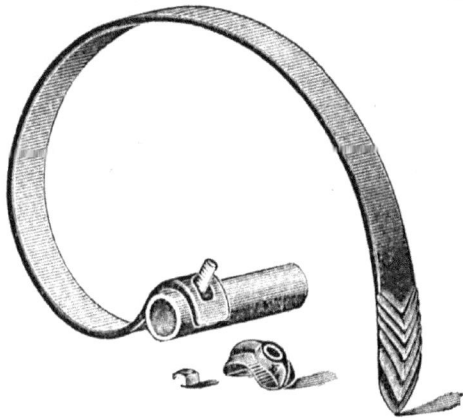

The Imperial spring tooth is held securely in position by a malleable friction clamp, which also prevents the loosening of the nut. The slotted hole allows for an adjustment of up to four inches at the point of the tooth.

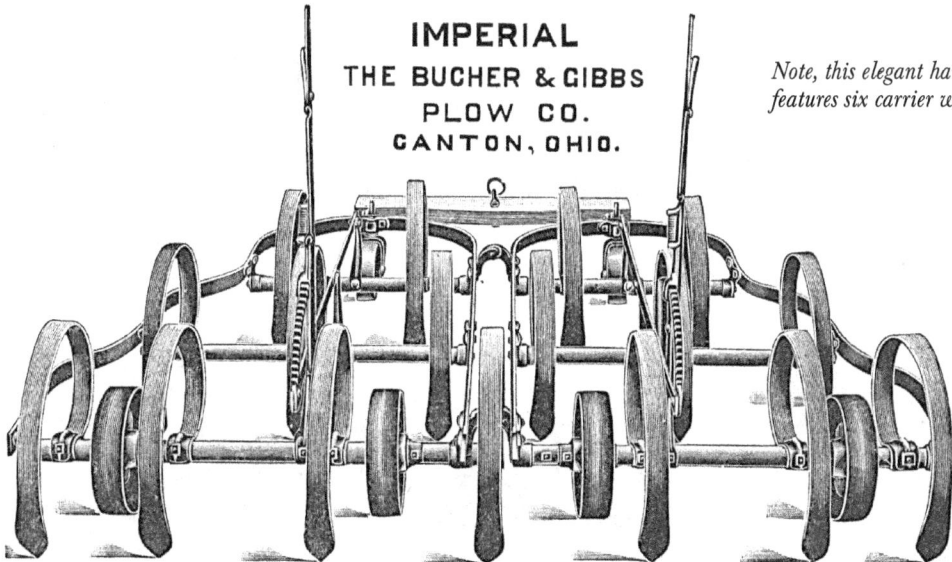

IMPERIAL
THE BUCHER & GIBBS
PLOW CO.
CANTON, OHIO.

Note, this elegant harrow features six carrier wheels!

Alfalfa Spring Tooth built with a frame high above ground to provide clearance and avoid clogging. Teeth are closer together than ordinary.

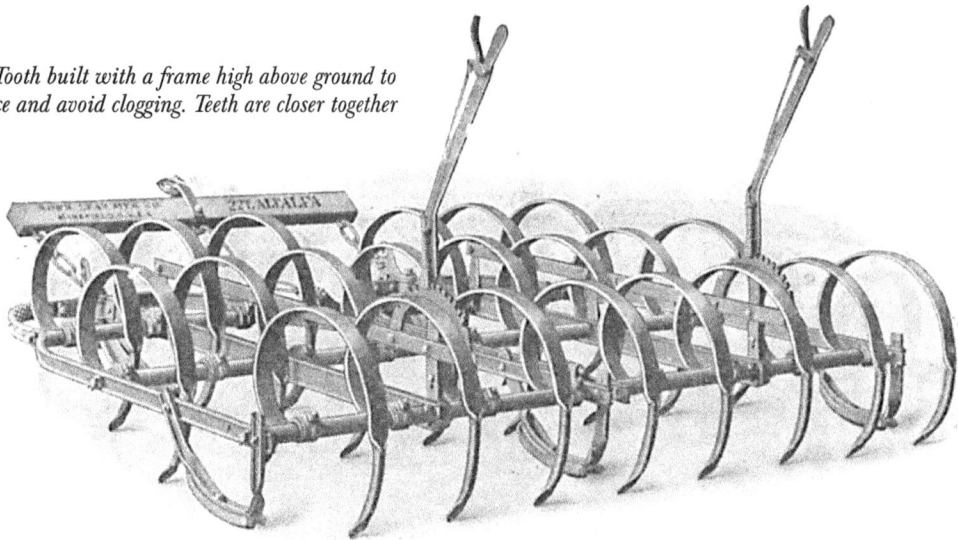

Roderick Lean

"R" Series One Horse

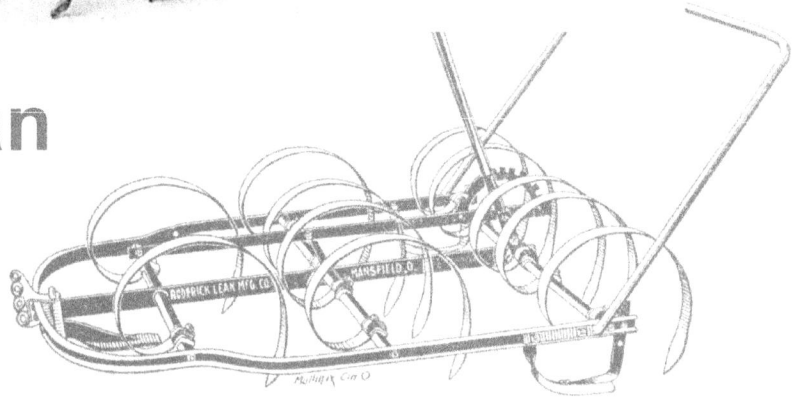

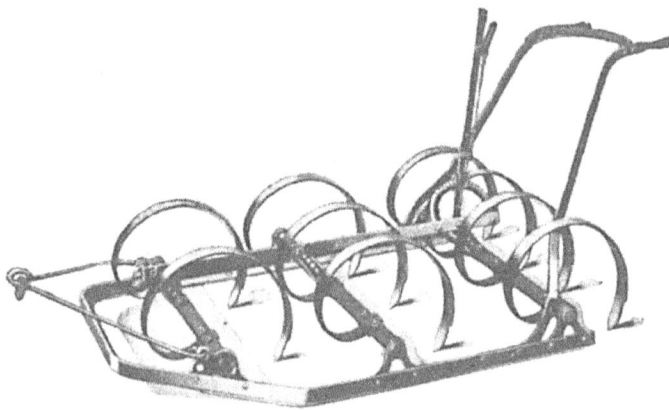

McCormick Deering

9-Tooth Spring Tooth Harrow with Channel Bars

19-Tooth Spring Tooth Harrow

McCormick Deering

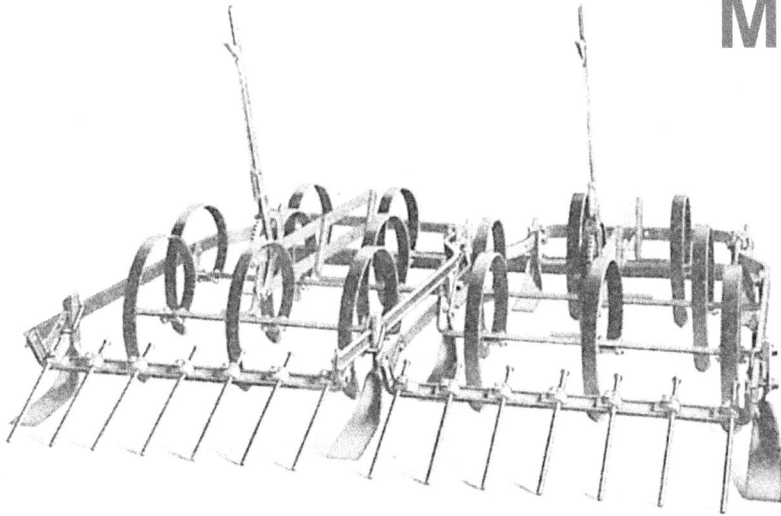

Combination Spring-Tooth Harrow (2-Section)

Oliver

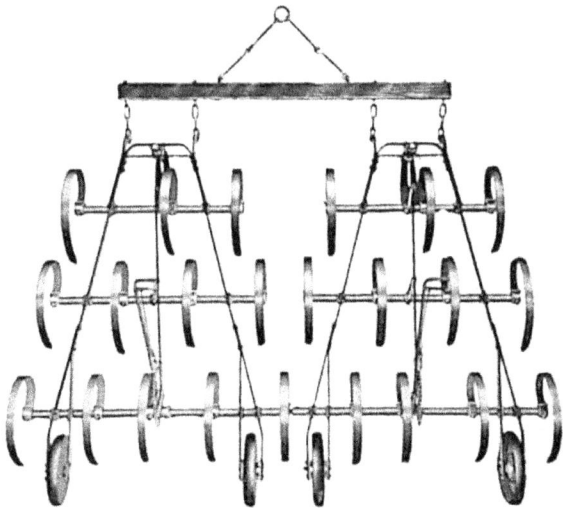

Top view of the AMJ shows how two sections are fastened together.

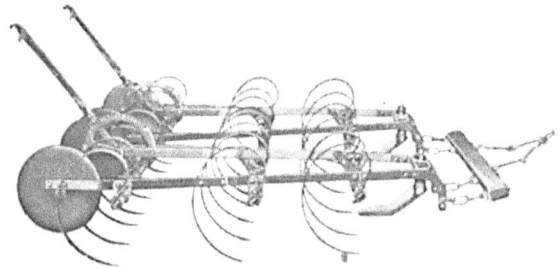

Oliver's AMJ Spring Tooth Harrow was specially designed to eradicate Quack grass and other noxious weeds. May also be used for general purposes. The sharp pointed spring steel teeth have a natural suction and penetrate hard soils quickly and positively. Their springy vibrating action tears out the roots and brings them to the top of the ground.

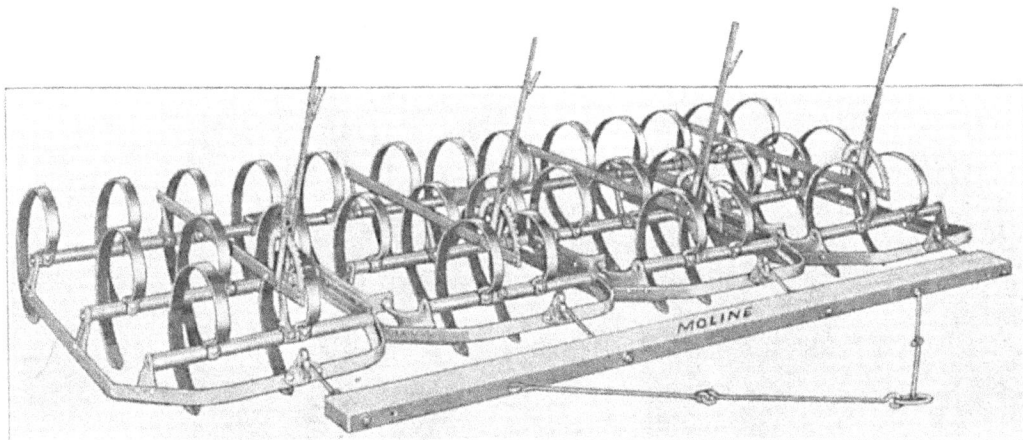

RS-33, 4-Section Harrow with Levers Set Forward for Tractor Use

Harrows

P & O

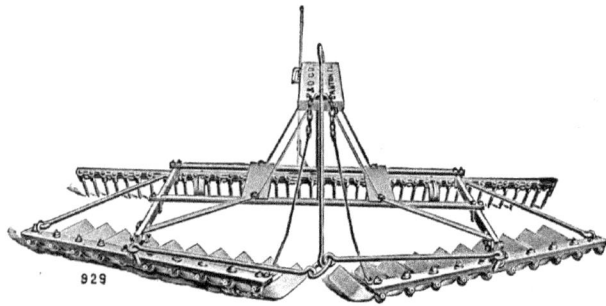

CANTON CLOD CRUSHER

grain is in, firming the soil so as to assist the seed to germinate, and at the same time leaving it loose above, to allow the young shoots to readily break through. Both the slant of the knives and the pitch of the teeth are readily adjustable to suit any condition of soil.

P & O writes: this is a tool which will quickly pay for itself on any farm. It not only thoroughly crushes all clods, but levels off the high places and fills up the depressions, leaving the surface level and smooth and in the very best of shape for any kind of seed. It does its best work if put in the field immediately after plowing, and will do good work in the hardest ground, although it might sometimes be necessary to go over the ground twice, the second time driving at right angles to the first. It can also be used to advantage in going over the field after the

John Deere

Deere Clod Crusher with angle frame 1912 style

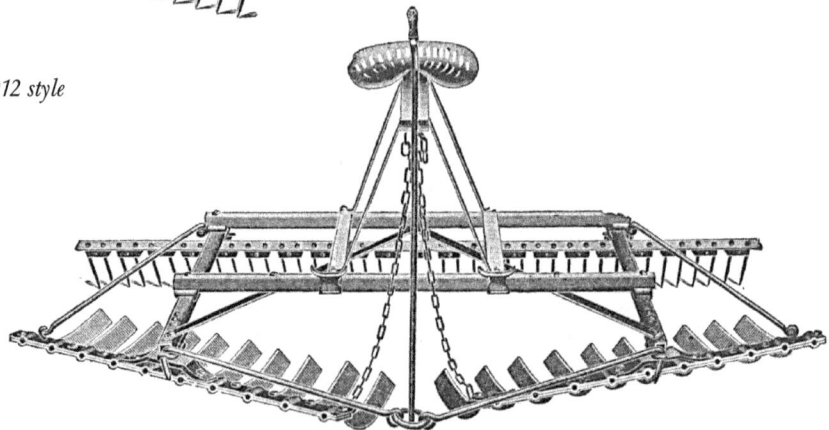

Deere Clod Crusher—Pulverizer and Leveler

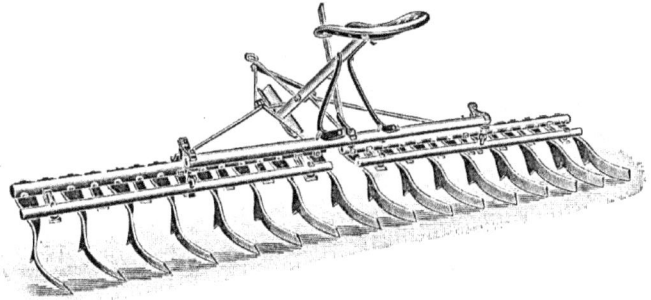

Acme Harrow, Clod Crusher and Leveler.
The forward half of each coulter is curved to the left, while the rear half is curved to the right, thus forming the equivalent of double gang.

Clod Crusher and Leveler *Size H - one horse with 8 coulters plus crushing spurs. Works 4 foot 4 inches wide.*

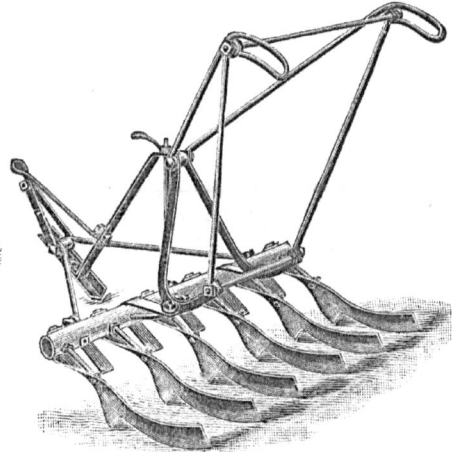

Size G - One Horse, 3 feet wide

Acme Double Pulverizing Harrow *seventeen and a half feet wide, arranged for four horses, and capable of covering 5 acres per hour.*

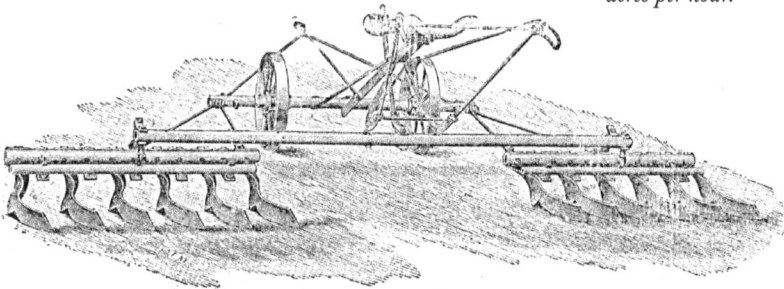

Acme Pulverizing Harrow and Orchard Cultivator *arranged to work 13 feet wide.*

NEW CYCLONE WEEDER

The Peoria Harrow came with the option of this button-on seeding attachment which drops seed ahead of the harrowing.

Positive chain drive that will outwear the seeder.

Showing simple method of attaching seeder to evener bar of harrow.

B.W. Macknair & Son

The Macknair & Son company of Lewistown, PA is building a handsome spring tooth harrow designed, handles back, for horses. See the back of this book for information on how to reach them.

Pioneer

THE PIONEER SPIKE TOOTH HARROW
FEATURES

Full Diamond Spike Tooth: This full diamond spike tooth is oil-tempered and hardened to avoid excessive wear, is reversible and easily replaced. Each tooth has a reinforced shoulder and a lock washer to prevent shaking out. The full diamond shape means a heavier, tougher tooth that stays sharp longer and does not wear round.

Open End Construction: Helps prevent clogging between sections, eliminates overriding, and allows complete flexibility for each harrow section.

Easily Adjusted Lever Assembly: Easy-to-handle lever sets tooth angle to fit any soil or crop conditions.

Extra-Strong Tubular Frames: For more wear on pivot points. Rigid enough to hold teeth in alignment yet flexible enough to follow uneven ground.

Tooth Bars: Steel tubes for extra strength and resistance to wear. Each bar is held in position firmly to assure every tooth is working the soil thoroughly and designed not to collect or drag trash.

Drawbars with Chain Hitch: Strong, tubular drawbars have adjustable section clevises and a chain type hitch.

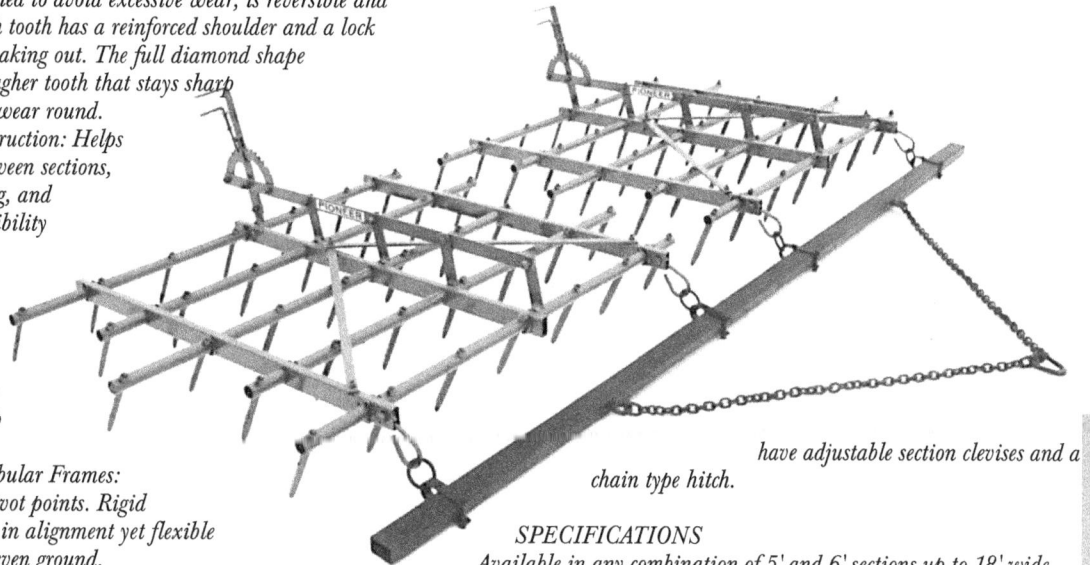

SPECIFICATIONS
Available in any combination of 5' and 6' sections up to 18' wide.
35 Teeth per 5' Section
40 Teeth per 6' Section

Harrows

PIONEER SPRING TOOTH HARROW

Designed and built for trouble-free performance and reliability using quality materials and PIONEER workmanship. The PIONEER Spring tooth Harrow is an all-purpose tillage tool, ideal for seedbed preparation and weed killing. One of the unique features of this harrow is the swiveling front runner shoes and rear raker bar which eliminates the need for full length runners and improves trash flow. Each section is an individual unit which makes for easy handling and storage. Also, additional sections may be added at any time by using a longer drawbar.

FEATURES

Spring Teeth: heat treated and tempered. Teeth are punched to take reversible point shovels after original points wear down.
Tooth Bars: rectangular steel tubes for extra strength and rigidity.
Tooth Clamps: bolt on clamp secures tooth in position, yet allows easy replacement or spacing of teeth.
Lever Assembly: easy-to-handle lever sets tooth depth to fit any soil or crop conditions.
Runners: extra wide swiveling runner shoes insure good flotation and easy turning. Shoes are bolted to swivel for easy replacement.

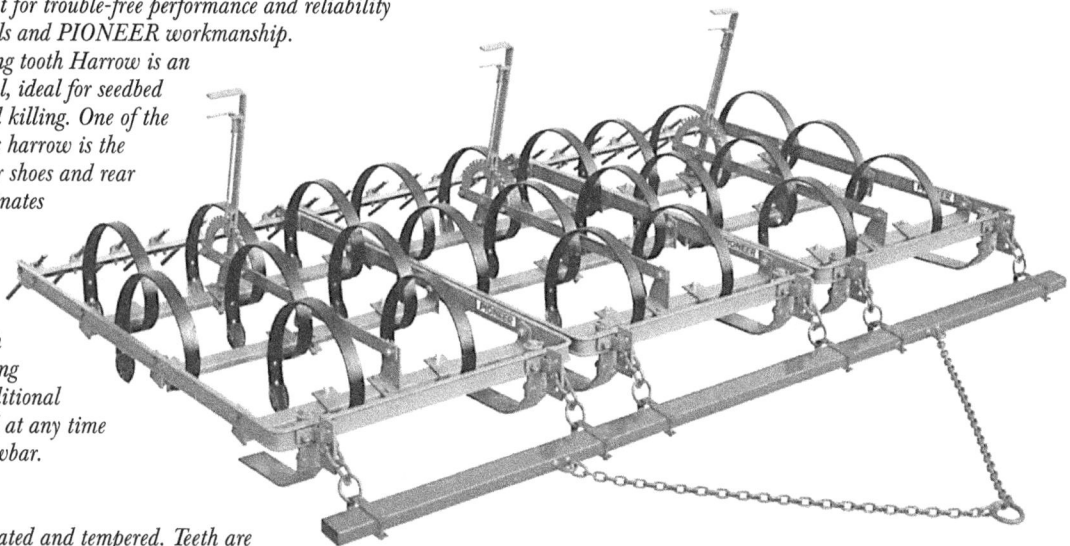

Raker Bar: exclusive raker bar helps break up clods and eliminates ridging. Raker rods are easy to adjust or replace.
Drawbars: strong tubular steel drawbars have a chain pull to eliminate bent rods. Section clevises are adjustable and designed to prevent unhooking when making turns.

SPECIFICATIONS
Available in any combination of 3' and 4' sections up to 16' wide.
8 teeth per 3' section
11 teeth per 4' section

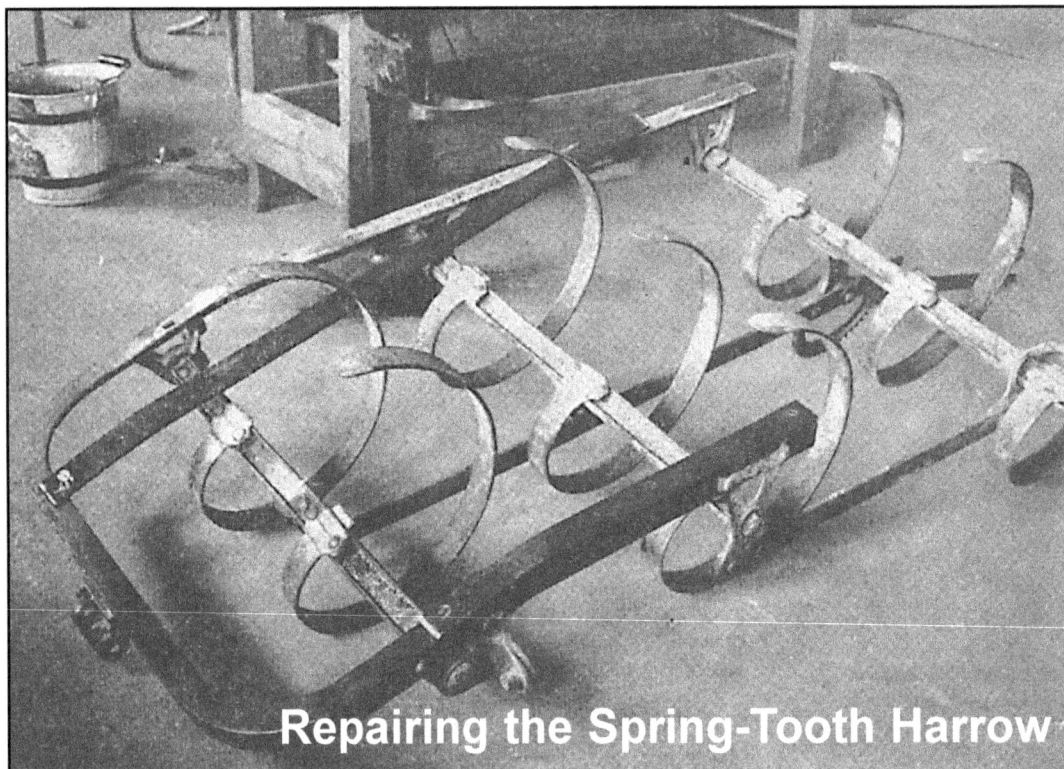

Repairing the Spring-Tooth Harrow

Figure 1. A discarded harrow frame: This harrow was discarded because the frame broke. It could have been prevented if the shoes had been replaced.

Edited from the original by B.A. Jennings, and appearing as Cornell Extension Bulletin #385

Many a spring-tooth harrow has gone to the fence row or junk heap years before its usefulness was past, merely for the lack of a few hours of labor and a few cents for parts. In fact, there is almost no reason for ever discarding a harrow unless it has been neglected. The harrow should be repaired each year. On farms where a warm shop is available, the work is a good excuse to spend a worthwhile day in a heated building.

Each year harrows are discarded because the shoes have not been replaced soon enough and the frame has been allowed to wear completely thru (figures 1 and 2). If the teeth have worn dull or short, new teeth or detachable points can be purchased. The cost of these is considerably less than the price of a new harrow.

To repair the spring-tooth harrow one needs to:

    Inspect and replace, if needed, worn shoes.
    Inspect and repair the depth levers and bars so

that they work correctly and easily.
So adjust the teeth that all are set at the same depth.

REPLACING HARROW SHOES

Harrow shoes should be replaced as soon as they are worn thin at the bends, or where the runner turns up.

To remove the old bolts, a cold chisel is used to split

the nut or a rivet buster is used to cut the bolts (figure 3).

Homemade shoes

Many farmers, objecting to the manufactured harrow shoes because they do not last long enough, either make heavier ones at the farm shop or have them made at the blacksmith shop. A bar iron 3/8 or 1/2 inch thick and 1-1/2 inches wide is commonly used and makes a satisfactory shoe. The length of the iron depends, of course, upon the length of the harrow. Usually, about 12 feet of iron is needed for the two sections. A 3/8- by 1-1/2-inch bar iron weighs approx. 2 pounds a foot.

Tools for fitting shoes

Expensive tools are not needed to fit shoes to a harrow. A solid vise, a large monkey wrench, a 7/16- and 3/4-inch drill, a drill press, and a stove or forge are all the necessary tools. A forge and an anvil are convenient, but a coal stove and a piece of railroad iron can be used.

Bending iron for shoes

Harrows that have tapered or narrowed front ends require two distinct bends for the shoes, one bend is up for the runner effect and the other is to the side for the narrowing effect. This means that the iron must be bent both edgeways and flatways.

These bends are made as follows: A piece of iron is cut the length of the shoe. The distance from the end of the shoe to the bend is laid off on the new iron and marked plainly with a prick punch or cold chisel. The iron is then placed in the fire and heated at the punch mark to a bright red color. It is placed in the vise with the short end down and with the punch marks even with the top of the jaws of the vise. A monkey wrench is clamped over the iron and, with the wrench set close to the vise, the iron is bent edgeways an amount equal to the angle of the frame (figure 4). The iron is then reheated and bent flatways to form the runner part of the shoe. This is done in the same way as for the previous bend. When the iron is nearly the correct shape, it can be

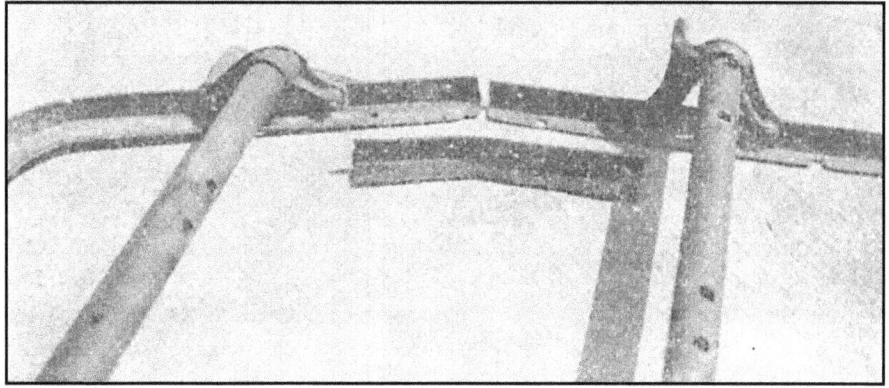

Figure 2. Repairing a broken harrow frame: If only one frame had worn thru or broken, it can be repaired with a short piece of angle iron fitted to the inside of the frame and riveted, bolted, or welded in place. Welding is preferred since it is more rigid.

Figure 3. Removing old harrowshoes: Bolts for the shoes are usually so rusty that they cannot be unscrewed with a wrench, and it is necessary to cut the bolts or to cut the nut with a cold chisel. A rivet buster works well for this job. See insert. This tool is similar to a cold chisel except that the lower surface is straight. The lower surface should lie flat on the iron when in use. The action is to shear off the bolt instead of cutting it as with a cold chisel.

reheated and held or clamped to the frame and correctly fitted by hammering (figure 5).

After the iron has cooled, the holes are carefully marked and drilled. The new shoes should be bolted on with new bolts. Flat-headed stove bolts, 3/8 by 1-1/2 inches, can be used for this purpose. A 7/16-inch drill is used to drill the bolt holes, and then the bolt head is countersunk at least 1/8 inch deep with the 3/4-inch drill. The 3/4-inch drill should be ground to fit the angle of the bolt heads (figure 6). The shoe can then be slightly rounded on the front edge, by filing or grinding, and bolted in place.

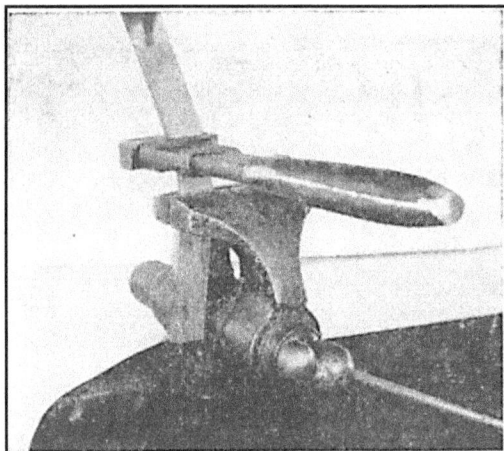

Figure 4. Bending an iron for a harrow shoe: The iron is heated, placed in the vise, with the short end down and with the punch marks even with the top of the jaws of the vise, and bent with the monkey wrench.

Figure 5. The final fitting of shoes: After the shoe has been fitted as closely as possible with the wrench, it should be reheated, one end clamped in place and the final fitting done with a hammer. Note that the short shoe is completed, the long shoe has had one bolt drilled in the front end and is ready for reheating and clamping in place and for the final fitting.

CHECKING AND ADJUSTING THE LEVERS

Very often the rod between the handle and the ratchet becomes stretched or bent so that the ratchet does not clear the quadrant sufficiently for easy adjustment. If so, rebending may help or a new shorter rod may be made. A good way to make the bends in a new rod is to heat the rod in fire and wrap a complete circle around a 3/8-inch bolt or punch held securely in the vise (figure 7). The extra end is then cut off with a hacksaw, the circle is spread, placed in the handle, and pounded closed again.

The quadrant must be tight on the frame and the uprights from each of the tooth bars must also be tight. Play in any of these places changes the depth of the setting of the teeth. If the bolts are loose and permit play, they should be retightened. If they are rusty or worn, they should be removed and replaced.

ADJUSTING DEPTH OF TEETH

If the harrow is to do a good job of fitting land, all of the teeth must be set at the same depth. One or two teeth set too deep pull up sods, leaving trash on top of the ground. Teeth running shallower than the others do little good and leave streaks of unfitted land.

The teeth of a spring-tooth harrow should be checked each season. The harrow should be on an even floor and the depth levers so adjusted that the teeth are just touching the floor. There should be a slight amount of tension on the teeth. The workman stands on the frame of the harrow to hold it firmly to the floor and tests the tension of each tooth. It is good practice to loosen the tension on the tight teeth and to

tighten the tension on the loose ones, rather than to attempt to set all of the teeth equal to the tightest tooth. Usually, changing the setting of two or three teeth will be all that is necessary (figure 8).

After one section is thus completed, the next section should be adjusted. The levers for the other sections should be in the same notch as for the first one, and the tension on the teeth should be set alike, for this gives equal setting for all sections by simply setting the levers alike.

A strong socket wrench with a long extension handle is good for loosening or tightening the bolts which hold the teeth. Such a wrench saves time and it may save skinning a knuckle or two.

HARROW TEETH

Harrow teeth wear dull and short until finally the levers cannot be set for good tillage. If the harrow has

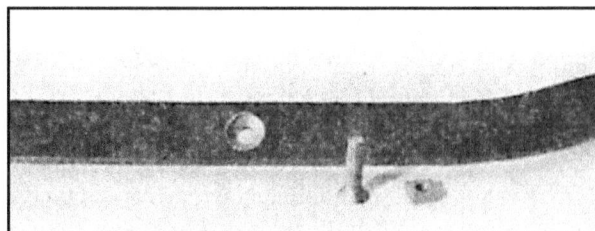

Fig. 6. Bolting harrow shoes: Flat-headed stove bolts (3/8 by 1 1/2 inches) can be used for fastening shoes. The holes should first be drilled 7/16 inch for the bolts and then countersunk 1/8 inch deep for the head. A 3/4 inch drill ground to fit the angle of the bolt head can be used for this purpose if a regular countersink is not available.

Fig. 7. Bending hooks in rods for handles: A good way to make a bend in a new rod for the depth lever is to heat it in a fire and wrap a complete turn around the bolt held in the vise. The extra part can then be cut off, the circle spread to fit into the handle, and pounded closed again.

been properly cared for, it will pay to buy a new set of teeth.

Many farmers have repaired a worn-toothed harrow by using clipped-on teeth or detachable points, and have been well satisfied with the results. An advantage of the detachable points is that they have two ends and can be reversed when one end becomes dull. The disadvantage is that some types of these teeth are not securely attached and occasionally one may come loose and be lost.

A tooth sometimes is caught and is straightened out. Since a new tooth often gives trouble in setting or adjusting because it is longer than the old or worn one, it may be simpler to repair the sprung or bent tooth. Again, the monkey wrench and vise are used for this job. The tooth is heated to be bent back to shape. In heating one must be sure that the tooth does not get too hot, for too much heat spoils the temper of the tooth. Not quite to a cherry-red heat is correct. In cooling, the tooth takes on its original temper if it is placed on a dry-dirt or concrete floor. No water should be used.

HARROW HITCH

Most harrows have a number of holes for changing the height of hitch. This should be changed according to the length of the draw bar and the chain. A low hitch pulls up on the front part of the harrow and makes the back teeth do more work than the front ones. A hitch that is too high wears the shoes rapidly and makes the front teeth go too deep. The height of hitch should be adjusted for each source of power so that all teeth work equally.

Fig. 8. Adjusting the depth of the teeth: It is essential that all of the teeth of the harrow are set for equal digging. A long-handled socket wrench is convenient for loosening and tightening the bolts that hold the teeth in place.

*The author with two Belgian mares, one in training, on
two sections of Pioneer Spike Tooth Harrow in 1984.*

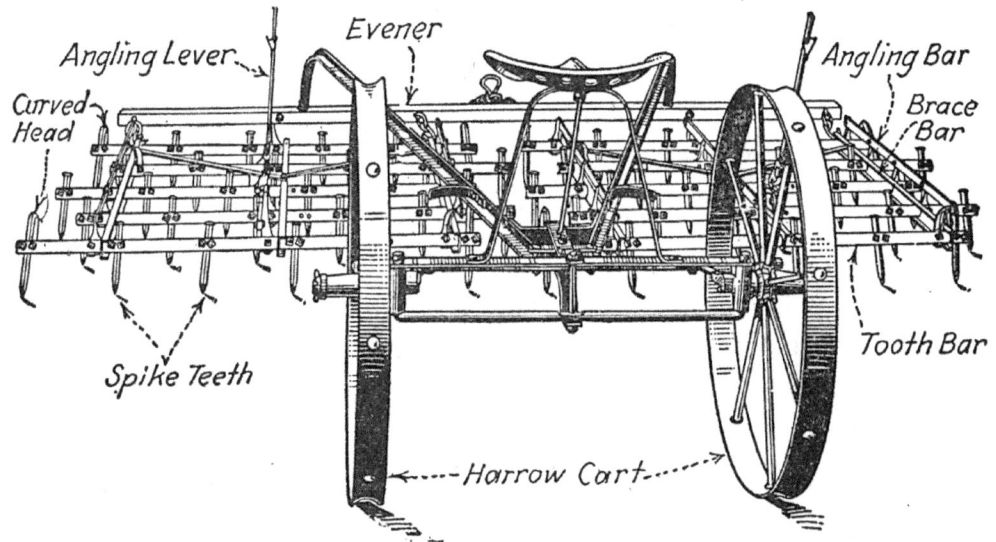

Angling Lever

Curved Head

Evener

Angling Bar

Brace Bar

Spike Teeth

Tooth Bar

Harrow Cart

Rigid spike-tooth harrow equipped with harrow cart.

Chapter Five

Harrow Carts & Forecarts

Case

A harrow cart, strictly speaking, is an implement designed to attach to a spike tooth harrow draw bar by a long framework which passes back over and across the harrow to a two wheeled carriage or cart. This cart carries the teamster, in relative comfort, behind the working harrow. The wheels, cart axle, and/or draw frame are made, differing ways specific to manufacturer's design, to turn in opposing direction to the cornering action of the harrow. This causes the cart to follow the harrow's travel pattern. The turning may be actuated by foot pedals or by a scissors action set to work by the tension from the drawbar.

The harrow cart puts the teamster in a safer position than standing upon an affixed board or riding ahead of the harrow on a 'forecart'. The harrow cart allows the

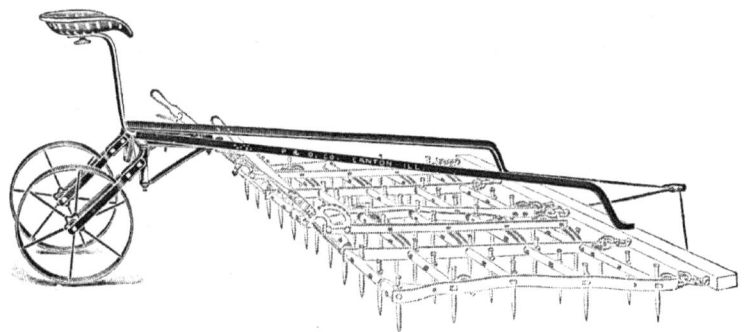

Parlin & Orendorff

129

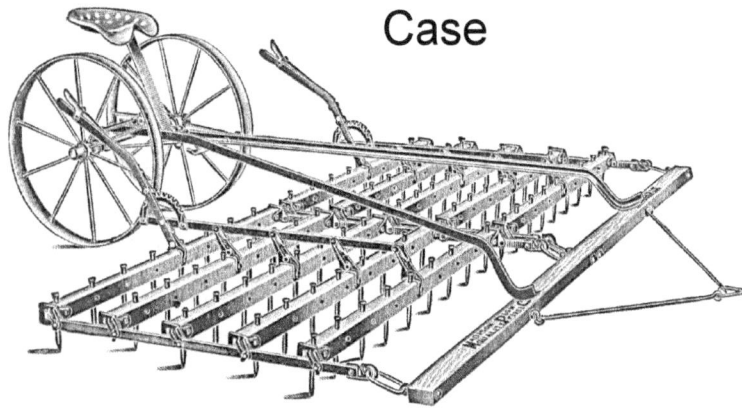

Case

added advantage that the teamster may see the ground being worked. As of publication date we know of no one building new harrow carts. This is surely an implement worthy of some farm shop creativity and sweat. The first individual or company to manufacture a good one for sale will likely do well.

Riding Attachment for Harrows

This attachment finds favor with many farmers, not only because it gives them a chance to ride after the harrow, but also because it tends to hold the harrow in line. It can be attached on any drag harrow, and when attached to a lever harrow the driver has the levers within easy reach.

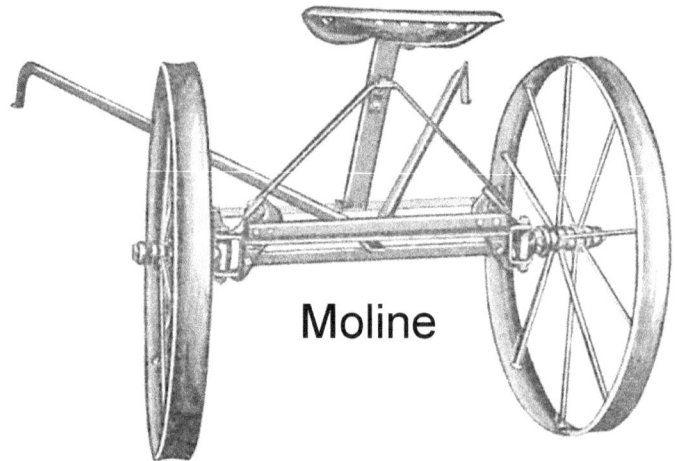

Moline

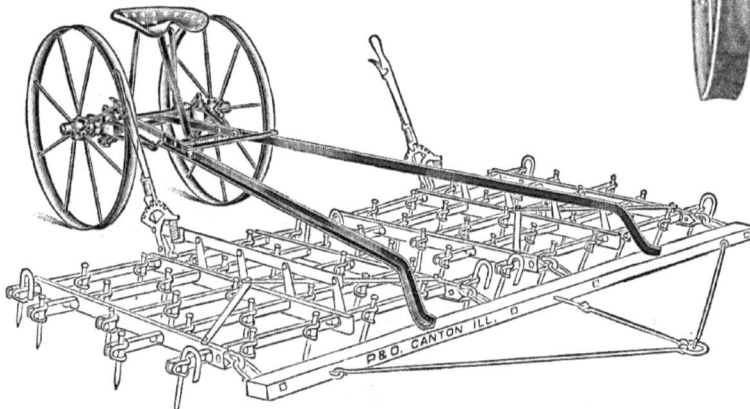

below and above

P & O

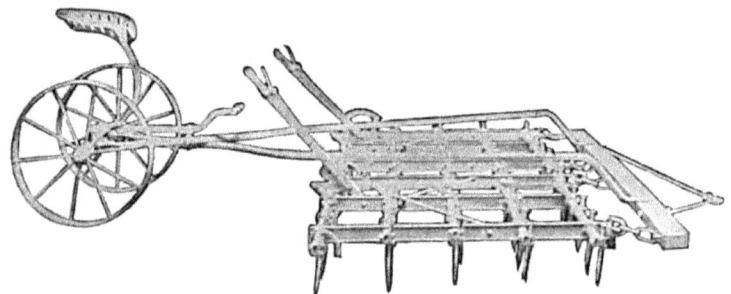

Rock Island

130

Rock Island

Demonstrating scissor action at the turn.

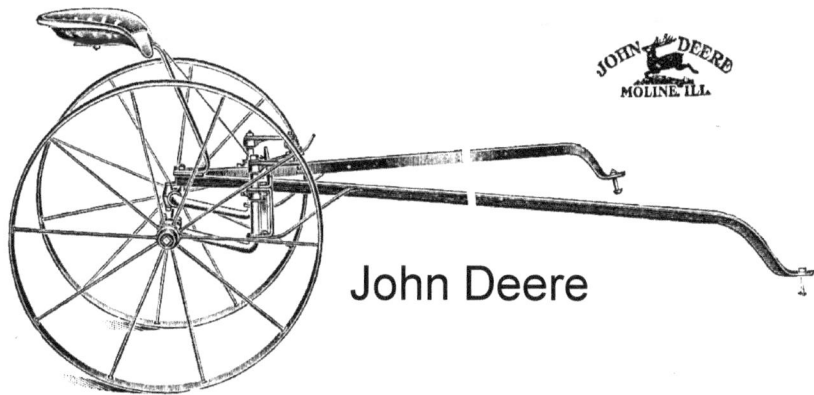

John Deere

Harrow Sulky

The Deerecart above features a seat back off the framework and no straight axle. The wheels are individually attached by right-angled steering spindles.

Moline

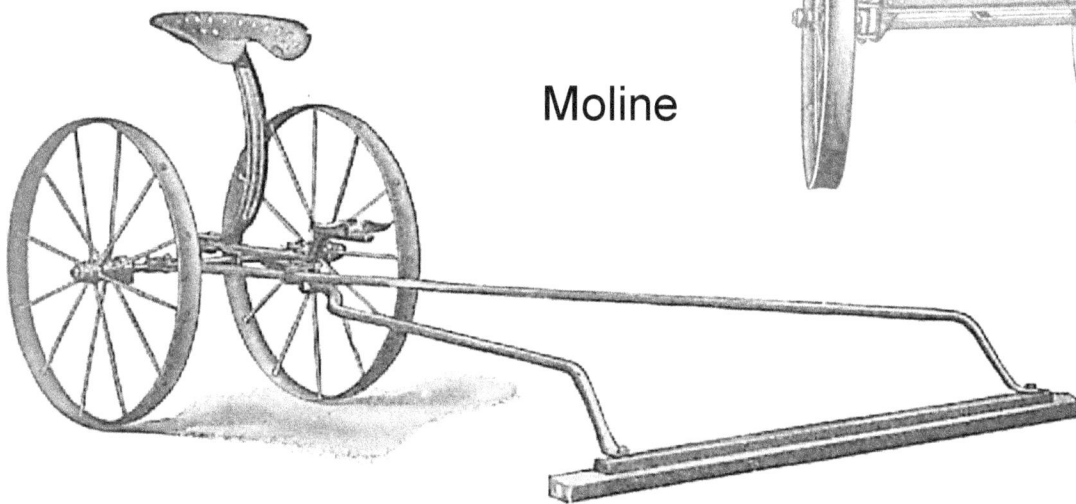

Showing Riding Attachment in Straight-Away Position

Vulcan

MITCHELL HARROW CART

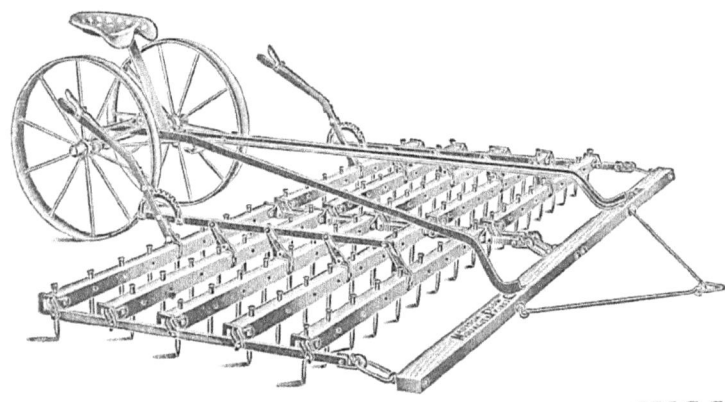

Attaches to any harrow.
Works up close.
Turns without changing position of
seat.
Guaranteed to carry the weight of
driver no difference how heavy.
Rider is up out of the dust.
A great labor saver.
Weight 75 lbs.

SUCCESS HARROW CART No. 3

Do You Need a Harrow Cart?

Strictly speaking, no. They are nice and add a level of improved comfort and safety to the horsefarming routine but if the budget is tight you can get by without one. However, should you happen on one at a cheap price and you don't spring for it, likely you'll kick yourself later. I have seen old ones sell at auction for anywhere from $40 to $400.

Moline

Forecarts

Hitchcart, implement truck, foretruck and forecart are all names for the same basic tool. Expanding on the practical aspects of the tongue truck or foretruck, as was seen on many disc harrows in Chapter 3, the forecart attaches to the front of the implement and gives a place to hitch animals. Through the pole or tongue of the forecart a braking system is provided. In its modern incarnation the forecart has become

A shop built pony-sized forecart

A shop built forecart with battery powered hydraulic cylinder.

Identifying the need to be able to apply hydraulics and drive shaft power, some inge-nious companies have designed and built motorized forecarts which feature full power-take-off, remote hydraulics, and three point hitch capability. These are expensive units. (But they still cost less than the loyalty of an elected official.)

Harrow Carts & Forecarts

arguably the single most useful tool in the horse and mule farmer's arsenal.

The standard forecart usually consists of a frame built upon a single axle with two wheels. Some varia-tions were developed including our own three wheeled bench seat forecart (as portrayed in the *Work Horse Handbook*).

Many a handy farm shop welder has built his own forecart. The more common approach has been to use an older front axle from a car or truck and secure the steering yokes, adding a platform, drawbar, and tongue. *(The less common approach incorporates a full stereo system and a lock box to hold the manicurist's tools.)*

We will not go into any detail on the variables of forecarts in this book. (That's subject for another.) Suffice it to say that harrows of all types, and indeed any pull-type tractor implement, may be drawn by animals with the insertion of a good basic forecart.

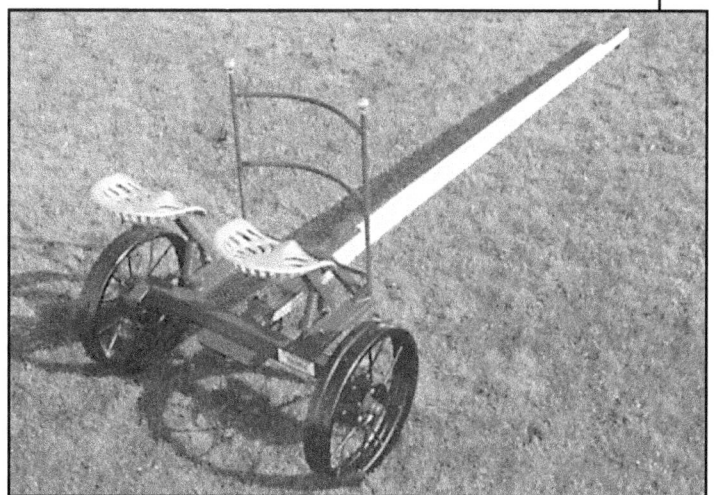

Standard cart with implement seat and steel wheels. Shown with optional mechanical brakes, guard, tongue, tongue cap, extra implement seat and bracket.

There are a few small manufacturing companies which build new forecarts. They include Midway Machine, White Horse Machine, Gateway Mfg., and Pioneer Equipment. For contact information see the back of this book.

(Below) Standard cart with bench seat and air tire wheels. Shown with optional mechanical brakes, guard, shafts, cushion seat and fenders.

Standard cart with implement seat and air tire wheels. Shown with optional mechanical brakes, guard, tongue and tongue cap.

THE Standard PIONEER FORECART
A partial list of the options available for this new cart includes;

Guard: *Bolt on guard for teamster protection. Also serves as a foot rest and line holder. Readily mounts on other implements as well.*

Seat bracket: *Spring mounted seat bracket swivels a full 360 degrees...locks in any position. Seat bracket mounts in any of three positions or mount two seats side-by-side for pleasure driving.*

Mechanical Brakes: *Economical and effective foot operated braking system. Easy to adjust and relatively maintenance free. Foot pedal locks in position for parking or load holding.*

Shafts: *For single horse driving. Shafts are easy to install or detach using only two bolts. Singletree is included. All steel construction.*

Fenders: *We recommend this option when using the bench seat. Bolts directly to the cart frame. Includes floorboard extensions and covers wheels for teamster safety.*

Engine Mount: *Allows mounting a small engine for PTO power. Engine mounting brackets installed for a motor platform.*

Hillside Steering: *Mostly used for row crop work on hillsides. Allows operator to steer wheels either right or left. Hand lever operation.*

PIONEER LIBERTY PTO CARTS
The PIONEER Liberty PTO cart enables the horse and mule farmer to use the latest, most up-to-date equipment available today. Choose the horsepower to fit your application from a range of 20 HP to 72 HP. Quiet running gas or diesel engines help you do more work in less time. Field ready and loaded with features, these PTO carts are proven performers.

These motorized PTO carts are the result of years of research and extensive on-the-farm testing. They have been successfully used on rotary mowers, haybines, rakes, tedders, square and round hay balers, manure spreaders, corn pickers, picker/shellers, plus others. In a class by itself, the PIONEER Liberty PTO cart will perform an endless variety

of tasks on the modern horsepowered farm.

All PIONEER Liberty PTO carts feature the exclusive adjustable drawbar stabilizer. Eliminates all the balance problems of a two-wheeled cart. Ends swaying and tongue weight problems. Offers better stability when making turns and working hillsides.

This is accomplished by first clamping or welding a pin to the tongue of the implement being pulled. Next the implement tongue is hitched to the cart drawbar in the usual manner. The cart stabilizer bar is now attached to the pin on implement tongue. Finally, the cart tongue latch is disconnected. The cart and implement have now become one single unit and will operate and steer like an auto-turn wagon.

Two stabilizer pins with clamp are included with each cart. Additional pins are optional and may be ordered separately.

FEATURES and SPECIFICATIONS

Gasoline Power: choose from 20 HP or 25 HP Kohler gasoline engines.
Diesel Power: choose from 27 HP, 41 HP, 54 HP or 72 HP Deutz air-cooled diesel engines
Wheels: air tire wheels or steel wheels with rubber bolted on. Brakes: foot pedal operated hydraulic brakes. Convenient brake lock valve. Operator Platform: easily accessible. Elevated platform provides excellent visibility. Self-cleaning expanded metal floor. Seat: spring mounted seat bracket swivels a full 360 degrees. Locks in any position. Guards: ample guarding provides teamster comfort and safety. Hillside Steering: hydraulically operated hillside steering allows you to steer the wheel on the go. A must for row crop work. Clutch: hand lever belt tightener clutch for positive engagement and full power transmission.

PTO Shaft: standard 1-3/8"-6 spline - 540 RPM PTO shaft.
Hydraulic Pump: 2 GPM on gas models. 5.5 GPM on diesel models. Reservoirs have filters and temperature/ level gauges.
Hydraulic Valve: convenient to control machine functions from operator's platform.
Hydraulic Outlets: 4 standard outlets. Pull to disconnect type for your convenience and safety.
Electric Start: all models feature convenient 12-volt starting.
Throttle: large hand throttle lever is easy to adjust from operator's platform.
Tachometer: makes it easy to adjust and track proper engine speed.
Hour Meter: included on all models. Helps to schedule maintenance.
High Temp/Low Oil Pressure Shutdown: included on all diesel models. Protects engines from overheating or oil pressure loss.
Tread Width: 61" center-to-center on wheels.

No. LK-25

No. LK-25

The author (left) uses a Pioneer forecart, with 2 seats, to allow a friend an opportunity to drive a Belgian team under supervision.

A classic Pioneer forecart set up with steel wheels and two seats.

The author dragging a chain pasture harrow with four Belgians and a Pioneer forecart.

Richard Chamberlain of Missouri uses a homemade forecart and four mules to pull a tractor implement.

Chapter Six

Roller Packers

John Deere

"The Culti-Packer Will Roll, Pulverize, Pack, Stir, Level, Cultivate and Mulch the Soil IN ONE OPERATION Better Than Any Other Type of Tool"

John Deere

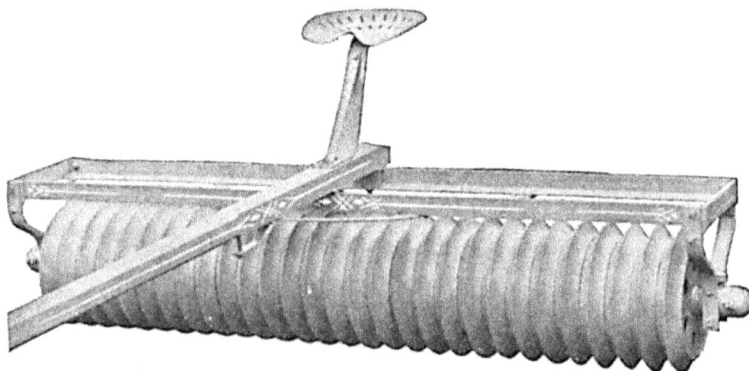

"ND" Single Gang Pulverizer

All through the history of the last 150 years of western hemishpere farming, manufacturers and ag historians have haggled and wrangled over the proper names for implements. In some circles all the tools in this chapter are referred to as "cultipackers" or "culti-packers." In other circles they are referred to as soil pulverizers. And to still another group, of which I consider myself a member, a "cultipacker" is a combination roller and spring tooth. The tools in this chapter we choose to call "rollers" or "roller packers."

This implement is ideal for finishing a seed bed. It is simple to use. There are no adjustments and few parts that ever need replacing with the possible exception of the older models which may need to have their oil-soaked wood boxings removed and replaced.

The effects of the use of the roller, either in seed bed preparation or after seeding, are far reaching. Well pulverized and packed soil resists blowing, conserves moisture, and is firm around newly planted seed.

Please consider a good corrugated roller, preferably a double, to be indispensible to your implement lineup.

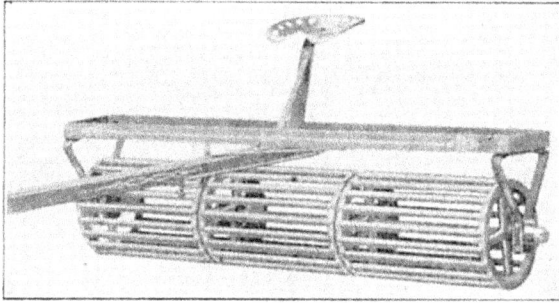

Tube land roller.

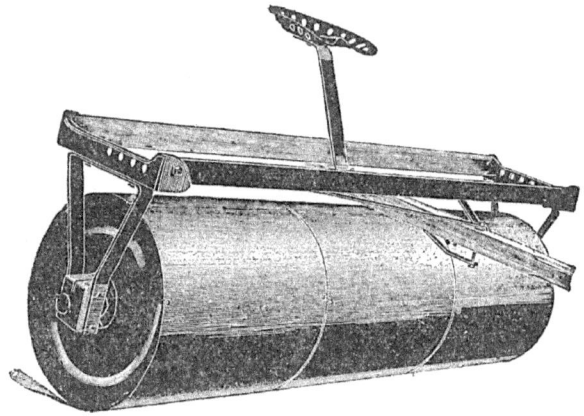

Cast-iron drum roller.

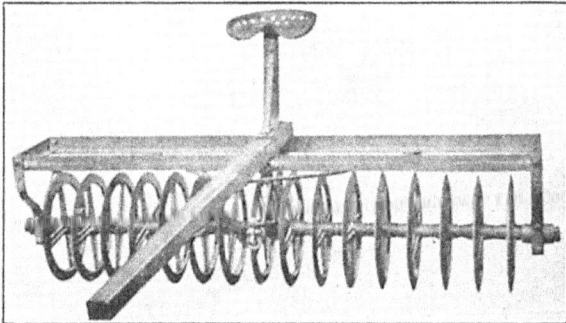

Subsurface pulverizer and packer.

Buch's wood roller.

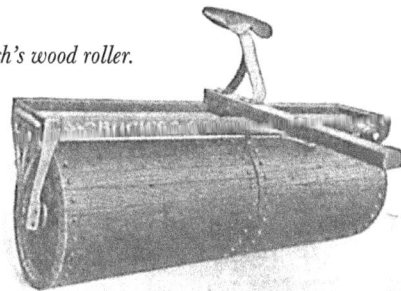

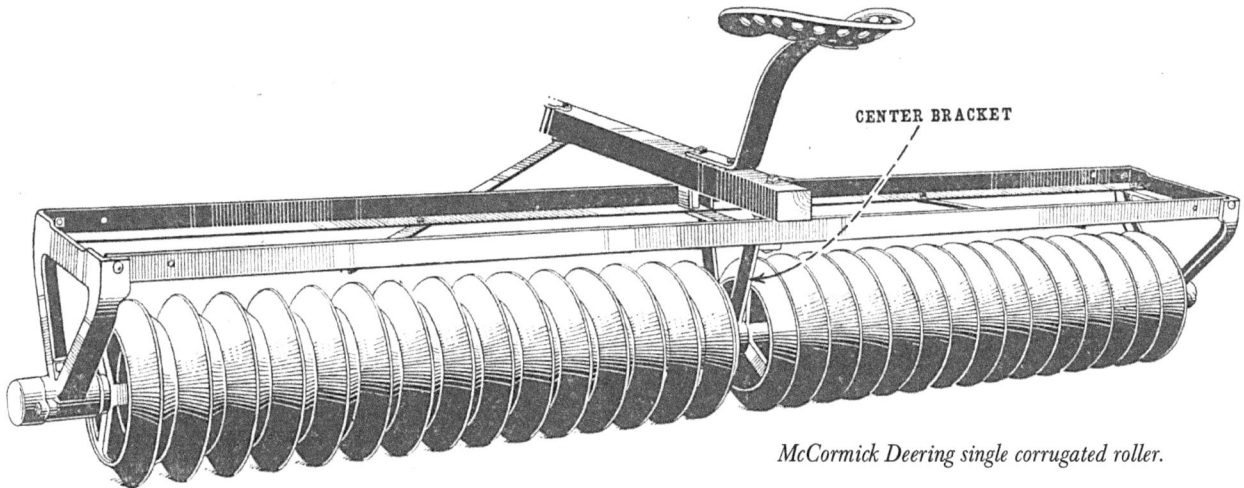

CENTER BRACKET

McCormick Deering single corrugated roller.

Rollers

Vulcan double corrugated double roller.

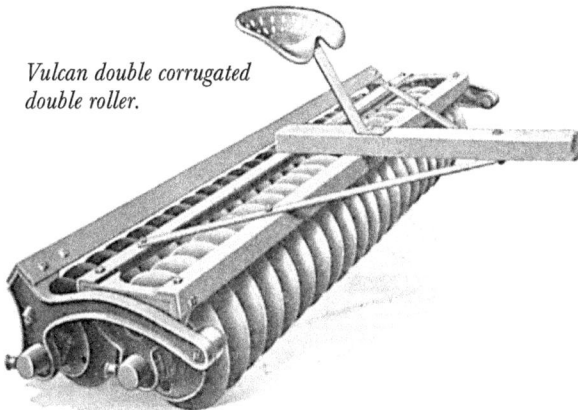

Unusual 'Crowfoot' roller.

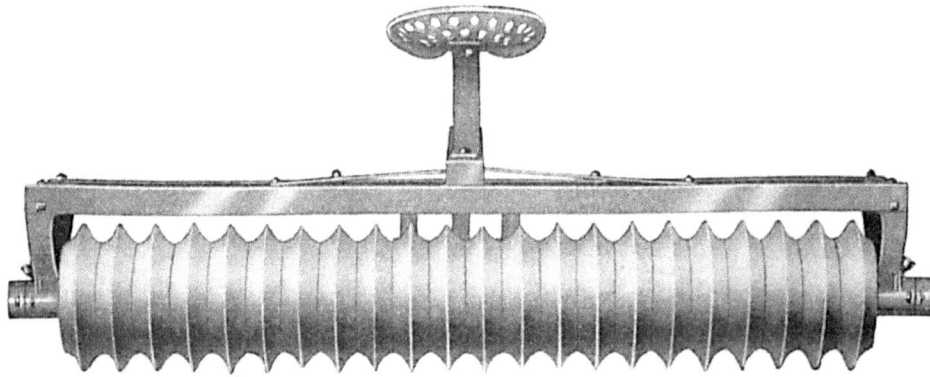

Rear of 8-foot NTSP-8 horse drawn single gang pulverizer with rigid pole ("B" equipment). While not as efficient in its work as a double machine, it is recommended where horse power is limited.

Oliver

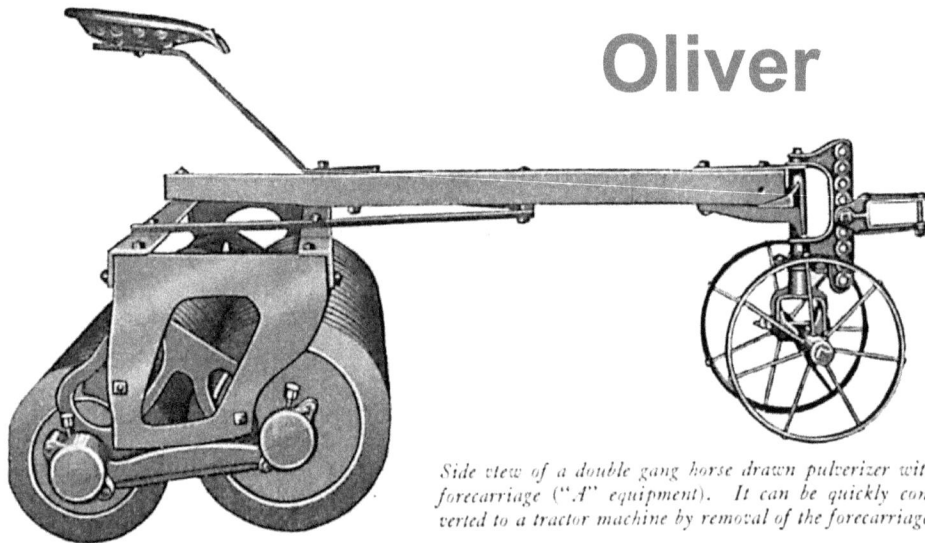

Side view of a double gang horse drawn pulverizer with forecarriage ("A" equipment). It can be quickly converted to a tractor machine by removal of the forecarriage.

Buch

New Buch Double Gang Crusher. *The company wrote these words in their catalog: "it is a well known fact that nothing increases crops so much as well pulverized seed beds. The question has always been the method of doing the work the best and most economically... This Crusher pulverizes every inch of the land, the rear discs track between the front discs, thus no ground is left unpulverized as is the case with a single gang pulverizer."*

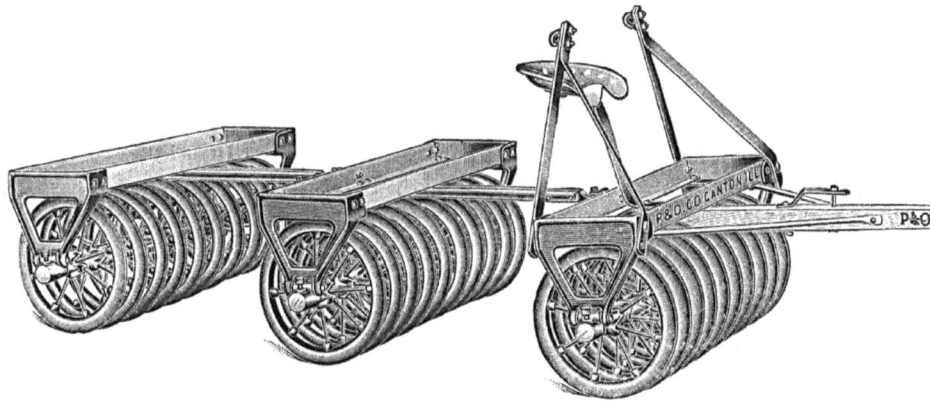

30-WHEEL PACKER READY FOR TRANSPORTATION.

Moline

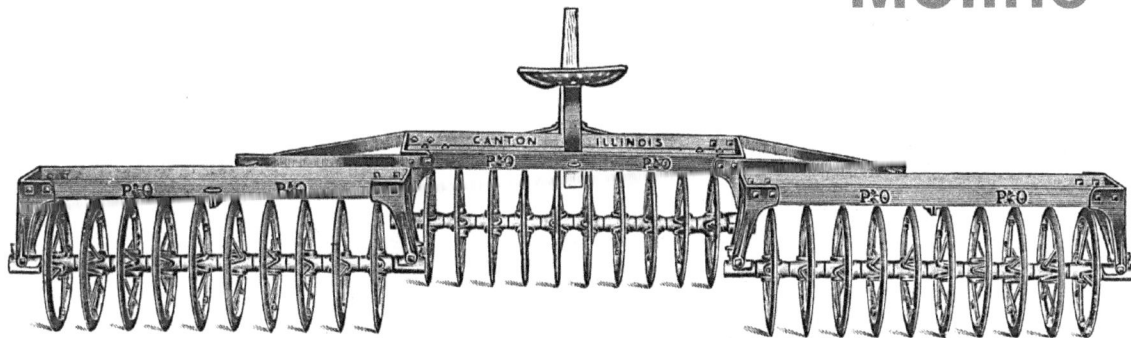

30-WHEEL NO. 1 FLEXIBLE SUB-SURFACE PACKER.

Text lifted from the Moline Implement Catalog:

MOLINE SUB-SURFACE PACKERS
The Moline Sub-Surface Packer is a most necessary part of every dry farmer's equipment - it saves moisture and increases yields.

Packs Sub-Surface
The action of the Moline Sub-Surface Packer is entirely opposite from that of the ordinary roller or packer. The wheels with V shaped rims sink into the soil, thereby packing the sub-surface or bottom of the plowed ground and establishing capillary contact with the sub-soil. It leaves the surface soil loose.

Aids Capillary Movement.
To secure capillary movement, or upward rise of soil water, it is necessary that the soil particles be in close contact, as the water is passed from one to the other in the form of a film which, when there are any open spaces in the soil, is broken and the flow is stopped.
In humid regions the soil is packed sufficiently by natural agencies to secure capillary action; in arid or semi-arid regions, however, there is very little rainfall and it is necessary to pack the sub-surface by artificial means, thus the value of the sub-surface packer is apparent.
The sub-surface packer also assists in the rapid decomposition of vegetative matter, making available plant foods and humus by establishing capillarity and removing air spaces.

Prevents Evaporation.
To prevent evaporation of the soil water, it is necessary that capillary movement be destroyed at the surface--where it was necessary to have the soil firmly packed, it is now necessary to have the opposite quality if the soil water is to be retained.

The sub-surface packer leaves the surface soil in a loose condition, forming a mulch which effectively prevents evaporation. This mulch is also valuable in absorbing any rains that fall on it.

P & O

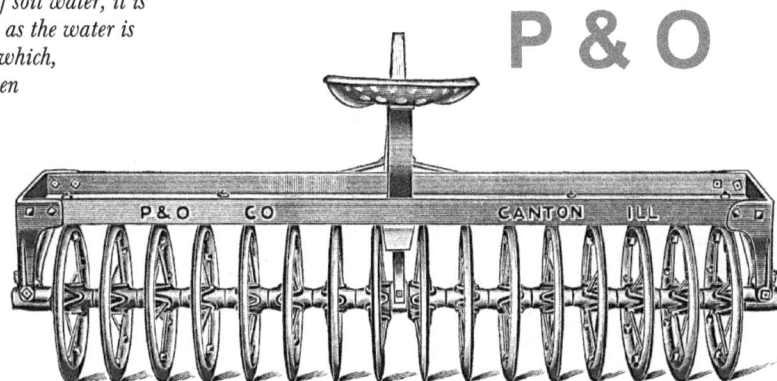

16-WHEEL NO. 1 SUB-SURFACE PACKER.

Rollers

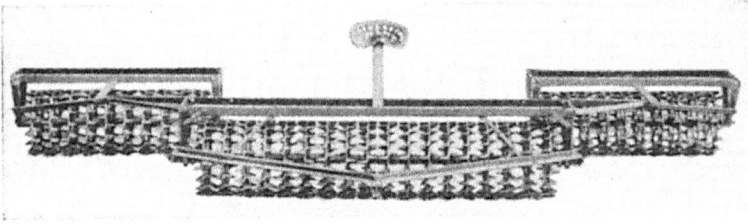

Subsurface packer.

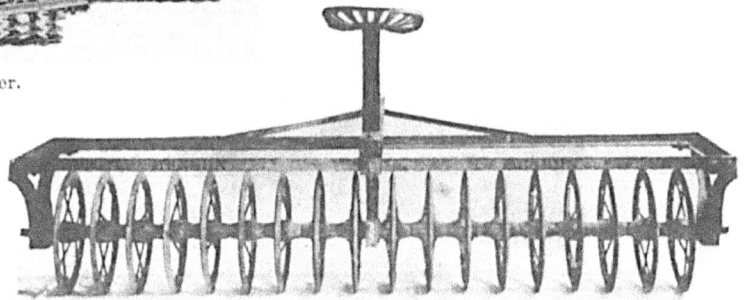

SUBSURFACE PACKER

Imperial

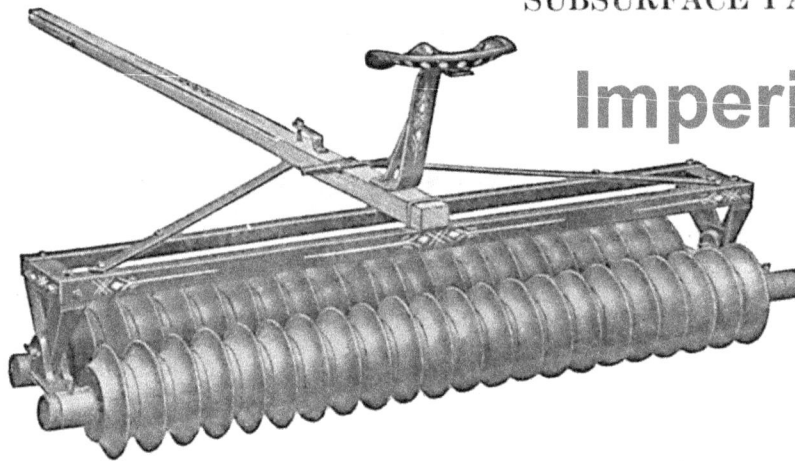

DOUBLE GANG SOIL PULVERIZER, CRUSHER AND PACKER

Hobson

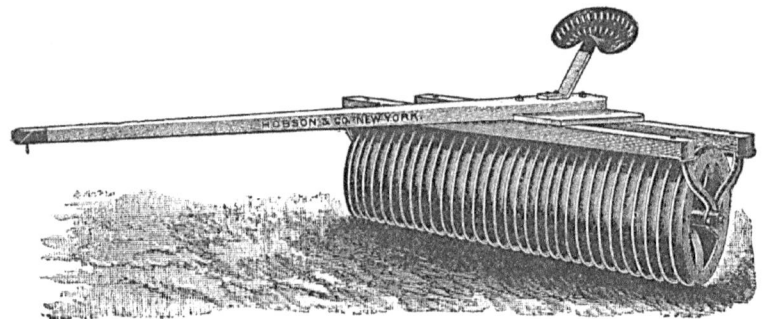

1890 BY HOBSON & CO.

English clod-crusher.

Clod-crusher. *Lowcock & Barr, Shrewsbury, England.*

Chapter Seven

Cultipackers

In keeping with Boswell's third rule of *biblio-dynamics,* which states that every book should have a really short chapter, here it is.

Combine a roller with a cultivating action and you have something which cultivates and packs, ergo: 'culti-packer.' Most of the time this has been accomplished by the farmer hooking the two implements together. Below are two recent examples of modern horsedrawn cultipackers.

Beiler's Machine Shop of Pennsylvania makes the above unit.

I & J Manufacturing, also of PA, makes the handy unit below from which the walking cultivator may be detached. Both companies are listed in the sources at the rear of this book.

Cultipackers

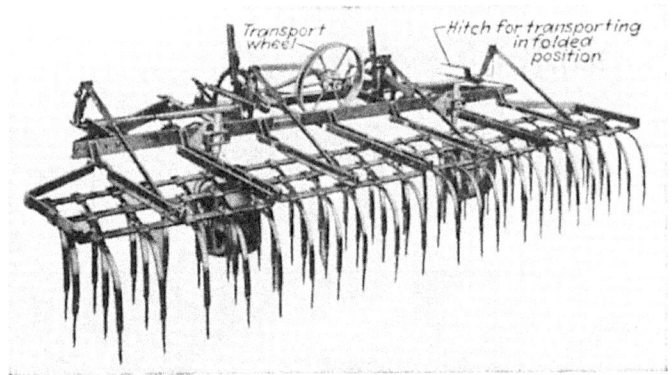

Weeder-mulcher.

Chapter Eight

Field Weeding

THE MITCHELL IMPROVED MOUNTED WEEDER

Riding and walking field weeders are adapted to a number of different jobs from cultivation of corn to the cultivation of sugar beets, cabbage, onions, potatoes, beans and practically any crop grown in rows any distance apart.

In the *old days* successful potato growers began using this weeder as soon as the potatoes were planted and continued once a week until plants were 15 to 20 inches high. They stirred the soil at least an inch to an inch and a half deep.

Using the weeder this way, in the row crops listed, destroys weeds, forces the plants to root deeper, stimulates the young plants to send up more main stalks, prevents the formation of surface crusts and produces a surface mulch which holds the moisture in the ground.

THE MITCHELL IMPROVED WEEDER

Mitchell

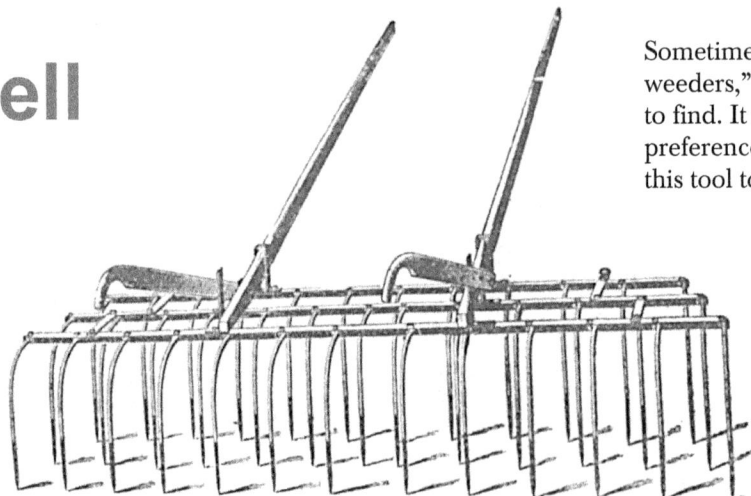

Sometimes referred to as "finger weeders," the old ones are difficult to find. It is a matter of personal preference whether or not to add this tool to the lineup.

Vulcan

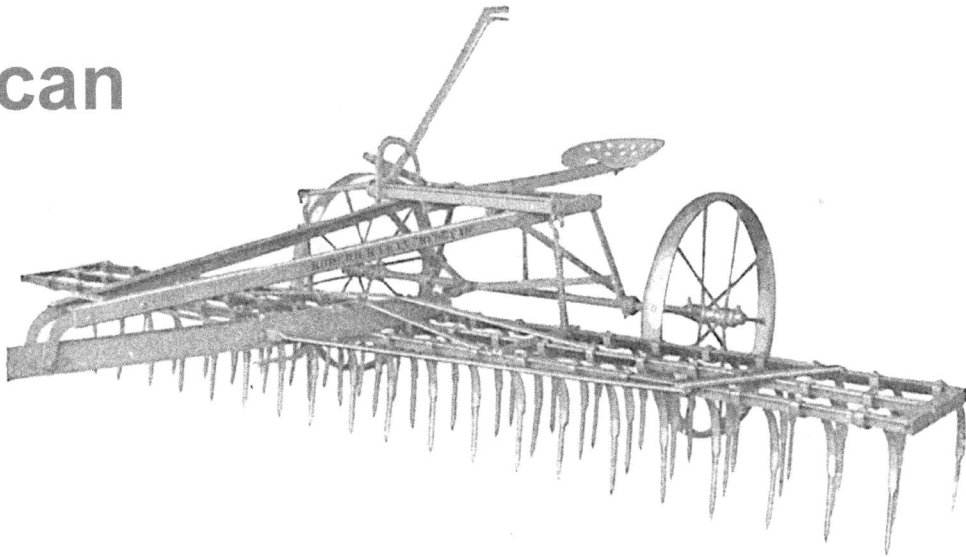

Oliver

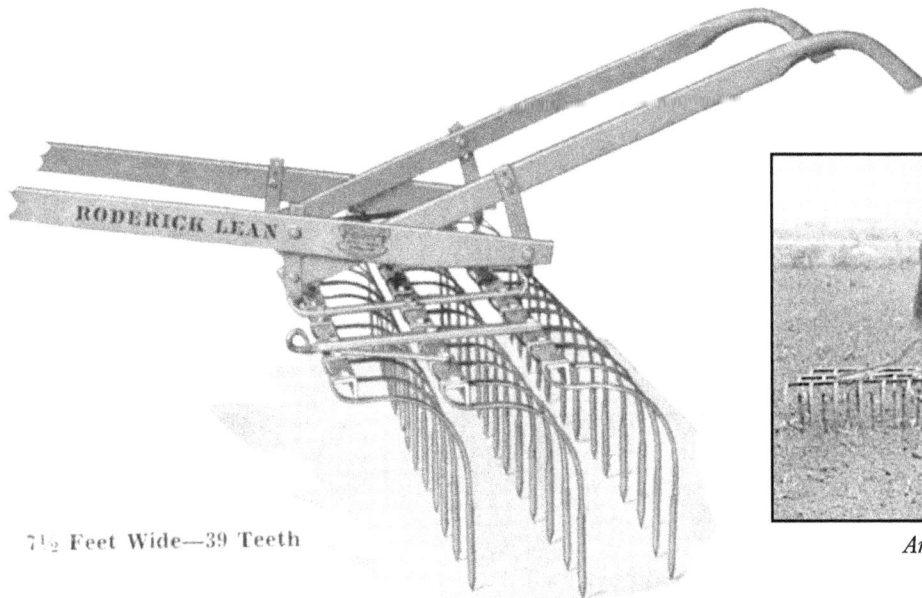

7½ Feet Wide—39 Teeth

An Oliver field weeder at work.

Roderick Lean

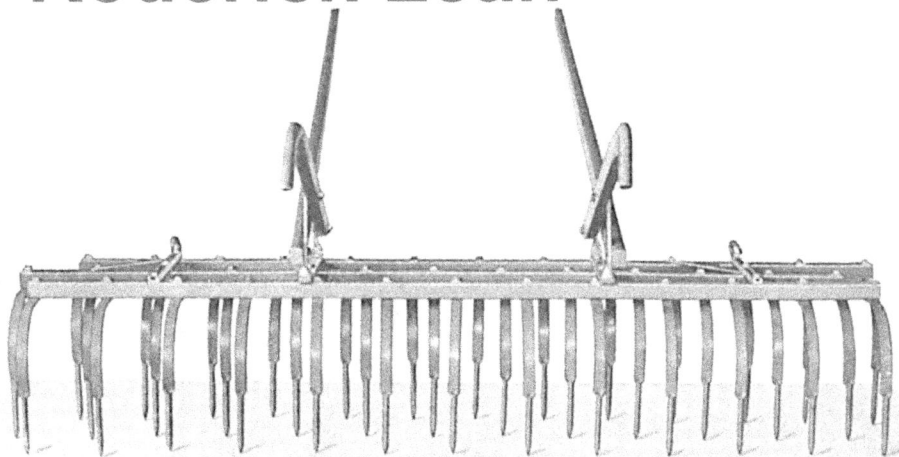

Rear View of Roderick Lean 7½ Ft. Weeder

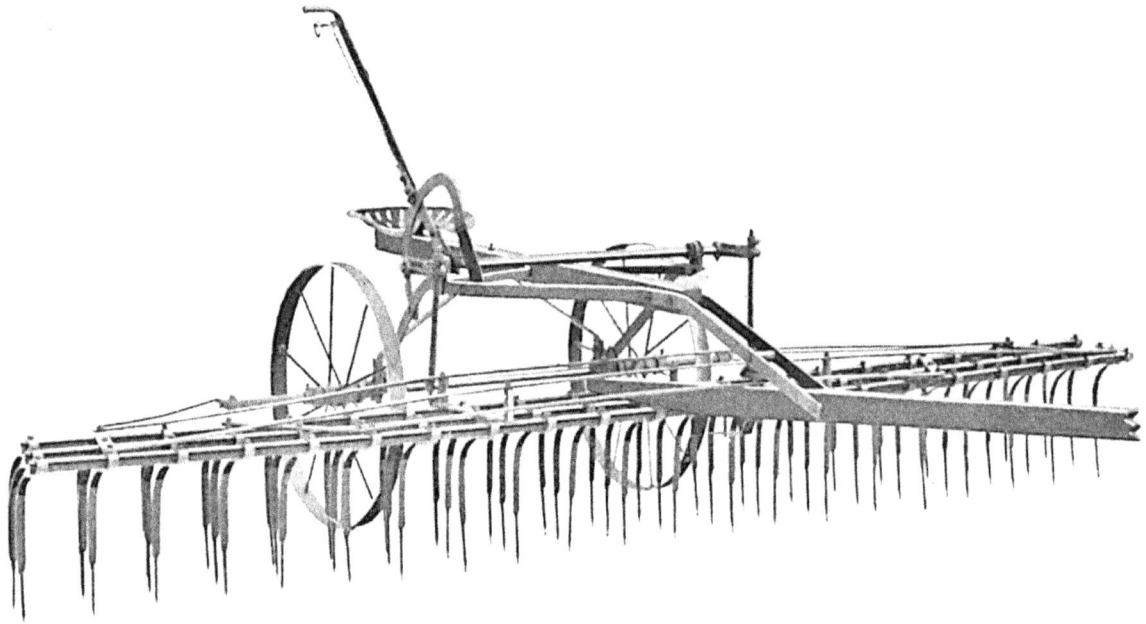

The No. 2 12-foot two-horse riding weeder—an all steel machine except the pole and seat support.

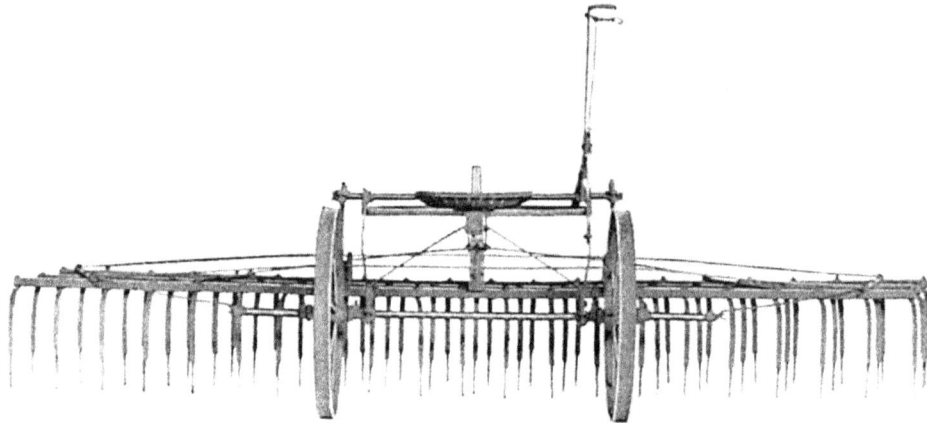

Rear view of the No. 2 showing how every part is well braced for great strength. The wheels are adjustable on the axle for different width rows.

Oliver

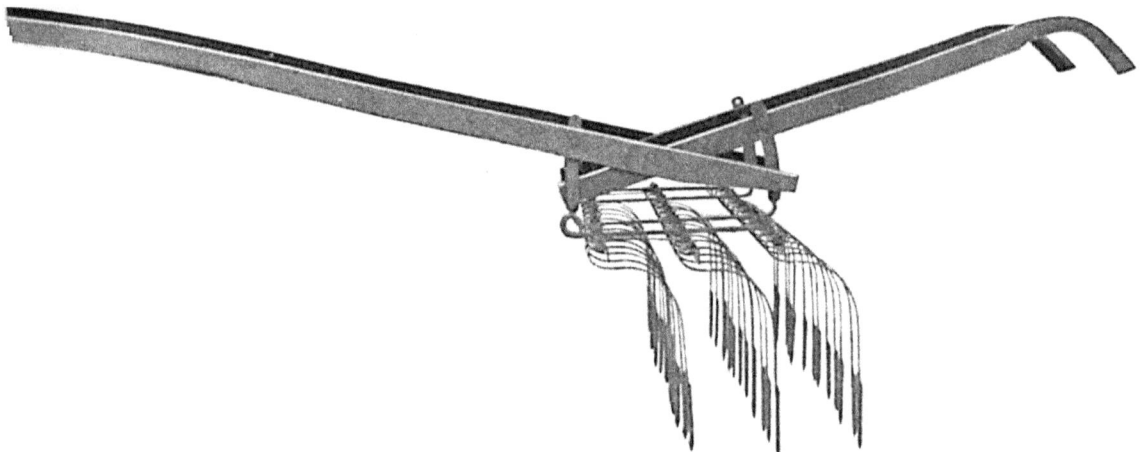

Oliver

Rear view of the No. 1 showing the flat shank, round pointed teeth which have proved the most successful kind for weeders.

CYCLONE WEEDER

Showing Weeder as with the teeth set for general field work.

Cyclone

These Cyclones are more akin to the Acme harrows but were sold by the maker as a field weeder. The teeth, as illustrated, could be moved. This implement was flipped over on its back for transport.

There is tremendous opportunity for the inventive and energized individual interested in the design and development of new horsedrawn implements. (Such a line of work and such a livelihood would be more satisfying and less prone to association with criminal elements than a life of elected office.)

These photos from Norway appeared in the Small Farmer's Journal and show off a European finger weeder attachment for a small forecart. For more information refer to SFJ volume 21 issue 4.

By name this chapter includes those implements which were designed to cultivate field crops as opposed to row crops, with at least two specialized exceptions; orchards and fallow lands.

The nature of the soil and any crop or crop residue resistance will have a bearing on the draft power required to pull these implements. Since these involve shovels or points which are drawn through the ground, it will greatly improve the draft if everything is properly set and sharp.

Chapter Nine

Field Cultivators

The Deere Alfalfa Cultivator—This Implement Breaks Up the Crowns, Stimulates the Growth of Plants, Removes Weeds and Thickens the Growth

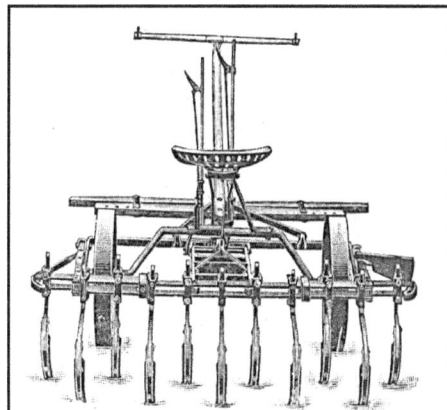

Field cultivator.

John Deere

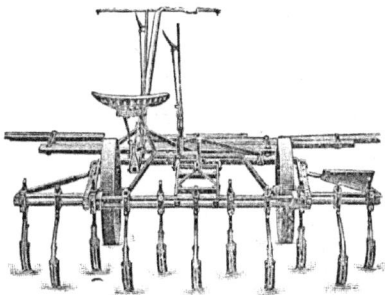

No. 42 Orchard and Universal Cultivator Seven foot nine inches wide. Two horse.

Planet Jr.

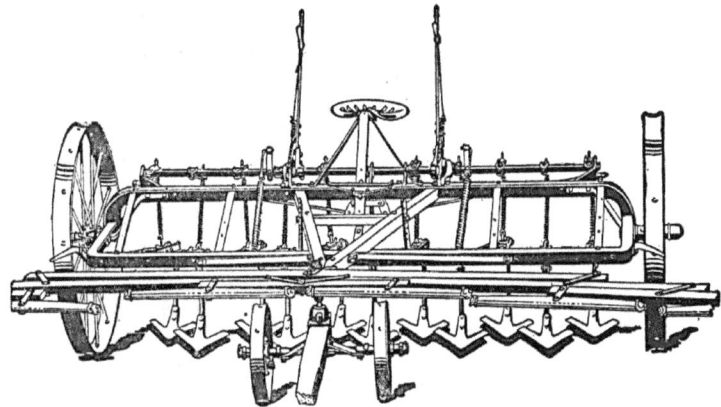

No. 41 Riding Harrow and Cultivator. Three horse.

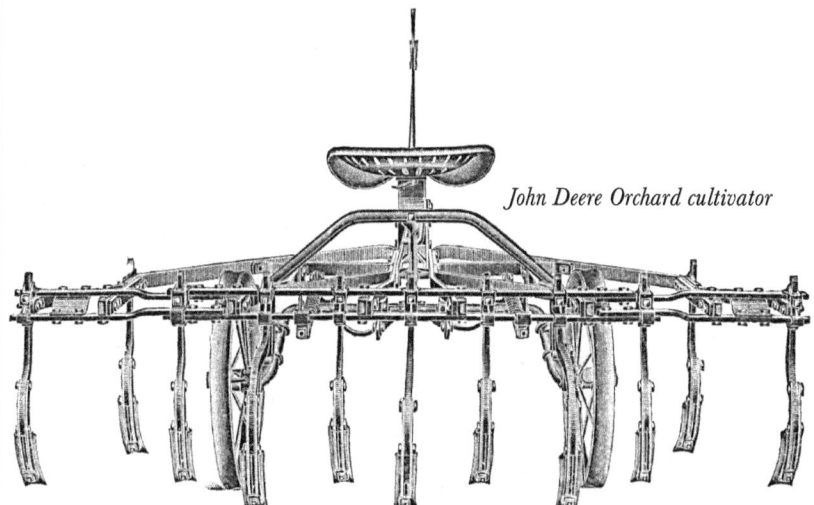

John Deere Orchard cultivator

MOLINE ALFALFA CULTIVATORS

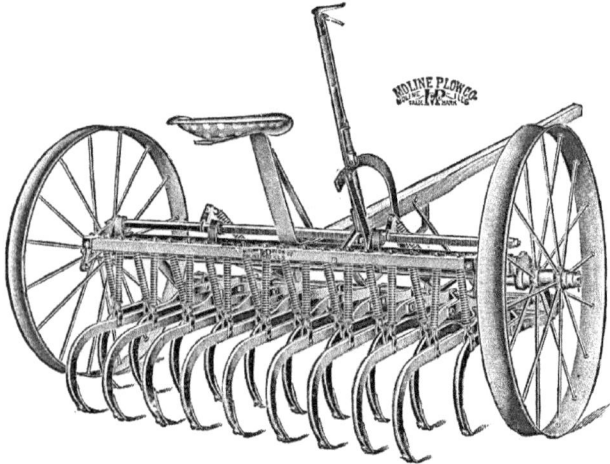

What follows is information provided by the manufacturer to entice folks to purchase their Alfalfa cultivator.

MOLINE ALFALFA CULTIVATORS

The Moline Alfalfa Cultivator is the most successful type of alfalfa cultivator made. It thoroughly cultivates the soil without injury to the plants; is light in draft and convenient to handle.

Advantages of Cultivation.

Cultivation loosens up the soil which becomes packed by rains, by tramping of horses, or by the weight of machines, and forms a mulch which prevents the evaporation of soil water. It also puts the soil in such condition that it will easily absorb any rains falling upon it. Cultivation aerates the soil, thereby aiding bacterial action, without which the alfalfa plants will not thrive, and nitrogen, an essential soil element is lost. Cultivation also destroys many noxious weeds and kills many

injurious insects. Without cultivation the alfalfa crop will die out and the yields obtained will be small.

Shovels.

The shovels of the Moline Alfalfa Cultivator are narrow and do not tear the alfalfa plants, but rather work around them. They are reversible, and when one point becomes worn, a new point is available.

One of the (available) points is sharp and the other blunt. The blunt point is better adapted for cultivating young alfalfa plants and for use in loose soils. The sharp point is more generally used for cultivating old alfalfa and in hard soil.

Beams.

The beams of the Moline Alfalfa Cultivator are very stiff and strong, and will withstand any twisting strain. They also have some play and permit the shovels to work around the plants instead of tearing through them as rigid beams will do. The rear shovels will not trail the front ones, and always keep to their work. These beams are connected by means of steel rods to a strong angle steel draw bar. The set of long beams has a connecting rod separate from the short set. This permits more play and individual action and gives a wider front support.

Pressure.

Any amount of pressure can be given to the beams. Each beam has an individual spring which permits it to follow any unevenness of the ground without interfering with the action of the others, thus all the ground is thoroughly cultivated. One heavy angle bar raises all the beams.

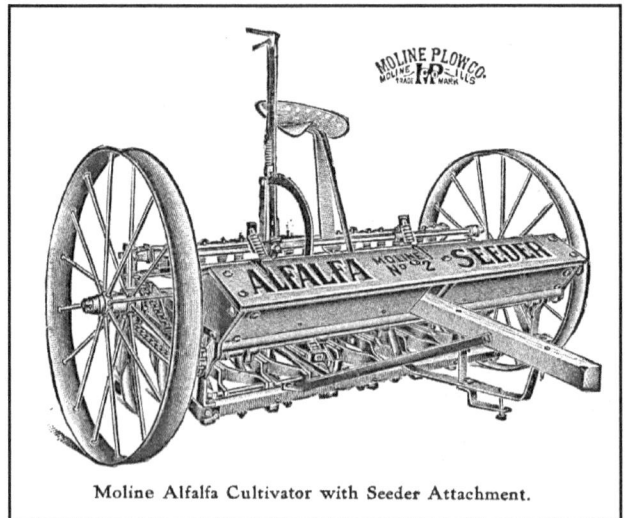

Moline Alfalfa Cultivator with Seeder Attachment.

Please note: Every company needed to believe that their particular implement design was the very best, or at least sell them to you in that regard. The manufacturer's information reprinted throughout this book should not be taken as an endorsement of one make over another.

John Deere

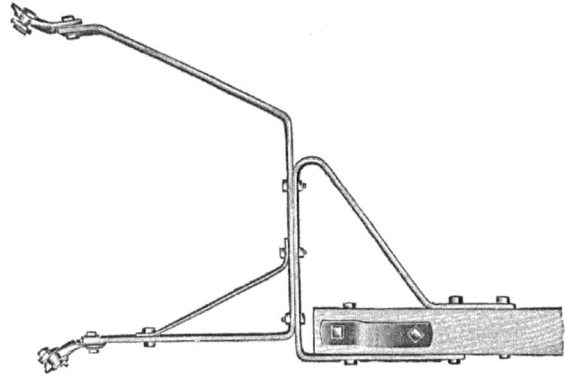

Pole Attachment—1910 Style

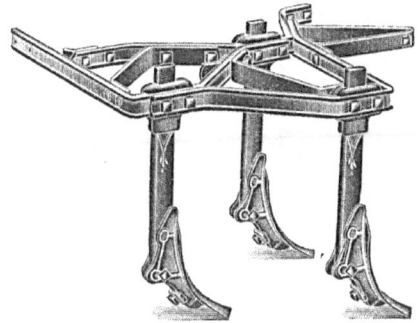

Shovel Extension

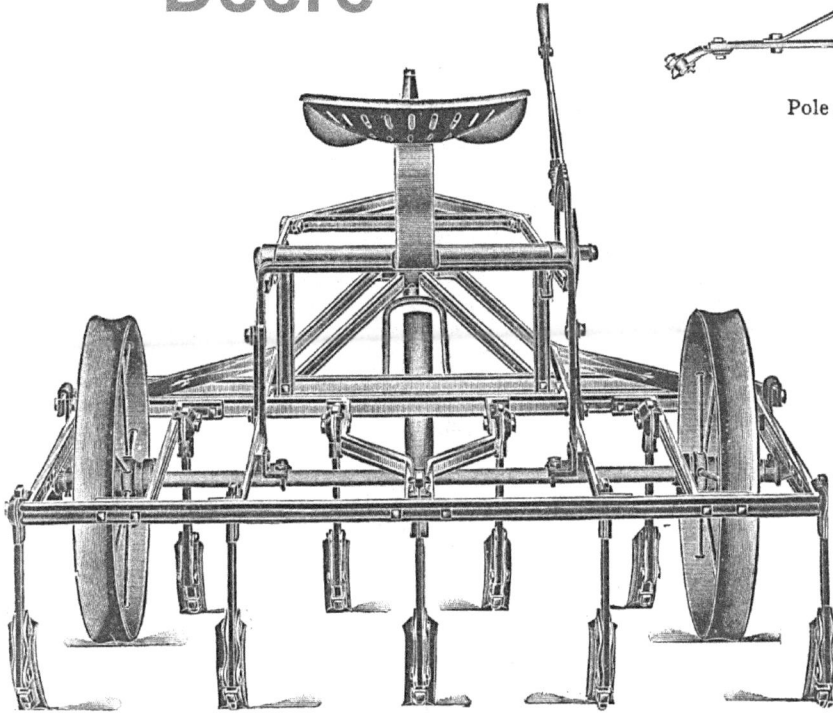

John Deere Orchard Cultivator

P & O

also McCormick / International

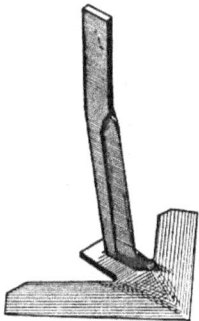

Sweep.

Alfalfa Steel
on Front Leg.

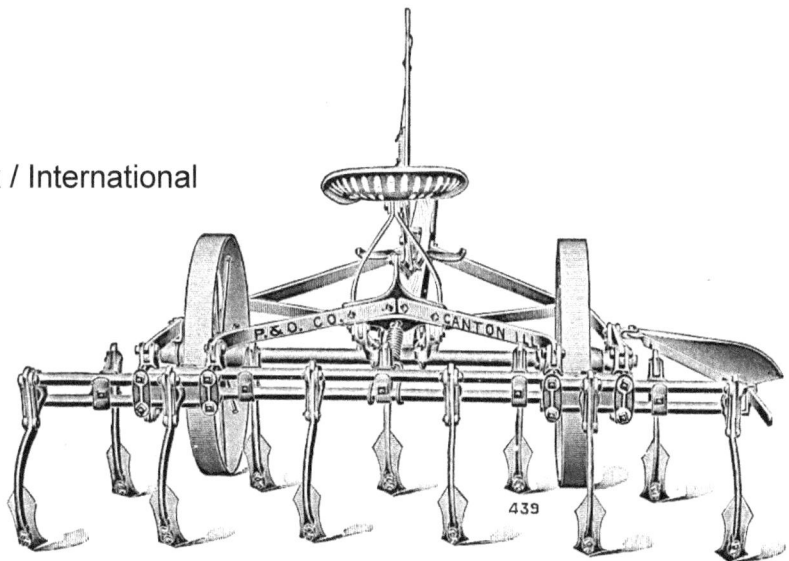

439

NO. 11 ORCHARD CULTIVATOR WITH SHOVEL GANGS.

Forkner

*No. 28 Forkner Spring tooth Tiller. Ten foot wide.
Two horses.*

These riding spring tooth harrows represent an implement which would find ready market for today's horsefarmers and, it would seem, pose little challenge for the energies of good shops such as White Horse Machine, I & J Manufacturing and Pioneer Equipment. Maybe if several of us sent them letters of encouragement...?

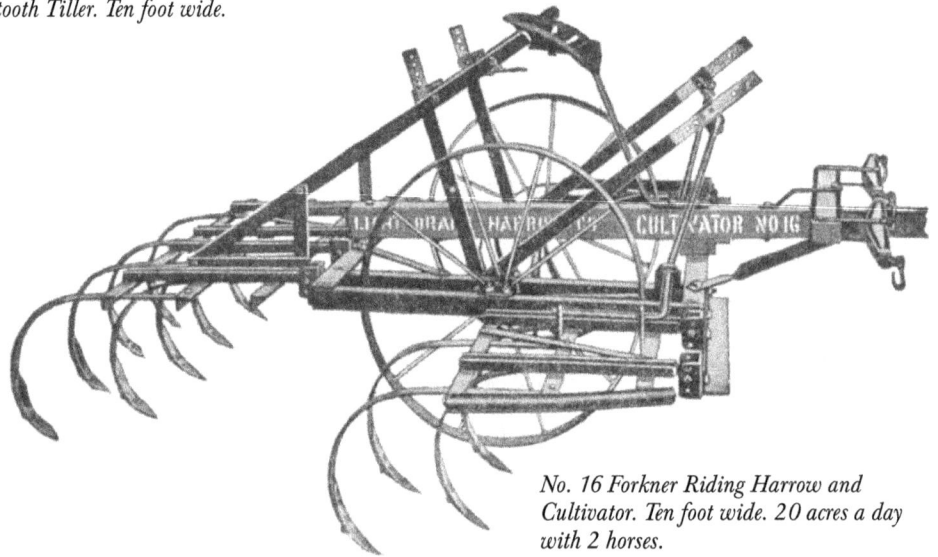

No. 16 Forkner Riding Harrow and Cultivator. Ten foot wide. 20 acres a day with 2 horses.

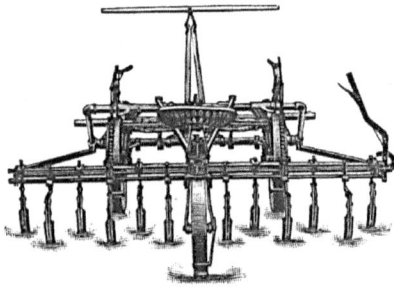

Planet Jr.

No, 3 Planet Jr. 3 wheeled Sugar Beet Horse Hoe for 1908 with foot actuated steering wheels.

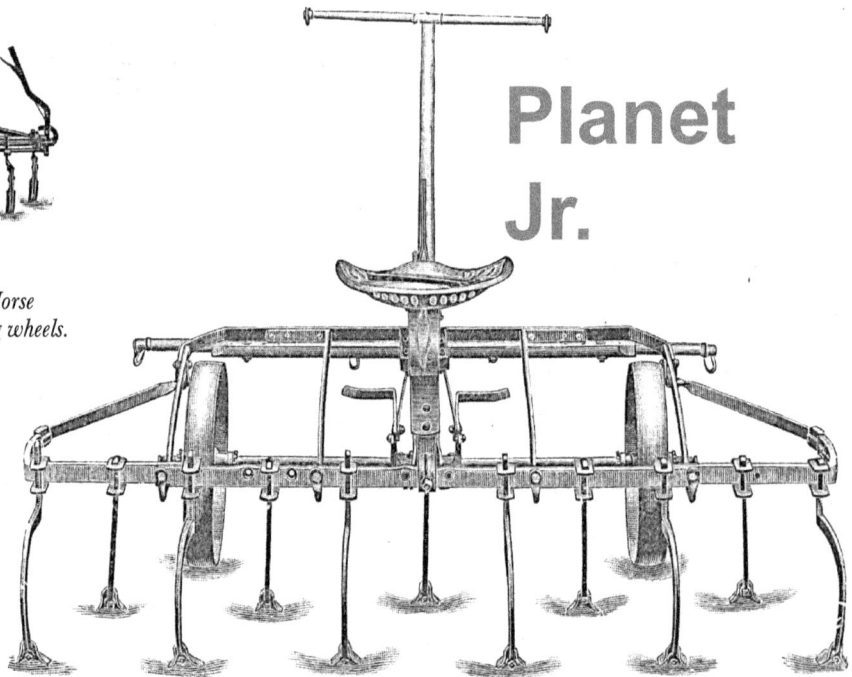

No. 45 Planet Jr. Riding Harrow and Cultivator. Especially adapted for orchards, hopyards, and vineyards

Forkner

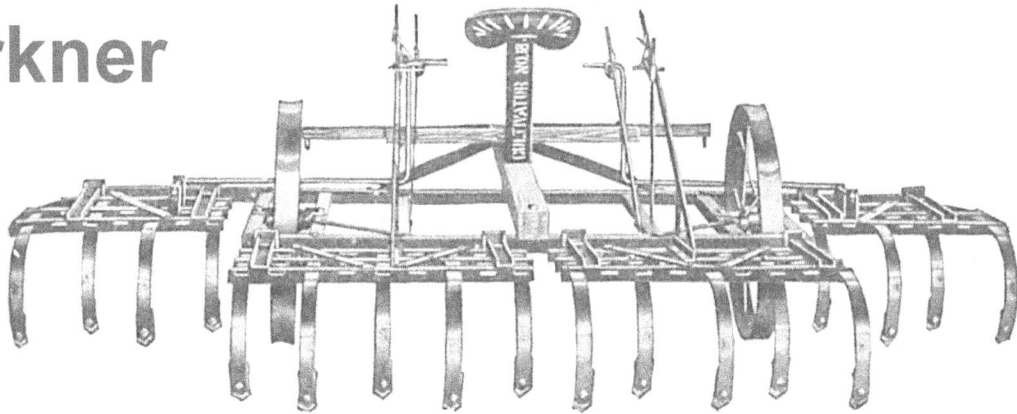

No. 18 Forkner "Light Draft". This implement was designed to leave the ground in slight waves, a condition that conserved moisture and tended to prevent blowing. Eleven foot 4 inch wide, the manufacturer claimed that 2 horses, 1,200 lbs each, could pull this unit across 25 acres in one day.

Planet Jr.

Fore Carriage for No. 41-42 and 46 Cultivators. 16 in. Steel Wheels, 2 in. ⅜ tire, 16 in. apart, 1 in. hinged axle, stub tongue, takes same braces as Nos. 41 and 42.

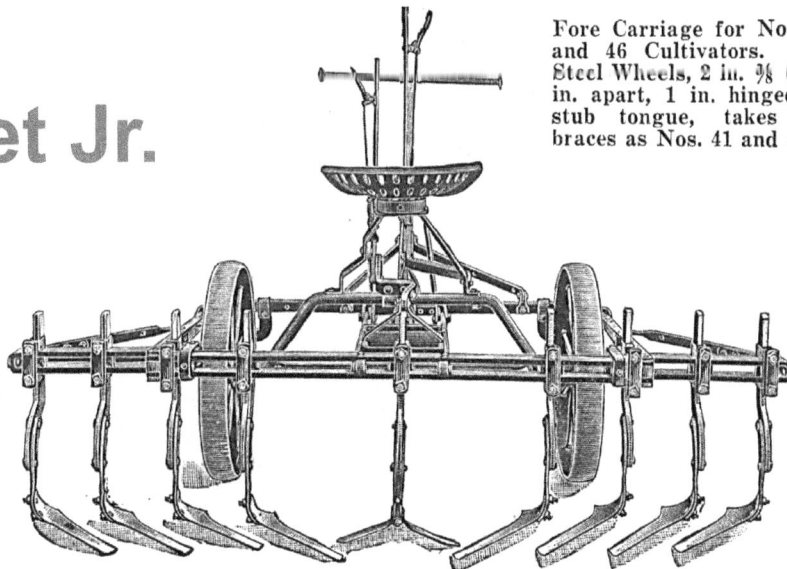

THE PLANET JR. ORCHARD AND UNIVERSAL CULTIVATOR
Equipped with Cyclone Weeder Attachment

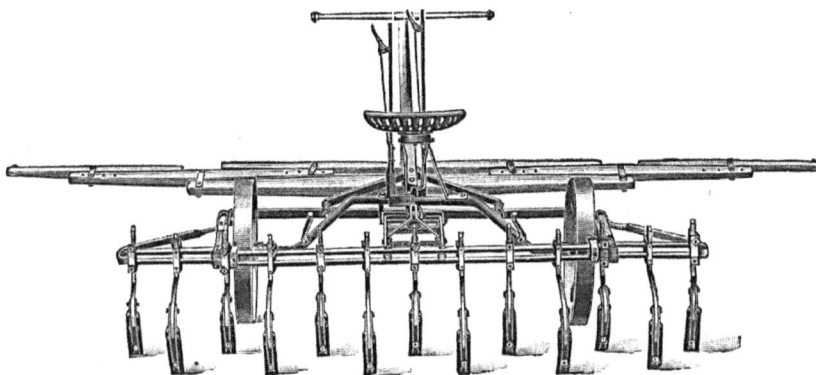

No. 42 Improved Planet Jr. Orchard and Universal Cultivator

Planet Jr.

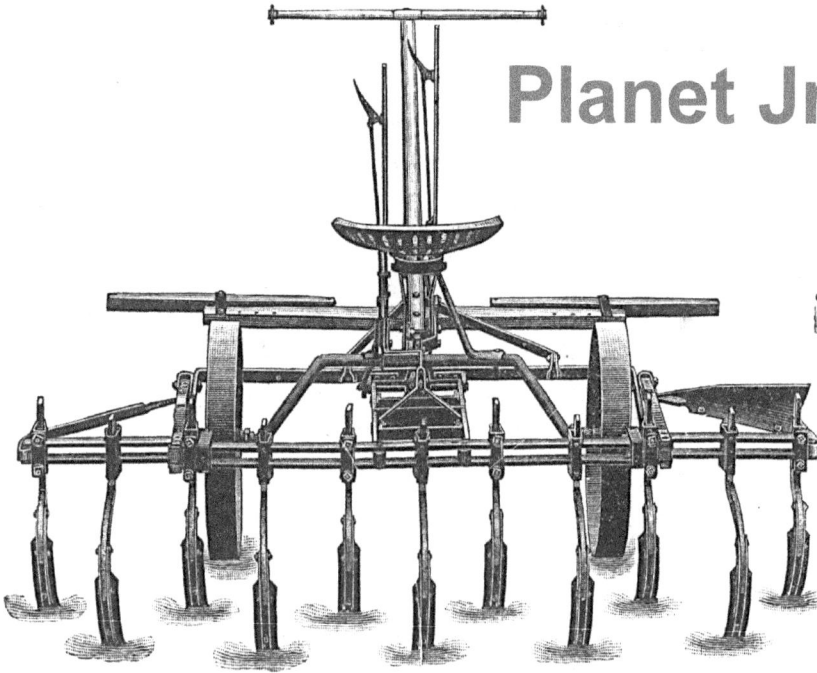

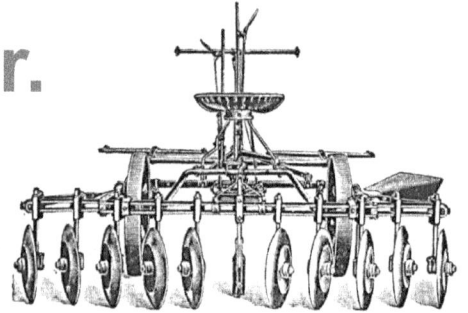

Planet Jr. No. 41 rigged with disc blades,

No. 41—Planet Jr. Orchard and Universal Cultivator

Cuts 6 Feet 6 Inches as Shown

P & O

CANTON ILL.

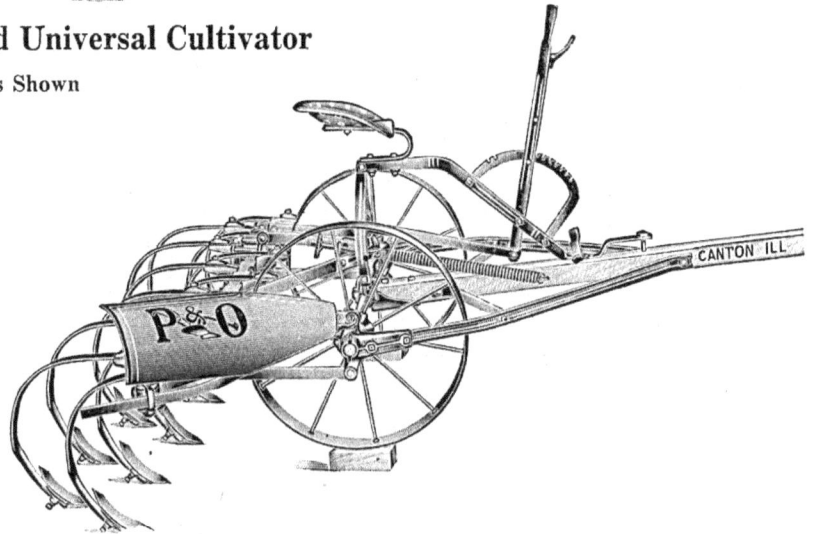

NO. 31 ORCHARD CULTIVATOR WITH SPRING TOOTH GANGS.

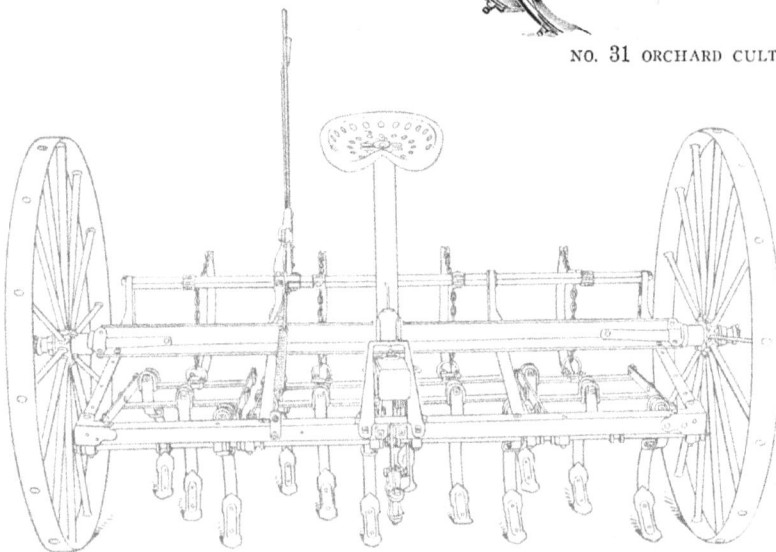

McCormick

Field Cultivator

Great Western

Originally designed for the arid Pacific Northwest.

No. 47 Great Western Weed Killer. Fifteen foot cut, raised seat. Leaves soil in wavy condition. Note that the teeth are arranged to offer no obstruction to trash working back out of the way. Good in trashy ground. Eight to ten horses recommended.

Pointers as to the Best Methods of Working Summer Fallow with the Great Western Weeder. *(Authored by the manufacturer.)*

The land should first be disced either in the fall or as early as it can be gotten onto in the spring. This discing will start the weed growth. As soon as this growth is well started, the land should be plowed and the weeds turned under. The second crop of weeds will then come, which is usually along about in May and this is where the Light Draft Harrow comes in. When this second crop of weeds is up approximately 3 inches, in other words, well started but still young and tender, the Light Draft Harrow should be put onto the land, going over it thoroughly and if in June or July the weeds again appear, the Light Draft should be used again and still again, if they appear the third time, say in August or the first part of September.

Emphasis must be laid upon the fact that there is nothing that will take moisture from the land as will the growth of these weeds, if permitted to continue. It is, therefore, as much for the purpose of conserving the moisture as for cleaning the land of weeds that the Light Draft Harrow is used and recommended.

Another great advantage in the use of the Light Draft Harrow is that the 4 inch shovels we use on the No. 34 and No. 47 machines ridge the land and prevent blowing of the soil. Furthermore, the proper method of using these weeders with 4 inch shovels is to put them down into the ground to the bottom of the furrow--in other words, as deep as the surface of the ground has been plowed, this will bring the clods to the surface, putting the pulverized or fine soil underneath, next to the bottom of the furrow, where it will have a tendency to attract or draw the moisture from below.

This method will conserve the moisture better than where open places are left next to the moisture in the bottom of the furrow by leaving the clods there.

No. 34 Great Western Weed Killer. Similar to No. 47 but in smaller size. Suitable for 6 horses. Eleven and a half feet wide.

Front view of No. 47 (same as top of the page.

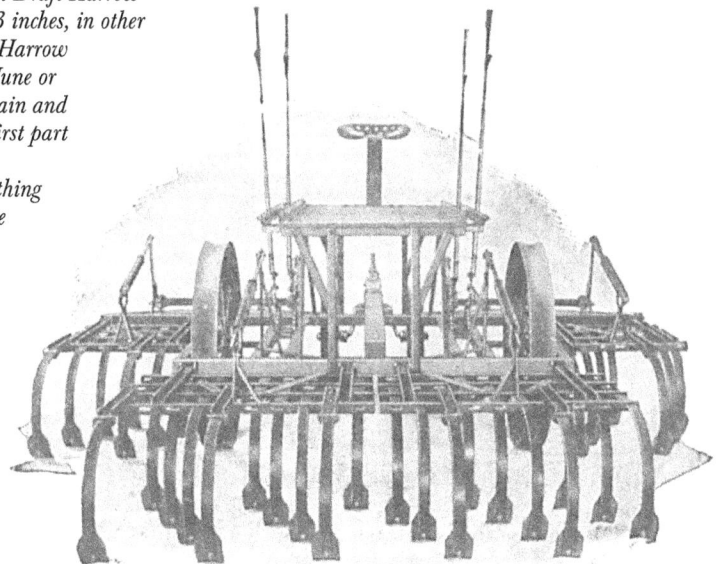

155

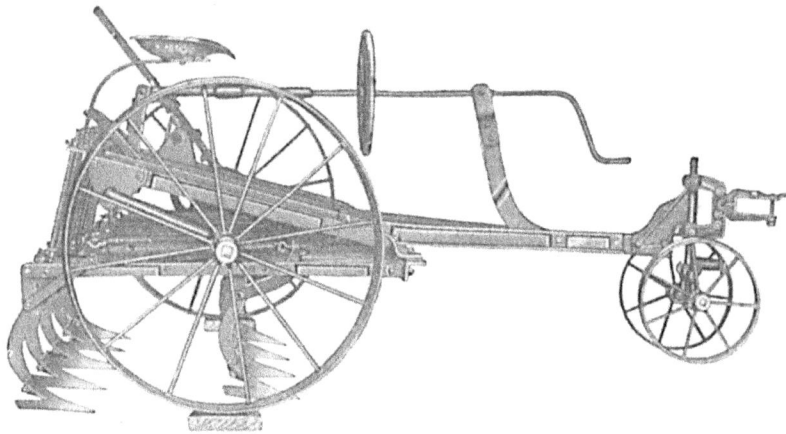

Side view of the No. 131-C 7-foot horse drawn duck foot Fallovator.

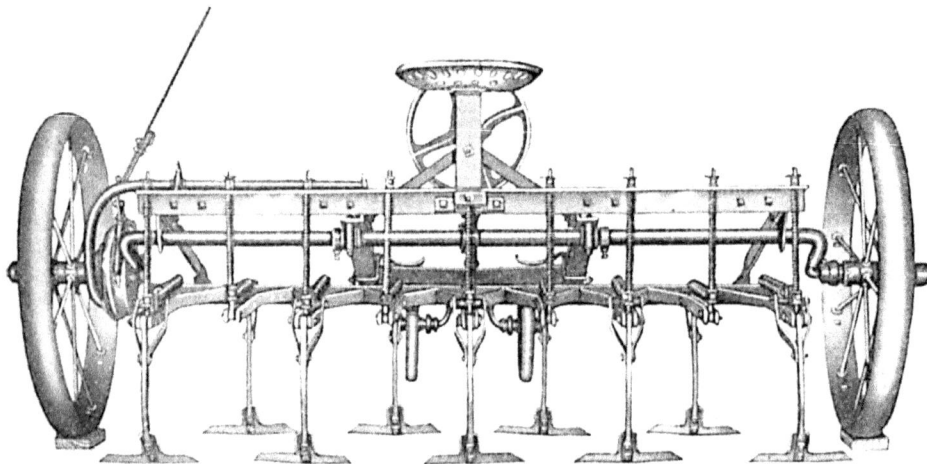

The 131-C 7-foot duck foot Fallovator is built for use with horses. A large wheel on the depth adjusting screw makes depth adjustments easier.

Side view of the No. 131-E 7-foot spring tooth Fallovator for use with horses.

Top view of the 134-B 10-foot horse drawn duck foot Fallovator showing the rigidly braced non-sag frame and the heavy wide tired wheels.

Do you need an orchard or alfalfa cultivator or a field cultivator? Too many variables to be able to answer that question. You can certainly start out without them and utilize disc harrows, and modified straddle row cultivators, in the beginning. But, should the amount of land warrant, these wider swath tools would certainly be appreciated when the difference between 10 acres a day and 20 becomes critical.

Chapter Ten

Rotary Hoes

Rotary Hoes and Rotary Harrows are fascinating tillage tools which, unfotunately, are difficult or impossible to find in a usuable older model. Vague references from the historical literature of the early 19th century manufacturers suggest that these tools never had the wide acceptance of disc harrows or straddle row cultivators and because of this far fewer were built.

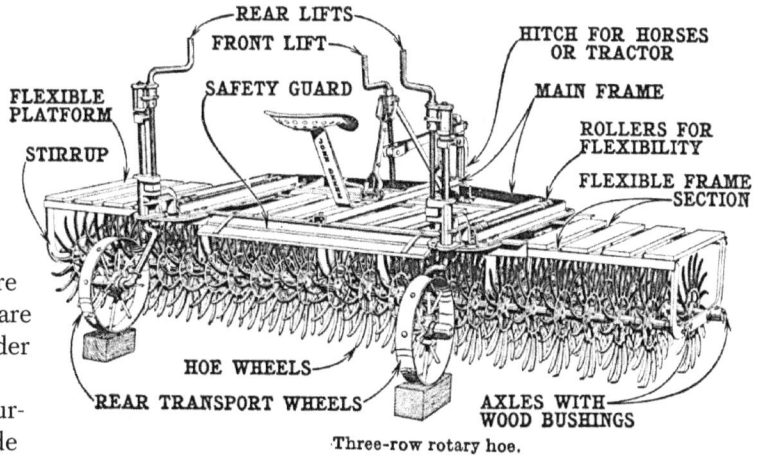

Three-row rotary hoe.

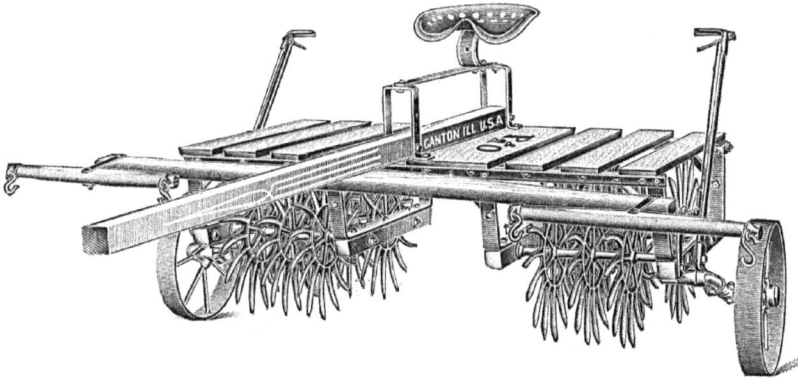

NO. 1 TWO ROW ROTARY HOE, OPEN TYPE.

John Deere

Within this chapter we again offer the manufacturer's comments on some of these units.

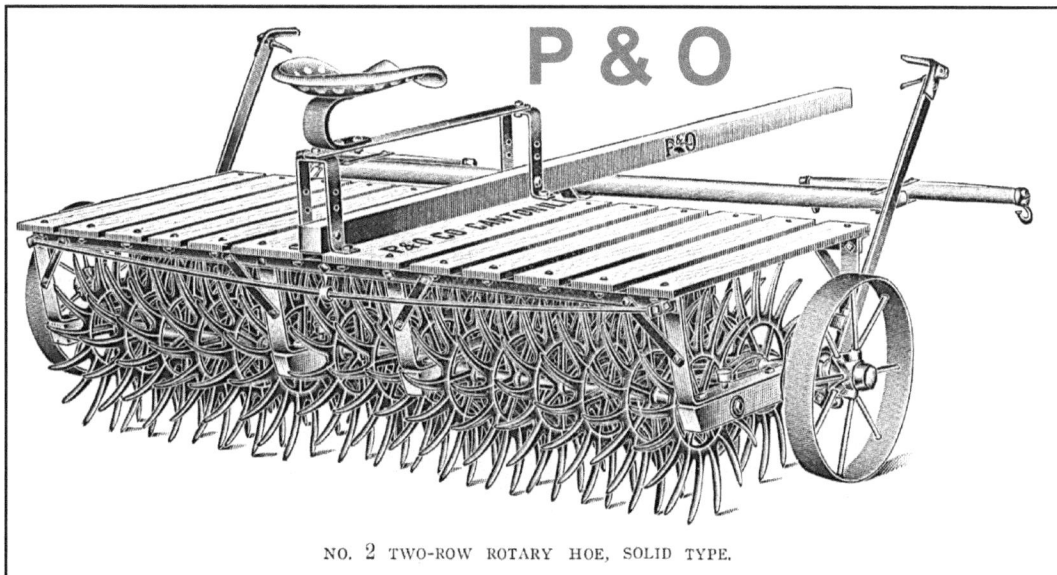

P & O

NO. 2 TWO-ROW ROTARY HOE, SOLID TYPE.

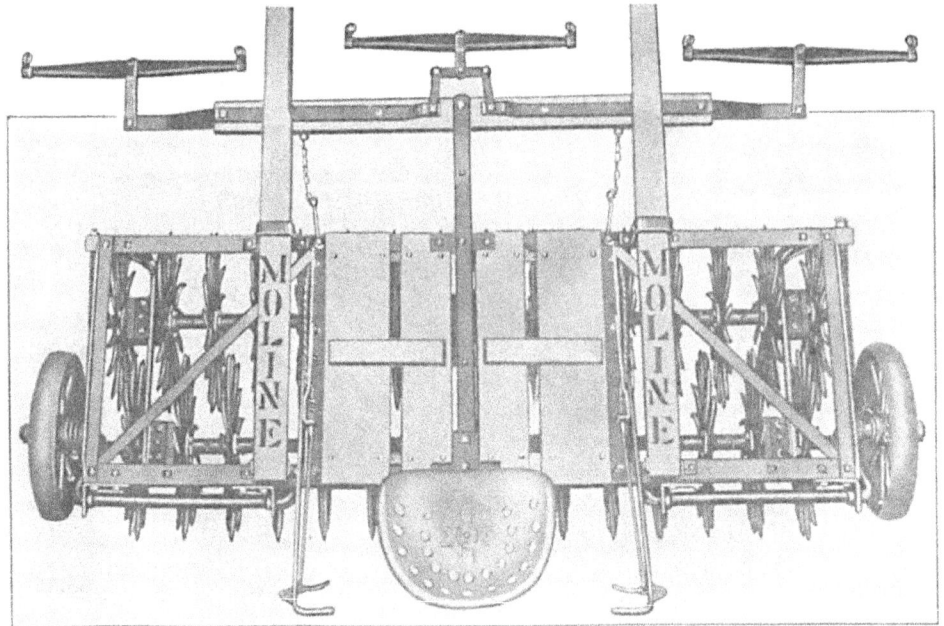

Moline Rotary Hoe

The Moline Rotary Hoe is an exceptional tool for cultivation of small corn, to break a crust, create a surface mulch and destroy weed seedlings. The hoe wheels thoroughly aerate the soil and cultivate closely which promotes the rapid growth of the young corn, or other crops, because it gives an opportunity to develop a healthy root growth right from the start. Weed seedlings are dislodged and exposed to the hot sun, thus being destroyed. The Moline Rotary Hoe makes possible successful blind cultivation of corn where the soil is inclined to pack after a rain. It also provides the only satisfactory method of cultivation of crops drilled solid, such as soy beans, or peas; and is suitable for renovating alfalfa and clover.

This machine is not intended to replace shovel cultivators but is a valuable supplement for early cultivations, because it permits earlier and closer cultivation than is otherwise possible. The hoe wheels do not injure the corn but pulverize the crust and dislodge weeds without disturbing the more deeply seeded plants.

The Moline Rotary Hoe cultivates to a width of 84 in. or two rows. It is provided with either two-horse hitch, or a three-horse two-pole hitch with spacing to

Moline

prevent the horses tramping on the rows. This arrangement allows the gangs to cultivate to the middle of the two outside rows, as in the case of a two-row shovel cultivator.

Cultivates Thoroughly

With the Moline Rotary Hoe, it is possible to do a most thorough job of cultivating. The front gangs consist of

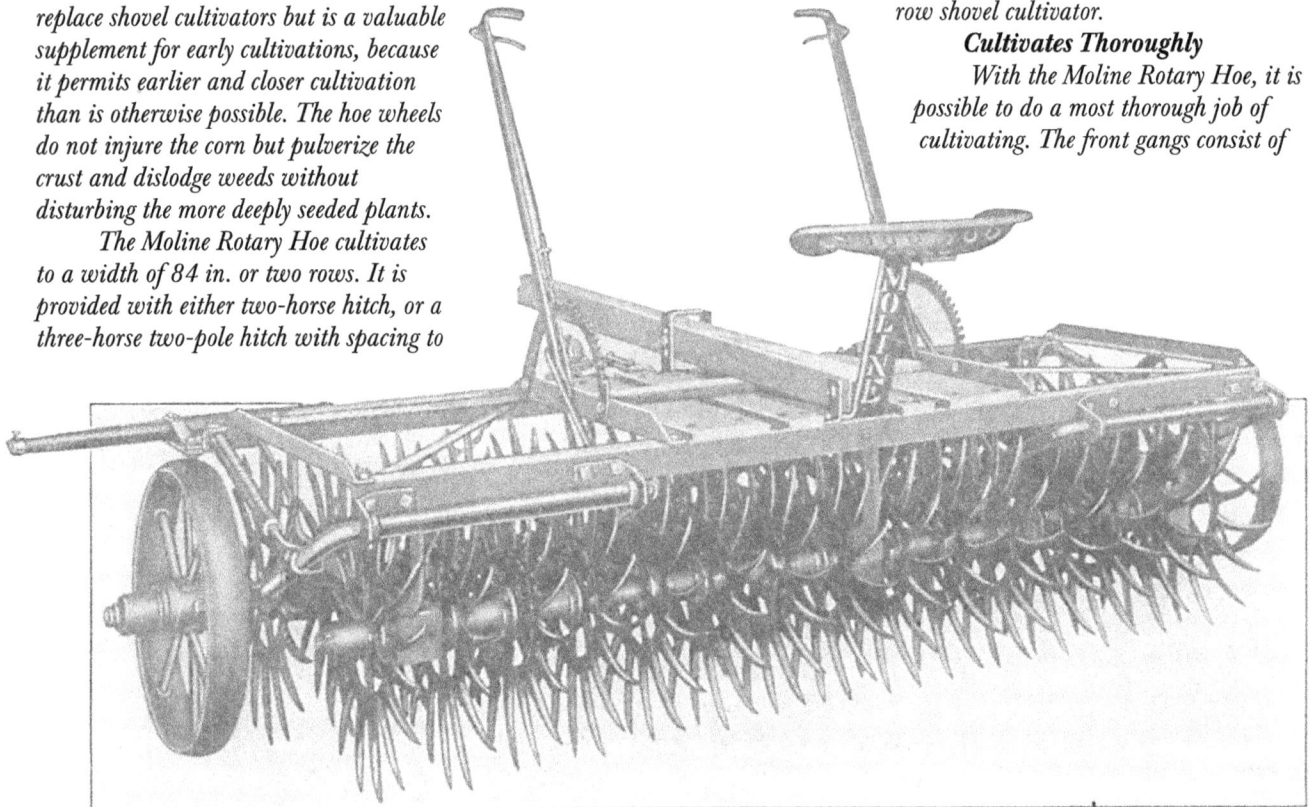

MOLINE ROTARY HOE

158

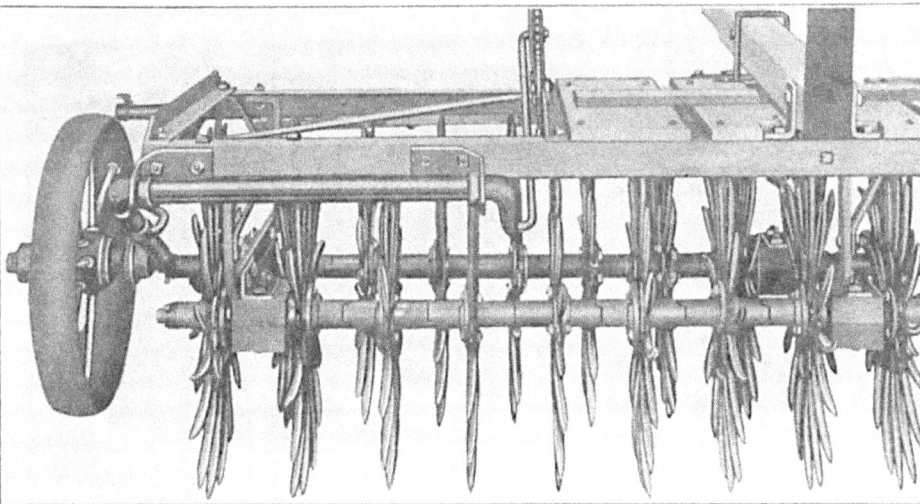

Detail of gang, showing section of four hoe wheels locked together—observe large bearings with Alemite oilers.

Moline

16 wheels; the rear gangs have 15 wheels which run between the front wheels pulverizing the centers. All together there are 31 wheels of 21 in. diameter and each having 16 points or fingers. These fingers are correctly shaped to pierce the soil and pulverize the crust without dragging effect or injury to the plants.

Hoe Wheels Operate in Sections

A feature of paramount importance on the Moline Rotary Hoe is the arrangement of the hoe wheels on the gang bolts. They are locked together in sections of four on both front and rear gangs by an ingenious design of the hubs. In this arrangement the wheels shed trash better than when mounted individually because the combined turning power of each section serves to clean individual wheels and thus avoids clogging. It is also a decided advantage when turning at the end of rows to have the hoe wheels in sections rather than all locked together on the axle for the sections make it easier to turn short and thus avoids any tendency to injure the crop.

Gangs Penetrate Uniformly

Since the gauge wheels on this rotary hoe are placed at the end of frame, instead of behind, it is flexible enough to follow the contour of the field and penetrate uniformly regardless of variations, such as gullies. The gauge wheels are set close to the frame, thus providing ample row clearance for narrow rows.

Low Hitch
--No Truck Required

On the Moline Rotary Hoe the hitch is attached to the frame directly in line of draft, and therefore is low enough to eliminate excessive neck weight on team. For this reason it is not necessary to use a tongue truck. Adjustment of the pole for height of team does not change the height of hitch.

Levers Operated from Seat

Observe the convenient placing of the levers on this Rotary Hoe. Both are operated from the seat, being in convenient reach of operator at all times. It is not necessary to get on and off the machine to raise or lower the gangs, for they may be adjusted while the machine is in motion. This new feature of convenience appeals to all farmers. It materially improves the operation of the Rotary Hoe. It is now possible for a boy to use the machine efficiently and safely. A large assisting spring is provided at each end of frame, which makes the levers operate easily.

Good Balance

The seat is placed in the correct position to balance the machine. The end wheels crank forward as the gangs are lowered; as the gangs are raised the end wheels are moved to the rear to balance it when transporting.

Turns Easily

It is easy to turn the Moline Rotary Hoe at ends of rows. By having the hoe wheels arranged in sections of four, they allow greater freedom in turning than would be true if they were all locked together on the axle. Another important factor in the operation of this Rotary Hoe is its ability to turn short due to the elimination of the tongue truck.

Team Spaced Correctly

The hitch is designed to allow the team to walk squarely between the outside rows without danger of tramping on the corn row; the pole being in center of middle row. The three-horse, two-pole hitch, which is optional, is quickly adjustable to a two-horse hitch by merely inserting two bolts in evener bar and changing position of pole.

Removable Platform

The platform is in two removable sections. It does not form a part of the frame and, therefore, gives free access to the hoe wheels at all times.

Rotary Hoe

With Depth Regulated from Seat

Vulcan

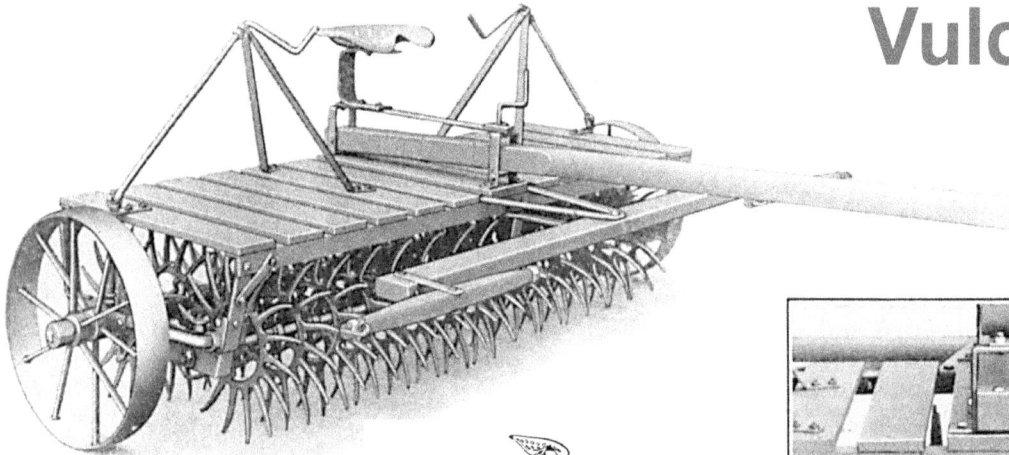

P & O NO. 1 TWO-ROW
ROTARY HOES

Two-row rotary hoe.

The No. 1 Rotary Hoe is designed for breaking the crust of corn ground before cultivating with the regular shovel or disc cultivator. It is unsurpassed as a crust breaker, as it digs into and loosens the soil, breaking the crust into fine pieces without dragging great chunks of it over the hills, with the consequent damage to the growing corn which often occurs when breaking crusty soil with the ordinary cultivator. A rotary hoe loosens the ground closer to the corn roots than any other type of implement. When using the No. 1 Rotary Hoe with the open center, it is driven so that the gangs go through the rows (not between) and the solid ground that is left in the center between the rows is afterwards plowed by the shovel cultivator.

HOE WHEELS. These are exceptionally strong and so designed as to effectually penetrate and loosen the soil 12 inches each way from the rows. Each side of the machine has a

Closeup of the Vulcan

set of gangs comprising two square shafts of solid bar steel, the rear shaft having four and the front shaft three wheels. The wheels are set five inches apart on the shafts and they are so placed that the front wheels cut half way between the rear wheels, having the same effect as would seven wheels set 2 ½ inches apart on one shaft. The wheels are held in place by spreader spools, and the axles revolve in hardwood bearings, which are equipped with compression grease cups for using hard oil.

CARRYING WHEELS. These constitute an entirely new feature and afford distinct advantages, as the depth at which the hoe points work can be regulated to suit the requirements. The carrying wheels operate on cranked axles, the levers being used for throwing the wheels backward or forward, an action which raises or lowers the frame and the hoe appliances up or down. The working depth is from nothing to four inches. When raised for transportation the hoe-wheels clear the ground four inches. The carrying wheels, in addition to regulating the depth of the points, also save the points from unnecessary wear when taking the machine from one place to another.

P & O

Two Row

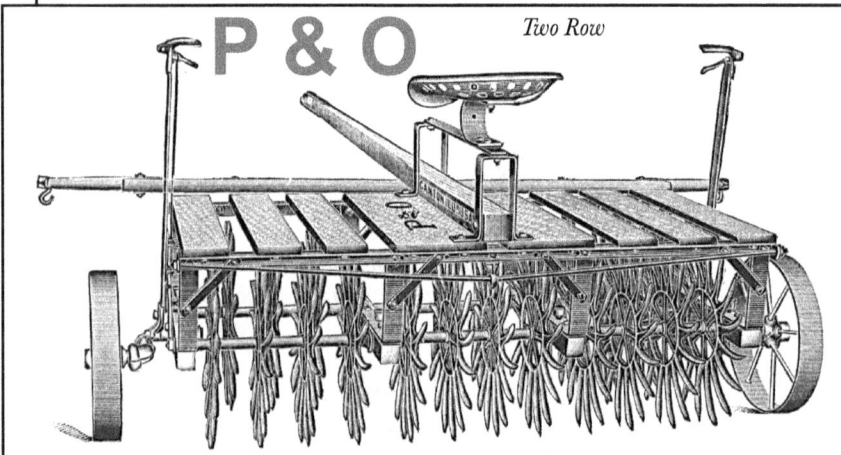

ALFALFA CULTIVATOR

the heads has 18 teeth, about 5 inches of each tooth being exposed, and as the ends of the teeth are only about 3 inches apart at the points, it makes the machine a thorough pulverizer. The inner ends of the gangs are held down by adjustable corrugated brackets on the frame. As the tendency of all harrows of this nature is to raise in the center, the brackets can be set so that the gangs will be held down at a uniform depth.

TEETH. The teeth are beveled on one side so as to thoroughly pulverize the soil and avoid heavy draft.

CLEANERS. These cleaners are bolted to the channel steel frame and are curved around the heads of each hub so that they are always set and require

Moline

P & O NO. 1 ALFALFA HARROWS

The object of an implement of this kind is for going over an alfalfa field after a crop has been cut in order to loosen up the hard baked soil. Experience has shown that the alfalfa will thrive better if the ground is pretty well stirred up after each cutting. As an ordinary disc harrow would simply cut up the field, the harrow with spiked teeth has been designed for this work, as it will tear up the ground and loosen it without damaging the alfalfa itself.

The use of this harrow in an alfalfa field keeps the ground fresh and alfalfa can be grown year after year on the same field without reseeding. This harrow might properly be called a soil renovator, as that is practically what it does.

FRAME. The frame is made of steel, braced to the tongue on both sides, and is very strong and rigid. The frame and hitch are low, to avoid neck draft.

SEAT. The seat can be set back and forth on the seat rail and the foot stirrups are adjustable up or down, to accommodate a tall or short driver.

GANGS. The tooth plates are made of two circular malleable iron plates, one of which is recessed to receive the teeth and the other is bolted up close, the two plates being held together by nine bolts. The heads of the teeth rest against the flanges on the inside of both the circular plates, and when the plates are bolted together the teeth cannot be pushed in or worked loose. Each one of

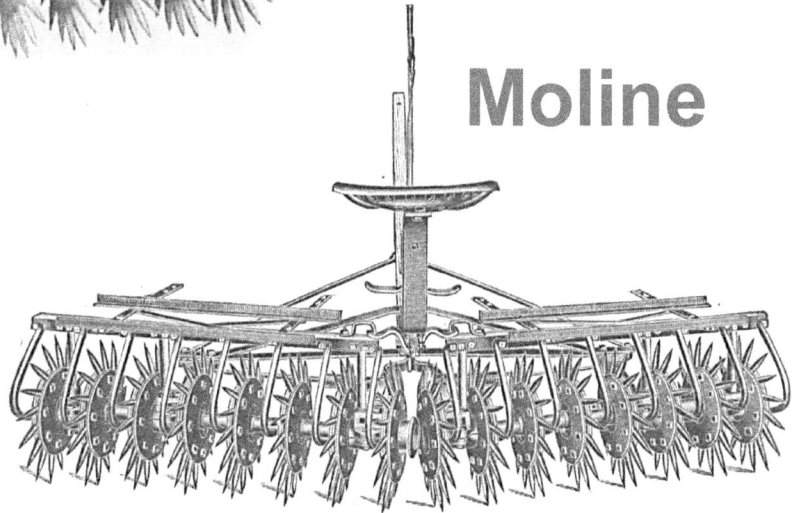

no trip levers. They are placed on the outside of each hub and as the gangs are invariably worked at an angle, the teeth are always kept clean.

BEARINGS. The bearings on this harrow are made of oil soaked hard maple.

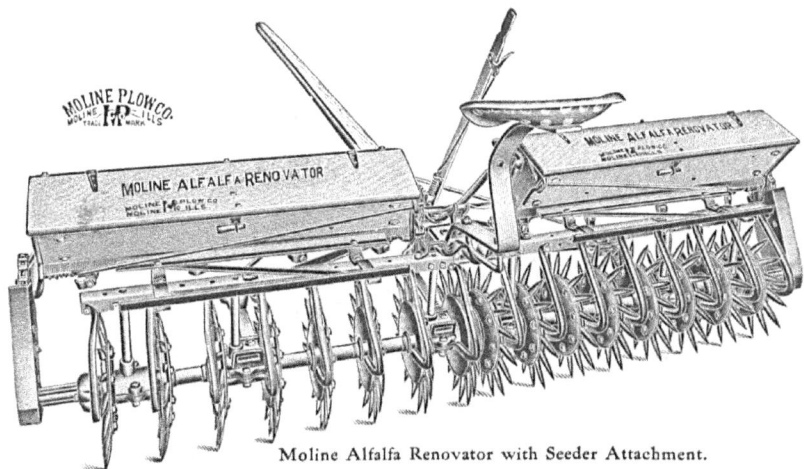

Moline Alfalfa Renovator with Seeder Attachment.

Rock Island

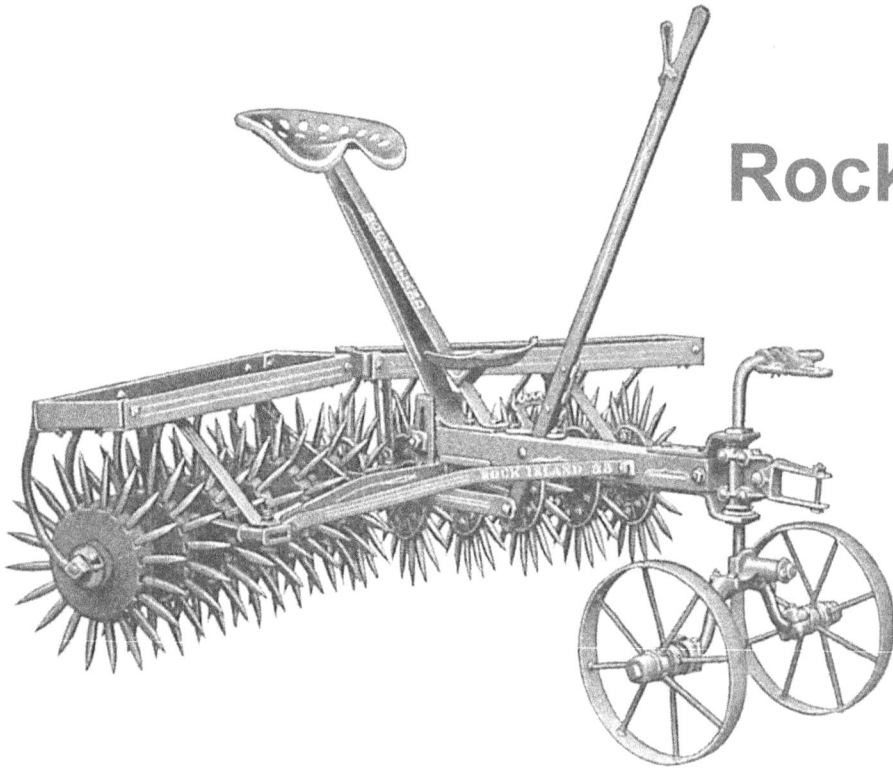

In Vol 21 No. 4 of the Small Farmer's Journal one horse cultivation practises in Scandinavia were looked at. The photo on the right originally appeared in that issue and demonstrates a small diameter rotary hoe behind a one horse forecart. This is a clear demonstration that applied appropriate technologies will expand the economy and productivity of the 'world's' farmers in affordable and exciting ways.

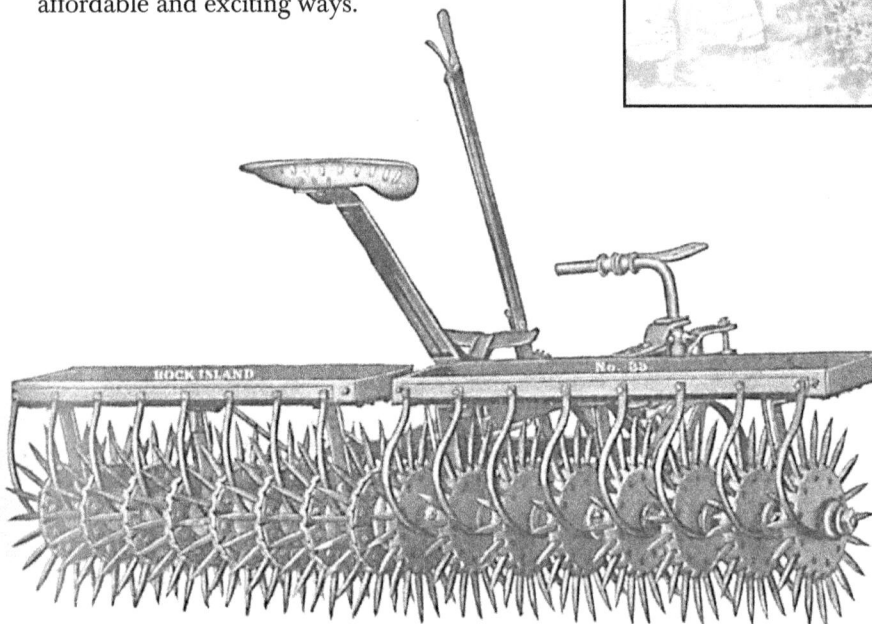

Moline Lay-By Lister Cultivator

Sled Lister

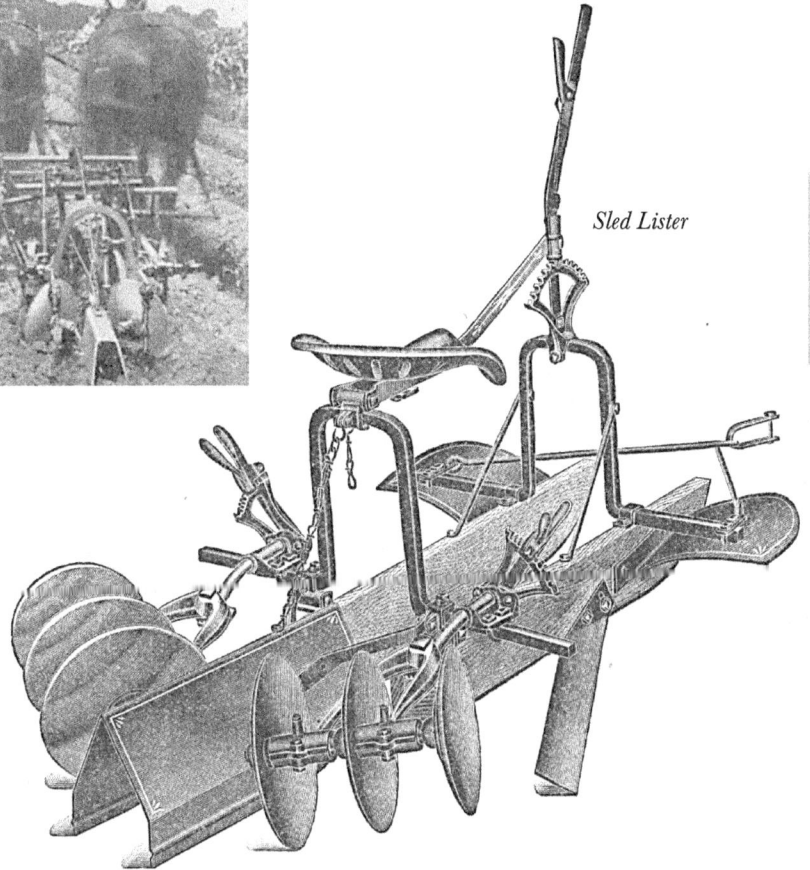

Chapter Eleven

Listers
Ridgers
Hillers

John Deere

John Deere Ridge Burster

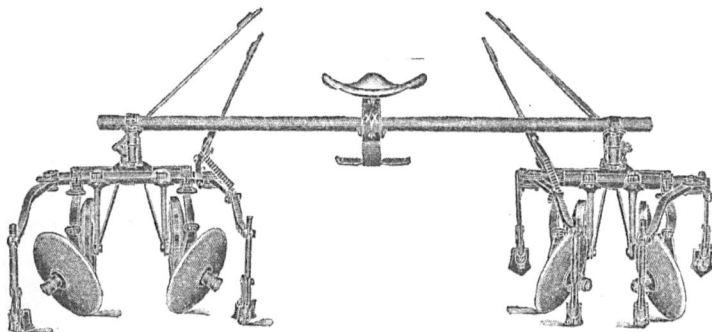

From the listers through all manner of the classic straddle row cultivators, old-time manufacturers had a hey-day when i t came to design variables. Fascinating solutions to lift, pressure, spread, shield, stir and coverage concerns resulted in some of the most beautiful and some of the strangest agricultural technologies devised before or since.

"Listing" is that action which lifts parallel ridges of soil which are meant to provide protection for the crop that is planted in the resulting trough. Seeds go down into the 'ditch' and soil

Avery Plainsman Lister

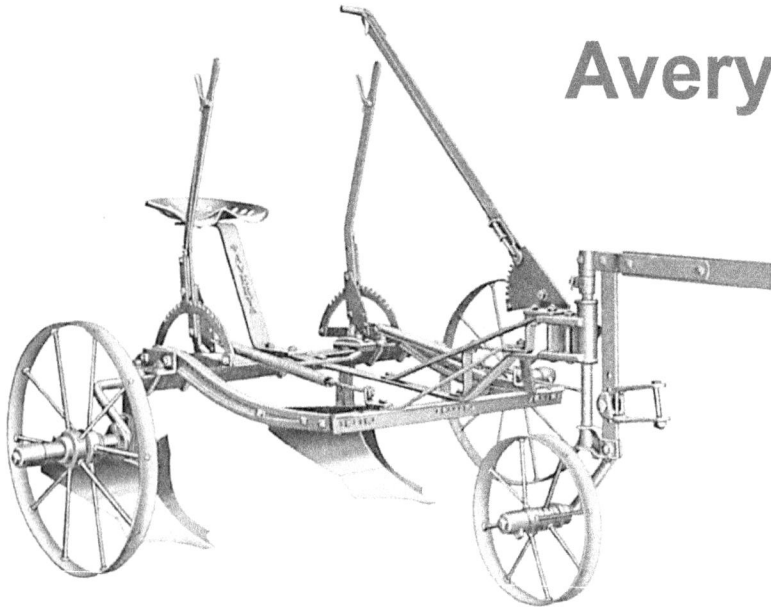

Avery

goes up into parallel ridges. The action is sometimes referred to as 'ridging.' And, in some rare instances, it is also referred to as 'hilling'. (This is confusing because a more accurate definition of hilling, in this context, would refer to throwing soil up against the planted crop row.)

In this chapter we have included only the straight ahead listers, deliberately leaving out the combination lister/planters which will be shown in an upcoming book.

Listing, especially in irrigated crops, can result in more or less hardened ridges all across the field. For this reason Ridge Bursters were developed which leveled out the field.

John Deere

Deere King Listed Cultivator

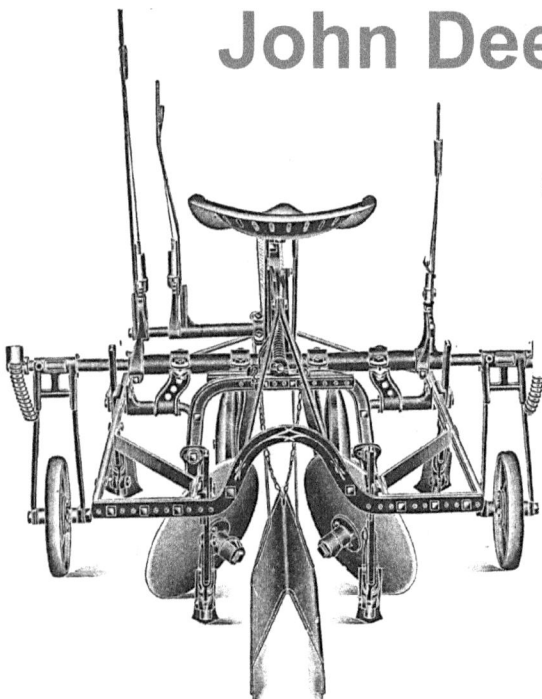

No. 116—Lister Cultivator Single Row

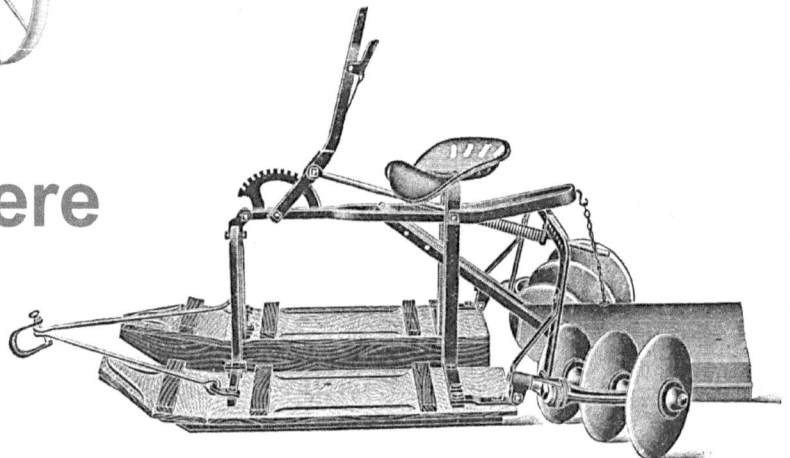

McCormick

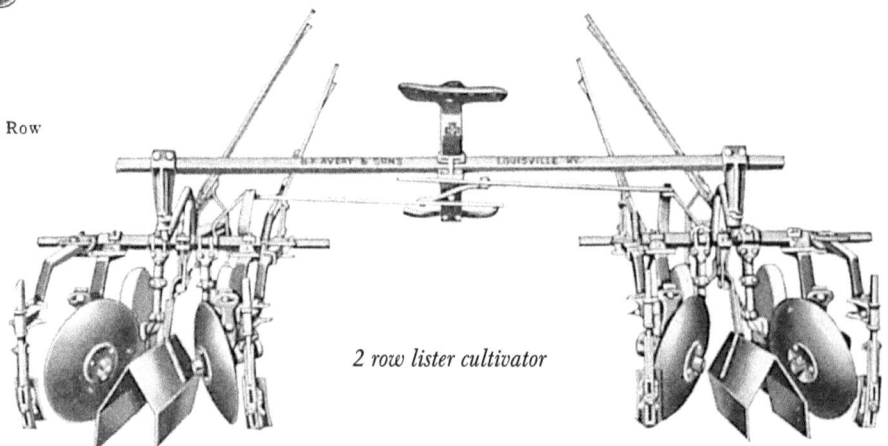

2 row lister cultivator

164

Nos. 315 & 316 prior to 1915

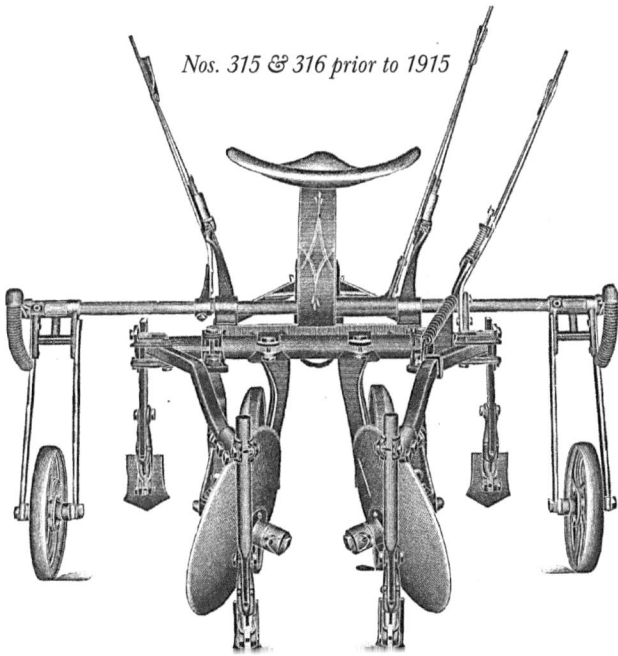

Oliver

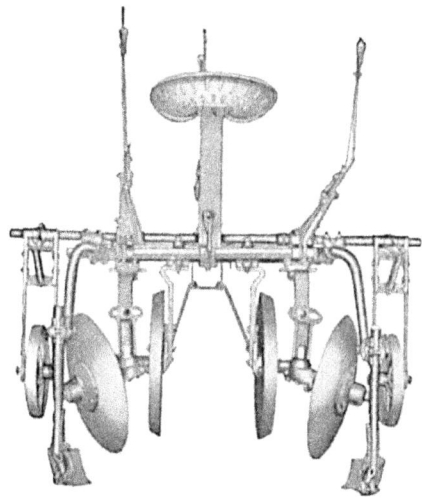

The 19-A one-row lister cultivator. The 19-B has spring trip shovels, otherwise it is the same.

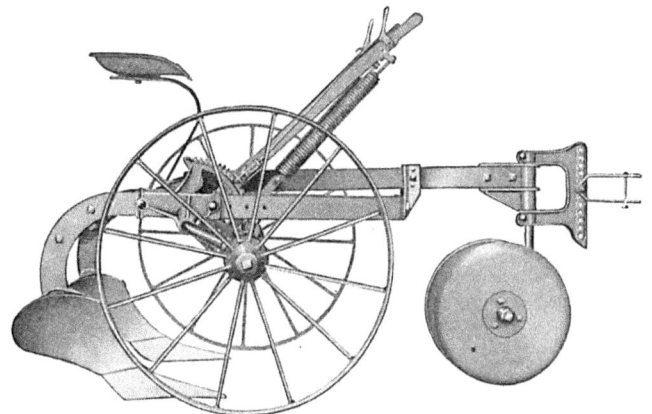

John Deere

The Oliver No. 38 Improved wheatland lister. Converted from the planting lister by the removal of the seeding attachment.

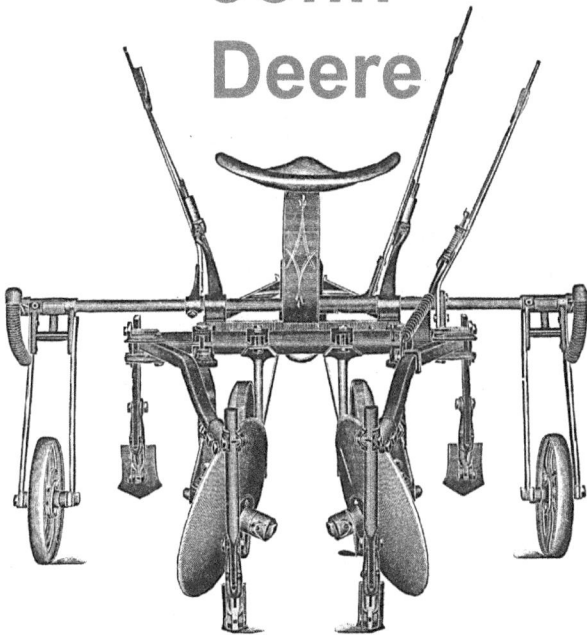

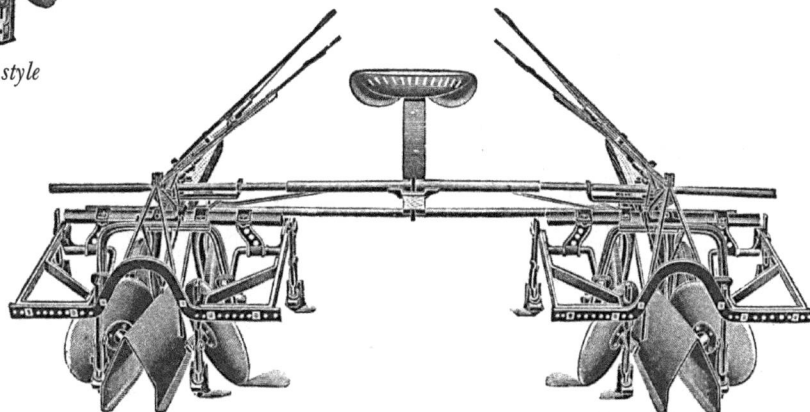

Nos. 315 & 316 1915 style

Above shows the subtle style changes these implements went through from year to year.

No. 118—Lister Cultivator Two Row

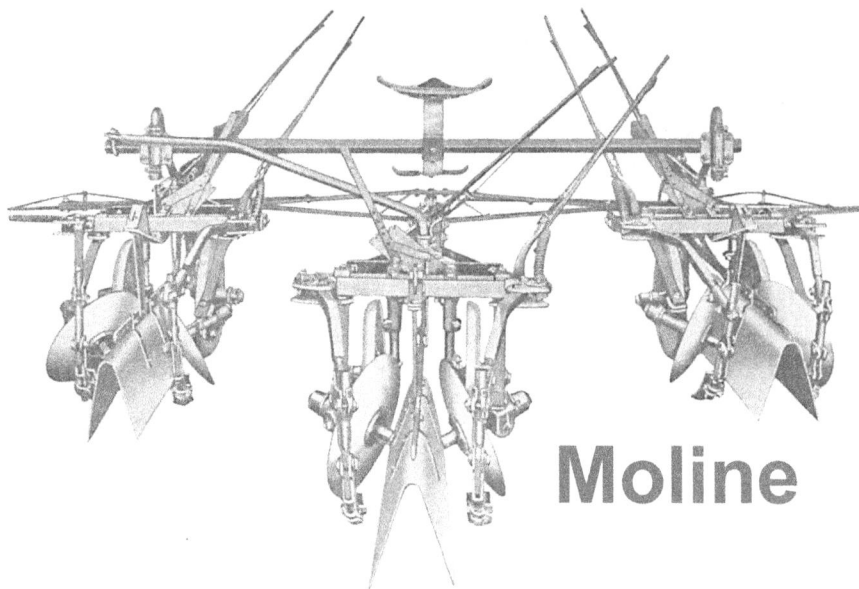

MOLINE THREE-ROW LISTER CULTIVATOR

the banks into the trench close to the corn. The discs are reversed and the frame and beams are spread by simply turning an adjusting rod on the frame.

The discs cut down part of the trench, and throw the moist soil up against the corn where it is needed. The shovels are set close to make a track for the wheel on the last cultivation. At this stage, the use of the ordinary lister cultivator is discontinued for there is no trench to guide the cultivator.

Laying By.
When laying by, the discs and shovels are set. The wheels follow the tracks made by the shovels on the previous cultivation. The discs cut down the trench and throw the soil up against the stalks, providing a supply of moist earth and affording protection against hot winds, while the shovels cultivate the center and leave the field in perfect condition.

From this it can be readily seen that the Lay By will cultivate corn in any stage, thus requiring only one cultivator--a decided advantage over the ordinary cultivator, the use of which must be discontinued after the first cultivation.

The Lay By Lister Cultivator was built to meet the demand for a two-row lister cultivator that would properly cultivate corn from the time it peeps out of the ground until it is laid out.

Construction.
The Lay By consists of two complete cultivators connected by a steel frame. The frame is made of steel pipes, thoroughly trussed. These are mounted on rollers to allow for any side motion of the two cultivators. The distance between the cultivators may be easily changed. An equalizing device also keeps the seat directly between the two cultivators, and the levers within convenient reach.

First Cultivation.
The wheels follow the trench, while the discs and shovels cultivate the bottom and sides of the trench. They are set close together. The first cultivation loosens up the soil and destroys weeds. The discs are set to throw the soil out, and the shields furnished prevent the small corn from being covered.

Second or Third Cultivation.
As the corn increases in size, it is necessary to keep the weeds down and throw some of the soil from

LAY BY LISTER CULTIVATORS

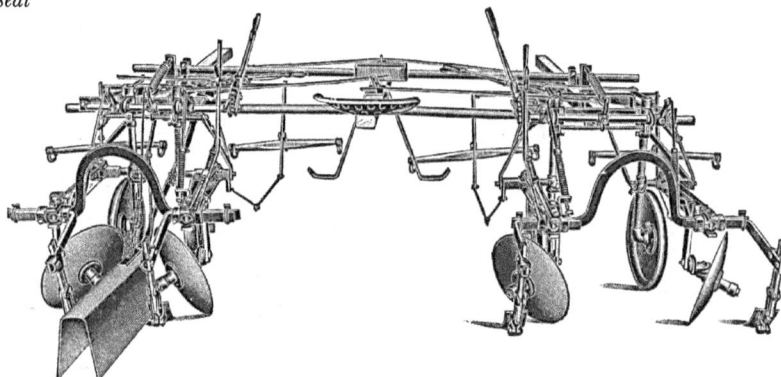

Three row lister cultivator

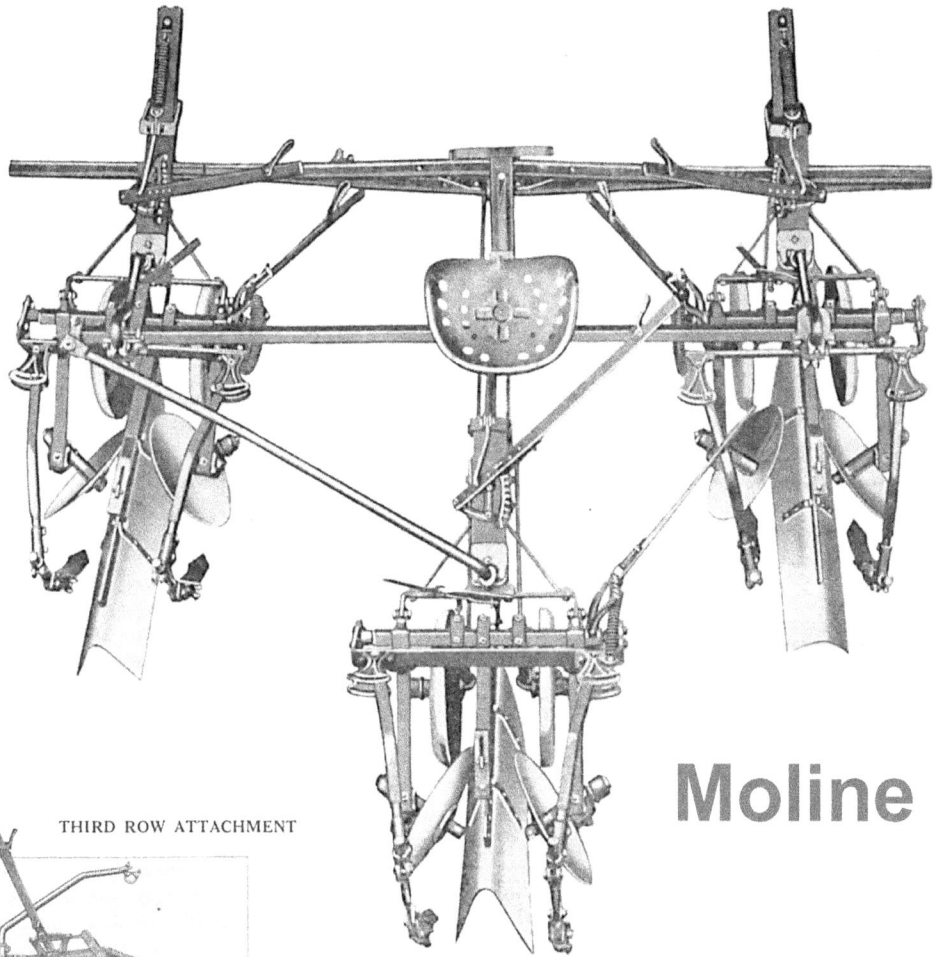

Moline

THIRD ROW ATTACHMENT

Moline Hooded Shield—Adjusted from Seat by
Lever—Held Horizontal and Parallel to
Row—Adjustable for Length.

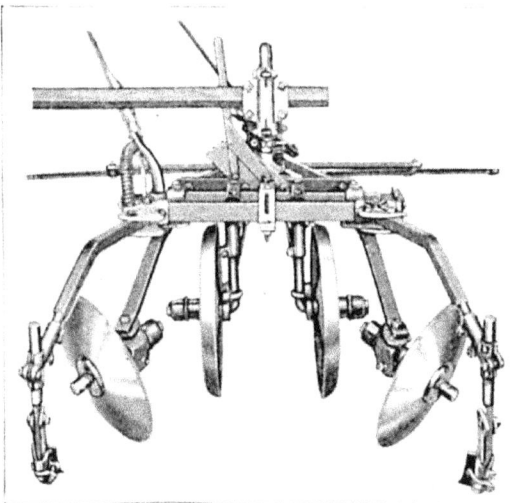

Shovels and Disc Set for Second Cultivation

SIDE VIEW OF OUTSIDE GANG
—OBSERVE EVENER ROLLER

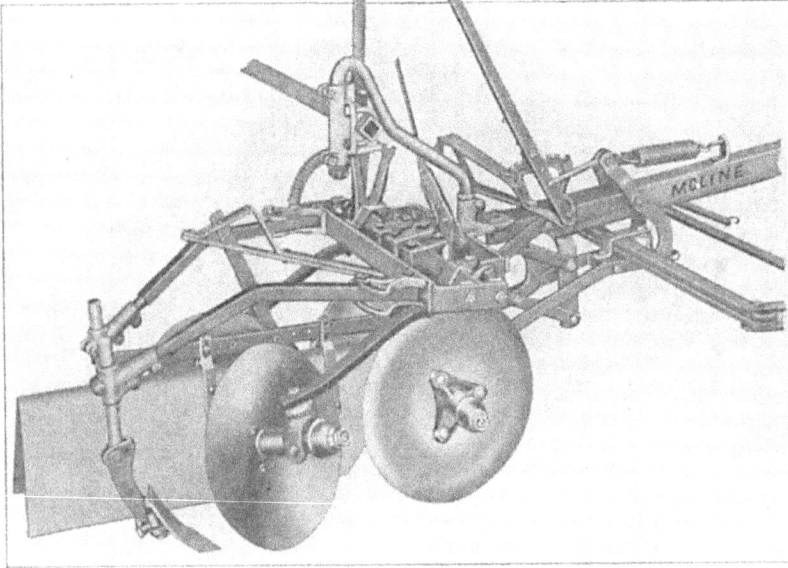

Moline

Bearing Detail - Observe long Maple bushing and Alemite Oiler

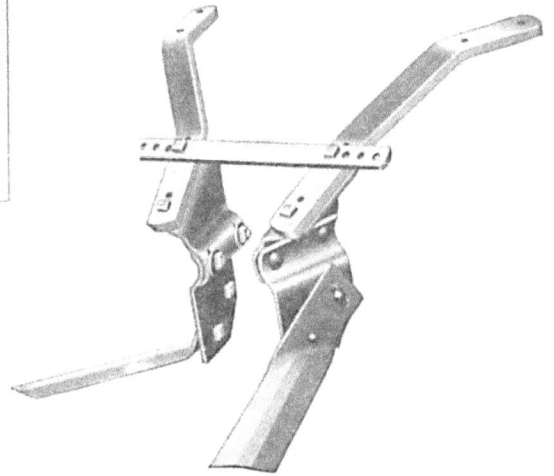

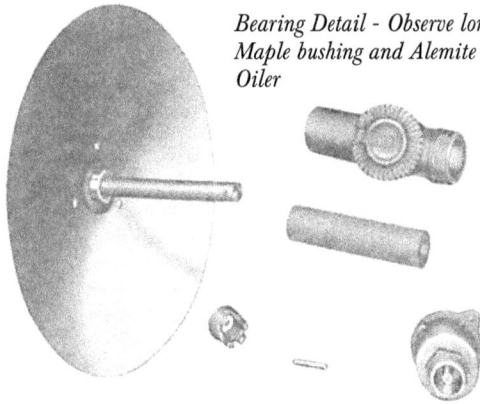

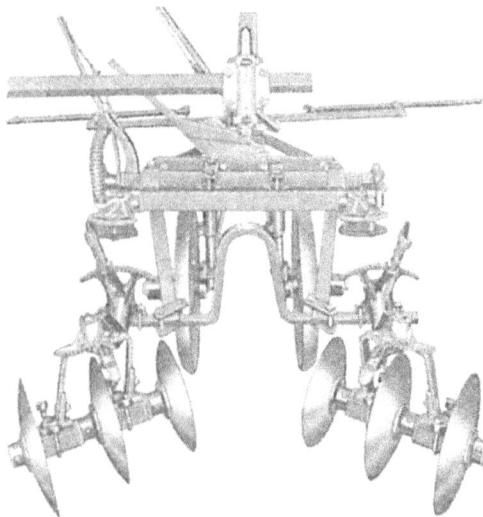

Moline Adjustable Weeder Knives.

Gang equipment with outside shovel attachment set for first cultivation.

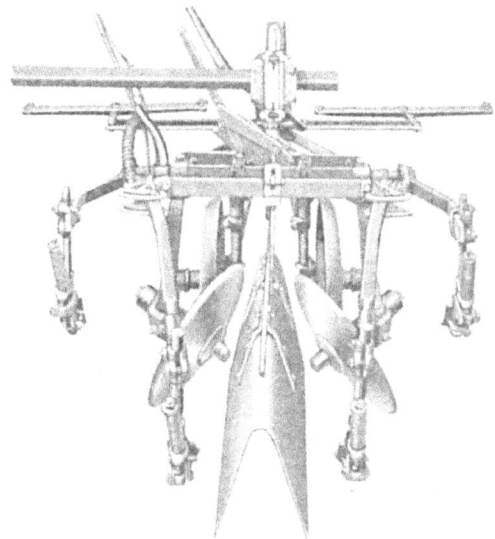

Six disc gang attachment - adjustable to or from row, for in or out-throw and for cultivation of ridges by tilting.

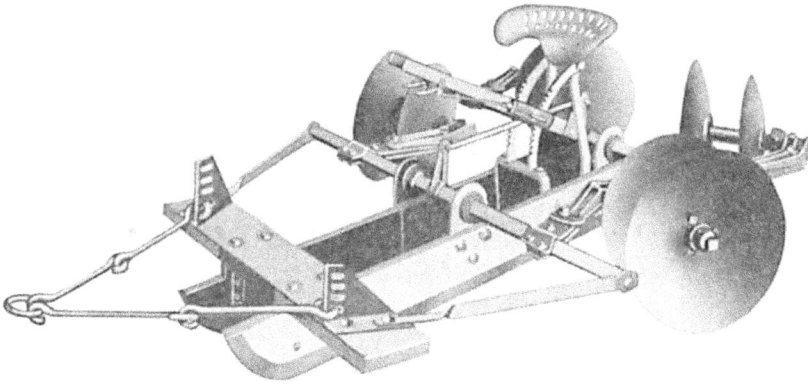

RB-2, Double Row Ridge Buster

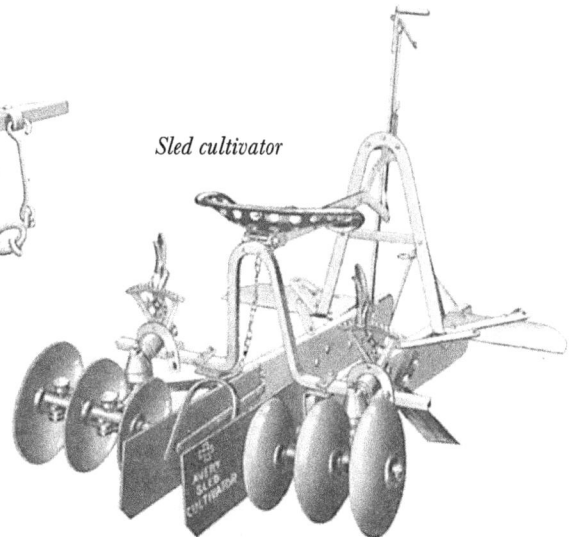

Moline

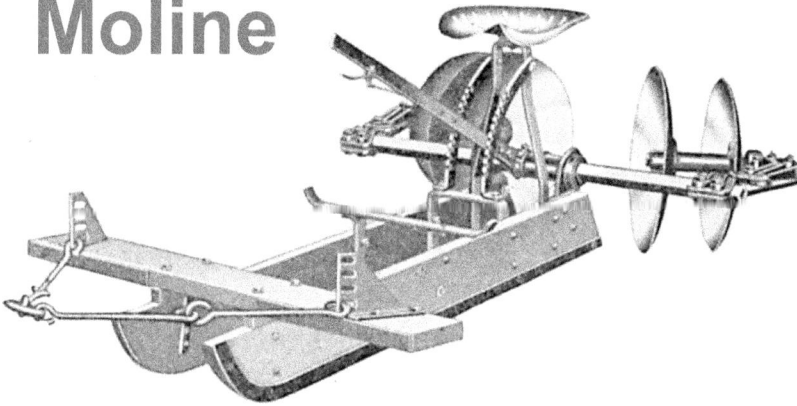

RB-1, Single Row Ridge Buster

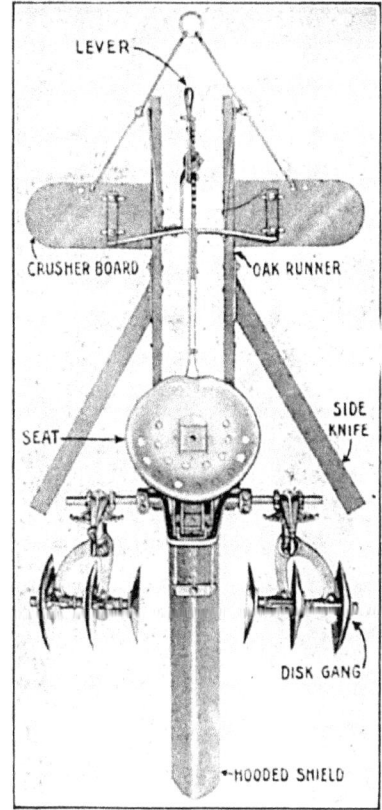

LEVER

CRUSHER BOARD · OAK RUNNER

SEAT · SIDE KNIFE

DISK GANG

HOODED SHIELD

Sled lister cultivator.

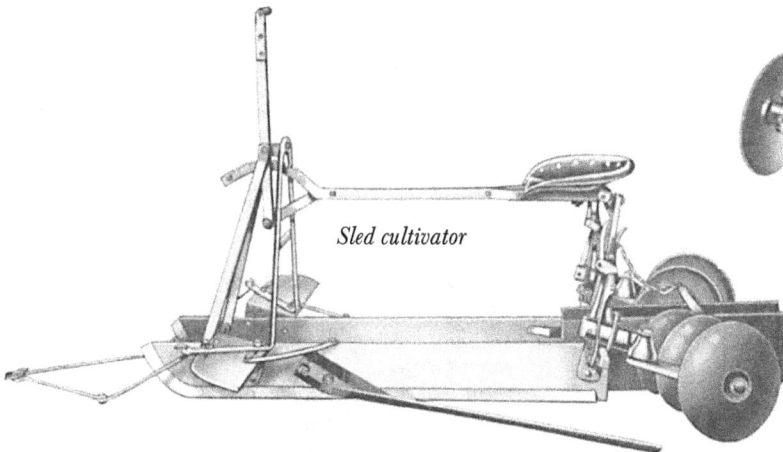

Double row Ridge Burster

Sled cultivator

Sled cultivator

McCormick

Levels Ridges and Prepares the Seed Bed

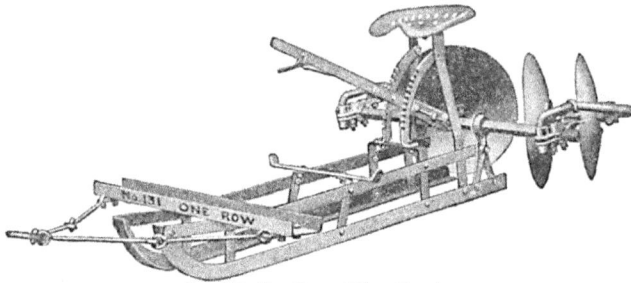

No. 131 One-Row Ridge Burster

Rock Island Ridge Burster

Rock Island

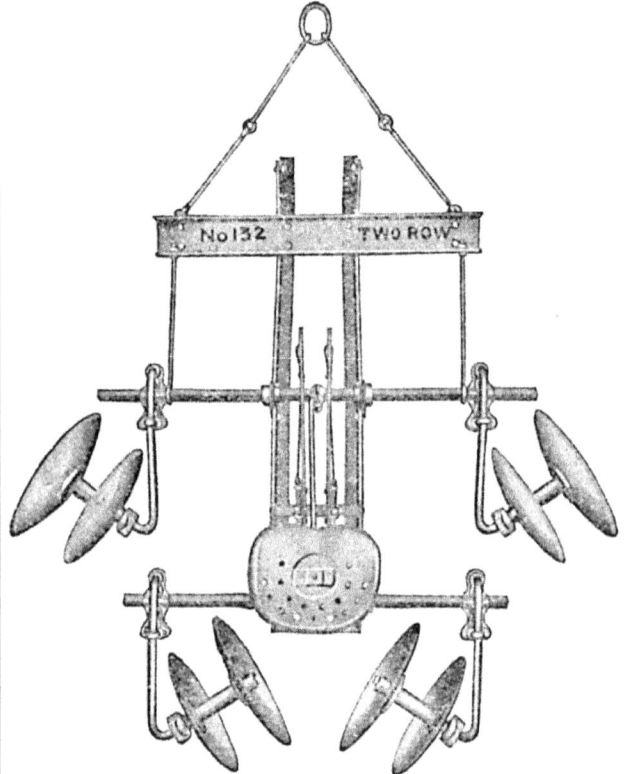

Top View No. 132 Two-Row

Moline

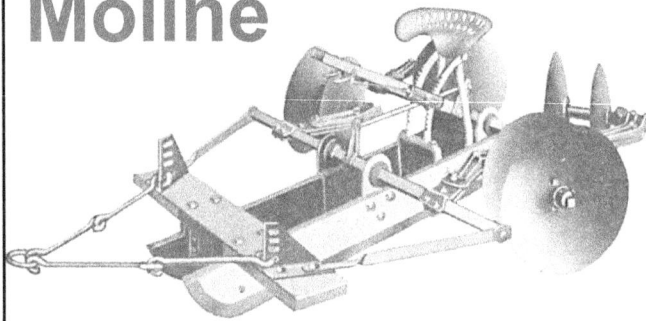

RB-2, Double Row Ridge Buster

RB-1, Single Row Ridge Buster

P & O

McCormick

No. 17 Sled Lister Cultivator.
This cultivator is a favorite in localities where the soil is light and the work is not of the severest character. It is a simple, substantial and useful tool, especially adapted for the first cultivaton of listed corn.

McCormick

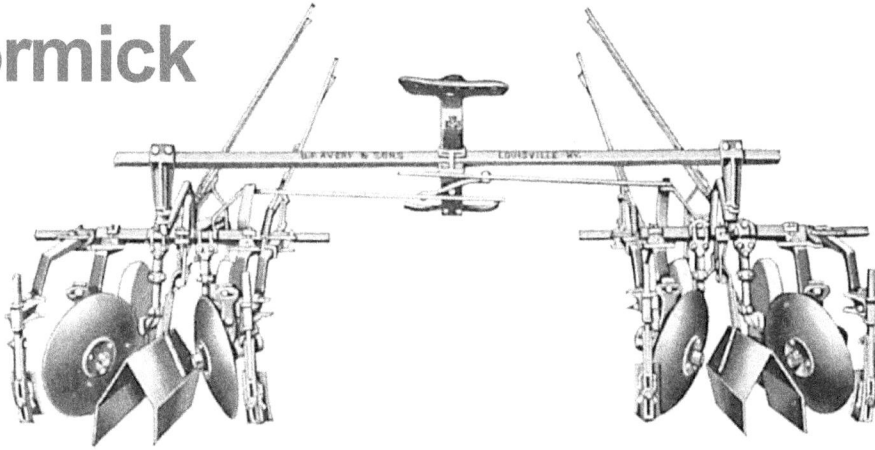

Canton Lister Cultivators

CANTON SINGLE ROW LISTER CULTIVATOR

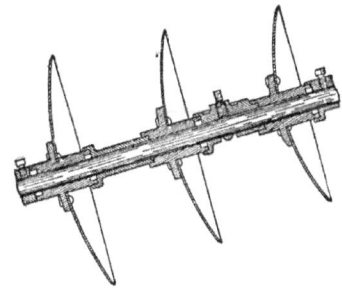

P.& O. CO. No. 761

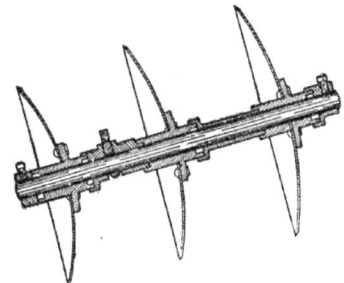

Detail showing construction of disc gangs and the manner in
which the spools are telescoped into the axle hub.

P & O

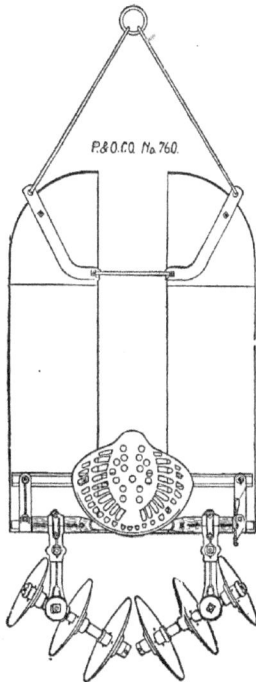

P.&O.CO. No.760.

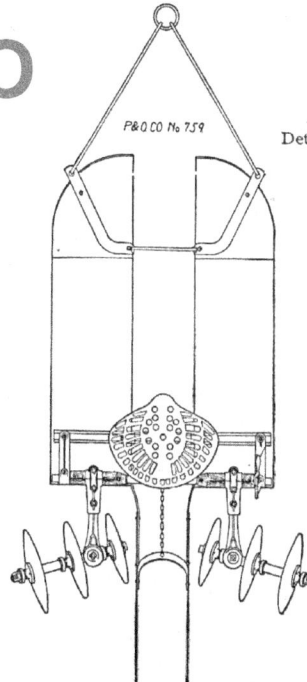

P&O CO No 759

Top view, showing the gangs used for first cultivation
in throwing the soil from the corn.

Top view, showing gangs reversed with shields in
place for throwing the soil to the corn.

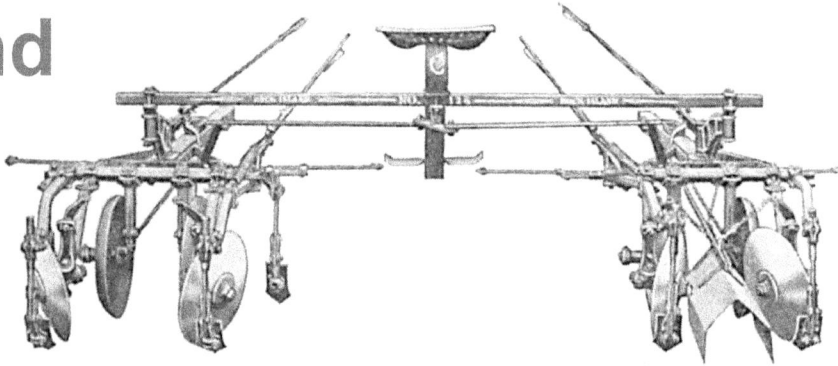

ROCK ISLAND NO. 126 TWO-ROW
LISTED CORN CULTIVATOR

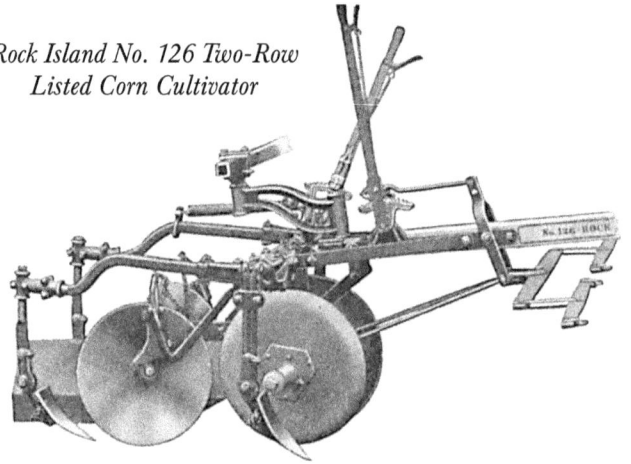

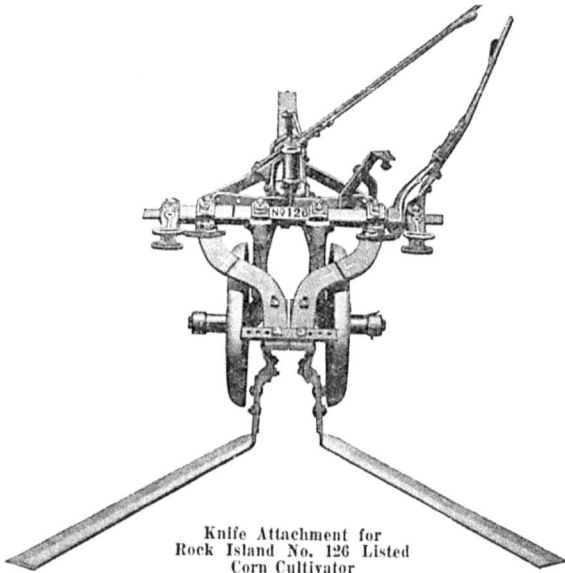

Rock
Island

*Rock Island No. 126 Two-Row
Listed Corn Cultivator*

**Knife Attachment for
Rock Island No. 126 Listed
Corn Cultivator**

**Runners
Protected Full
Length by
Sheet Steel
Plate**

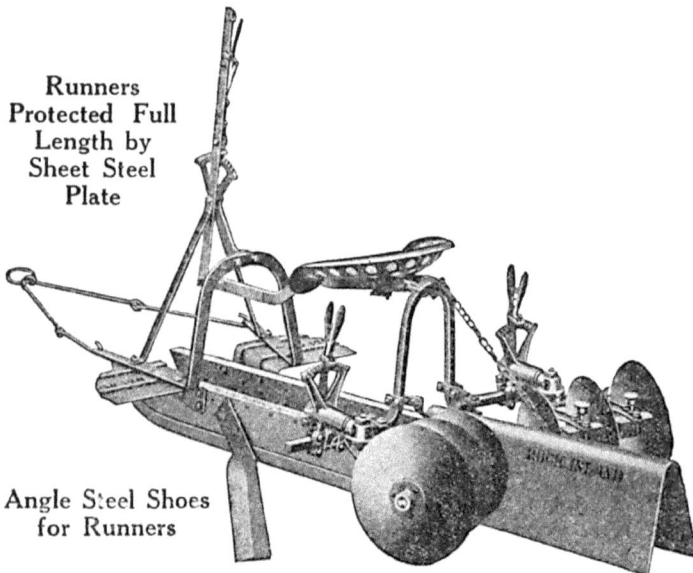

**Angle Steel Shoes
for Runners**

P
&
O

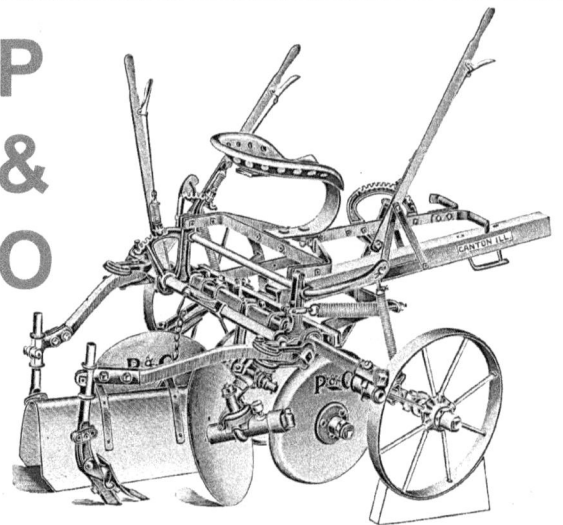

No. 16 Disc Lister Cultivator.
*The essential difference between the No. 16 and our sled
lister cultivators is that balance wheels are used instead of
runners, eliminating the friction of the runners and thereby
greatly lightening the draft. It is also built higher, placing
the driver far enough above the ground to avoid the dust
and in a better position to handle his team and watch the
field ahead.*

P & O

Moline Two-Row Lister Cultivator

Canton Two Row Lister Cultivators

Twelve Discs

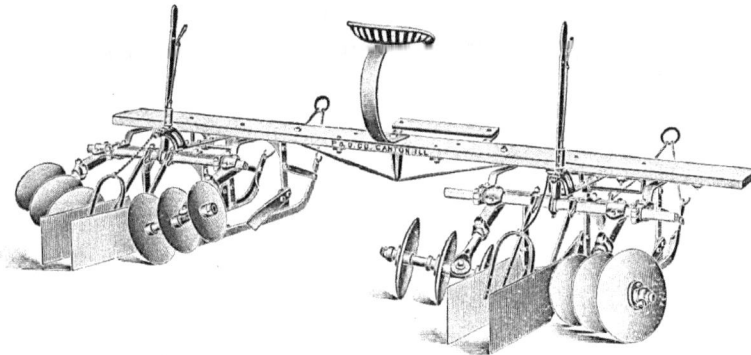

Four or Eight Discs

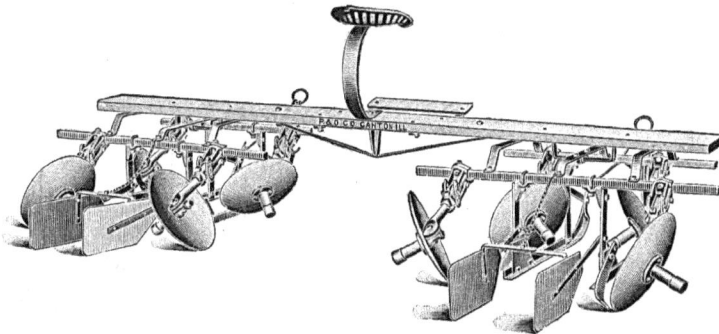

LISTER SHIELDS FOR P. & O. CULTIVATORS.

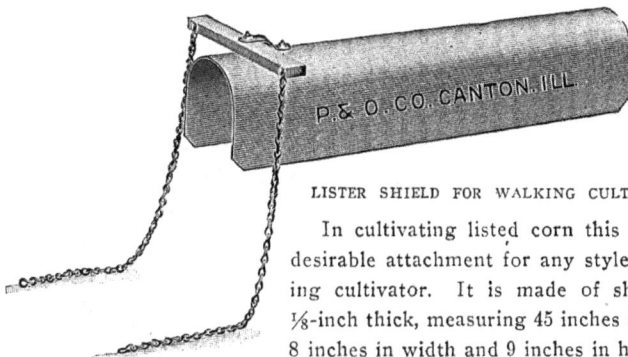

LISTER SHIELD FOR WALKING CULTIVATOR.

In cultivating listed corn this is a very desirable attachment for any style of walking cultivator. It is made of sheet iron, ⅛-inch thick, measuring 45 inches in length, 8 inches in width and 9 inches in height.

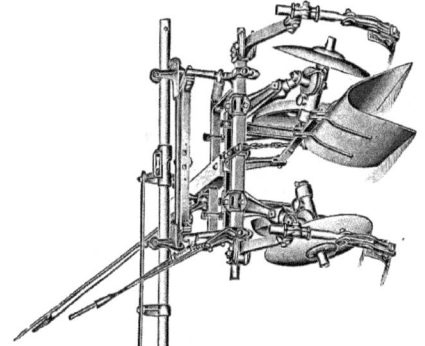

P. & O. TWO-ROW LISTER CULTIVATORS.
No. 27.

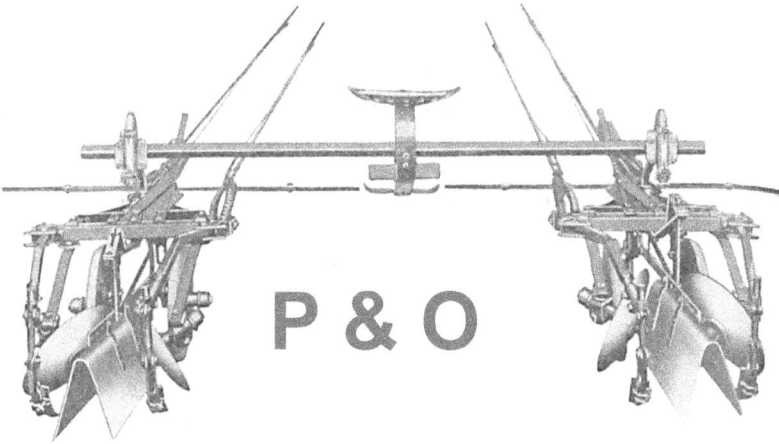

Set for Second Cultivation.

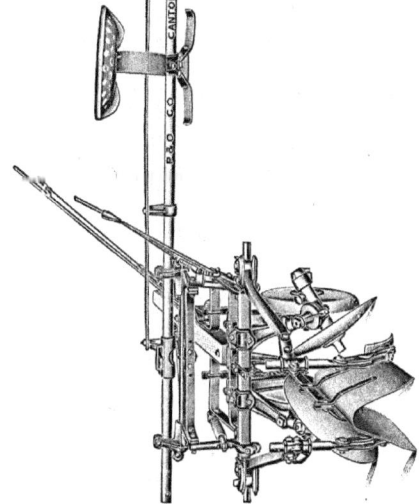

No. 27 TWO-ROW LISTER CULTIVATOR.

Set for First Cultivation.

Listers
Ridgers Hillers

Rock Island

Holds to the Furrow at All Times

For Cotton

For Corn

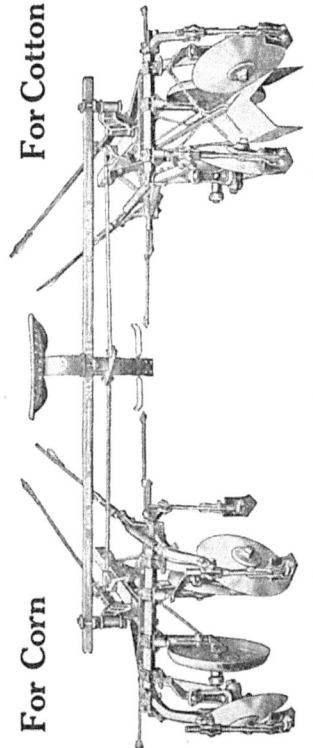

Rock Island No. 126 Two-Row Lister Cultivator

This photo by John Nordell shows the coulter slicing through the cover crop residue on the top of the ridge just ahead of the exta large sweep. An example of ingenious adaptation of the conventional riding cultivator in a bio-extensive* market garden.

Chapter Twelve
Cultivating

Good yields of all row crops are a source of real satisfaction as well as a source of profit. These objectives may be realized by having good soil -- good seed -- and good equipment. It helps to understand procedural timing (the advantage and disadvantage of WHEN things are done) and to continuously work at learning what value may be locked up in the soil.* Good yields may also rest on the assumption, of course, that plowing and other seed bed operations were proportionately well done. The cultivation work done can only be as good as the soil working tools attached to the cultivator, such as shovels, sweeps, etc. and then too, this equipment should be

* *No better literature exists on this very subject than the exhaustive articles by Eric and Anne Nordell published over several years in Small Farmer's Journal. Please see references at rear of this text.*

Nordell's McCormick cultivator rigged with extra large sweeps

matched as to size and shape, or assorted in accordance with the system required for the particular crop to be cultivated and the soil requirement.

Each of the many sizes and shapes of shovels and sweeps has an important part to play in the art of cultivation and will prove productive of good results if used in accordance with its purpose and design. Each size and shape is a masterpiece and has a fixed function to perform in the destruction of weeds and towards a satisfactory production from row crops.

The most carefully prepared and fertile seed bed cannot utilize its ability to produce a good crop, and may fail entirely, if the progress of cultivation is hampered or improper shovels and sweeps are used. There are good reasons for the various sizes and shapes and the quantities used.

The Nordells customized their cultivator for special cover cropping and ridge till applications.

They are styled and sized to fit soils and crops. Cultivating equipment, matched or assorted, that successfully cultivates one crop may be damaging to another crop. The soils, seasons, and crops need proper systems of cultivation and the equipment used should include the proper shovels, sweeps, spring teeth, discs, spike-teeth and rotary hoe attachments. (Don't mistake this to mean there is an only way.)

Equipment that is right for the first cultivation may be wrong when the crop is up strong. In some crops the equipment required will go through the cultivating season without any change. Then again, there are crops and soils and weather conditions that require a change each time through the field. In some hybrid corn territories the rotary hoe has been known to start and finish the job of cultivation--no other equipment used.

Alfalfa is a crop that requires cultivation, no special implement is needed. However, correct soil tools are necessary. They are alfalfa double-pointed spring teeth used on the field cultivator, and regular alfalfa spring teeth used on the spring-tooth harrow. They both do excellent work.

Soils that crust over and interfere with the growth of small-grain crops should be spike-harrowed or rotary-hoed. It is just as desirable, and profitable too, when a top crust prevails, to break it up and give the young plants a chance to thrive and grow as it is to break up this crust in row crops. Spike harrow attachments are also used for the cultivation of tobacco. This attachment is usually a locally made unit, because of the limited demand which comes from just a few of the tobacco growing sections.

It was generally conceded that there are two reasons for cultivating crops, and that takes in all crops that are seeded and grown. First--to destroy weeds, that is, control weed growth. Second--to conserve moisture. And modern creative horsefarmers have discovered a third purpose - to enhance fertility. Over a long period of time modern experiments have concluded that the major purpose of cultivation is to kill weeds. It has been concluded, too, that it is of no real benefit to corn to loosen the soil. The seed bed for corn roots must be kept compact and firm. A deep mulch may have some value until the ground is shaded by the corn, but very little or nothing at all is gained by cultivating after each rain in a normal season.

It takes skill to select the proper shovel and sweep equipment because of the many sizes and styles, and the many crops for which they are designed. The important settings are: Enough tilt or "suck" to enter the ground quickly, and to lift the soil and lay it to each side; to set them to pitch the soil; and to turn them to throw the dirt toward or away from the plants.

The one thing that assures high corn yields is to prevent weed growth. This has been generally proved by experiments conducted in all corn belt states. The real purpose, then, is to set the earth tools for killing weeds, and the depth to set them is entirely determined by the setting that will destroy the most weeds while protecting roots.

The first cultivation is the most important because at that time it is necessary to pulverize and pitch the dirt well in the corn row so as to cover grass and small weeds peeking through the ground. Those between the rows will be cut out and covered--and some exposed to the sun for destruction. But, the fine grass and small weeds in the row can be killed and growth retarded only by a good coverage of fine soil.

The straight shovel, on the middle bracket of the Nordell cultivator, proved adequate for mixing compost into the furrow which the disc hillers covered and levelled with the dry soil. A very small glimpse of how the creative horsefarmer, with these magnificently adequate older implements, is able to add 'fertilization' to the many advantges of the technology.

Unless this first job is done well, it probably never will be, for these small weeds and grass grow fast and are too tall and too well rooted to kill or cover the next time through the field.

The skill of selecting proper shovels or sweeps is determined by the results of cultivation. Weeds must not take on a second growth after cultivation. However, they will do so if they were not thoroughly dislodged or completely covered. The soil has to be turned over or shaken up enough to dislodge and expose the roots. Seldom will they take on a second growth if this is done.

For cultivation purposes the walking speed of horses and mules, between 2 and 3 mph, is excellent with fast walkers sometimes preferred. Sweeps and surface blades can be traveled much faster than ordinary shovels. Weeds will start a second growth if lifted too gently and the soil layed down with little or no disturbance. Therefore, speed should match the sweep and shovel equipment for a good job of weed killing. Good equipment and a thorough knowledge of the soil tools are very essential, for we all know that it is just as easy to destroy a crop by cultivating as it is to make a crop. The method and degree of cultivation affect the quality and yield.

Fitting the soil for the growing of a crop means to have a well pulverized and compacted seed bed for holding moisture and permitting free growth of the root system. Fitting the seed to the soil best adapted to its growth is an entirely different subject. When more thought is given to soils and seasons and the ability of the various soils to grow crops--then and only then will it be possible to control growths and production. The

best cultivator, or shall we say a perfect cultivator, and all the shovels in the world cannot produce a crop if the seed is misplaced. That is, planted where it hasn't a chance to grow and produce.

There are three systems of planting--namely, on the hills or beds, in flat rows, and in trenches. These three and soil varieties are the factors that definitely determine the cultivating tools. While it would be entirely possible to pull the ridges to the center in listed crops with most any kind of earth tools, the results are not as good as when the definitely fixed tools are used.

Many personal ideas enter into the picture of cultivating. Farmers in adjoining fields, with soil and crops the same, will often use different tools. They are determined by personal experiences and each farmer is firmly convinced that his selection is correct. If production is about equal with each one, it is fair to say that the selection of tools, in spite of the varied opinions, was correct in either case. This proves conclusively that the destruction of weeds and a loosened surface (also called soil mulch) are the two essentials and it matters little, it would seem, which tools are used to accomplish these essentials if they fit in between the rows, loosen the soil, and do not prune the roots.

Because of the wide diversity of practices and soils, it is almost impossible to formulate a standard by recommending fixed styles. To illustrate: In the same crop where soils vary, the soil determines the equipment. In crops where the root systems vary, the root system determines the equipment. Row spacings also determine the tools to use. Then again there is the requirement of meeting both soils and root growths.

Pruning biomass!? To understand what Eric and Anne Nordell were up to with this procedure you simply must read their excellent article, entitled Ridge-Till Vegetables, in the Winter 1999 Small Farmer's Journal.

For instance: the root growth determines the tools to use but these particular styles are not suitable to work the soil, or the tools are adaptable to the soil but are not correct for the root system. This problem is answered by the individual in his or her specific case. The root system or root growth is usually favored. Shallow cultivation is always preferable. Deep cultivation is particularly bad in dry periods and should only be done to get deep rooted weeds that will eventually get the crop.

Corn, beans, grain, peas, sugar beets, sorghums, hops and melons demand thorough surface cultivation with tools such as shovels, sweeps, discs and knives, and seldom, if at all, are furrowing tools used. In fact, these tools are regular for all flat or surface planted crops. Cotton is planted in all three systems. Corn, potatoes and some vegetable crops are planted in two of the three systems--in flat rows and on the hills or beds. Lister bottoms, large discs or plow sweeps are used to form the beds.

Where there is plenty of rainfall, or where the land is irrigated, the system of planting on the beds is the general practice. Only in the semi-arid sections is planting seed in the trenches a regular procedure. The purpose of trench planting is to put the seed down in the moist ground for quick germination and growth, and so that the deeply planted crop may better withstand the hot and dry weather. As each cultivation progresses, by pulling the beds or ridges into the trenches until the surface is level again, the root system is lowered considerably below the root system from seed planted in flat prepared seed beds. When the crop is laid-by the soil is generally hilled or pitched to the

rows leaving the field again in a more or less ridged condition. Very little change takes place in bed planting. The beds remain from first to last cultivating. The middles or trenches are kept clean by using sweeps, blades or hillers.

It is generally agreed that level cultivation is more desirable than hilling to the crop as hilling does not increase the yield. Hilling is tradition, a common practice, and has always been the custom, therefore, continues to be consistently practiced. Hilling requires much dirt to be thrown on rows to cover the weeds. Too much dirt thrown to the crop may expose the plant roots between the rows, especially in flat planted corn, and if cultivation is very deep, damage to the crop will result. Some weeds may require extra hilling to cover, but it should be done with care and caution. Potatoes grown on heavier soils require some hilling.

Also, crops planted on beds, whether irrigated or not, necessarily require dirt thrown to the plants, but these crops do not have the roots extend across from row to row. In many hilled crops the middles, that is, the centers between the rows, are cultivated for re-hilling and can be made deep as there is no root interference and no danger of exposing the roots. The middles (furrows) are watered in irrigated areas and must be furrowed from time to time to deepen the channels to receive the water. This furrowing pitches the dirt to the rows and up on the ridges.

HILLING, CULTIVATION ON BEDS AND FLATLAND

Blanching is necessary to produce celery or white asparagus. This is the bleaching or whitening process. Special hilling tools are generally used for this work. Potatoes in irrigated sections are planted on the hills and as water is turned into the trenches, the soil washes from the hills and settles in the trenches demanding small lister bottoms, discs or hillers to go through and furrow them out. Shovels, sweeps and spring teeth are used in flat-planted potatoes. These are followed by special hillers or blades to pitch the earth to the rows. This hilling procedure to form bedded rows offers better exposure of potatoes after digging and assures a cleaner job of gathering.

It is desirable to ridge or hill the rows in many localities because of the drainage this method provides. The hilling helps to warm the land during cool season. Hilling should not be too high or overdone, especially in the sections where moisture is lacking in mid-season. It leaves loose friable soil for the potatoes to form in, but outside of this advantage the extra high hilling may be of doubtful value.

CULTIVATING IRRIGATED POTATOES

Weeds that spring up soon after the first cultivating should be killed by the use of spring teeth. The centers should then be furrowed out throwing the dirt to the beds. Irrigated potatoes are cultivated to control weed growth, keeping the beds built up, and keeping the soil loose. To conserve moisture by cultivating may or may not result. However, if the soil is the type that crusts badly, then moisture conservation is a result that can also be considered.

Weeds do more to take moisture away from productive plants than any other influence, and if soil is mellow and does not crust over it is doubtful whether cultivation would be responsible for holding any great part of it from getting away. It may be more apt to aid in its escape if shovels are used. Cultivation should then only take place if weeds have come on and need to be destroyed. If there are no weeds and the surface is loose, not crusted, it would probably be better not to cultivate in order to save moisture. This holds true in any row crop, bedded or in flat rows. In irrigated potatoes the cultivator follows irrigation as soon as it is dry enough to work the soil. This holds true in any irrigated crop.

While the first cultivation of bedded potatoes is usually deep, each subsequent time through the field the depth is shallowed and the distance away from the plants is increased so as to avoid root damage. If beds are wide it is also possible to early cultivate on the row to kill weeds. It would be a hazard to do this later on and must, therefore, be done when plants are small.

Spring teeth, hoe blades, and shovels are used in the various cultivating operations. Wide beds give potatoes plenty of room and as dirt is thrown to the beds there will be much less exposure of potatoes in the hills as they push themselves through the surface. If uncovered or left exposed they are bound to sunburn.

Potatoes and some other irrigated crops may be flat-planted with only a shallow trench in the centers. It is really a level planted field, but later on as irrigation ditches are formed for flooding the centers, and as each cultivation progresses, the flat-planted rows become hills which get higher and gradually widen, leaving deep trenches in the center. In some localities the hills are formed first and then planting is done on the hill.

Single Row Asparagus Cultivation

Large Disc Hillers and Spring Teeth for crowning and cultivating after discing tops into the soil.

Front equipment for general cultivation—six to eleven Spring Teeth on each side.

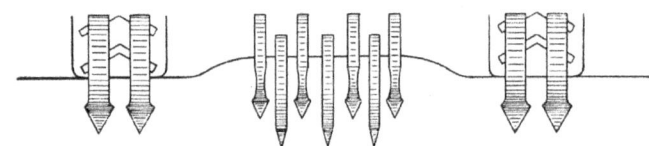

Rear equipment — Two Regular Spring Teeth back of each tractor wheel — Four Alfalfa Spring Teeth in center to work on the row -- Three Spike Spring Teeth in between Center Teeth for harrowing.

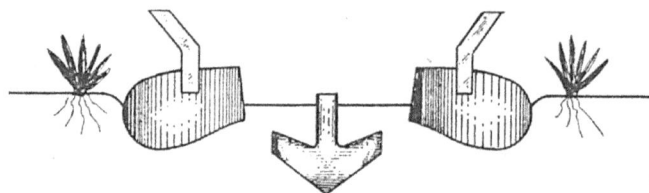

Hoeing beans, beets — sweeping center, rear view.

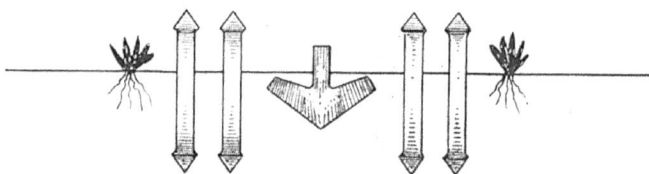

For deep cultivation — vegetables, front view.

Beet and Bean Cultivation

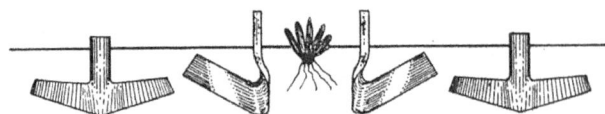

Hoes and Sweeps

CULTIVATING TOMATOES

It is recommended, when tomato plants are set in the field, that they should be set deeply and ridged as soon as they have begun to grow. The soil should be worked up to the lower exposed stems. Ridging facilitates drainage, cuts down erosion, and protects the plant by increasing root development higher on the stems.

This question comes up frequently just before tomato planting time. "Will it pay to stake tomato plants, or is it just as well to let them go flat on the ground?" The general opinion seems to be that more tomatoes are produced when they are not staked, but they are earlier and nicer when staked up off the ground.

If tomatoes are grown and not staked, a mulch of straw or grass under the plants will keep the fruit clean. There will be less danger of rotting and the mulch will also tend to conserve the moisture in the soil.

If it is true that more are produced by not staking it would seem that the rotting problem is solved by the mulching process, and also to let them go flat would give the production increase.

A tomato field should be kept clean of all weed growth. Tomatoes are usually picked from the plant about 8 to 10 weeks after the last cultivation. If weeds are allowed to grow during this period they may become a serious factor, not only as a nuisance to pickers, but from the standpoint of available plant food and moisture. Avoid deep cultivation after the roots begin to move out to the middle of the row. To derive the most benefit, follow good horticultural practices in handling and working this crop. This is very important.

Level cultivation is always more desirable. Hilling to flat-planted crops does not increase the yield. In some localities it is a common practice because it seems to have always been the custom. Permitting the weeds to grow and get a head start before putting the cultivator in the field has encouraged the practice of hilling to throw more dirt to cover the weeds on the rows. In some localities crops are hilled--in others they are planted on the hills or on the beds, which is done for irrigating, drainage, or perhaps blanching. Each system has it place.

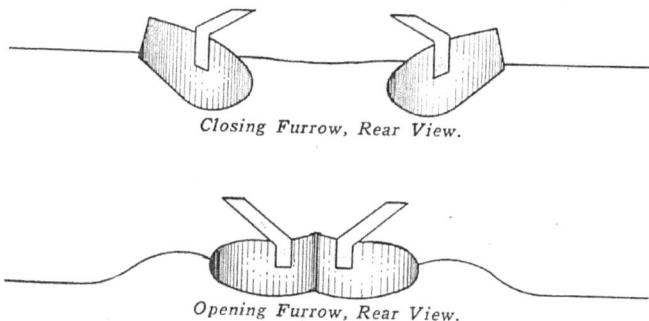

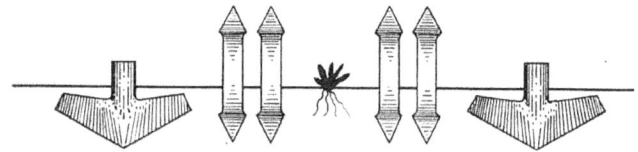

Group 2—Flat Cultivation

First, second and third cultivation of beans, sweetcorn, potatoes, celery, peas, peppers, rutabaga, etc. Shovels can be substituted for sweeps where soil tends to pack hard. Sweeps are best for light soil.

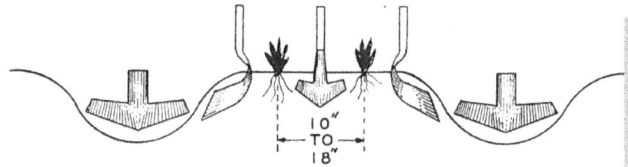

Group 3—Flat or Bed Cultivation

For cultivating carrots, lettuce, onions, parsley, spinach, turnips, table beets, and sugar beets. Most styles of hoes are suitable. For first, second and sometimes third cultivations, followed by double pointed shovels, teeth or sweeps for final cultivation. Beds are 36 to 42 inches apart.

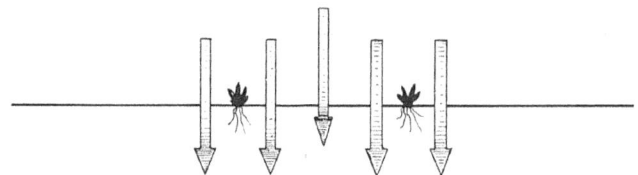

Group 4—Flat Cultivation Chisels

For first and second cultivation used in pairs close to the row. For deep cultivation when crops are small. Used in later cultivations to break up hard middles. The numbers used, one to five, depend on row spacings. Used to good advantage back of tractor wheels to break hard ground.

Potato Hilling Blades and Shovels

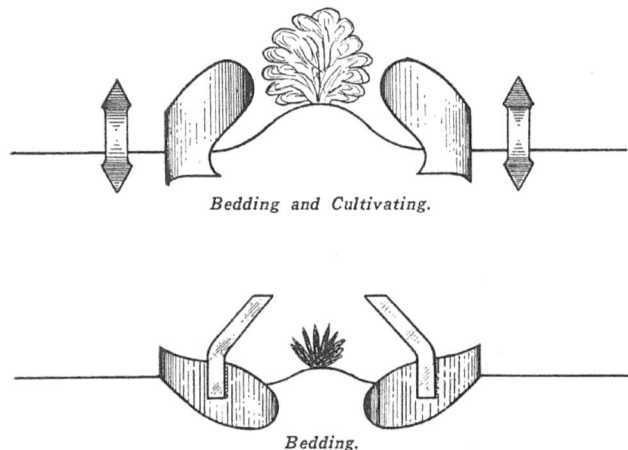

Bedding and Cultivating.

Bedding.

Vegetables

Closing Furrow, Rear View.

Opening Furrow, Rear View.

Soy Bean Steels

Double-Pointed Shovels.

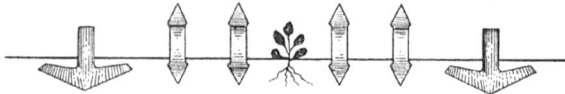
Double-Pointed Shovels and Sweeps.

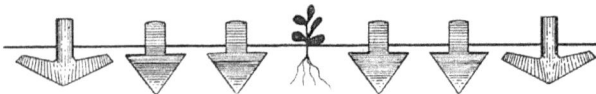
Spear Point Shovels and Sweeps—All Spear Points or All Sweeps

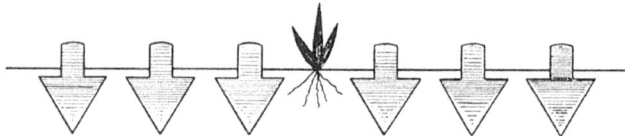
For first cultivation use Six Spear Point Shovels.

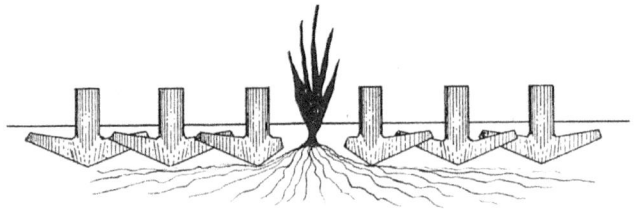
For second cultivation use Half Sweeps and 5-inch Spear Point Shovels.

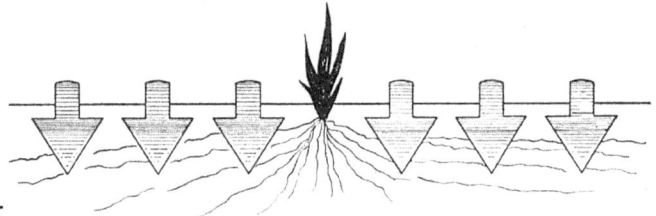
Weeding between rows.

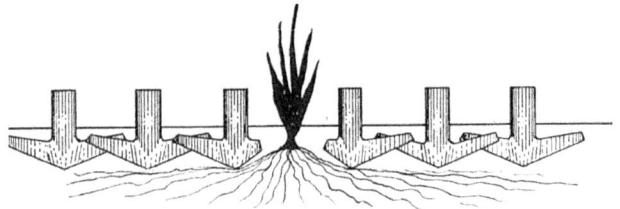
Hoeing single row, rear.

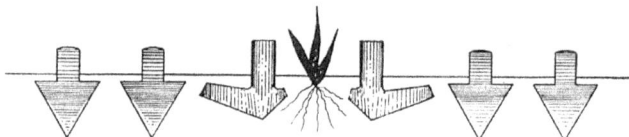
Cultivating nursery stock, front view.

For third and following cultivations use Half and Full Sweeps for shallow work. Notice lapping for full width cultivation. No blank strips.

Pruning the roots is a hazard.

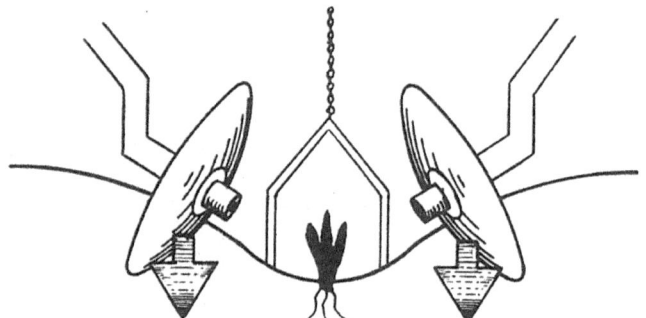
For third and following cultivations use Half and Full Sweeps for shallow work. Notice lapping for full width cultivation. No blank strips.

Listed Crop Cultivation

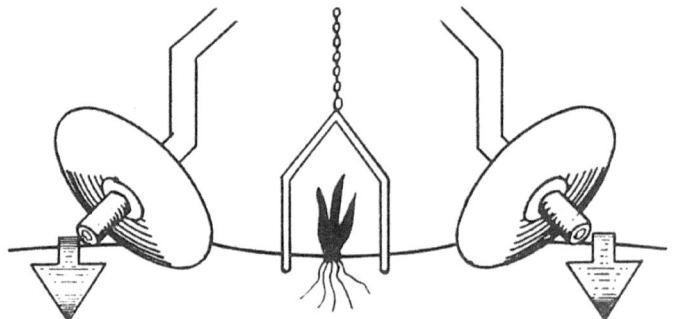
Throwing Out.

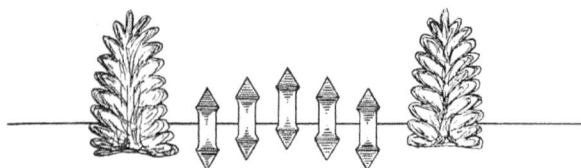
Throwing In.

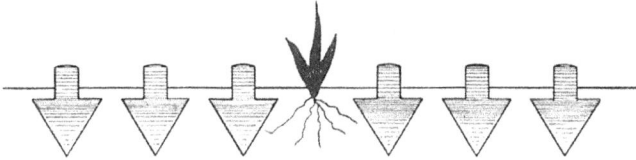

Full width cultivation — Six 5" Spear Point Shovels.

Four large Spear Point Shovels leave some unloosened soil in wide rows but the surface is smooth—no trenches.

Plain Shovels leave deep trenches and unloosened strips in wide rows.

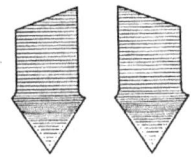

Spear Point Shovels and Sweeps. Nearly full width cultivation.

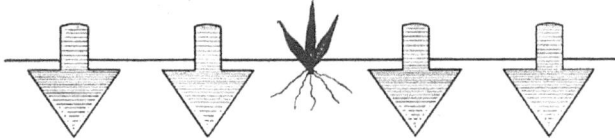

Six Full Sweeps

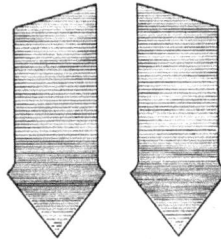

Two Half and Four Full Sweeps

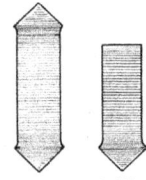

Two Large Sweeps

Hoes and Sweeps.

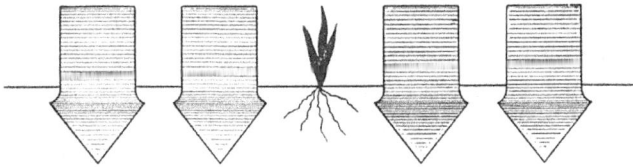

Half and Full Sweeps.

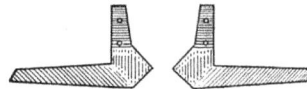

Spear Point

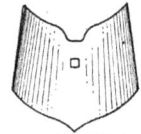

Plain Shovel

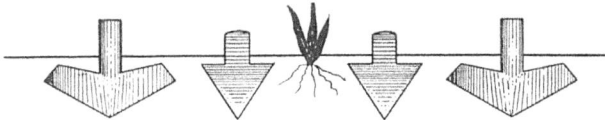

Twisted Shovels

Potato Shovels

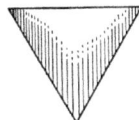

Double and Single Pointed

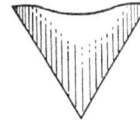

Bull Tongue

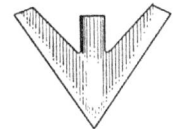

Hoe Blades

Irrigating Blade

Sweep Shapes

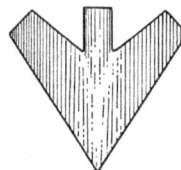

Plow Sweep

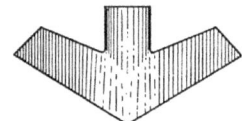

Planter Sweep

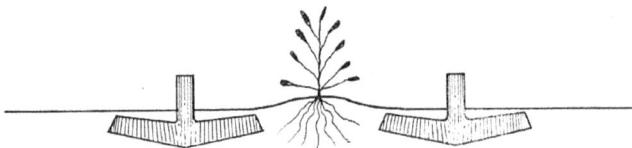

Blackland Sweep

Mixedland Sweep

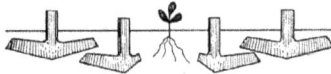

Duckfoot Sweep

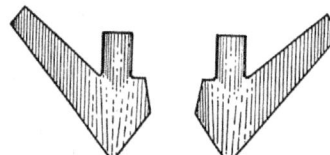

Half Sweeps

TUNING UP THE CULTIVATOR

You know that cultivating your crop is a mighty important operation. You can make or lose many dollars according to the kind of work you do while you are in the field with your cultivator.

Much time is spent with the cultivator in corn fields, cotton fields, potato fields and in the fields with many other crops requiring cultivation. It is important, therefore to have a satisfactory cultivator in the best possible condition before going into the field for the first cultivation. You can save time and money if your cultivator is working right.

As with the plow, you should know your cultivator and understand it; then you can take proper care of it and prolong its usefulness.

If your old cultivator would just lie down and refuse to work when some of the important adjustments are not correct, as will your binder, mower, tractor and other farm tools, the points brought out in this article would be considered more than they are today. But the fact that the old cultivator will just creep along and keep on doing some kind of job as long as the wheels stay on, is one reason why it is many times so sadly neglected. Tune up your cultivator this year before the cultivating season begins. Locate it where you can work comfortably for a few hours, and check over the following points:

First--Examine your shovels. If they were not taken off the cultivator and thoroughly greased after the completion of last year's cultivation, they will be badly rusted. They must be polished and sharpened if you expect first-class work when you go into the field this spring.

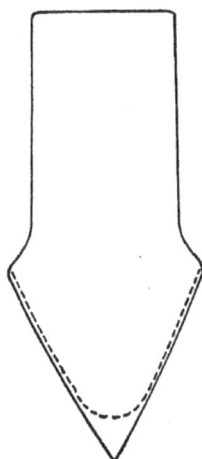

There are up-to-date blacksmith shops in most communities, equipped with polishing wheels. In having your shovels sharpened, borrow a new shovel of the make on your cultivator and ask your blacksmith to retain the original shape, comparing it with the new shovel during the sharpening process. If the shovels are so badly worn that they cannot be sharpened again satisfactorily, replace them with new after market shovels.

Second--Look to the condition of your shields, for next to the shovels, the shields are the most important part of your cultivator, especially during the first and second cultivation. Shields are often badly rusted and bent out of shape, and the bolt holding the shield adjusting washers or clamps becomes rusted tight so that adjustments cannot be properly and easily made. So the next easy thing for the operator to do is to take hold of the shield arm and try to bend it to get the shield located where it should be. A little light oil and patience will usually loosen the nut on the rusted bolt so that the shield can be properly set.

Third--Examine the sleeve bolts and break-pins. Break-pins often become badly cut through, giving the sleeve to which the shovel is attached the wrong pitch, so the shovel does not do as good work as it should. Replace worn break-pins with new hardwood pins so the sleeve will set up snugly as it should on the shank. Loosen the adjusting bolt and set the sleeve for shovels at a forty-seven-degree angle. The rig frame should be level when the sleeves are so adjusted.

In the cotton country, where sweeps are used extensively, there are so many different styles and shapes of sweeps that no angle can be recommended for setting the sleeve. Farmers skilled in the use of sweeps, however, make this adjustment frequently enough so that they are familiar with it.

Fourth--Take a look at your rig couplings--if they are loose and "sloppy," take up the wear. You will find means for making this adjustment on practically all makes of cultivators. Set the couplings up so they will work freely without being loose enough to permit the rig to wobble when in operation.

Fifth--Remove the wheels; clean both the axle and box with kerosene, and apply a good quantity of fresh grease. If wheel box is worn badly, replace with a new one. It will pay you to do this, because the cultivator will be so much easier to control.

Sixth--With the pole level, examine your lifting springs; grease the connections and adjust them so

View below shows, at left, hoof shovel with dotted lines indicating how shovel looks when worn. This shovel, however, has a slip point, as shown below at right; thus, the point, when worn, can be removed and replaced with a new one at less cost than that of re-sharpening an ordinary shovel.

View above shows cultivator shovel with dotted lines indicating how point looks when shovel needs re-sharpening. Obviously, a shovel in this condition penetrates poorly and pulls heavy.

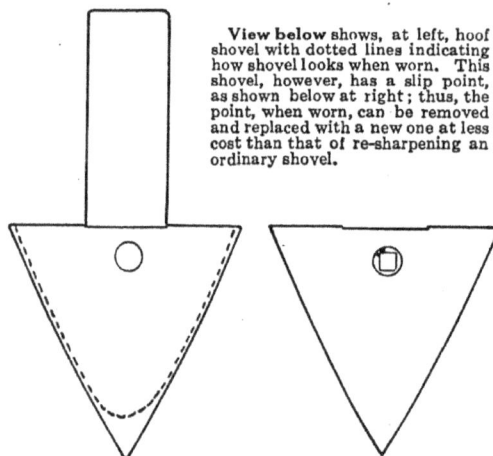

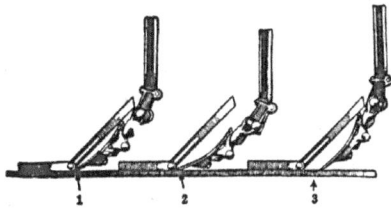

Showing correct and incorrect pitch of cultivator shovels: 1, shovel properly adjusted; 2, shovel set too flat; 3, shovel set too straight, will not penetrate or run steadily.

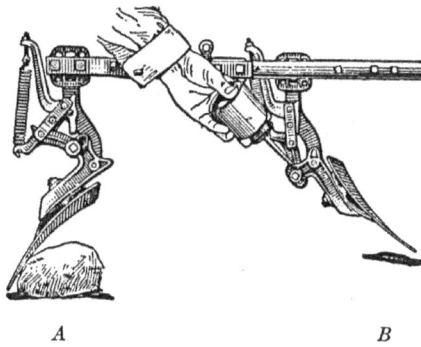

Cultivator spring trip: *A*, trip in action; *B*, trip should be oiled.

that the rigs balance to suit you. Proper adjustment of the lifting spring saves much hard work in the field. On most cultivators, means are provided for increasing or decreasing the leverage exerted in lifting the rig through the spring. A short lever and tight spring will give you better results than the long lever with all loose springs, because in the latter case, with the lifting spring set near the end of the lifting arm, it will have to be fairly loose to permit the rig to penetrate properly; then, in lifting the rig, when the rig is about half-way up, the spring lets go, leaving the operator to lift the dead weight of the rig the remainder of the way.

Seventh--Examine the pole connections where the pole is bolted to the frame or arch of the cultivator. Many times this bolt has been allowed to remain loose, with the result that the pole is badly chewed up, either by frame bars or other connections, so that the cultivator is really loose on the pole connection. Arrange to get this back tight again; if necessary, use a heavy washer between the frame bars. Sometimes a heavier bolt will go through the metal parts, taking up the wear

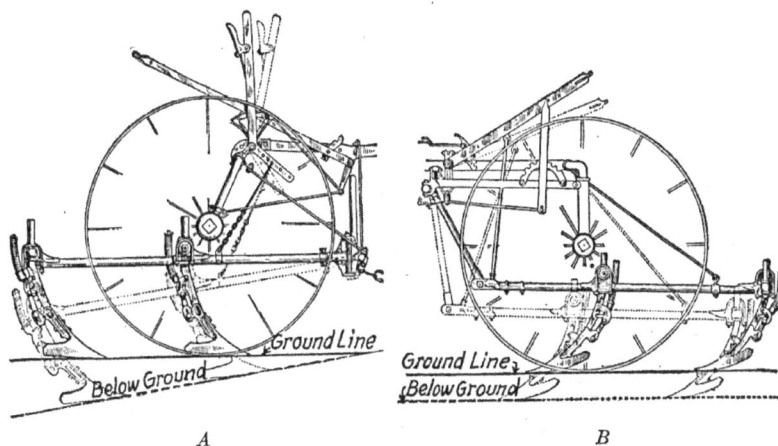

that has taken place in the wood pole.

If your cultivator has spring-trip rigs, oil every joint of the spring trip thoroughly. Limber up each trip by hand-tripping to insure every joint being in good working order. See that each spring is adjusted just tight enough to hold the trip when the shovel is at work. Notice picture on left which illustrates the oiling of spring-trip rigs.

And last but not least, in the case of riding cultivators, have a look at the seat and seat bars. A loose, poorly-adjusted seat is an annoyance to the operator every day he is in the field. Locate the seat where you want it, tighten it securely, and it will repay you in comfort many times during the season.

The foregoing directions are the result of years of experience in the field with cultivators. Perhaps you have had trouble with your cultivator; if not, you can save time and trouble by observing the directions given here.

FIELD OPERATION OF CULTIVATORS

The following remarks on field operations and adjustment of shovel cultivators will apply to these three types:

Single-row Riding Shovel Cultivators,
Double-row Riding Shovel Cultivators,
Two-horse Walking Tongue Shovel Cultivators.

All three types of cultivators should be operated with the pole as nearly level as possible. It is important, therefore, when hitching to the cultivator, to see that the breast straps carrying the neckyoke are adjusted so as to carry the pole practically level. If the pole is not level, the following trouble will be encountered:

(a) If the point of the pole is below level, the front shovels will penetrate deeper than the rear shovels. All shovels will stand straighter than they should, and will not penetrate easily.

(b) If the point of the pole is higher than level, the rear shovels

A, shows how shovels enter the ground when the gang is hinged at the front and lowered by the lever; *B*, shows uniform depth of shovels on entering the soil.

will penetrate deeper than the front shovels. All shovels will set too flat, and will not penetrate properly.

All cultivators are provided with an adjustment, either on the shovel shank or sleeve, whereby the pitch or angle of the shovel may be changed. This adjustment, when the cultivator leaves the factory, is correct for average soil conditions. It is well, however, to understand this adjustment so that it may be properly made when conditions necessitate. A shovel standing too straight will not penetrate readily; it will not run steady. There will be a tendency to skip and jump, and it will require unusual pressure to keep the shovels in the ground. If set too flat, the under part of the shovel will ride below the extreme point, and the shovel will not penetrate unless forced into the ground.

Many cultivators are equipped with levers that change the "set and suck" of shovels or sweeps when plowing up or down slopes.

This type of shield and the adjustments are very important during the first cultivation of corn. Two types of shields are used quite extensively throughout the corn belt; the solid sheet-iron shield, and the open rod-wire shield. They have been in use for many years, and their operation and adjustment are quite familiar to all.

As in the case of most other farm implements, there is more or less talk of cultivators being heavy draft. The draft of the cultivator is determined entirely by the amount of soil the shovels are stirring. If the shovels are properly adjusted and running at a reasonable depth, the cultivator is not a heavy draft tool. It is possible, however, with the shovels improperly adjusted, requiring unusual weight and pressure to keep them in the ground, to make the machine quite heavy draft. This condition also tends to make neck weight. If neck weight exists when the machine is running level with the shovels properly adjusted, note the following suggestions:

(a) Most straddle seat cultivators are equipped with what is termed a pendant type of hitch. On this type of hitch there is a series of holes that permit the singletree to be raised or lowered. Lowering the singletree will relieve neck weight on this type of cultivator.

(b) Hammock-seat cultivators do not have the pendant type of hitch. Most of them, however, are provided with a balance level whereby the wheels are moved forward or backward, slightly affecting the balance of the cultivator and relieving neck weight.

A good many times the operator finds it hard to keep the rigs running the proper distance from the row on account of a tendency to crowd sidewise. In a level field this is due entirely to the position of the shovel, as to whether or not it is straight or twisted. If the shovels are all set straight on the rig, there should be no tendency to crowd sidewise. If it is necessary on a swinging rig type of cultivator to twist the front shovels, either to throw the dirt away from the row the first time or toward the row during later cultivation, it is advisable to twist the rear shovel in the opposite direction to overcome the tendency of the rig crowding. Where a spread arch is used, and on the parallel type of rigid rig cultivators, the position of the shovels does not affect the operation of the cultivator.

And don't forget, too, that your cultivator needs attention while it isn't in operation. Leaving the machine out in the weather for even a short period may harm it much more than using it would. When the operating season is over, get the cultivator under good shelter as soon as possible. In case you do not have an implement house big enough to store all of your cultivators, be sure, at least, to remove the shovels, grease them thoroughly and hang them in a sheltered place. Vaseline is the best grease you can use.

Avery Pivot Axle Cultivator

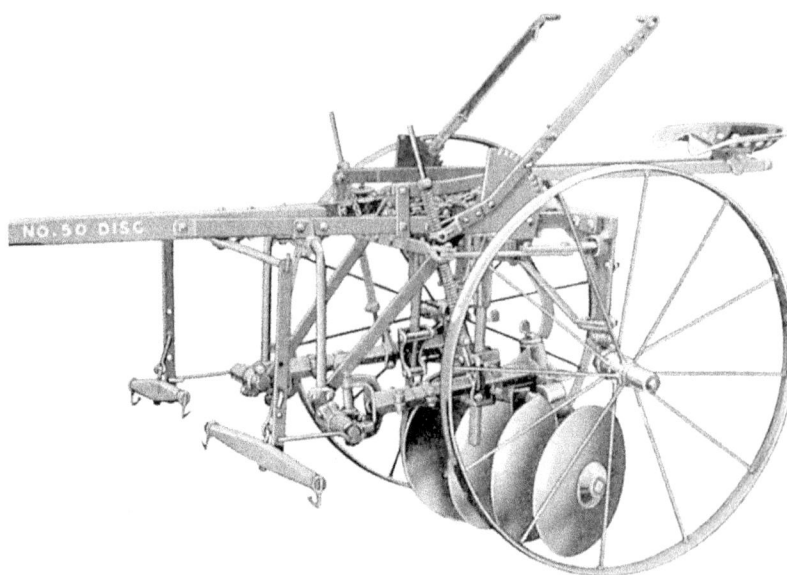

JOHN DEERE METHOD OF CULTIVATION

Increases the Yield—Improves the Quality

A simple method of cultivating corn that requires but one cultivator with two types of shovel equipment.

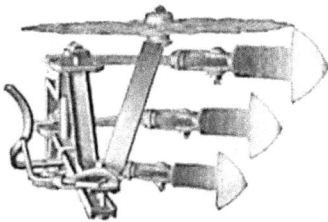

Sweeps

Hoof Shovels and Rotating Shields

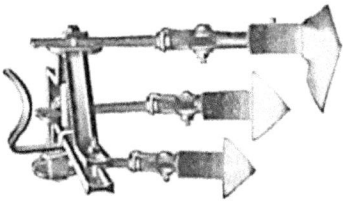

Combination Rig—Hoof Shovels and Sweeps

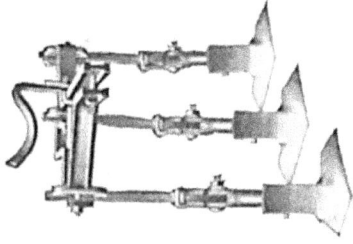

Third or Later Cultivation—Use Sweeps no danger of injuring roots, and a fine surface mulch is formed which retards evaporation of moisture in the soil.

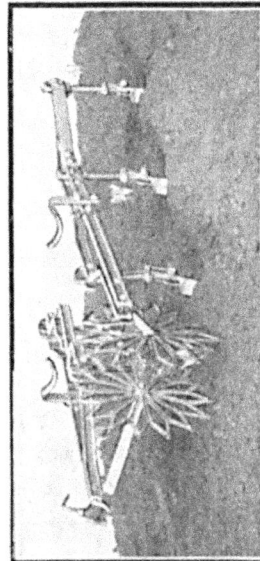

First Cultivation—Use Hoof Shovels and Rotating Shields. Ground is left level, well pulverized, and the growing plant is not covered by dirt.

Second Cultivation—Use Sweeps on the front shanks, hoof shovels on the balance of the rig.

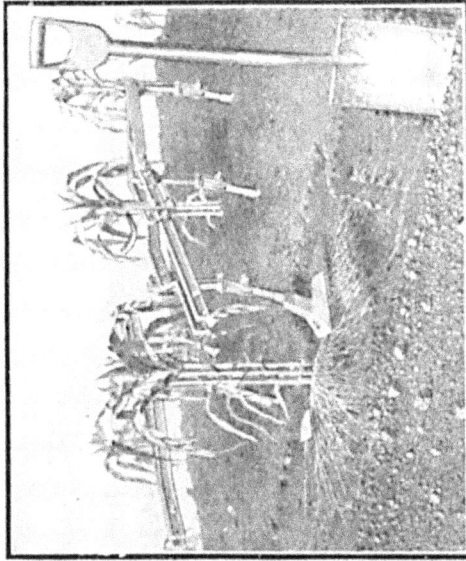

Each implement manufacturer had his or her focus or approach to farming procedure. In those early days it was essential because many of the specific innovations were designed to match a plan or method/idea and thereby offer a unique approach towards a farming challenge. Sometimes the differences between variations were so slight as to be difficult to see. Oft times the copywriters for the companies were successful at making a standard approach seem like a brand new cat's-meow full-chrome gizzy of amazing potent. Even so the literature contains information often helpful to those of us knee-deep by choice in the mystical particulars of a finer tuned farming.

Cultivating

Distribution of corn roots sixty days after planting. Notice the mass of roots that would be cut off by a cultivator running four inches deep

Corn Cultivation

The Moline Company recommended these approaches to corn cultivation in its 1921 implement catalog:

First Cultivation

Corn is cultivated primarily to eradicate weeds; and secondly to maintain a surface mulch. If weeds are allowed to get a start, it is well known that it is practically impossible to thoroughly clean the field. Therefore, in the first cultivation, weeds must be eradicated. At this time the corn roots are small, permitting close cultivation and penetration to a depth that will effectively destroy them--rather than merely retard their growth.

It is also desirable, during the first cultivation to allow the shovels to penetrate sufficiently to loosen and pulverize all soil around the plants. While a careful seed bed may have been prepared there are seasons that cause the ground to pack greatly before the corn can be cultivated. It is at just such times that it is most important to loosen the soil for the small tender roots. They can then spread out in search of food. Any method of cultivation that does not provide for thorough pulverization of the ground at this stage of plant growth handicaps the corn; for careful cultivation makes possible healthy, productive roots. Straight, twisted or wing shovels are very satisfactory; detachable point shovels are especially recommended.

Second Cultivation

Once the corn roots have begun to spread out and usually after the first cultivation it is impractical to use deep cultivation next to the row. To do so, destroys many of the feeders; consequently impairing the food supply of the plants. Half sweeps should be used next to the row to thoroughly cultivate the surface and destroy weeds. However, it is very important, that the middles be cultivated to a reasonable depth to avoid soil packing and to keep it in the best condition for the spreading roots. The regular shovels should be used on the balance of the gang.

Third Cultivation

After the second cultivation it is largely a matter of mulching the soil. The roots of the corn will, by this time, be spread practically across the row for by keeping the soil between the rows thoroughly stirred in first and second cultivation, it is in excellent condition to permit a healthy root growth. Shallow cultivation is therefore the most productive and can be done efficiently with sweeps or detachable point sweeps. They will carefully mulch the soil and destroy young weed growth without disturbing the corn roots. The roots may then reach a full growth and adequately nourish the plant, assuring a maximum yield. In a dry season there is less danger of the crop being reduced.

Cultivators and Equipment

This proven method of cultivation does not require special implements. Most any cultivator might be equipped with either, straight, twisted, wing or detachable point shovels, as well as sweeps and detachable point sweeps. For particular field conditions or for especially careful cultivation other equipment is provided. Many farmers will find it advantageous to use hilling shovels, disc blade hillers, rotary shields, side harrow attachment, gopher blades, cotton scrapers, etc.

Rotary Shields

Many farmers prefer this type of shield because it permits the pulverized soil to sift through the openings in the shield as it revolves and yet it protects the young plant from being covered with clods. It is hardly possible to secure such favorable results with ordinary shields.

Fig. No. 1 illustrates the first cultivation. Either regular or Quick Detachable Point Shovels can be used without danger of injuring the corn roots.

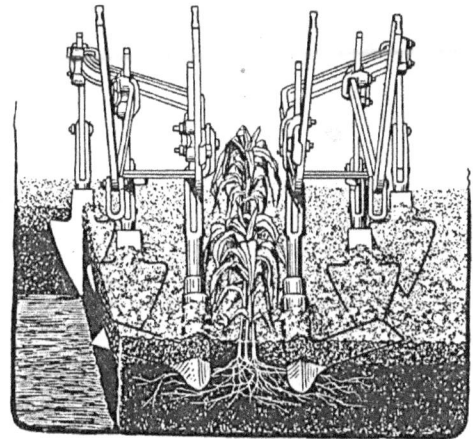

Fig. No. 2 illustrates the second cultivation. To avoid injuring the roots, half sweeps are placed next to the corn. If deeper cultivation is desired in the center of the row, the regular shovels can be left on.

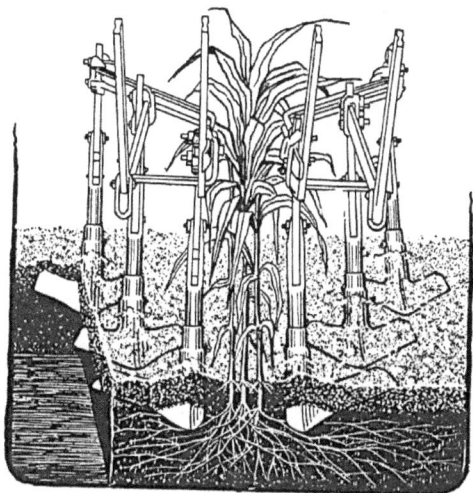

Fig. No. 3 shows how sweeps can be used for surface cultivation when the roots are near the surface of the ground. In this manner weeds can be destroyed and the soil kept in a well pulverized condition without danger of injuring the roots.

Checked corn planted in hills permitted cultivation in several directions and theoretically resulted in superior coverage.

Drilled corn in straight rows which can only be cultivated along the line.

The fine, pulverized soil thrown around the young corn plants forms a mulch to hold moisture. It also covers and destroys many small weeds close to the row, giving the corn every opportunity to make a healthy vigorous growth.

The Moline rotary shield is durably made of heavy wire loops fitted into an inner pressed steel rim. The wire loops, coming in contact with the ground, cause the shield to revolve. It will fit any Moline cultivator.

Disc Hillers

Disc hillers are especially designed for hilling crops and barring off cotton.

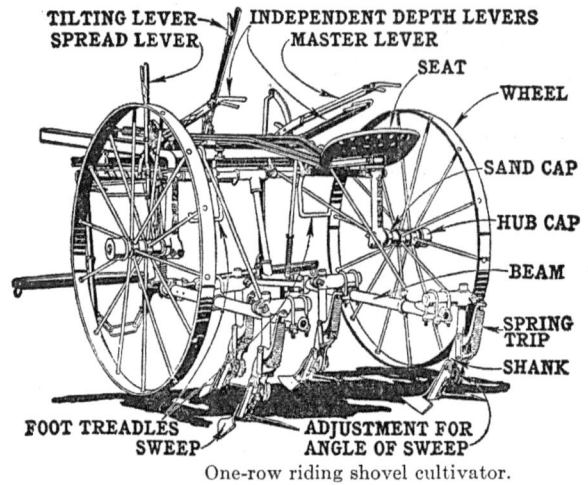

TILTING LEVER
SPREAD LEVER
INDEPENDENT DEPTH LEVERS
MASTER LEVER
SEAT
WHEEL
SAND CAP
HUB CAP
BEAM
SPRING TRIP
SHANK
FOOT TREADLES
SWEEP
ADJUSTMENT FOR ANGLE OF SWEEP

One-row riding shovel cultivator.

A young Amishman drives three Haflinger draft ponies hitched to a new I & J cultivator with side dressing tank during the Horse Progress Days 2000 at Kinzers, PA.

Chapter Thirteen

One Horse Cultivators

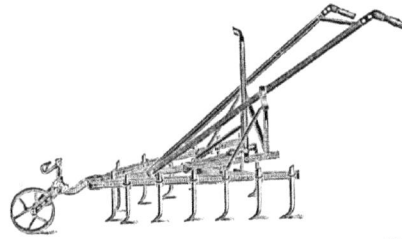

7-Tooth cultivator

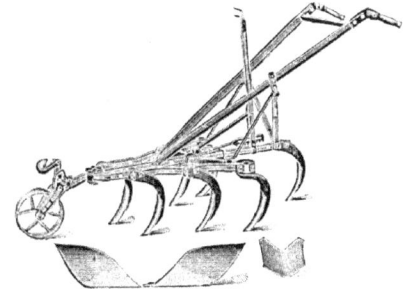

14-Tooth Cultivator

John Deere

5-Tooth Cultivator

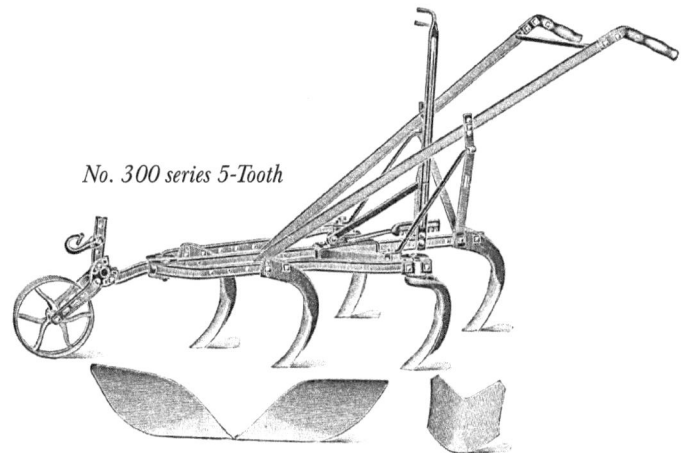

No. 300 series 5-Tooth

A Norwegian Fjord horse pulls a Scandinavian cultivator.

For the purposes of this book we have elected to separate cultivators into three groups; one horse, two horse and two rows or wider.

For hundreds of years farmers have planted certain crops in rows. In the beginning they probably worked on hands and knees pulling the competing weeds until the hand hoe was developed. Then came the wheel hoe pushed by hand or pulled by an agreeable or subpoenaed draft animal. More than sixty percent of the world's fresh fruits and vegetables are still produced in this manner in spite of the unholy lies of agribusiness and store-bought corporate governments.

Perhaps it was because of the direct simplicity of the job assignment but the walking one horse cultivator saw far less innovation than the two horse riding cultivators as evidenced by these two succeed-

ing chapters.

The modern horsefarmer will need to make his or her own mind up about whether a walking cultivator is a necessary implement. If the rows crops maintained come to less than an acre the one horse walker can do the work. Over an acre of row crops and the modern well-layered farmer may covet a tool he can ride on.

Most of the better cultivators allowed an adjustment for the width of the cut and permitted the farmer to move the shanks and interchange shovels. The basic walking cultivator frame was used by some companies to carry specialty applications such as celery hillers and even planters (covered in a coming text). Creative farmers have used the basic principle and framework of the cultivator for all sorts of jobs.

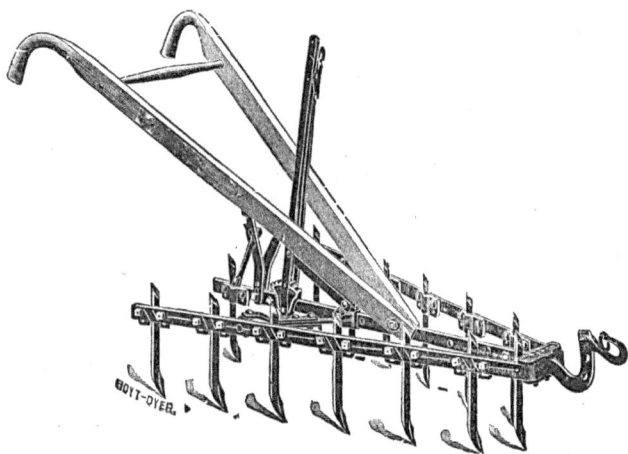

FOURTEEN-TOOTH HARROW AND CULTIVATOR

John Deere

Hitching to the one horse walker can be as simple as dropping a singletree ring into the waiting hitch hook on the implement. Care should be taken to hitch long so that the tug angle doesn't lift up on the front of the tool.

The experienced animal will walk down between rows as if it had no choice but this will take time and patience. Work animals enjoy the familiarity of routine and will surprise the teamster on how quickly they learn the job. It was common to allow a child to ride the cultivating animal and even help steer. The anxious green broke animal is no fun on the walking cultivator.

Celery Hiller

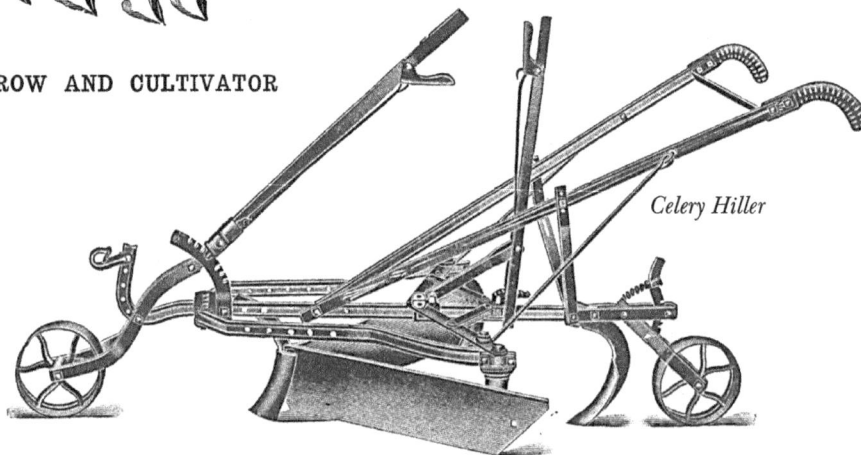

Three different I & J cultivators on display at the Pennsylvania Horse Progress Days 2000.

190

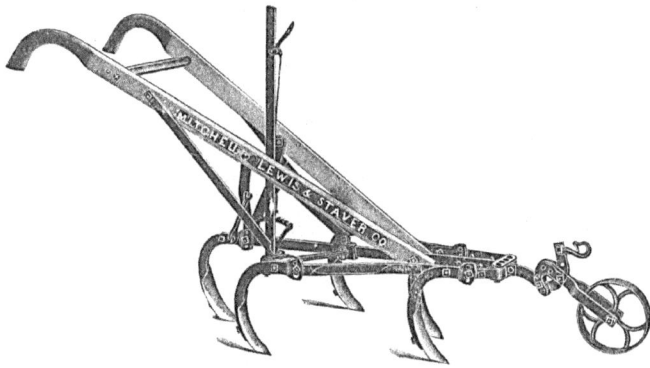

FIVE-TOOTH CULTIVATOR

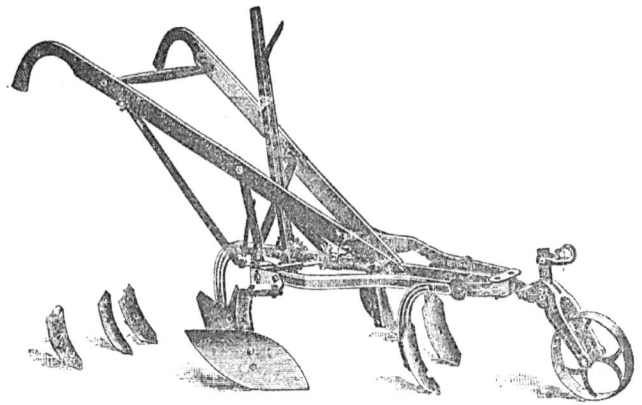

FIVE-TOOTH CULTIVATOR WITH HORSE HOES

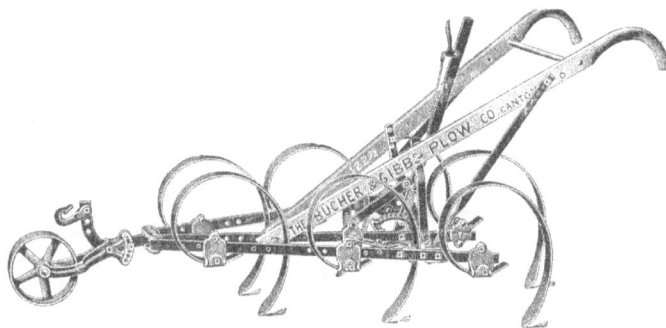

B & G Imperial Spring
Tooth Lever Cultivator
with runners

**Bucher & Gibbs Imperial Spring
Tooth**

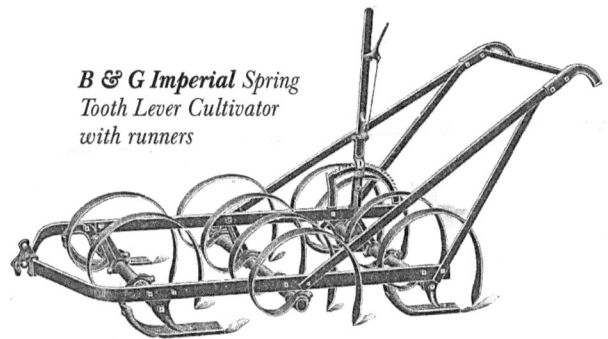

Misc.

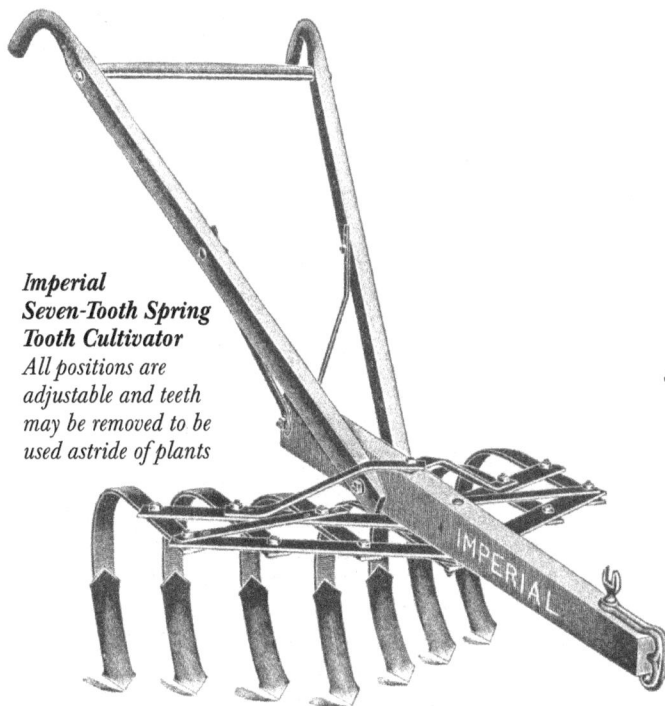

**Imperial
Seven-Tooth Spring
Tooth Cultivator**
*All positions are
adjustable and teeth
may be removed to be
used astride of plants*

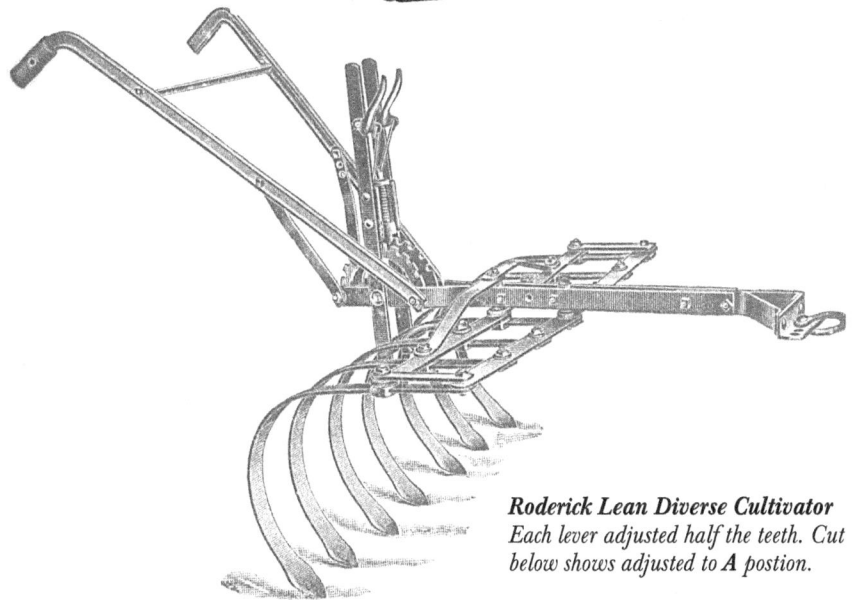

Roderick Lean Diverse Cultivator
*Each lever adjusted half the teeth. Cut
below shows adjusted to A postion.*

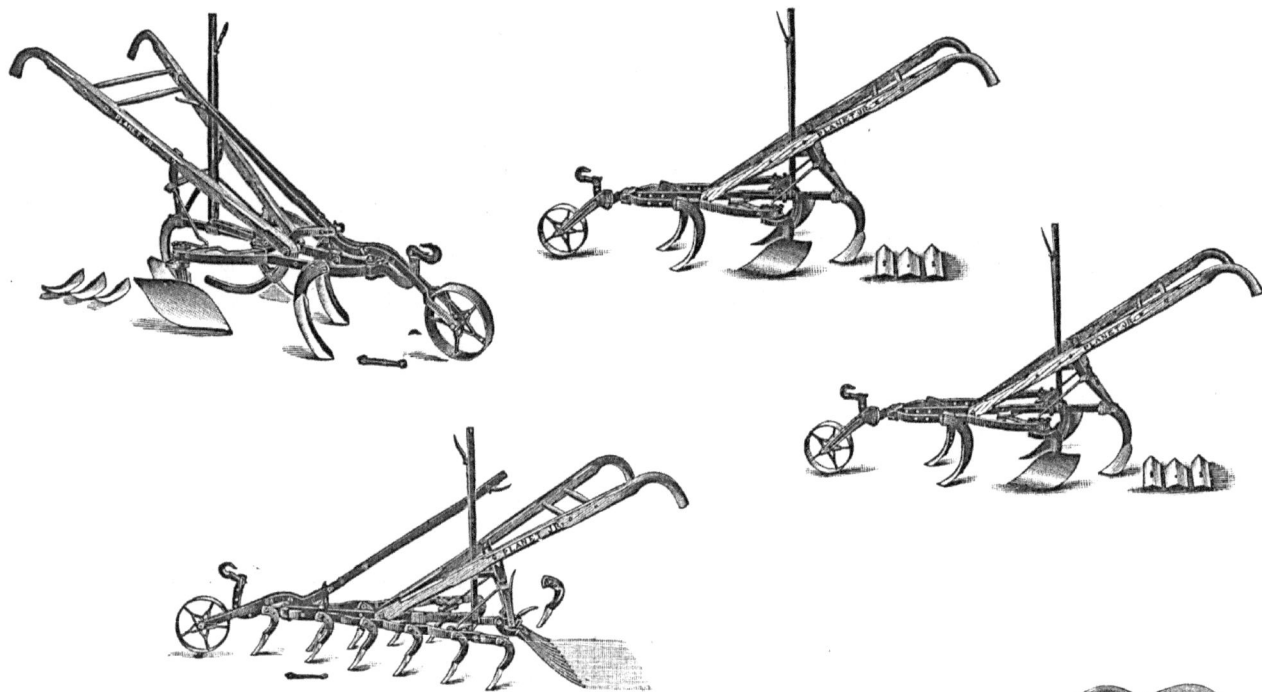

Furnished with Steel Handles
When So Ordered.

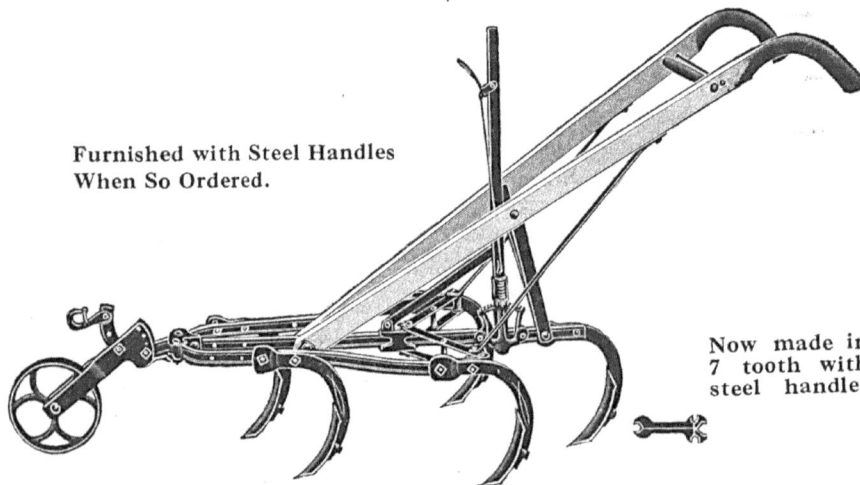

Planet Jr.

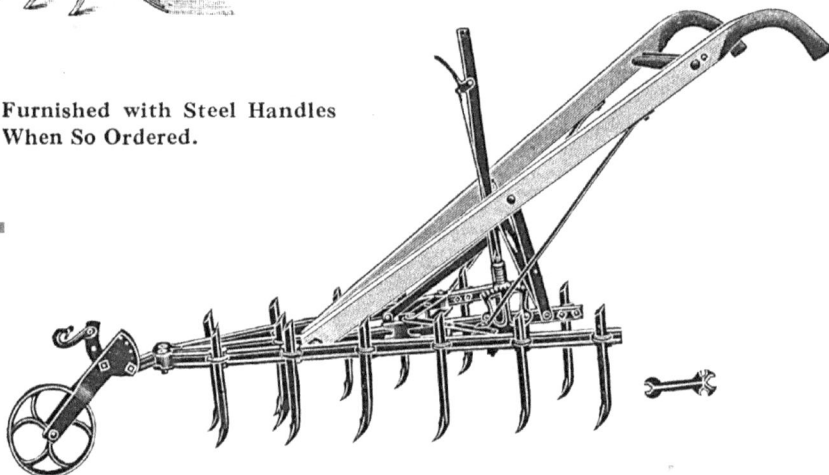

FOURTEEN-TOOTH HARROW AND CULTIVATOR

Furnished with Steel Handles
When So Ordered.

Now made in
7 tooth with
steel handles

FIVE-TOOTH CULTIVATOR

Planet Jr.

FIVE-TOOTH CULTIVATOR WITH HORSE HOES

Furnished with Steel
Handles When So
Ordered.

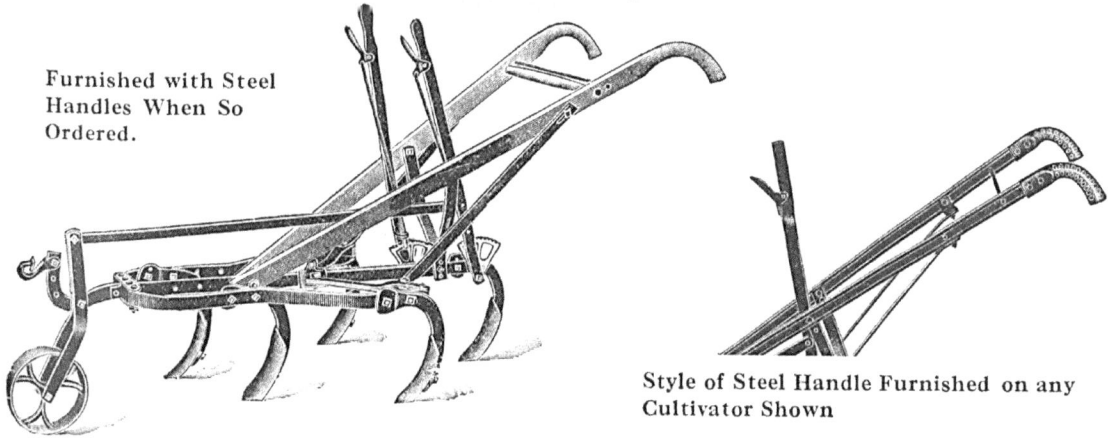

Style of Steel Handle Furnished on any
Cultivator Shown

FIVE-TOOTH CULTIVATOR WITH HORSE HOES

Equipment
4 3x8 Cultivator Steels
1 4x8 Cultivator Steel
2 6-inch Tillers
One 7-inch shovel
Lever Wheel
Lever Expander
Depth Regulator

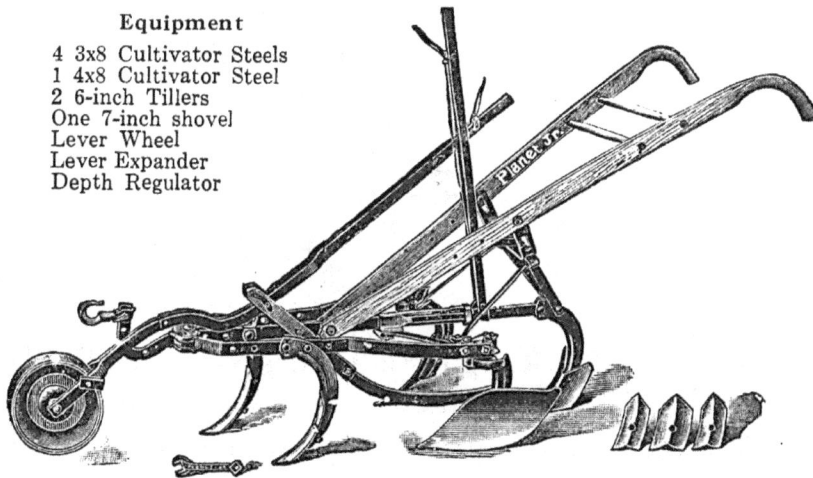

No. 8 Planet Jr. Horse Hoe and Cultivator

Planet Jr.

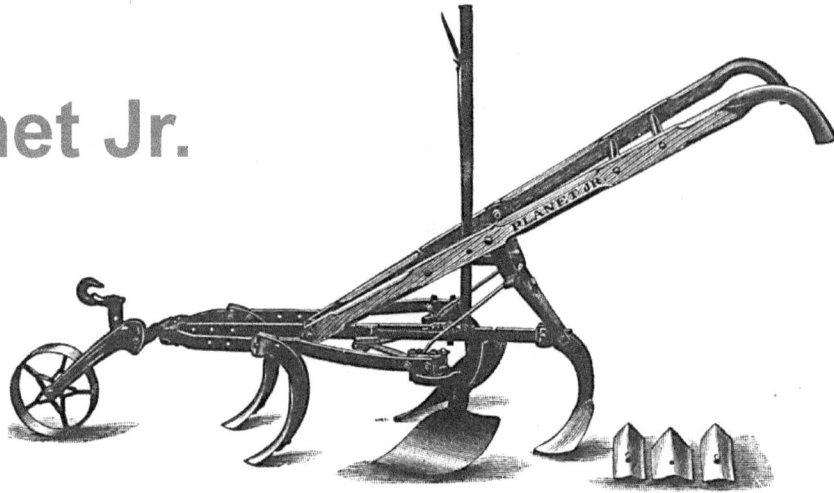

No. 9 Planet Jr. Horse Hoe

No. 10 THE GREAT PLANET JR. COMBINATION FARM AND GARDEN HORSE HOE, CULTIVATOR, PLOW, FURROWER AND VINE TURNER.

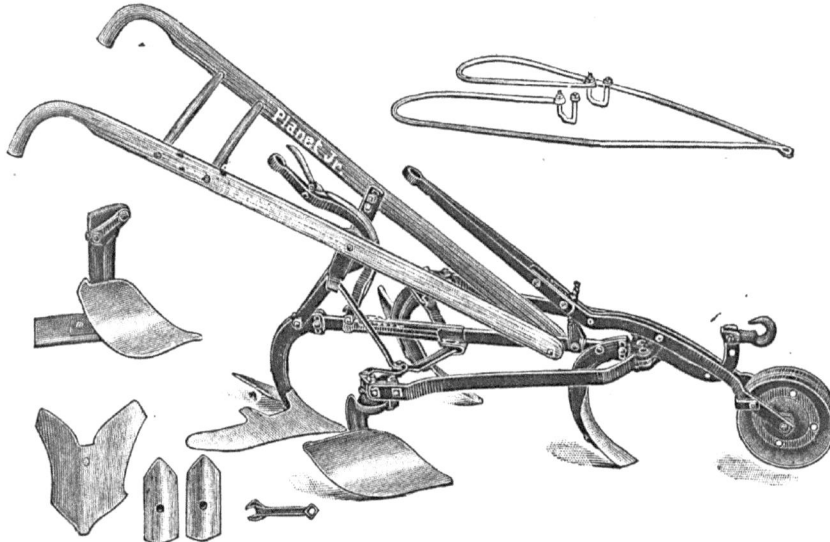

Equipment:

Three 3x8-inch cultivator teeth.

One pair 6-inch hillers.

One plow attachment.

One 15-inch fingered sweep.

One 10-inch furrower, one vine turner, lever expander and lever wheel.

The King of Planet Jr. Horse Hoes

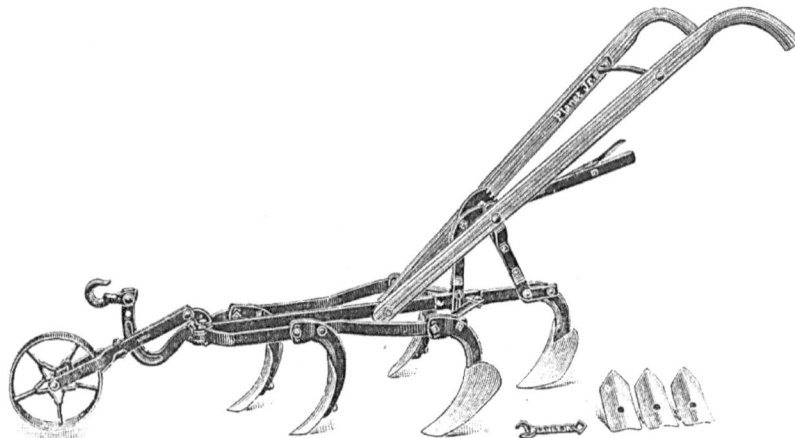

NO. 103 PLANET JR. HORSE HOE AND CULTIVATOR

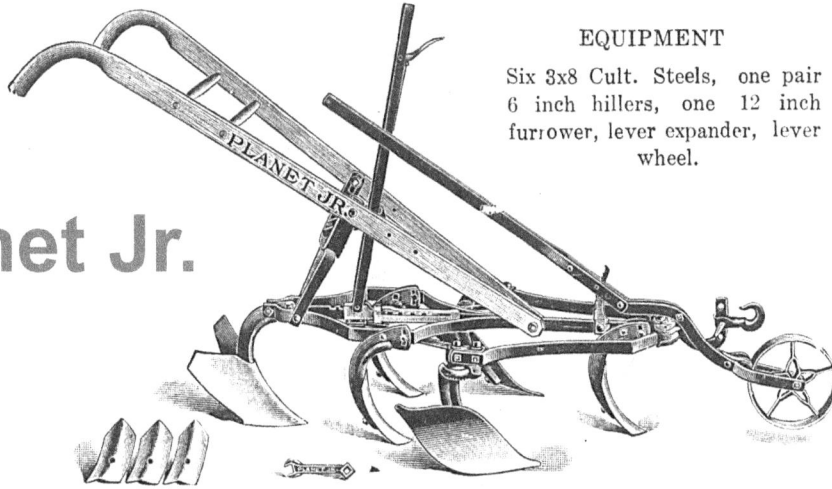

Planet Jr.

EQUIPMENT

Six 3x8 Cult. Steels, one pair 6 inch hillers, one 12 inch furrower, lever expander, lever wheel.

No. 81 Planet Jr. Horse Hoe, Cultivator and Hiller Combined

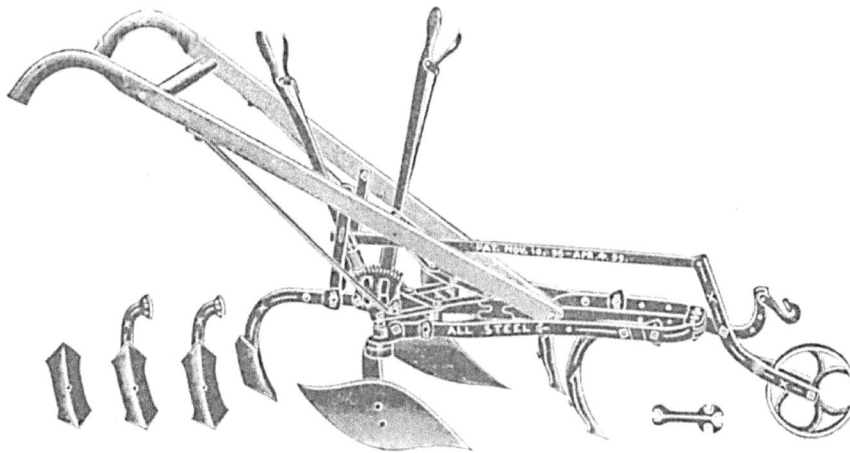

ROYAL BLUE 14-TOOTH CULTIVATORS

Rock Island

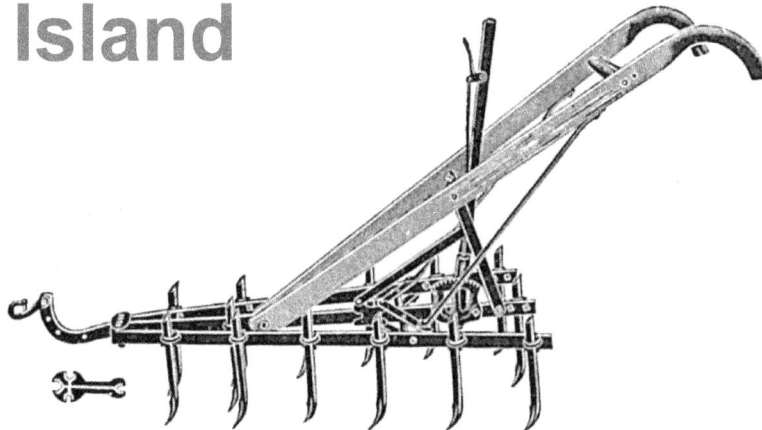

Fourteen Tooth Cultivator

One Horse Cultivators

Gee-Whiz Cultivators

THE most popular and generally used tillage implement in the Southern States. Its various adjustments adapt it to so many kinds of work that any farmer who has used one of these versatile implements will say he can hardly keep house without the Gee-Whiz.

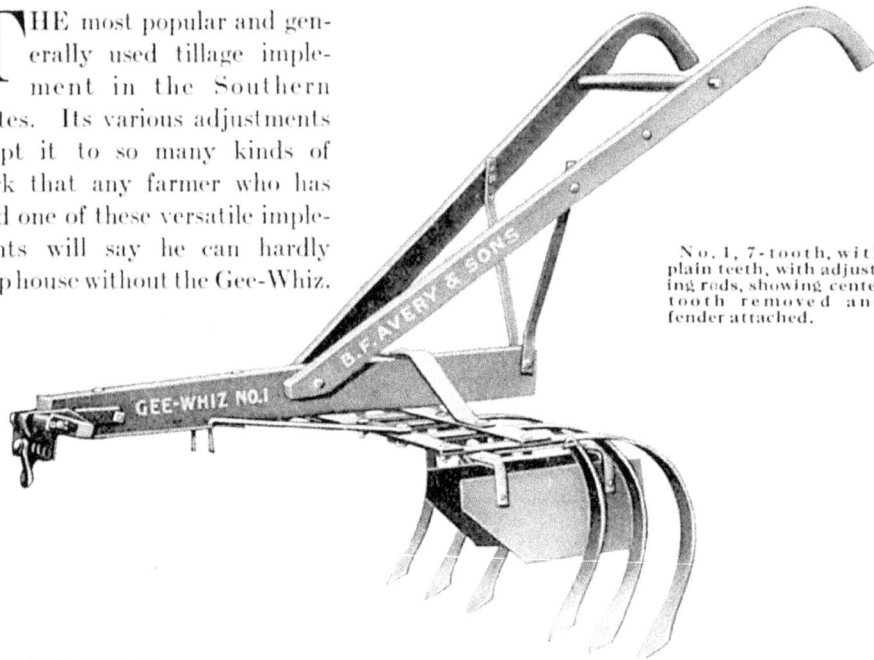

No. 1, 7-tooth, with plain teeth, with adjusting rods, showing center tooth removed and fender attached.

Avery

The Avery 101 Three-Shovel Cultivator
Avery's most modern cultivator design

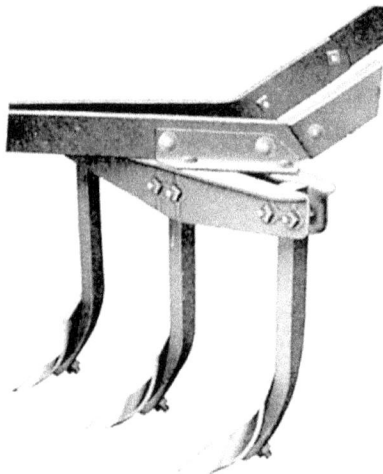

Detail Frame Construction

Detail Construction Floating Fender

The Avery fender was unique as it was constructed to move freely in an up and down direction between certain well defined limits.

No. 153, 7 tooth *equipped with adjusting levers and detachable blades.*

The jerky kicking action of the spring teeth was said to tear up weeds, shaking them free of soil and leaving them on the surface.

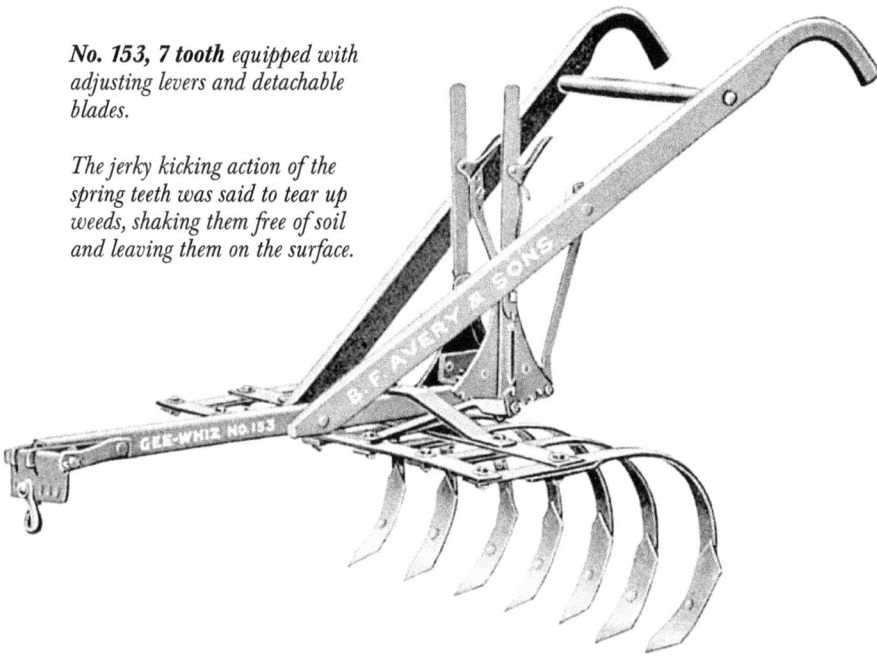

Avery

Avery One-Horse Trucker C

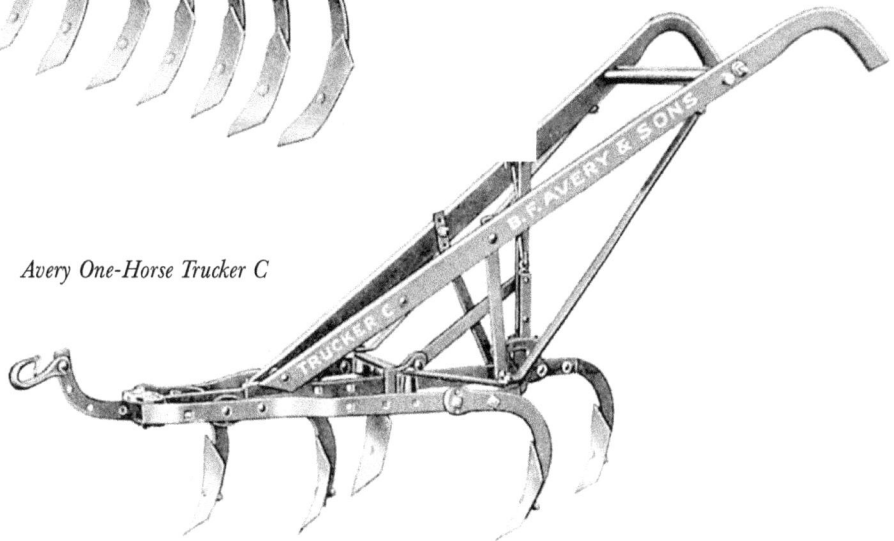

It is furnished with 5 or 7 teeth with wood or steel beam, with adjusting rods or adjusting levers with plain teeth or detachable blades.

It can be converted into a straight spring tooth cultivator, right or left hand side harrow, "A" shaped Cultivator or "V" shaped Cultivator. By hitching in the outside hole of the clevis and removing the center tooth, the Gee Whiz will straddle the row.

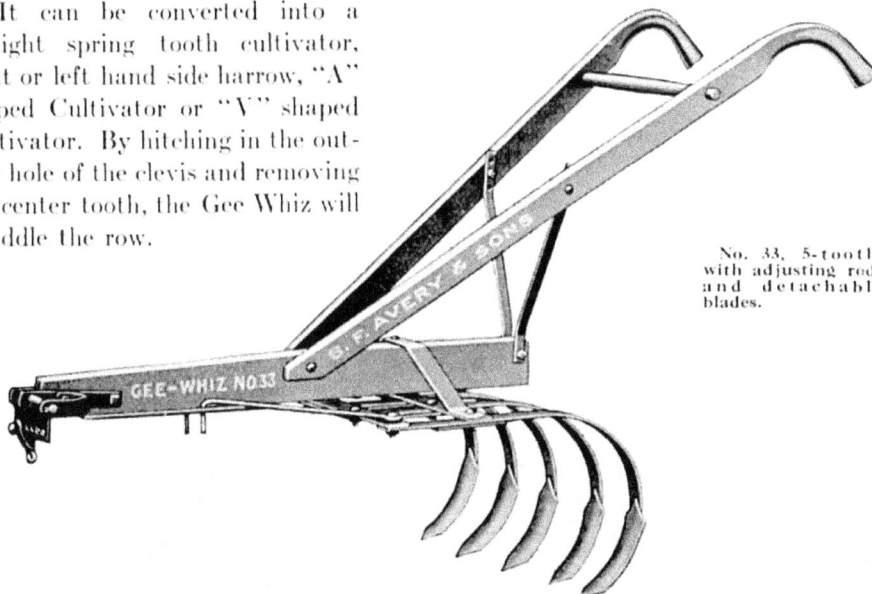

No. 33, 5-tooth, with adjusting rods and detachable blades.

One-Horse Trucker Cultivators

Avery

No. A, with Hand-Set Screw

Fourteen Tooth Avery Orchard Harrow No. G

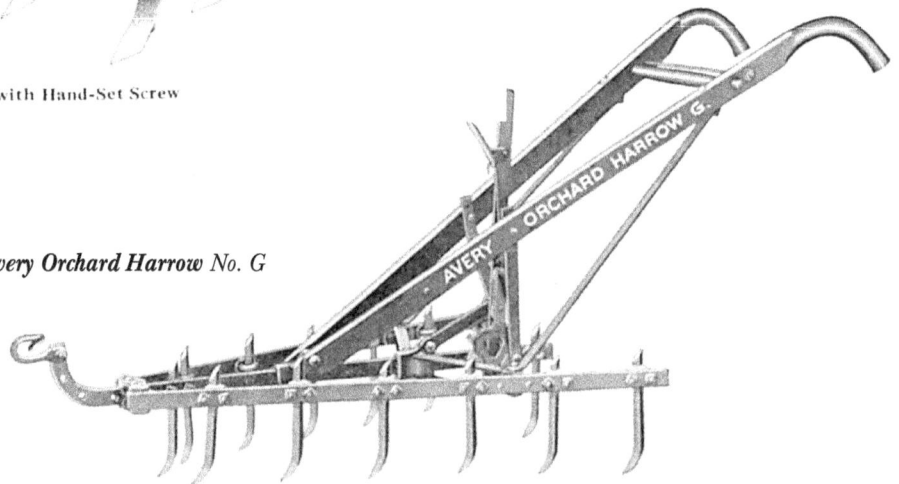

Cultivating scene in Norway with Fjord horse hitched, via shafts, to a walking single row cultivator.

One-Horse Trucker Cultivators

Avery New Trucker

AVERY New Trucker Cultivators are of superior construction, having hollow pressed steel standards. The standards are interchangeable, there being no rights or lefts.

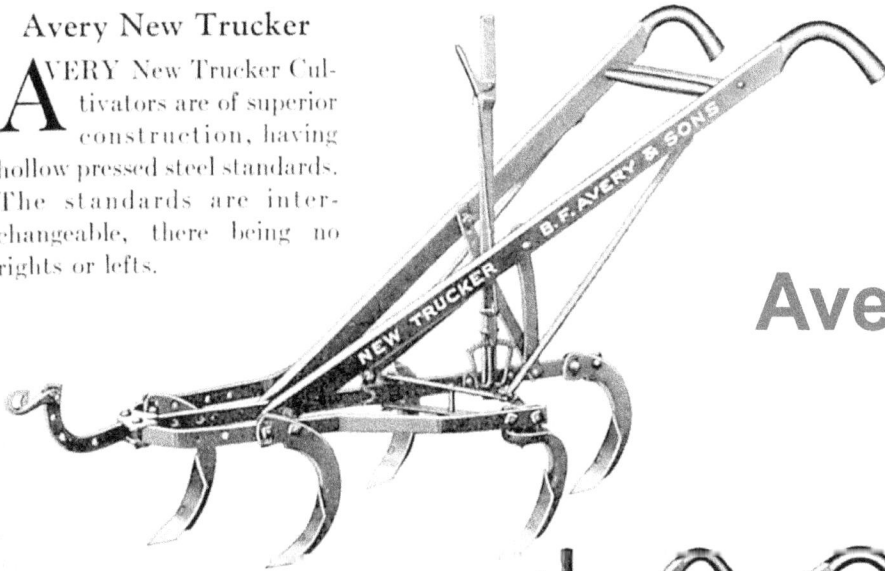

Avery

The horse hoe standards differ from the other three. They have a rotary adjustment and take either the regular cultivator teeth or hilling attachments. They have two holes to accommodate hillers, and may be turned all the way around, thus giving any sideway angle desired.

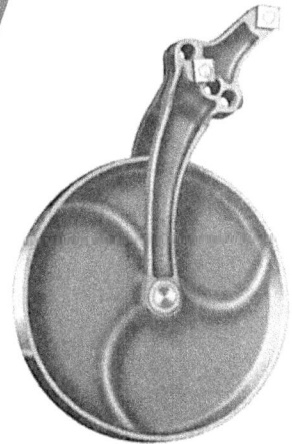

Gauge Wheel

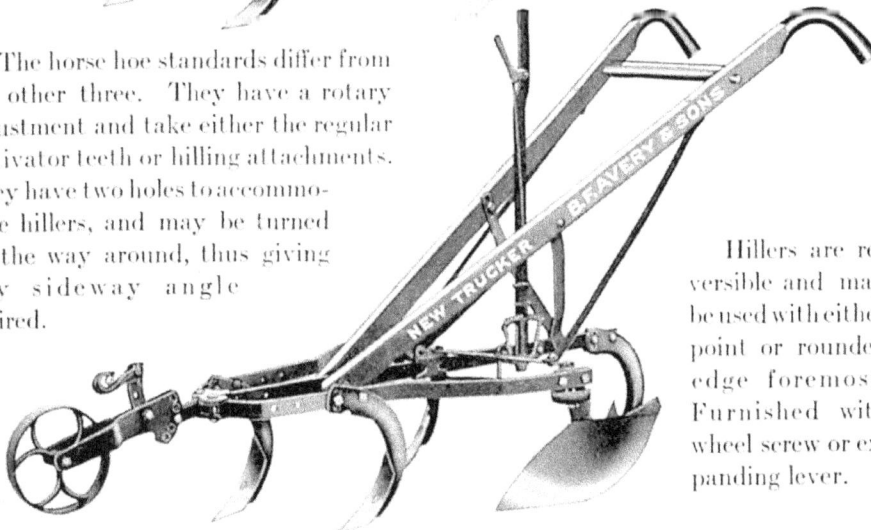

Hillers are reversible and may be used with either point or rounded edge foremost. Furnished with wheel screw or expanding lever.

Showing Gauge Wheel and Horse Hoes Attached.

Avery Orchard Harrows

No. F

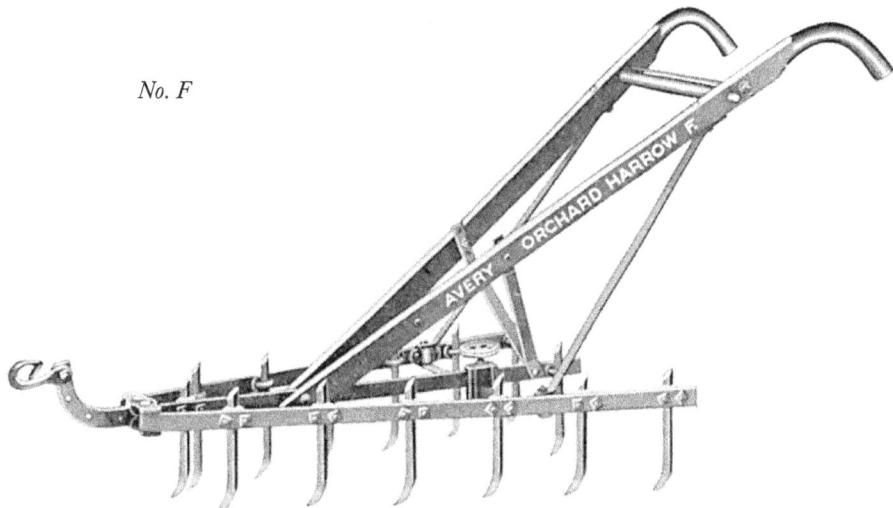

Avery One-Horse Cultivators

THE Delta Cultivator is an exceptionally convenient and useful implement. It is staunchly built to withstand hard work. It is furnished with five diamond point reversible blades 2x8x¼ inches and equipped with adjusting rods by which it may be readily converted into six different implements—"A" or "V" shaped cultivator, square cultivator, right- or left-hand side harrow or a four-tooth harrow for straddling rows.

Avery

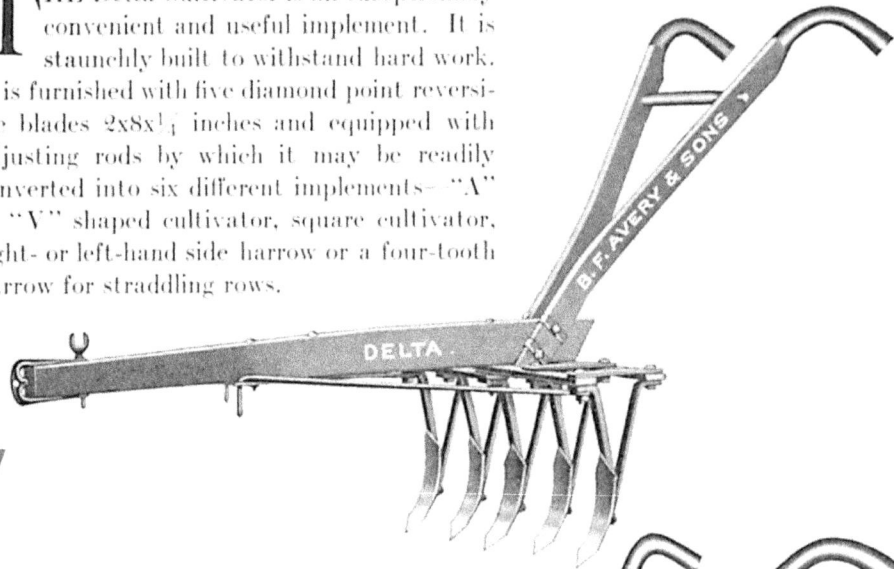

The Memphis is similar in construction to the Delta cultivator. It is equipped with four standards with diamond point reversible blades 2x8x¼ inches. By changing the adjusting rods, it can be readily converted into a right- or left-hand side harrow.

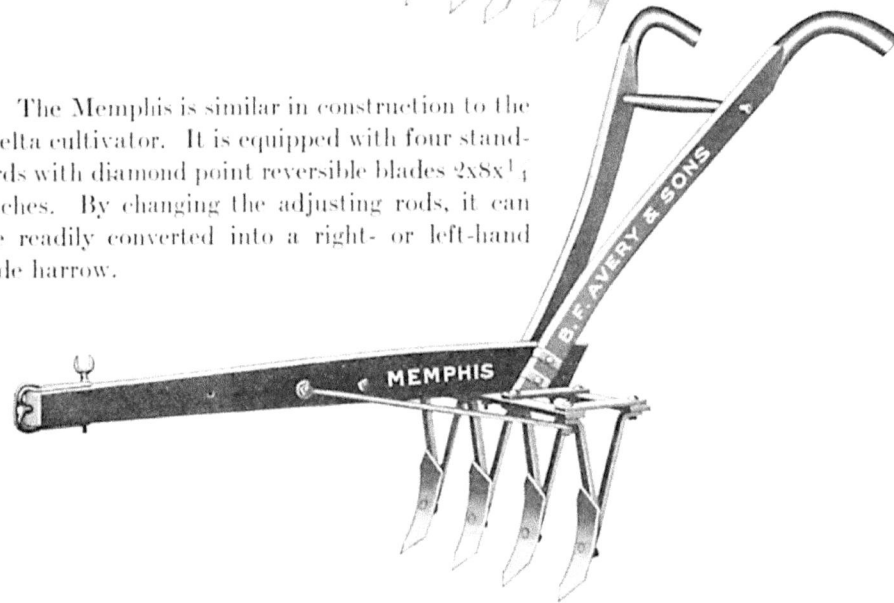

BLOUNT TIP TOP CULTIVATOR   BLOUNT DANDY CULTIVATORS

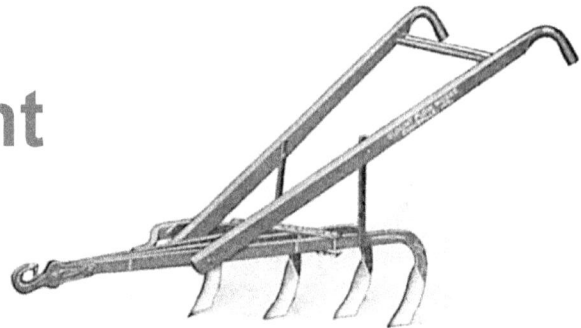

Blount

"TRUE BLUE" RANGER SIDE HARROWS

BLOUNT
"AIR-PLANE"
CULTIVATOR

Blount

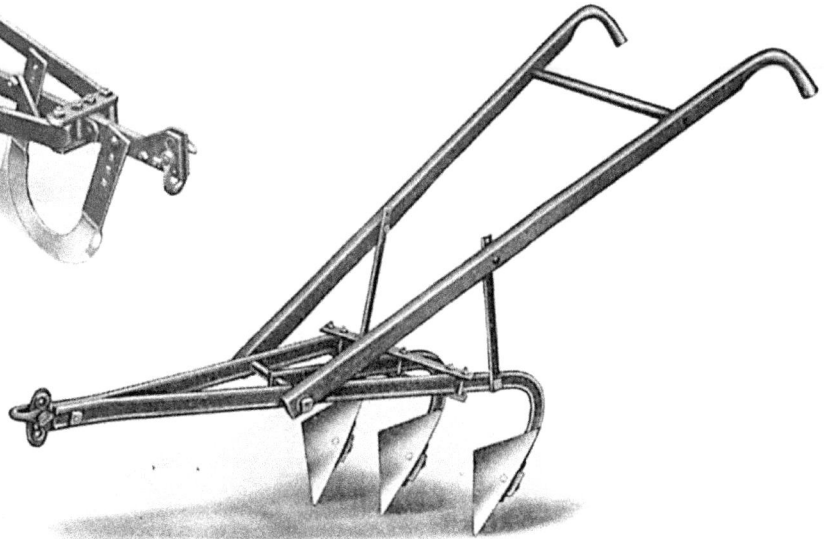

Blount B. Barton Cultivator

Blount "Hercules" Heavy Duty One Horse Walking Cultivator

Blount

Universal One Horse

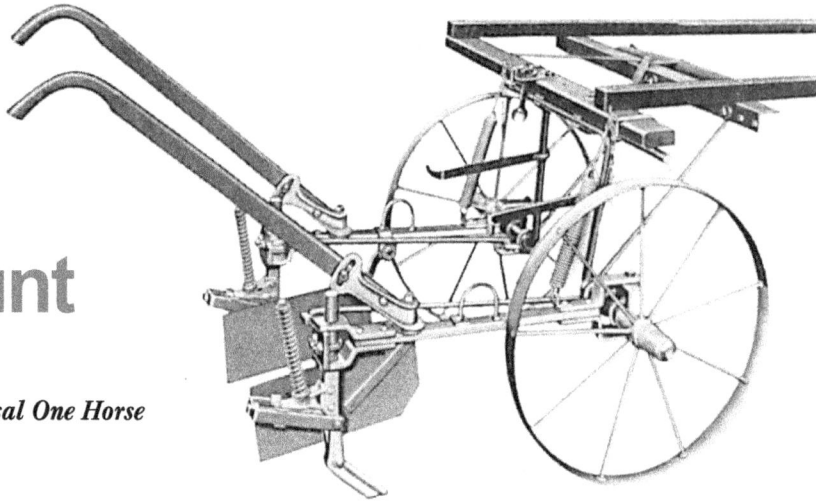

The Oliver Superior No. 1 and 2 Walking Cultivator

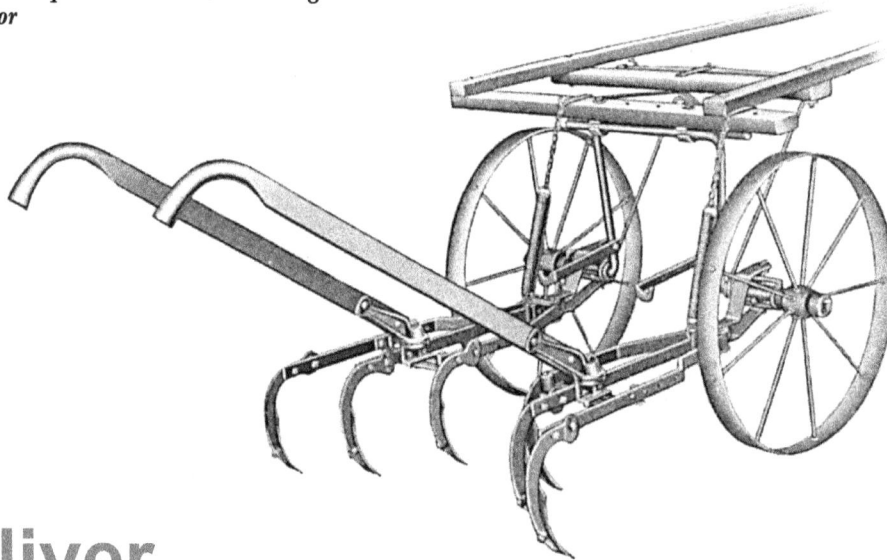

Oliver

The bridge between the walking cultivator and the riding cultivator came when wheels and shafts were added to the one horse implement ostensibly for the purpose of allowing greater depth control of the teeth especially for corn roots.

Right: A rear view of the No. 2 one-horse adjustable arch right spring tooth walking cultivator.

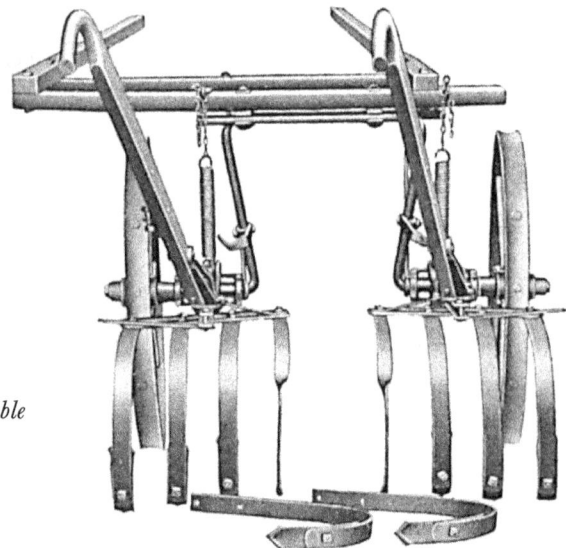

P & O

STEEL FRAME FOURTEEN TOOTH CULTIVATOR

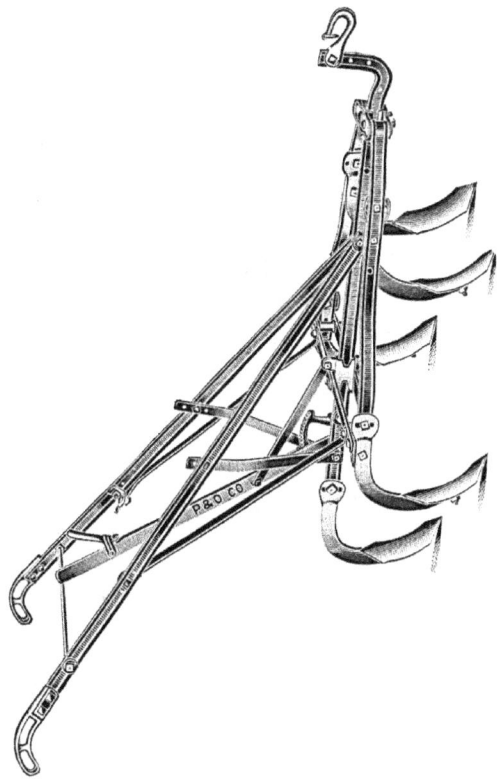

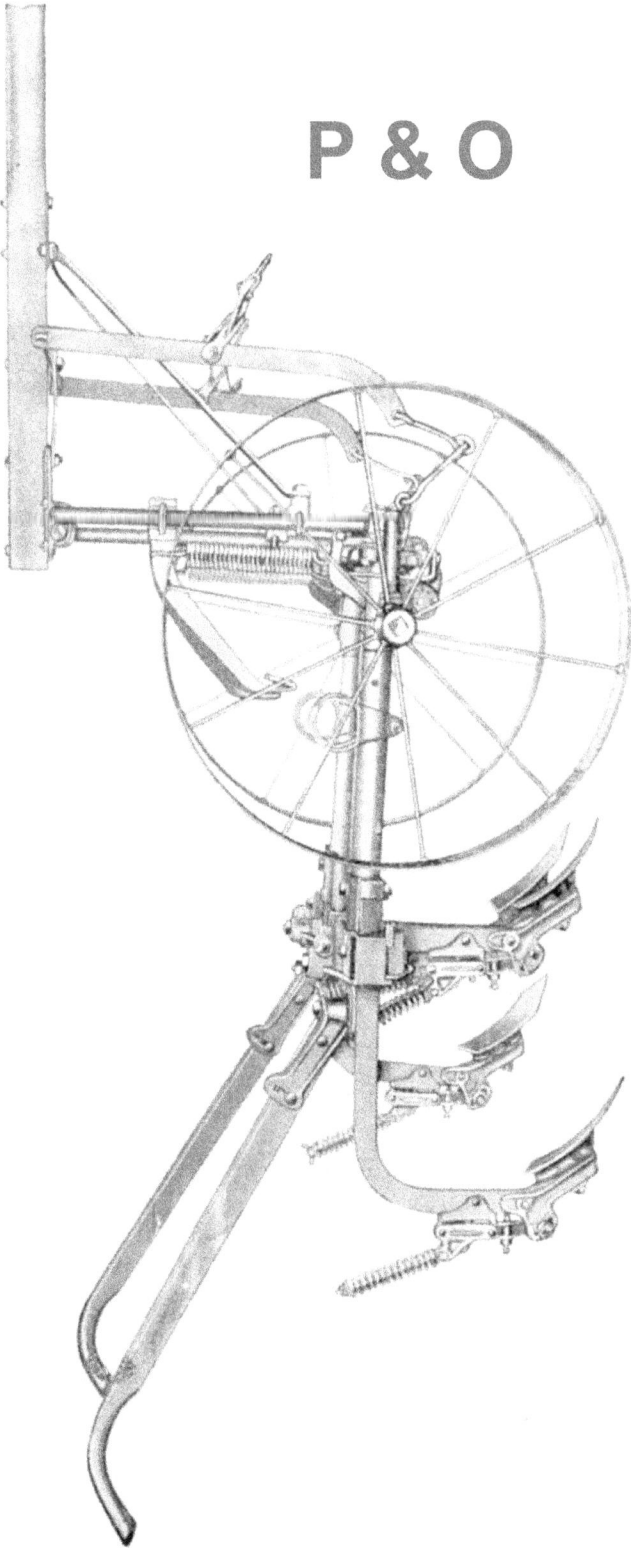

P. & O. ONE-HORSE CULTIVATORS.

NO. 3 SENIOR 5-TOOTH CULTIVATOR.

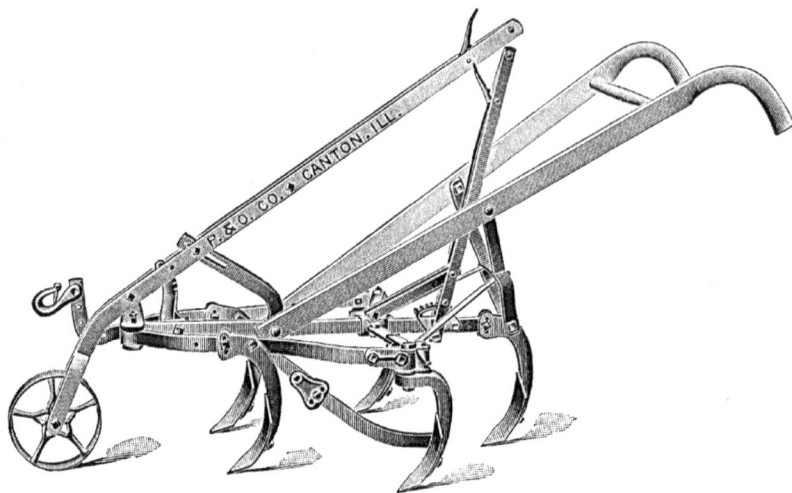

NO. 6 JUNIOR 5-TOOTH CULTIVATOR.

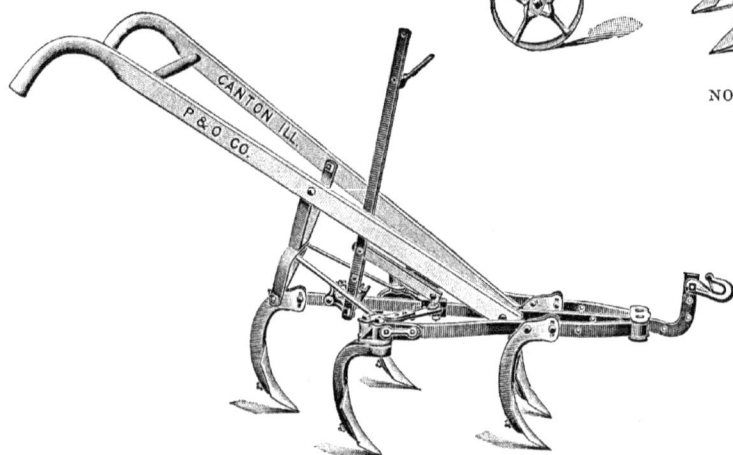

NO. 3 JUNIOR 5-TOOTH CULTIVATOR.

P & O

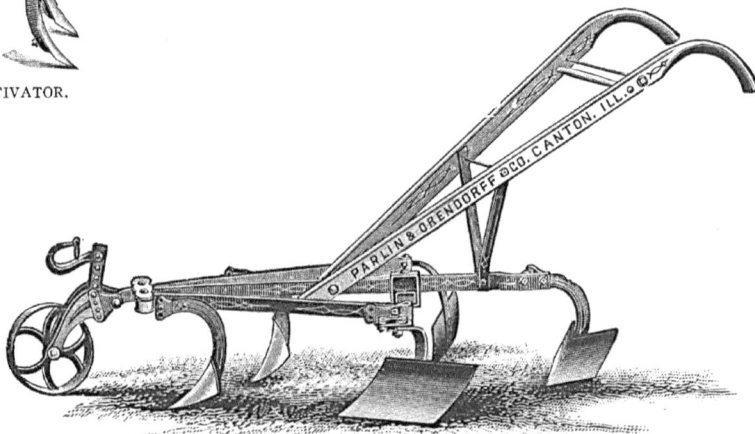

STEEL FRAME FIVE TOOTH CULTIVATOR EQUIPPED AS A HORSE HOE

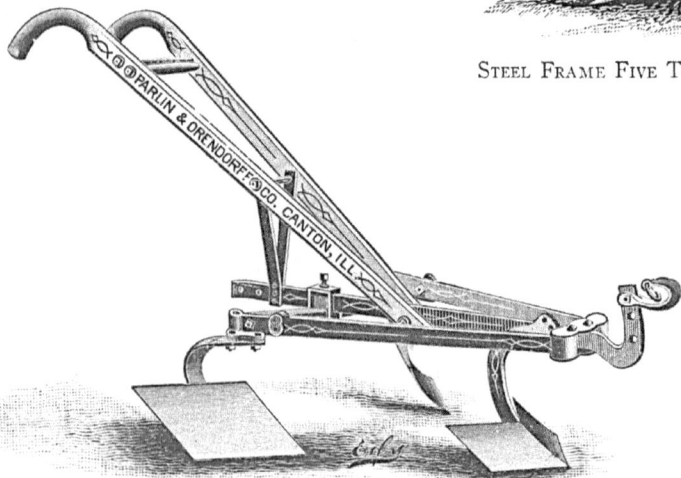

STEEL FRAME FIVE TOOTH CULTIVATOR AS A FURROW OPENER

P & O

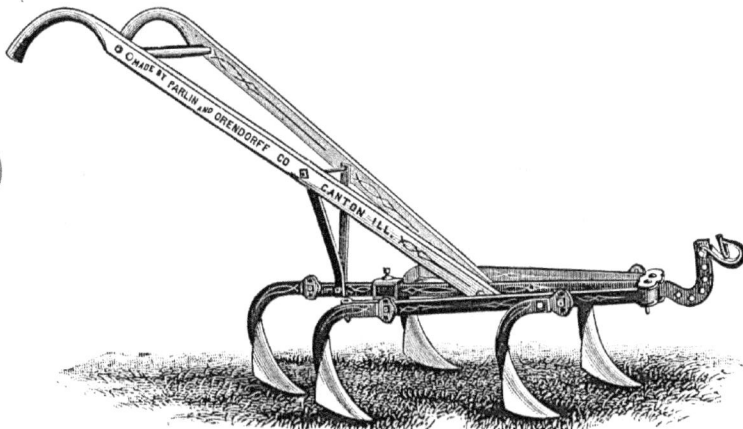

STEEL FRAME FIVE TOOTH CULTIVATOR

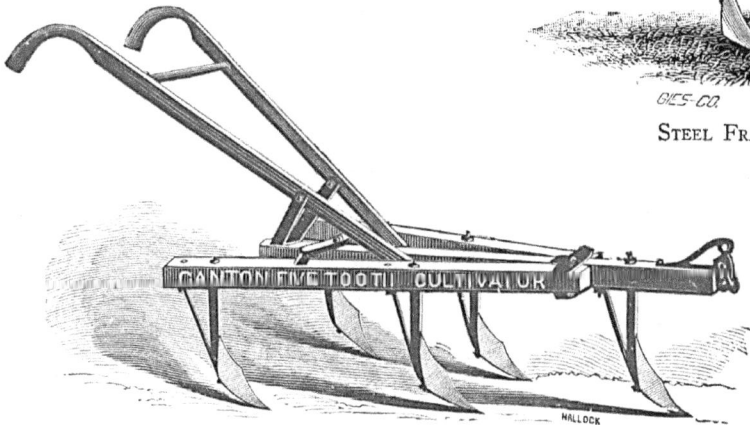

WOOD FRAME FIVE TOOTH CULTIVATOR.

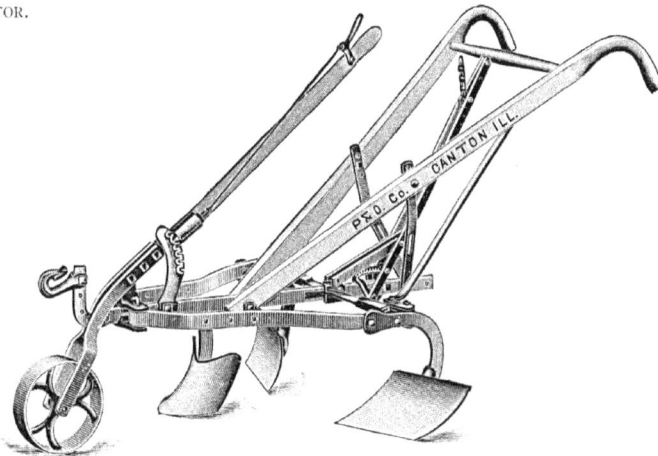

STEEL FRAME CULTIVATOR, WITH TWO LEVERS, GAUGE WHEEL AND SWEEPS

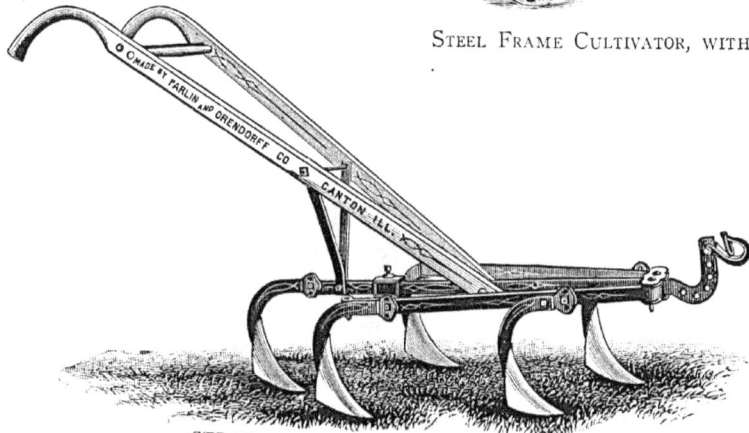

STEEL FRAME FIVE TOOTH CULTIVATOR

P & O Co 815

DIAMOND FLAT SHOVEL FOR ONE HORSE CULTIVATOR

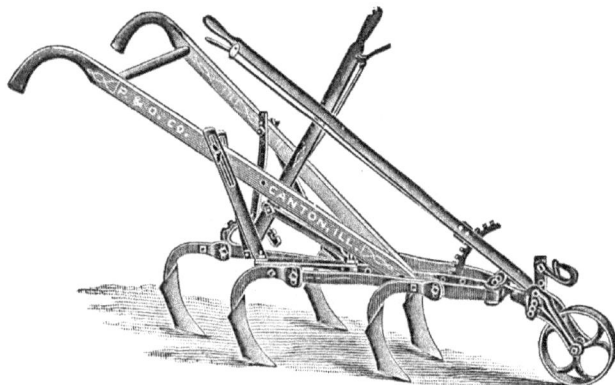

STEEL FRAME FIVE TOOTH CULTIV.

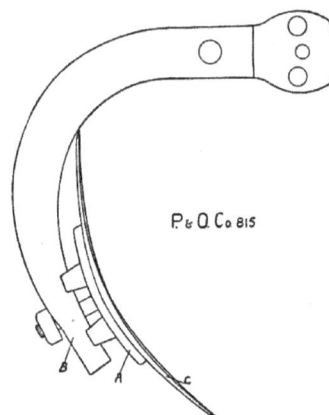

P & O

NO. 10 14-TOOTH CULTIVATOR.

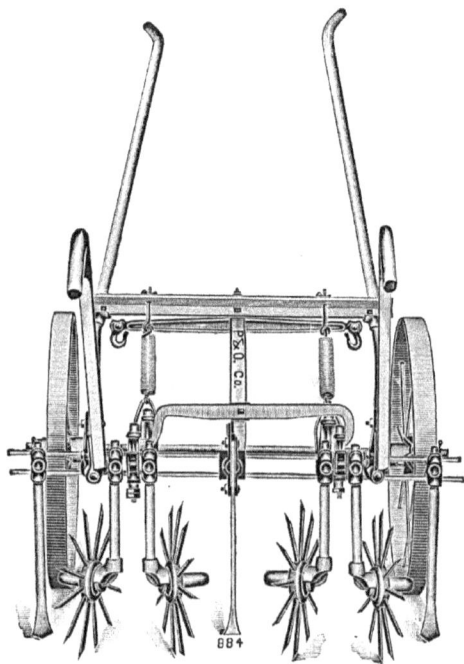

Canton Beet Cultivators

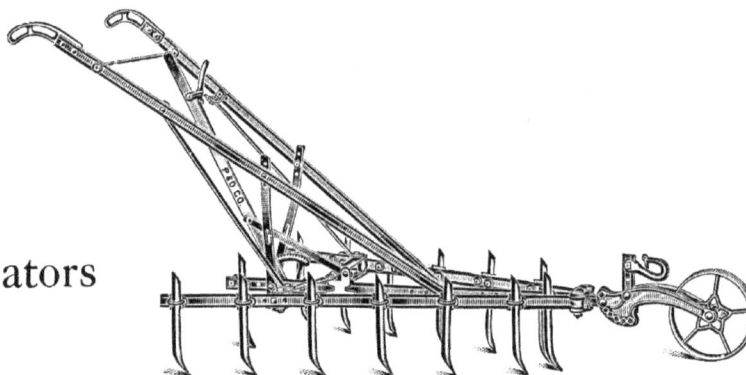

CANTON TWO ROW BEET CULTIVATOR

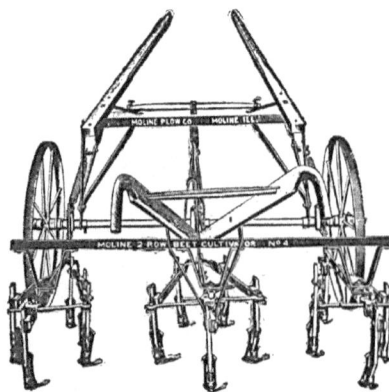

A most unusual one horse two row Moline cultivator designed for close planted beets.

I & J Mfg.

Five Tine Cultivator with adjustable row width.

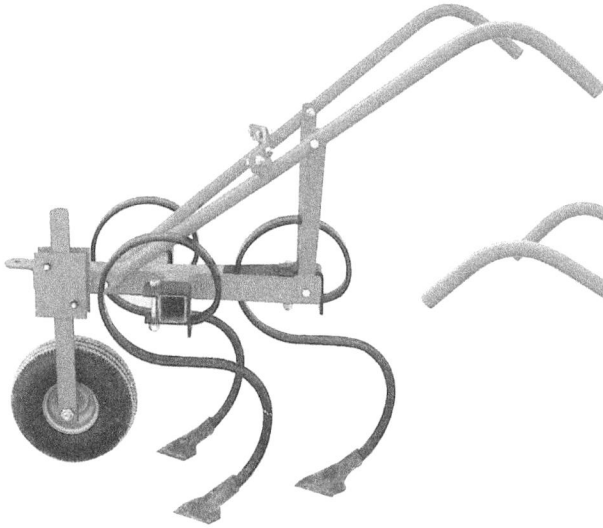

I & J Three Tine Cultivator for use with a horse or pony or oxen.

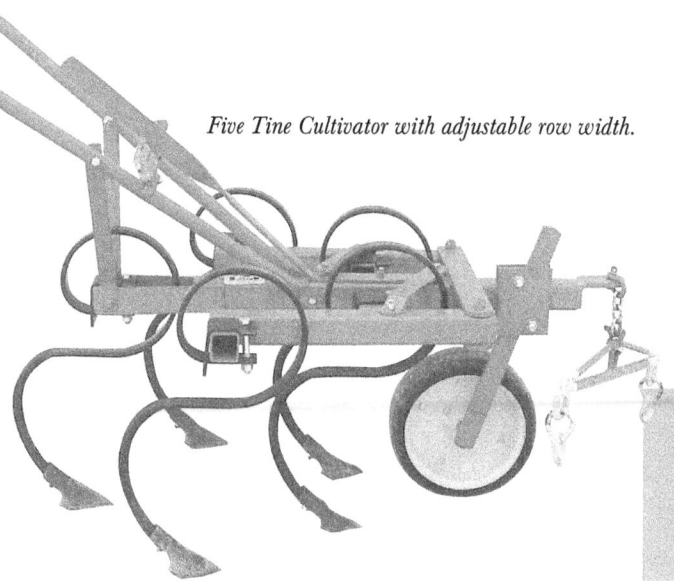

A riding version of the five tine is available

The I & J Manufacturing Company, in Gap, Pennsylvania, builds a handsome line of modern horsedrawn cultivators including the ones on these two pages and the ones at the end of the next two chapters. For contact information please go to the back of this book.

A buggy horse pressed to service pulling an I & J Five Tine Cultivator

I & J offers a detachable roller to add to the five tine and build a culti-packer or 'Culti-roller' as they call them. Here a Norwegian Fjord horse demonstrates the attitude and posture that makes the breed so popular with small farmers.

Eric Nordell and team on cultivator in Pennsylvania. Photo by John Nordell.

Chapter Fourteen
Two Horse Cultivators

Within the category of two horse cultivators the reader will discover more subtle yet significant variations than in any other horse drawn farm implement. This great variety was the result of vigorous competition between many excellent companies.

In the nineteenth century farmers were discovering en masse, that specific cares taken with row crops would result in dramatic increases in quality and yield. And that those cares, with regard to cultivation, were made infinitely easier with the advent of highly customizable cultivators drawn by horses or mules. From 1910 to 1920 the amazing design and procedural variety of factory-made cultivators reached a zenith which has never been paralleled or duplicated since. In this chapter you will see several hundred variations in design and setup, and these represent a fraction of

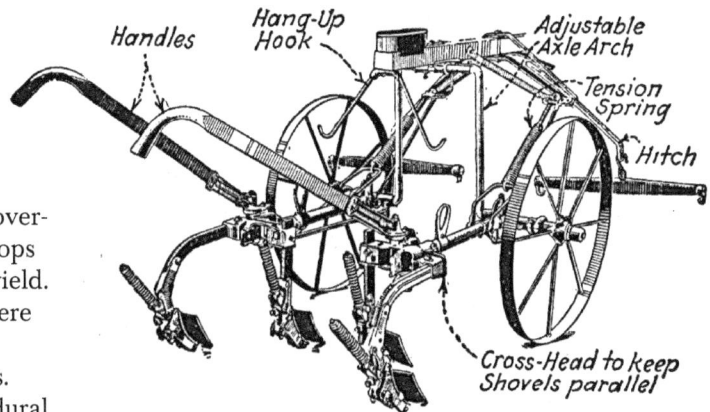
John Deere Straddle Row 2 horse walking cultivator with parts named

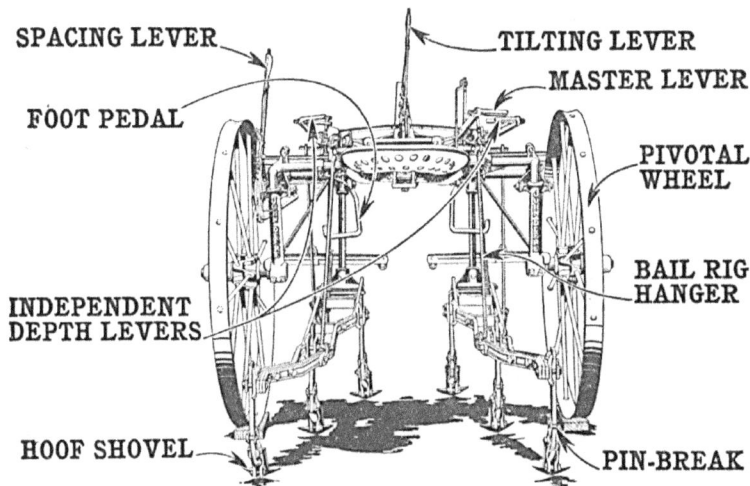
John Deere Two Horse Straddle Row Cultivator with parts named

what was available by 1925.

Our agriculture was well served (to vastly understate the condition) by the implement companies at that time and engineering design and procedural invention has only gone down hill.

Some would scoff and offer that the engineering and chemical advances of the latter half of the twentieth century far outstrip those of 1920. Their most common argument is that it takes fewer people to grow the same amount of food. Yet they cannot deny the physical evidence of these last 70 years, loss of topsoil, loss of natural fertility, loss of biological diversity in farm environs, loss of productivity, loss of product quality, loss of markets, loss of farms, loss of small towns, loss of heritage, and loss of farming craftsmanship. And what a silly mistake to proudly announce we've succeeded at needing less farmers? That's nothing to be proud of. When we can once again proclaim that we have gained X number of new small farms we'll have reason to be hopeful and pleased.

Meanwhile we make our important if feeble attempts to preserve the richness and specific design triumphs of that earlier way of working by picture books such as this.

This chapter features those more common cultivators which employed the draft power of two animals. Most of these are "straddle row" design. All but a few featured a tongue or pole. They are hitched to in much the same manner as one would hook to any wheeled vehicle.

Cautionary Note: While almost any implement may be a hazard when using green or nervous animals, the straddle row riding cultivator has a risk aspect somewhat unique to its design. Twenty years ago this author had occasion, for part of the spring, to enter a few pulling matches with a fine well broke team of farm geldings. On one Monday morning we hitched to our Oliver riding cultivator to go out to the young corn rows. The last thing the horses remembered was a Saturday night winning pull and all the rhythms of stop and go that entailed. When I climbed "into" the basket-like position of the seat and spoke to those horses they raised up anxious, as if getting prepared for a difficult pull. I pulled back on the lines and said, 'Whoa.' The horses kept on coming up, I was causing it. Feeling like the whole apparatus was coming over backwards, I pulled on the lines even more but for balance. We narrowly avoided an accident that would have damaged horses, harness, cultivator and one young idiot teamster. It taught me several important lessons one of which was that there is no quick and easy way to gracefully and quickly exit that cultivator seat.

Avery

Avery Queen Cultivator

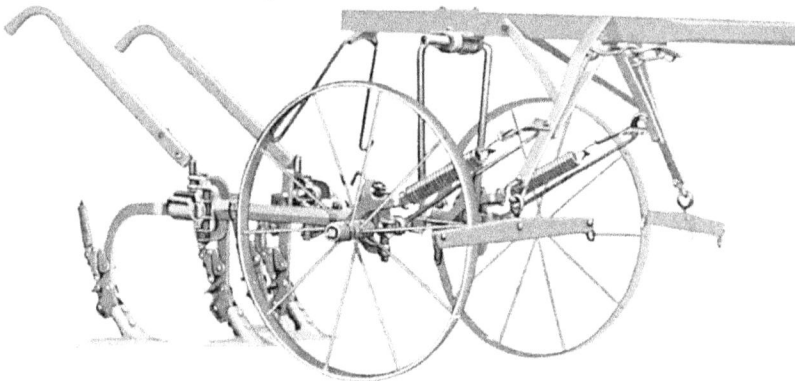

Majestic Cultivator

The two horse walking straddle row cultivator was not just a logical bridge to the riding cultivator. In many instances it remained the first choice because the farmer could see down the row to what variations were coming and make adjustments with the handles. For many farmers familiar with walking plows and single horse cultivators this procedure was comfortable and reassuring .

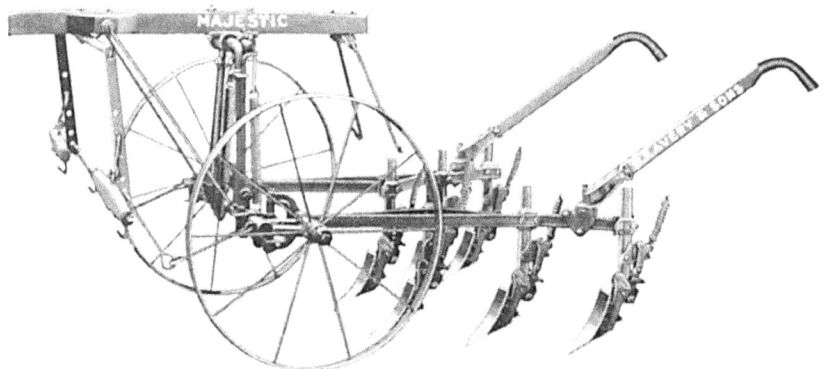

No. 6 Majestic, with Channel Beam, Equipped with Adjustable Spring Trip Standards, with Open Foot

Majestic Jr. Cultivator

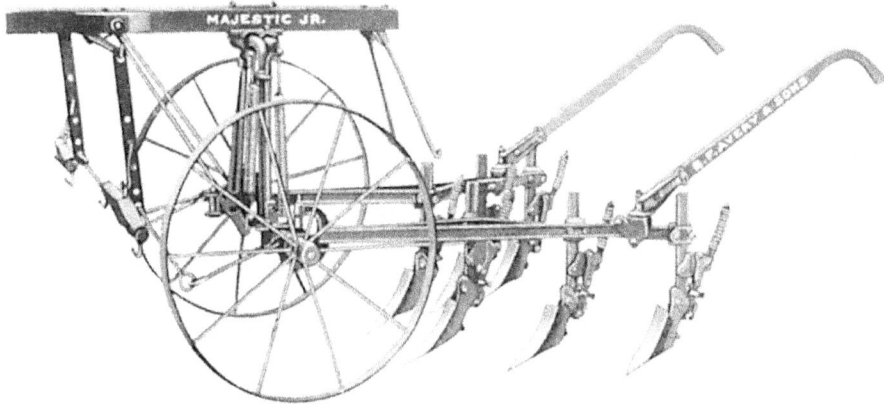

Avery

No. 6—Six-Shovel. Channel Beam, with Adjustable Spring Trip Standards, Open Foot

Bob White Cultivators

Showing the slide arrangement for adjusting gang position on the Avery.

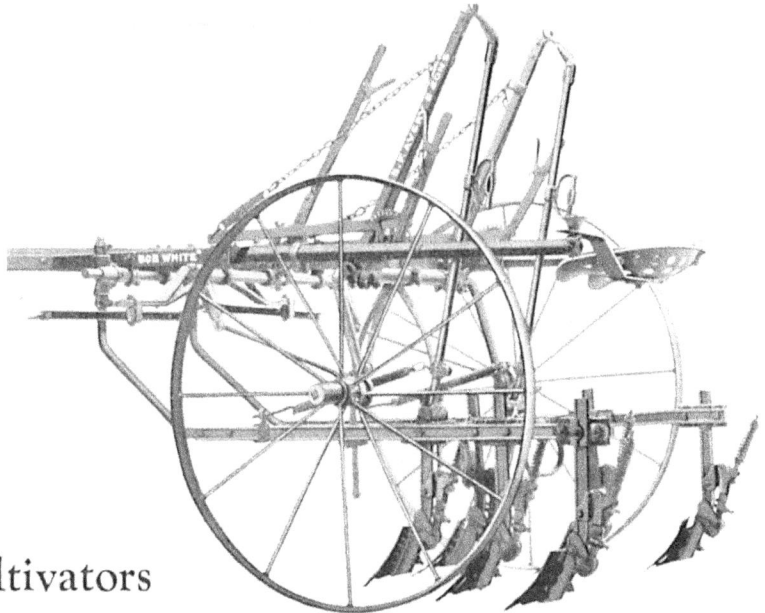

Avery Jack Rabbit Cultivators

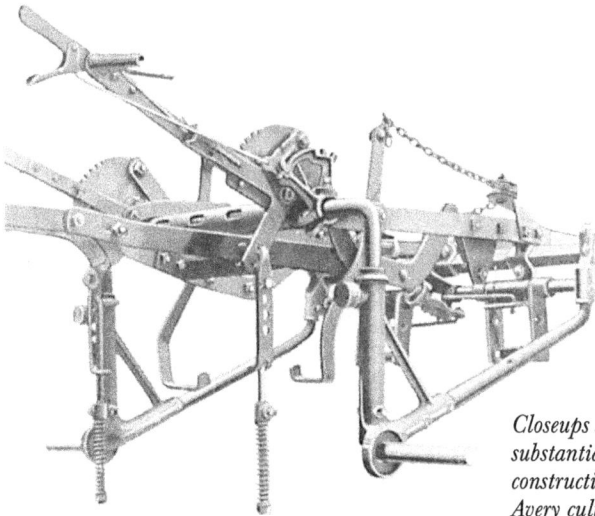

Closeups showing the substantial and impressive construction elements of the Avery cultivator frames.

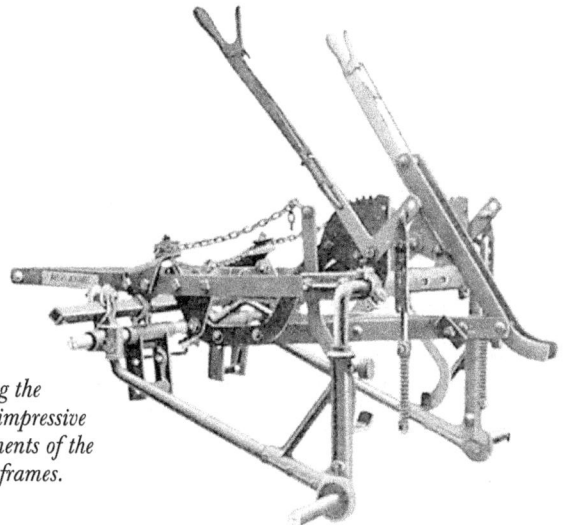

Avery

John Deere

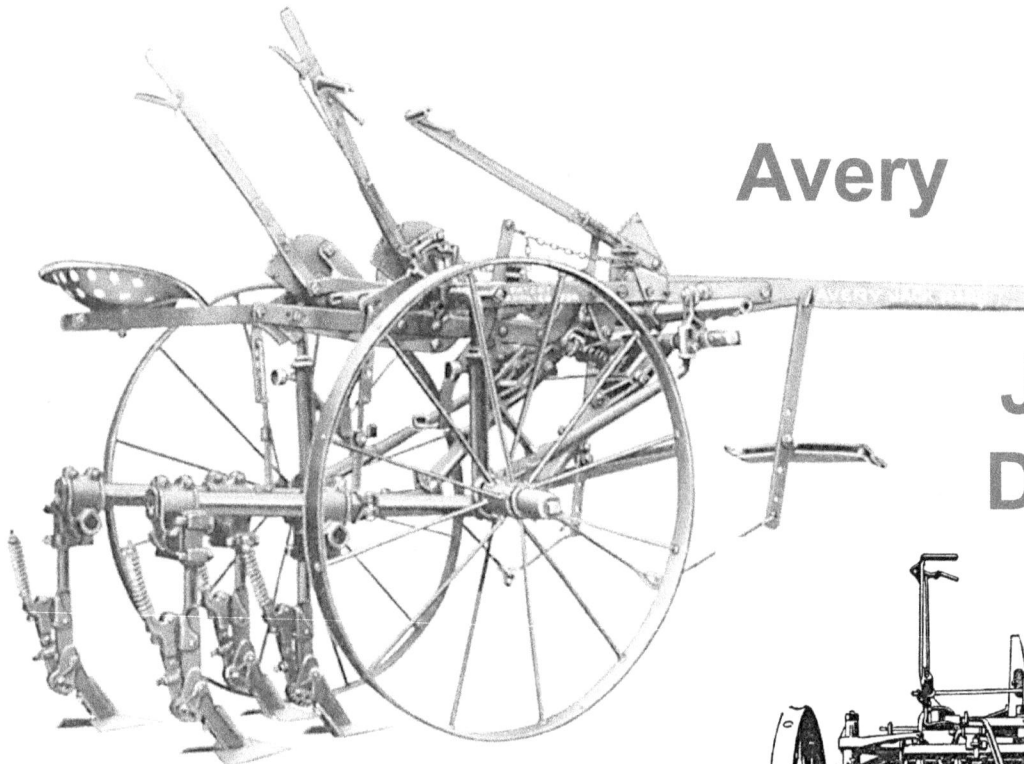

Avery Jack Rabbit Cultivator

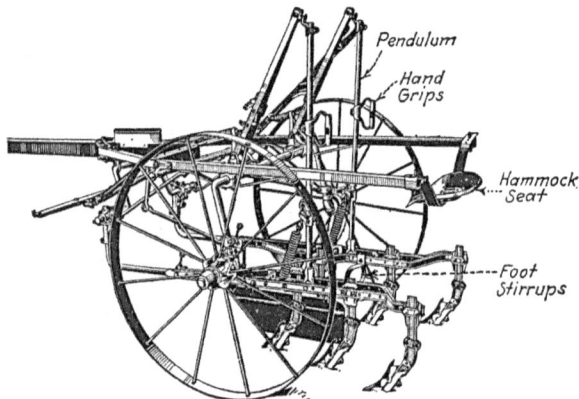

Pendulum

Hand Grips

Hammock Seat

Foot Stirrups

·Direct foot-controlled type of cultivator.

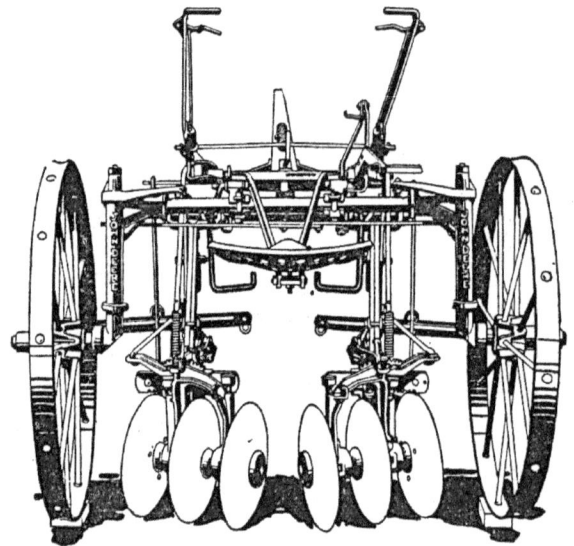

Disk cultivator.

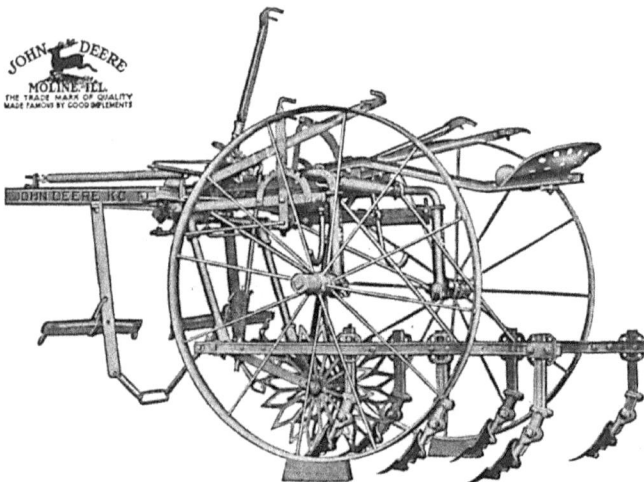

Hoof Shovels and Sweeps Can Be Used on this Corn Cultivator—Sweeps Are Splendid for Surface Cultivation

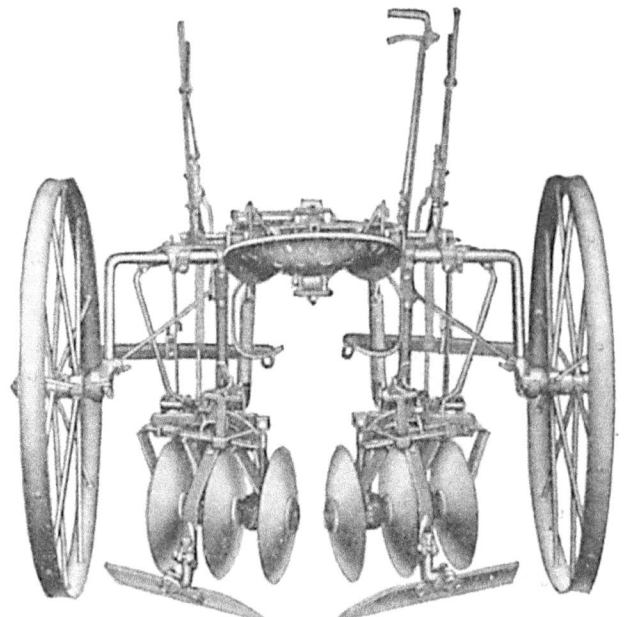

An Excellent Tool for Shallow Cultivation of Corn

212

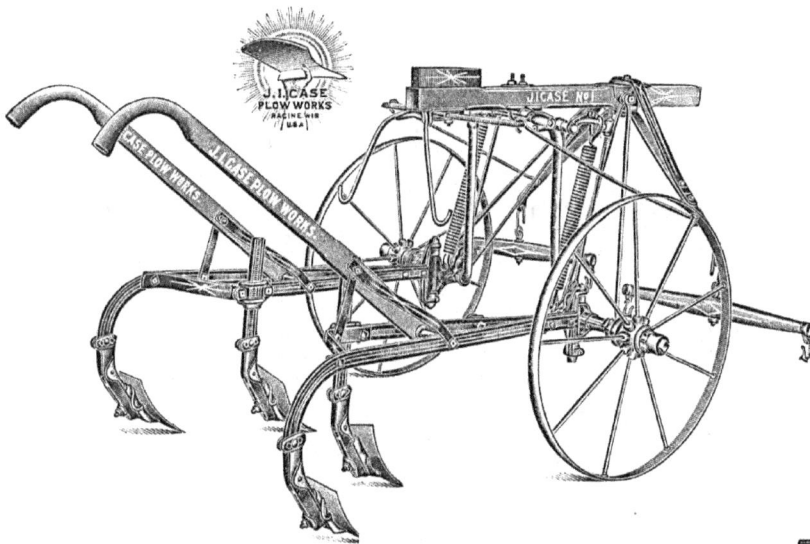

Case

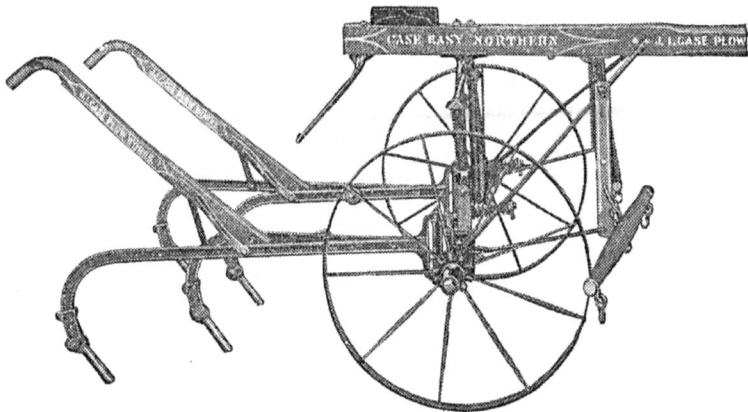

CASE EASY NORTHERN
WALKING CULTIVATOR

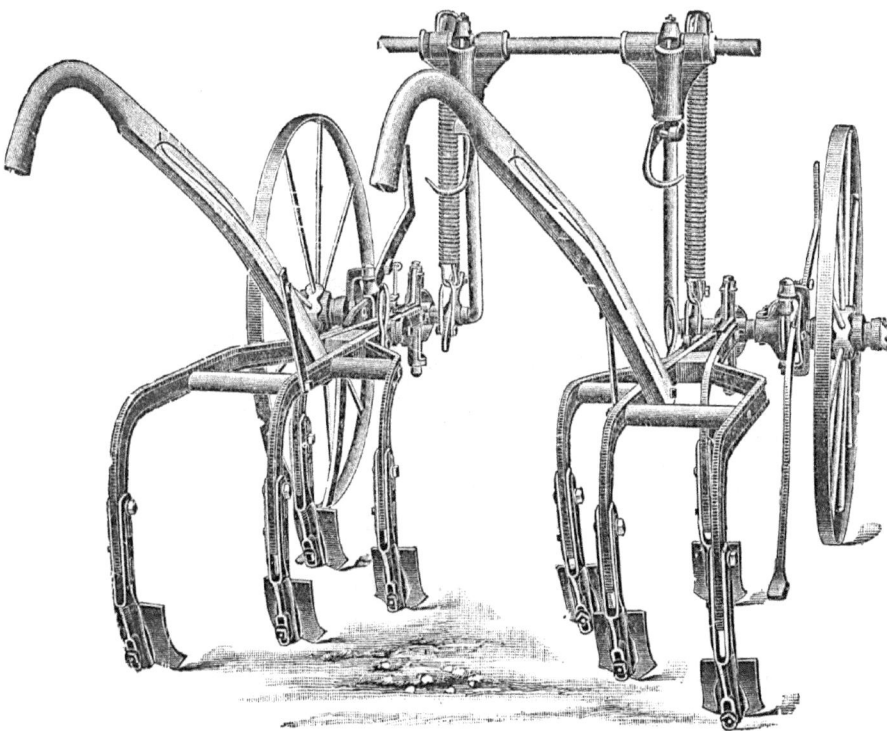

J. I. CASE RUBY TONGUELESS CULTIVATOR

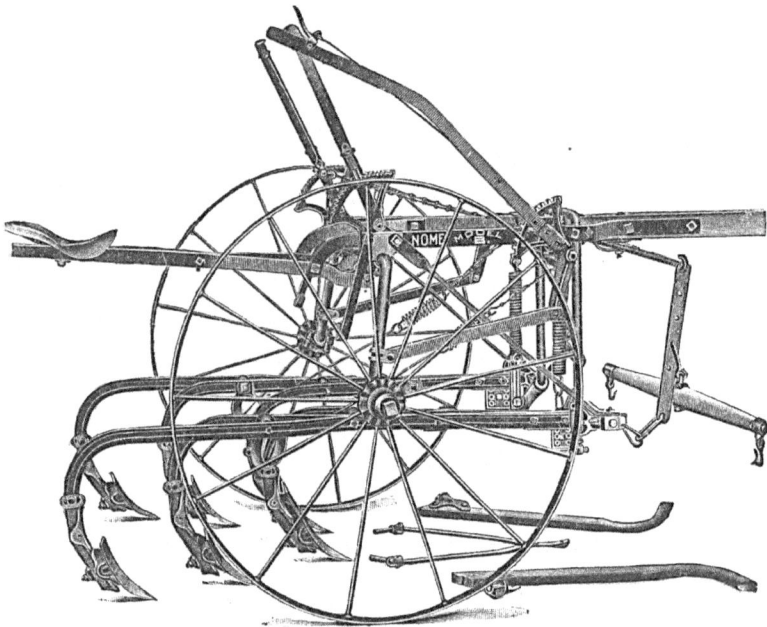

J. I. CASE NOME CULTIVATOR

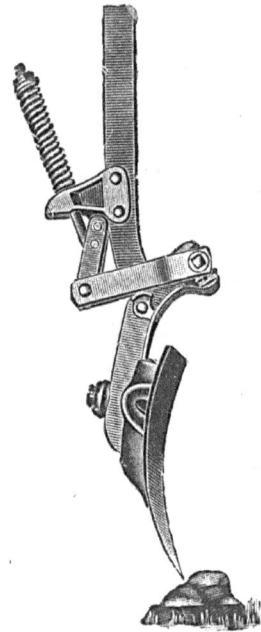

Illustrating the J. I. Case cultivator shank trip mechanism

Case

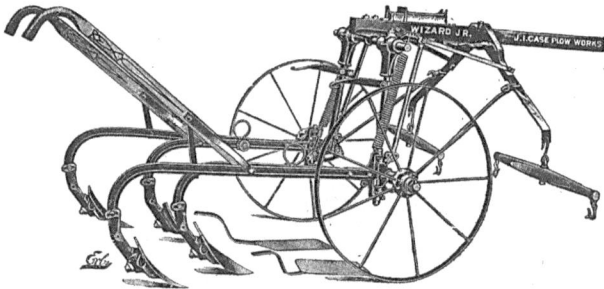

WIZARD JUNIOR CULTIVATOR

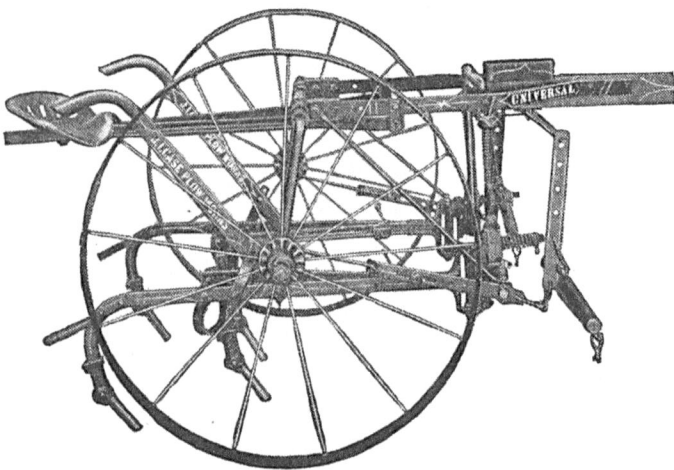

J. I. CASE UNIVERSAL CULTIVATOR

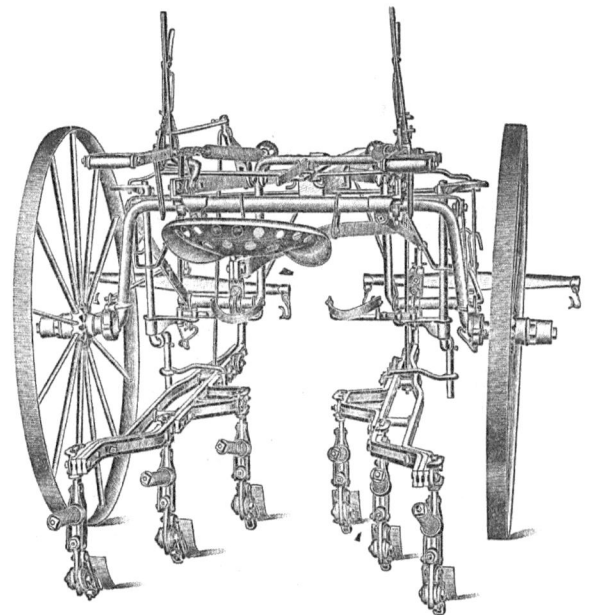

J. I. CASE PIVOT FRAME CULTIVATOR

NOME COMBINED RIDING AND WALKING CULTIVATOR

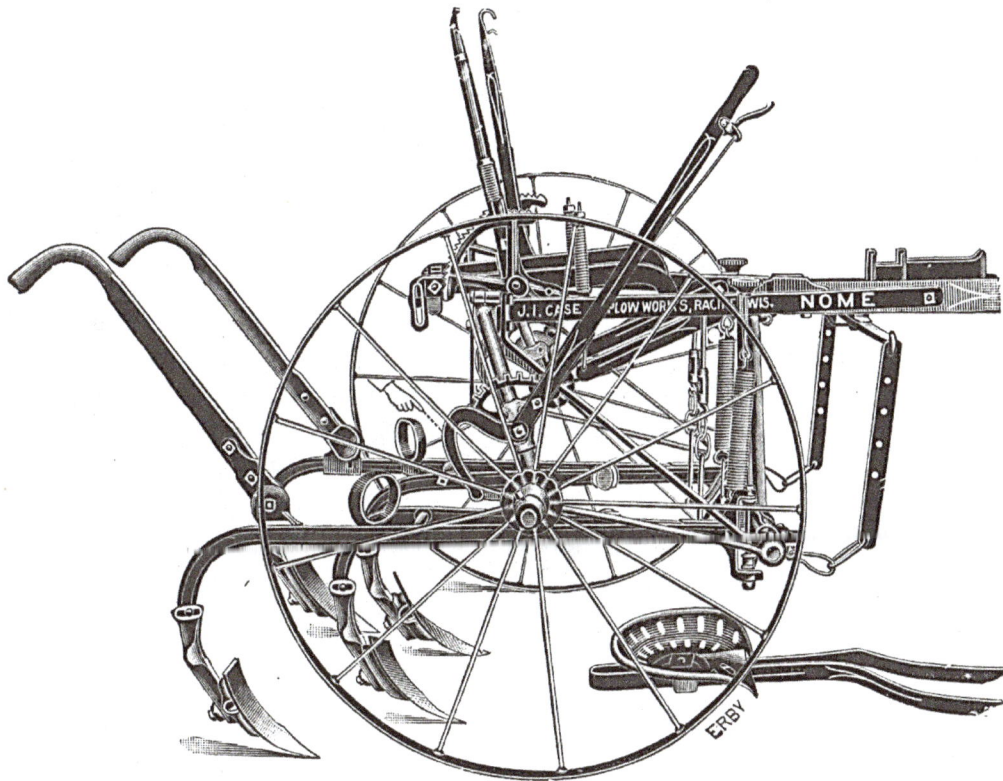

Case

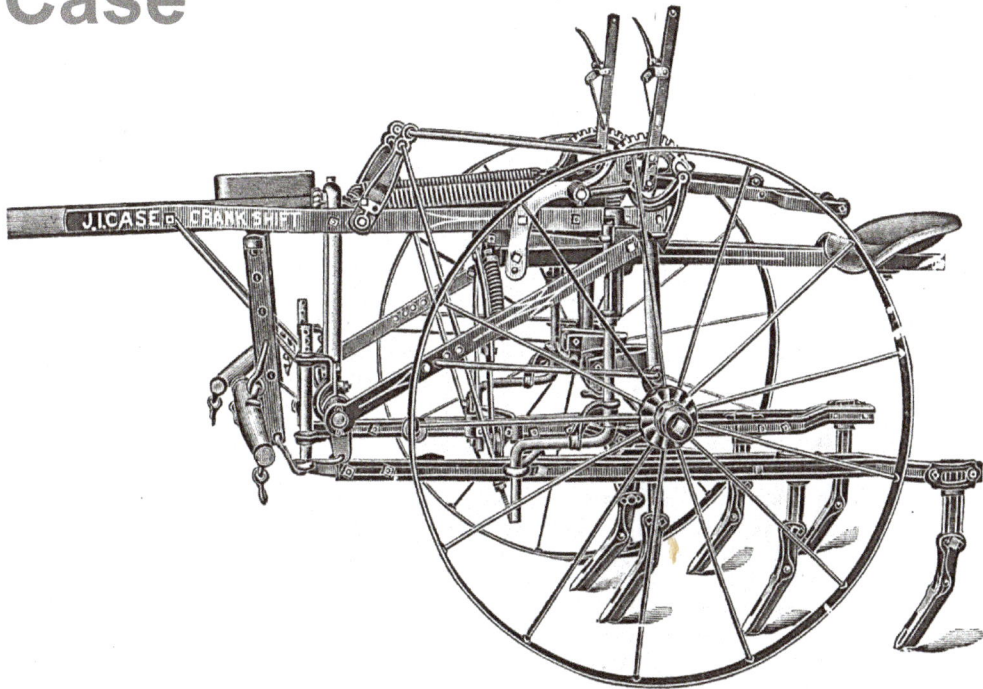

THE J. I. CASE CRANK SHIFT CULTIVATOR

Case

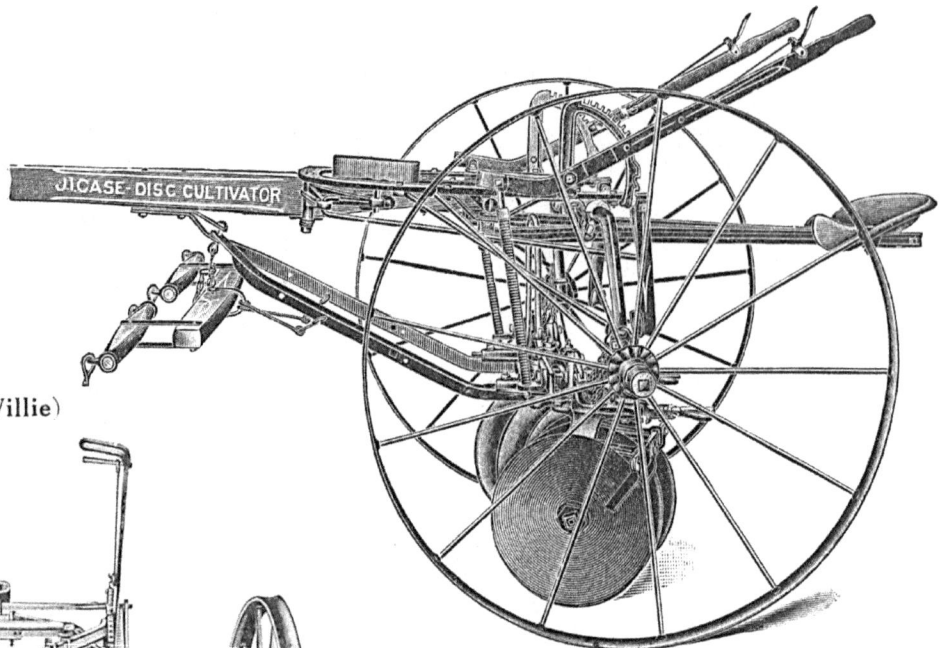

(Big Willie)

Side View Showing Hitch Direct on Gangs

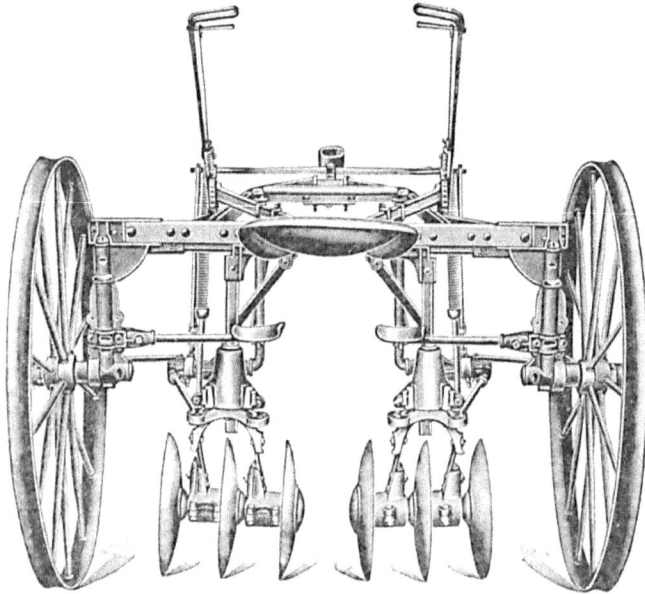

BW-16 Cultivator

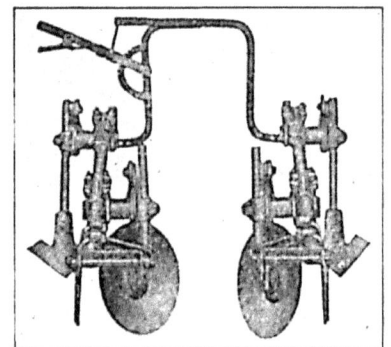

Disk-hiller attachment and jockey arch.

Moline

A-16 Disc Cultivator Equipped with Scrapers

Dustproof
Malleable
Parts and
Steel Axles.
St'l Ratchets
and Pins

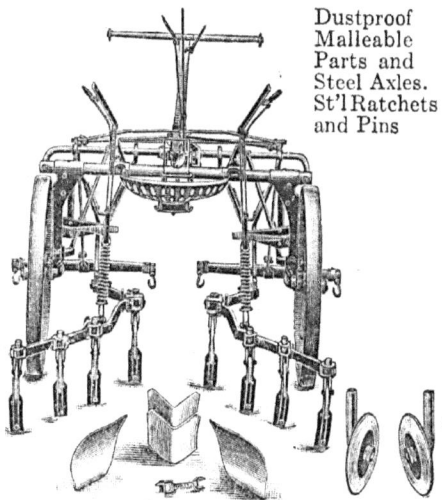

Cuts from Planet Jr. literature illustrating the package deals with optional shovel, points and discs thrown in.

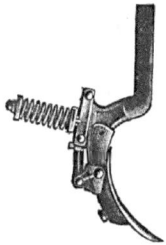

Planet Jr.

Roderick Lean No. 3 "New Century" Leverless Cultivator

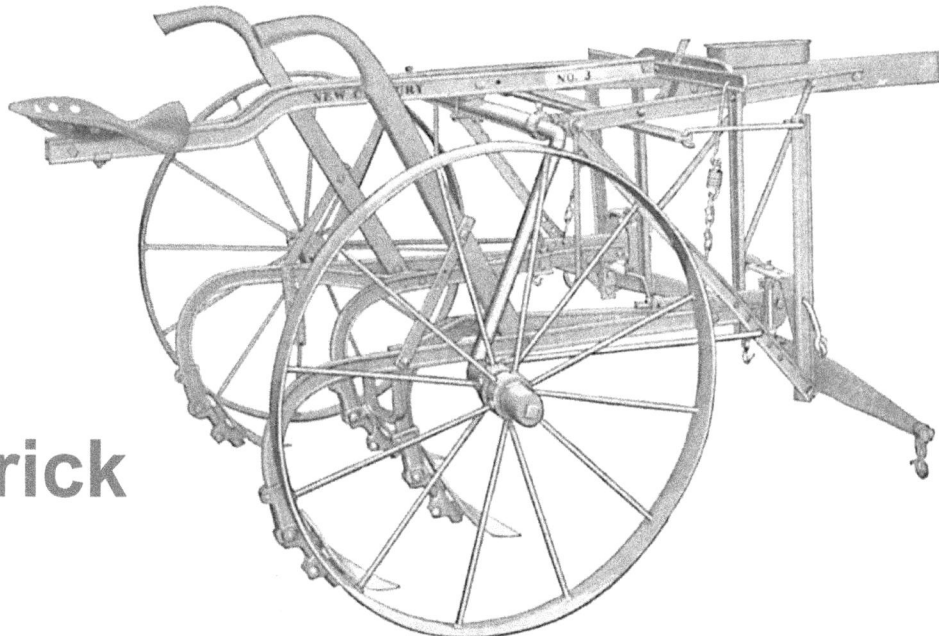

Roderick Lean

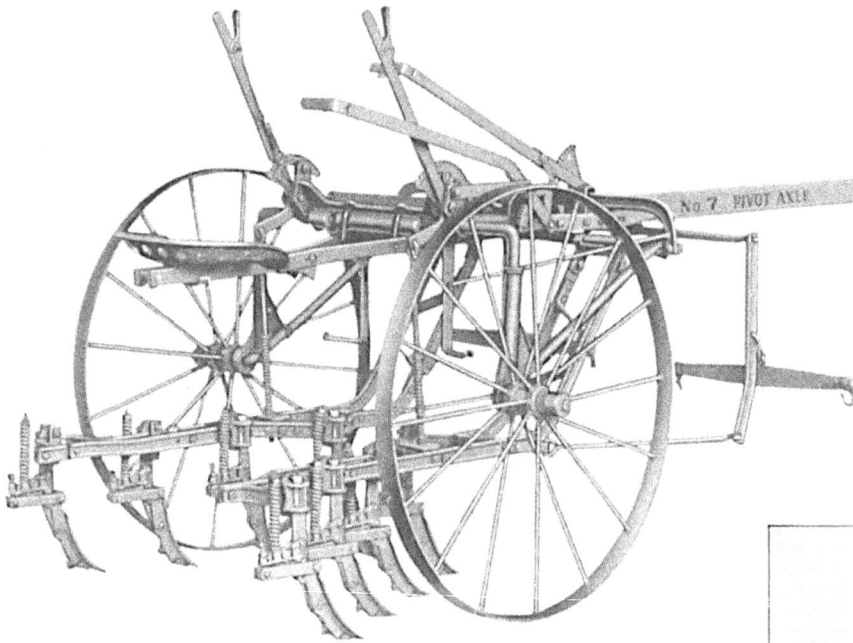

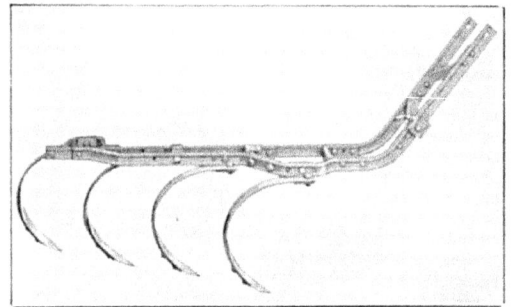

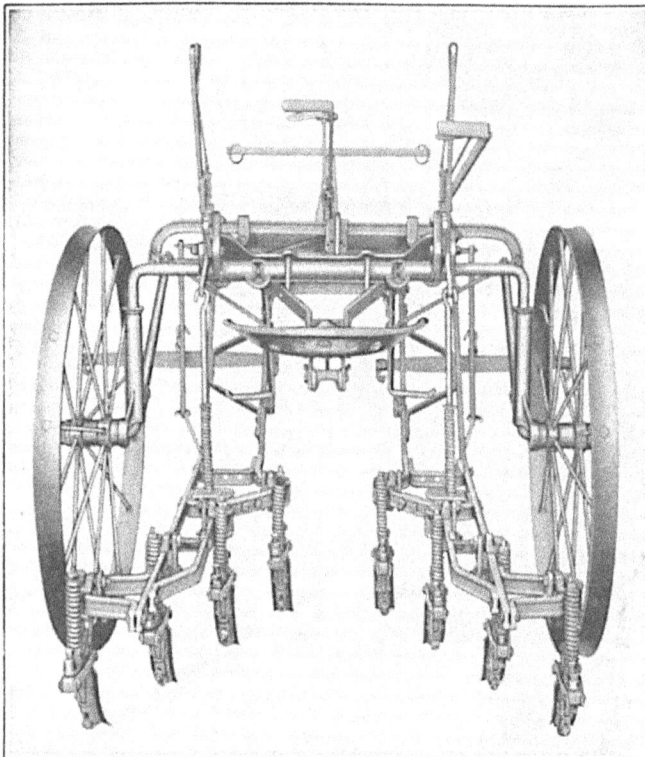

Roderick Lean

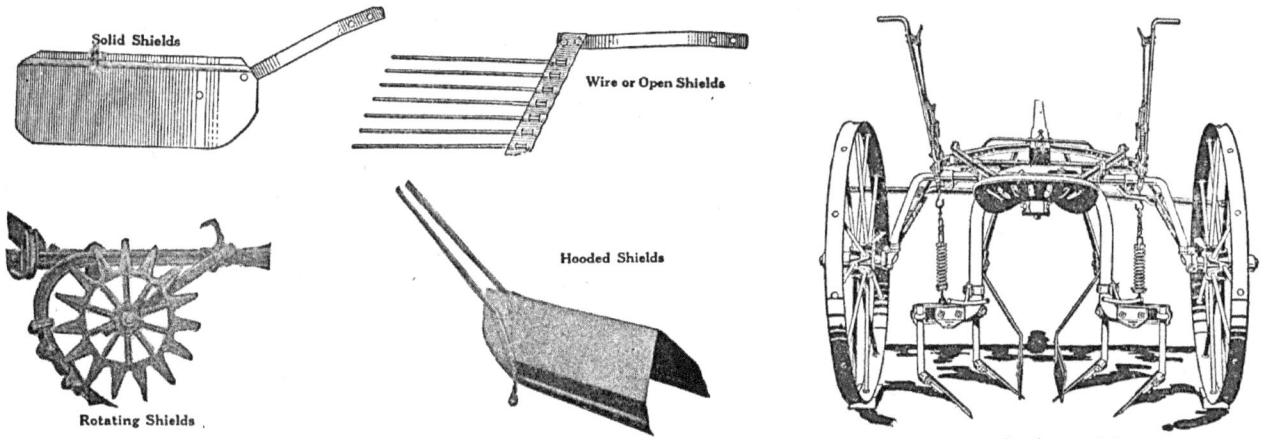

FIG. 323.—Shields or fenders used on cultivators.

Surface cultivator.

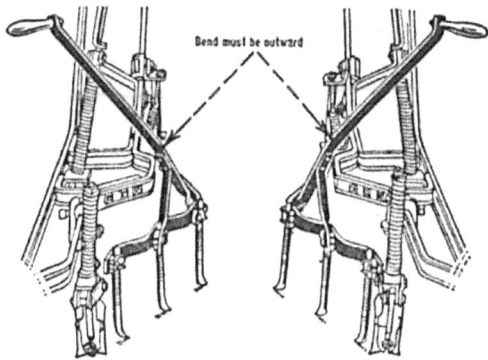

Tobacco attachment.

John Deere Rotating Cultivator Shields

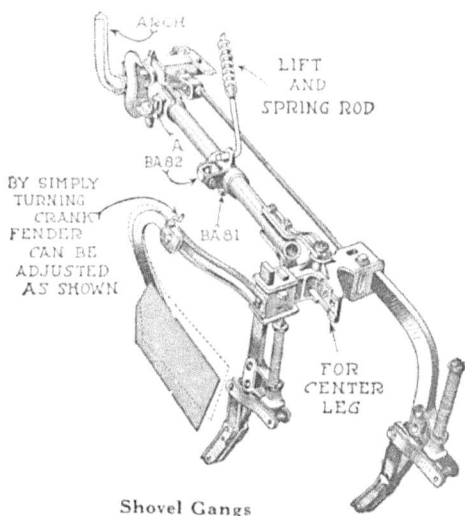

Shovel Gangs

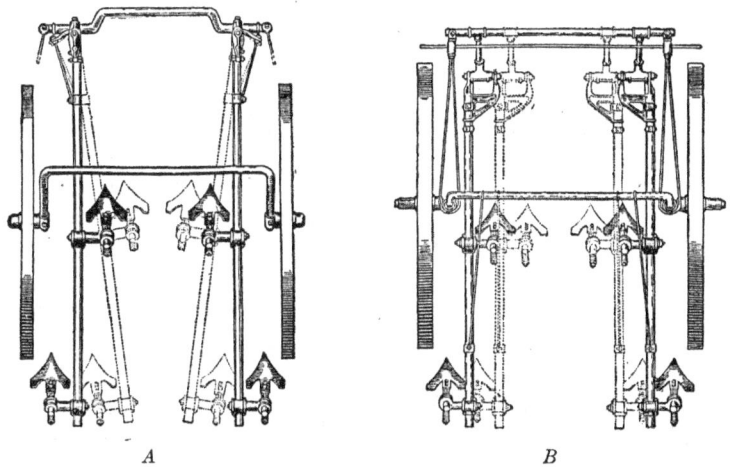

A, shows the action and movement of gangs and shovels when pivoted at the front; *B*, shows the movement when the gangs are shifted parallel.

Style "A" Walking Cultivator

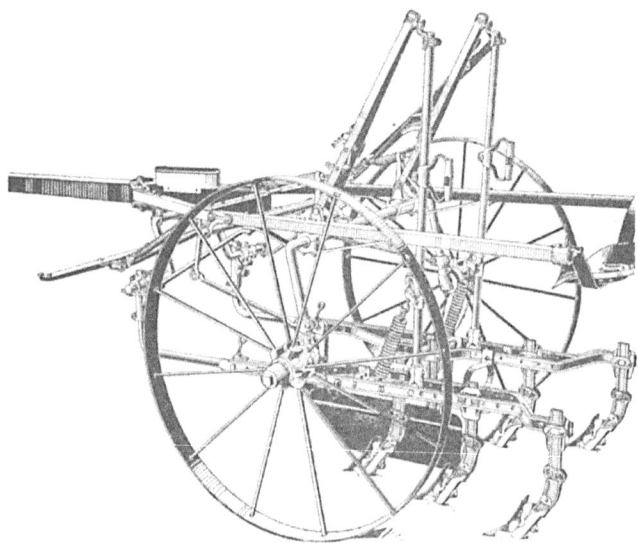

Style "ZD" Riding Cultivator

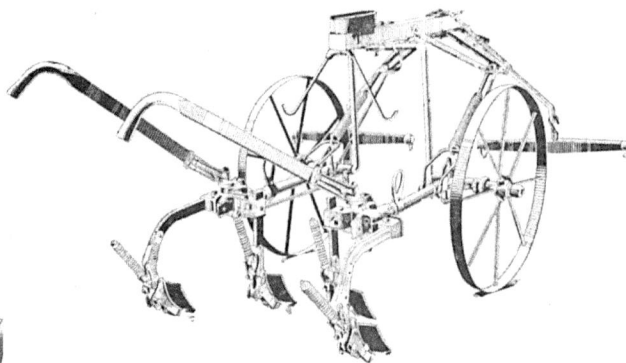

McCormick

No. 8 Surface Cultivator.

Rock Island No. 70 Perfection Cultivator

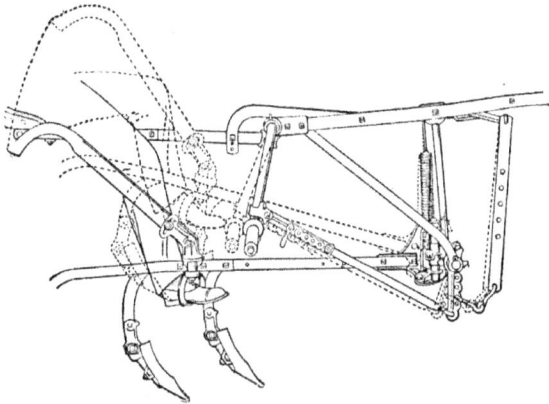

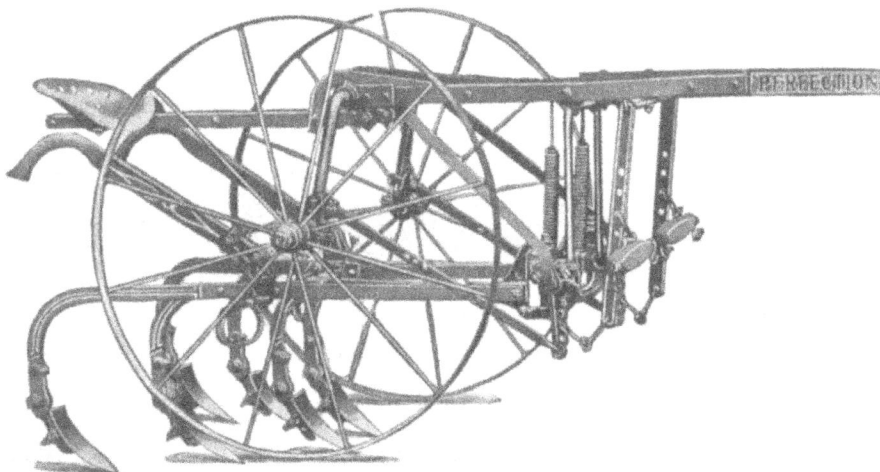

A detail of the Rock Island No. 70 Perfection shows the self-balancing aspect. "If you come to a place where you want to plow deeper, or the ground is hard, you can press down on one gang without affecting the other. Your own weight doesn't keep bringing one gang up when the other goes down. You don't have to overcome any spring tension, for the weight of the gang does that. When you start to lift the gangs with your feet, the spring acts and will bring the gang up so it requires very little effort or work on the part of the driver... Notice the dotted line and you will see how the wheels are shifted back as the gangs are raised and forward as the gangs are lowered... (the drawing on the right) shows how the arm to which the lower end of the spring attaches has several different adjustments, so that in a moment's time you can adjust the spring to any tension you want in order to give you all the spring action you want.

You will see that the connection between the arm on the coupling sleeve and the crank axle is adjustable to suit the weight of the driver, from a small boy to a heavy man. In this way the weight

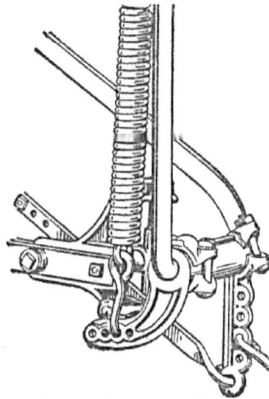

Gang Balance Spring

Rock Island

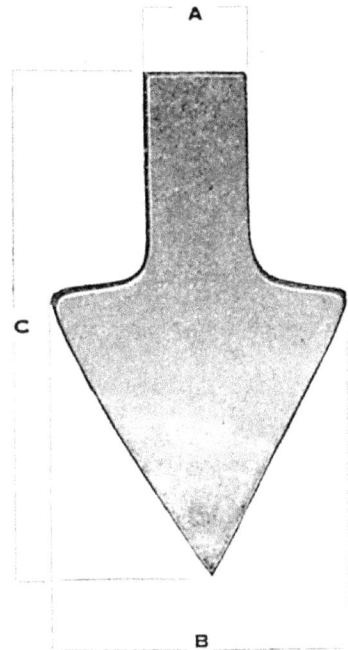

of the driver balances the weight of the gangs, and when the gangs are in the ground at work, the hitch pulls the shovels into the ground.

You can shift the front end of the gangs in or out any amount, from a fraction of an inch up to several inches. This gives you a wide scope to fit different conditions.

You have no set position for adjustment, so that you can put the gangs any place you want, and this is a very important feature."

ROCK ISLAND NO. 70 PERFECTION CULTIVATOR

The Original Balance Frame Cultivator

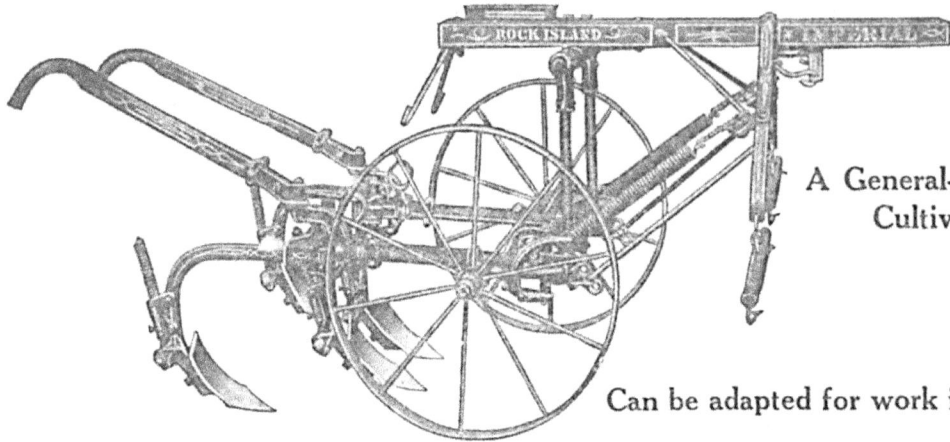

A General-Purpose
Cultivator

Can be adapted for work in cane

Rock Island No. 66 Imperial Cultivator

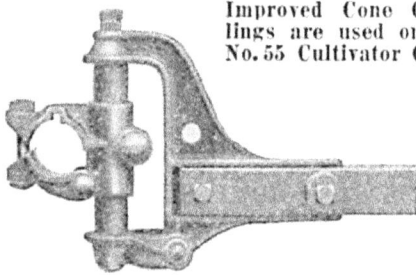

Improved Cone Coup-
lings are used on the
No. 55 Cultivator Gang.

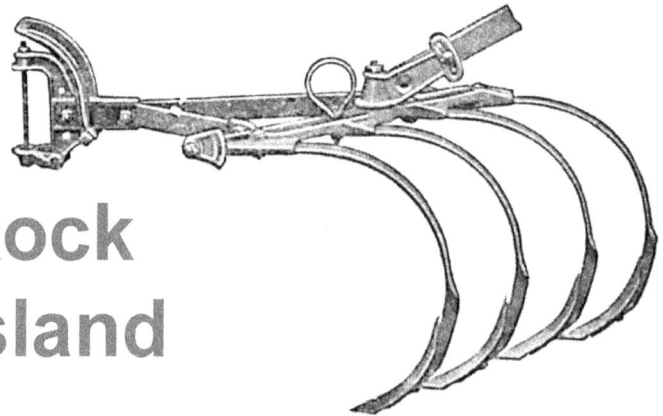

Rock Island

Spring Tooth Attachment for Rock Island Cultivators

The Walking Cultivator With the Easy Lift

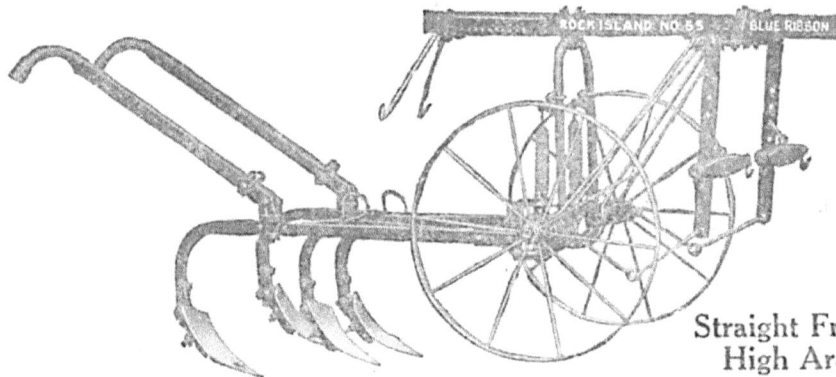

Straight Frame
High Arch

Rock Island No. 55 Blue Ribbon Walking Cultivator

Fits Any
Rock Island Pipe
Beam Cultivator

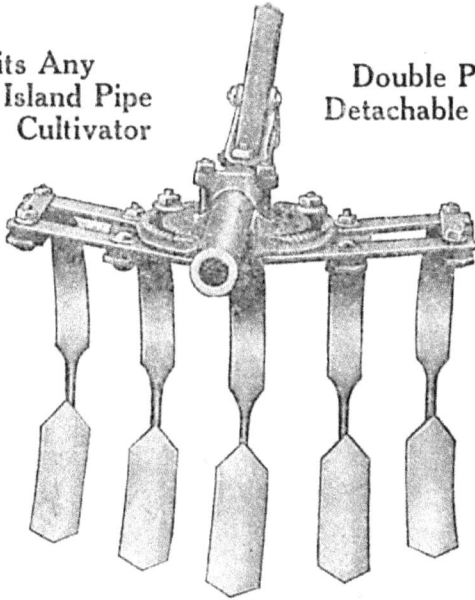

Double Pointed
Detachable Shovels

Rock
Island

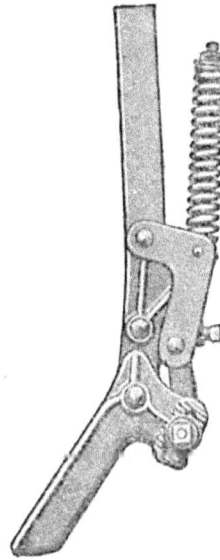

Rock Island
Spring Trip

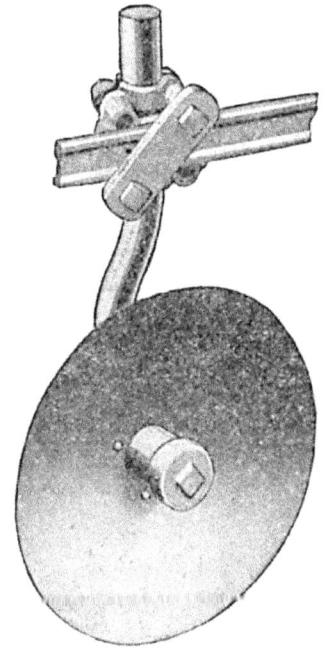

Disc Hiller

A Seat-Shift Disc Cultivator

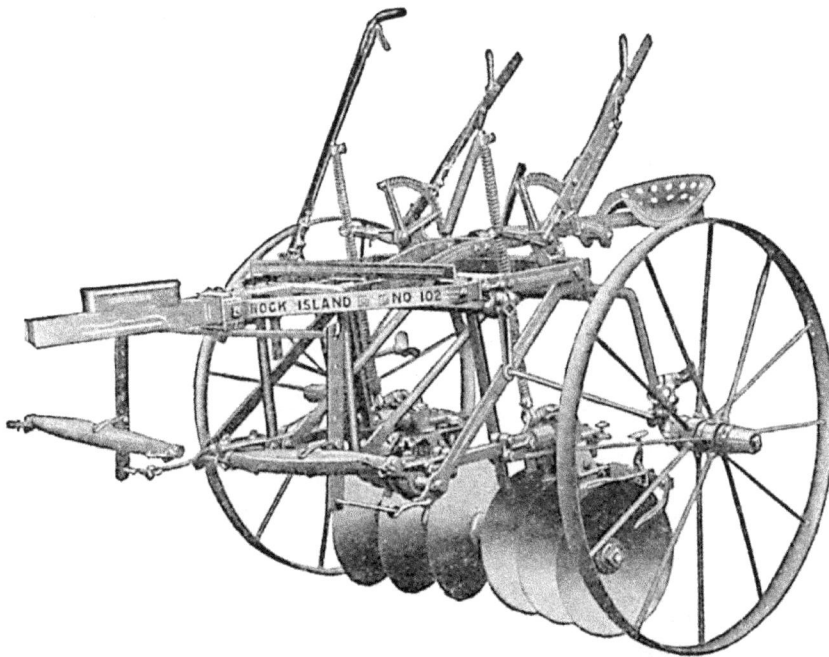

Rock Island No. 102 Disc Cultivator

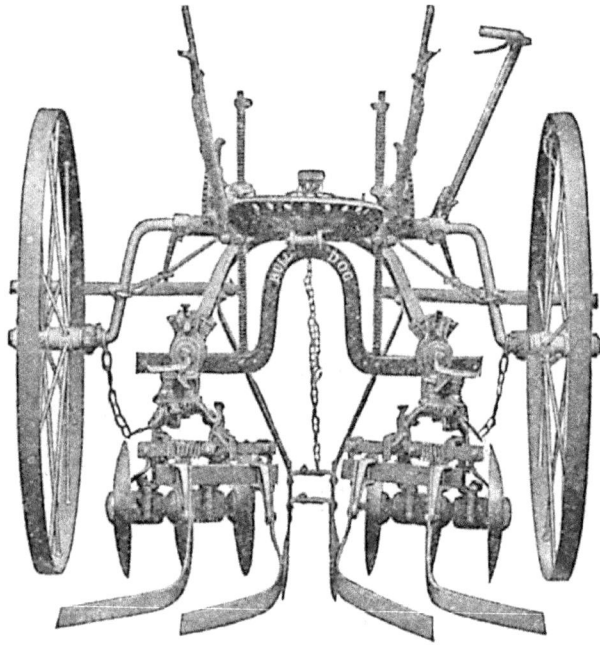

Rock Island

Rock Island No. 100 "Bull Dog" Cultivator

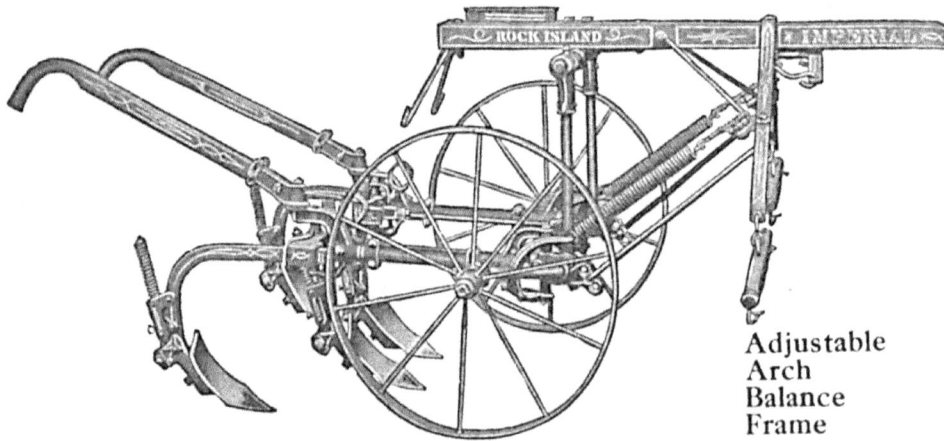

Adjustable
Arch
Balance
Frame

ROCK ISLAND NO. 61 WALKING CULTIVATOR

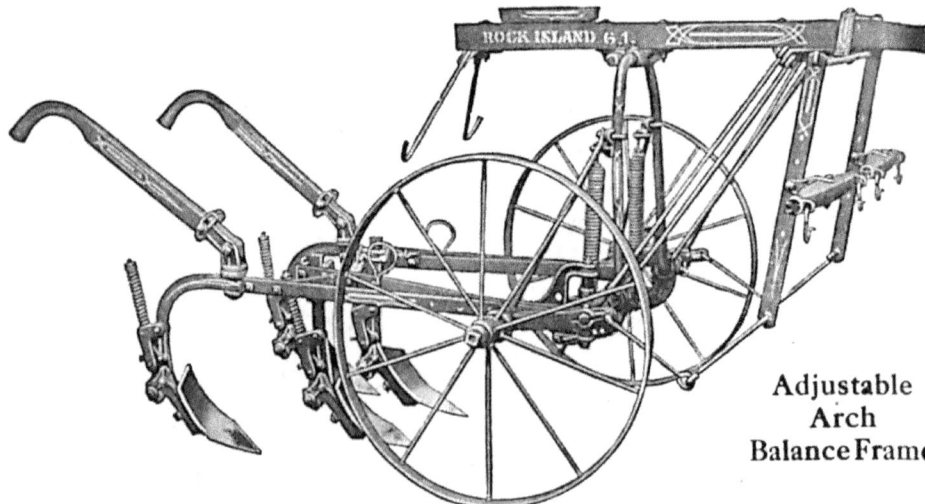

Adjustable
Arch
Balance Frame

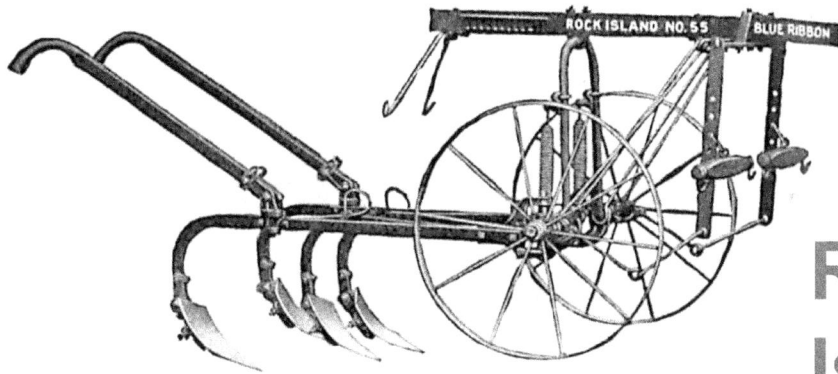

Rock Island

ROCK ISLAND NO. 50 TONGUELESS

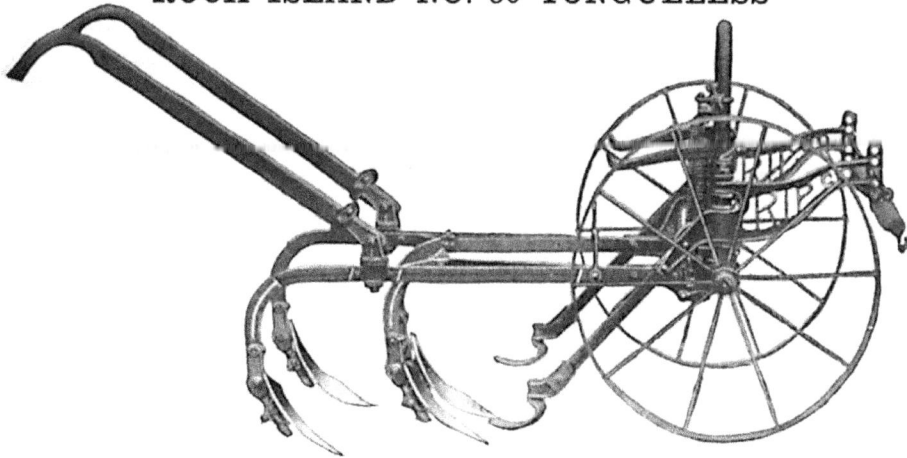

Short Coupled

Light Draft

Cultivates at Uniform Depth
Guaranteed Pipe Beams

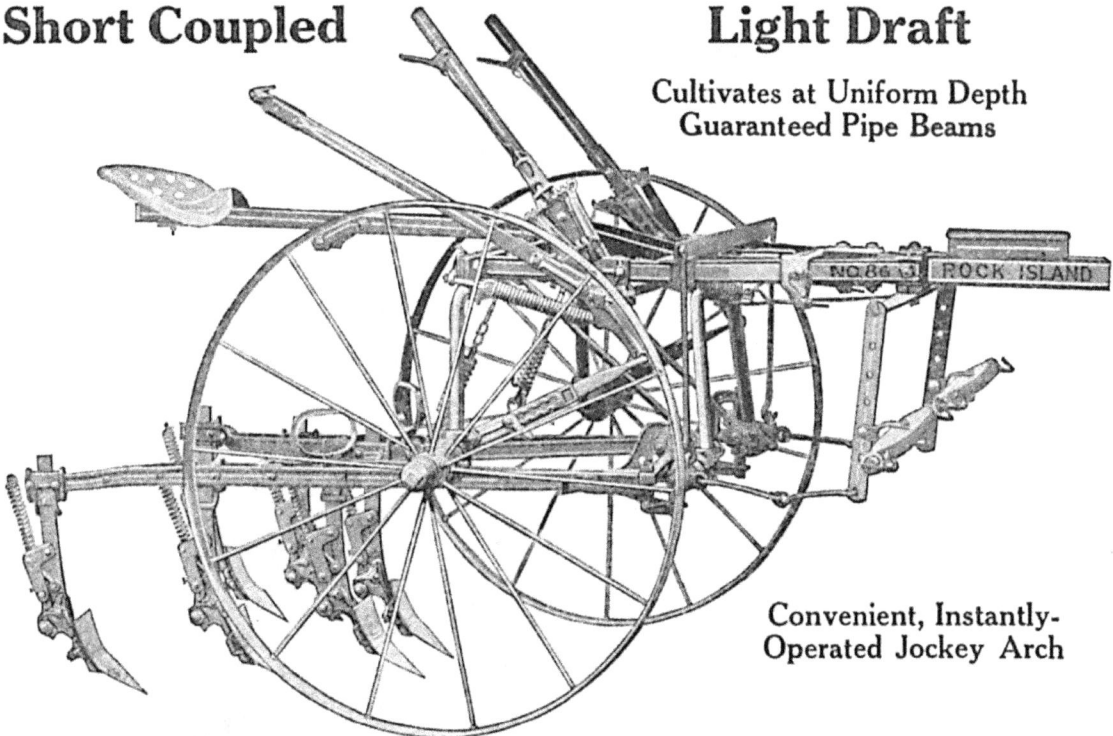

Convenient, Instantly-
Operated Jockey Arch

Rock Island No. 86 Seat Shift Cultivator

Rock Island

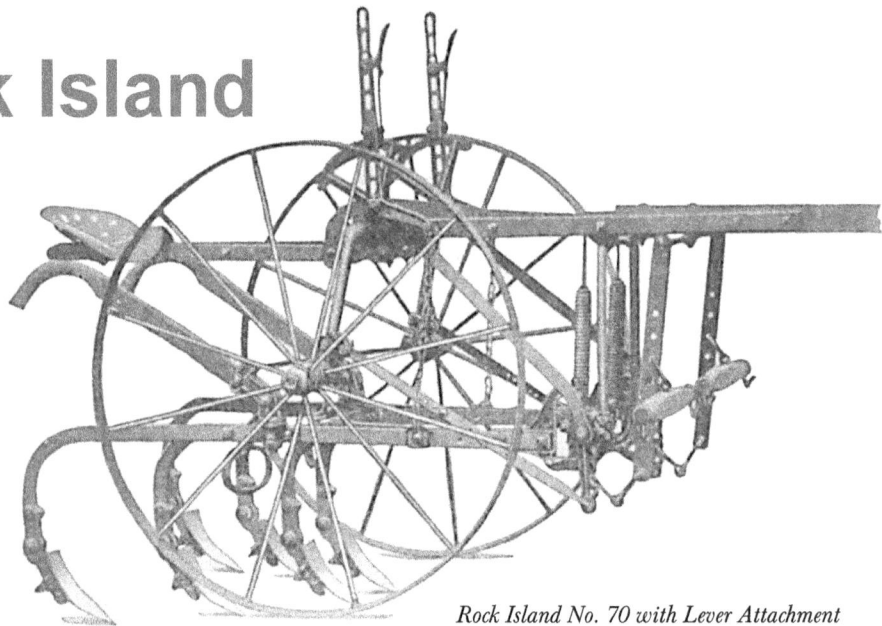

Rock Island No. 70 with Lever Attachment

Independent Levers
For Each Gang

ROCK ISLAND NO. 75 PIVOT AXLE CULTIVATOR

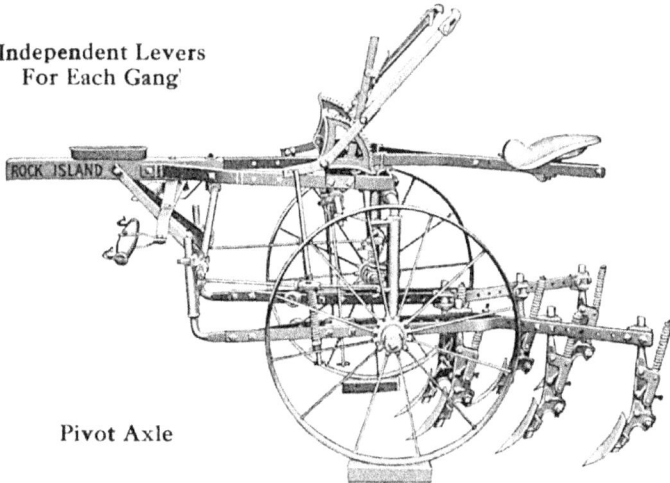

Pivot Axle

Rock Island No. 78

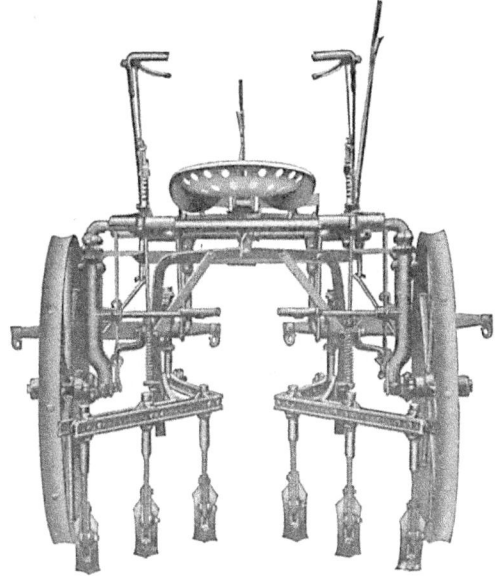

Balance Lever
Controlling
Both Gangs

Independent
Depth Levers

ROCK ISLAND NO. 75 PIVOT AXLE CULTIVATOR

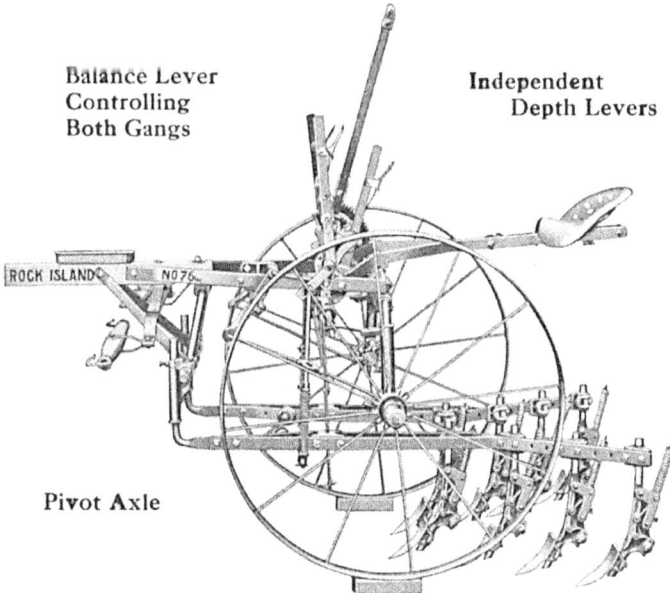

Pivot Axle

Rock Island No. 76

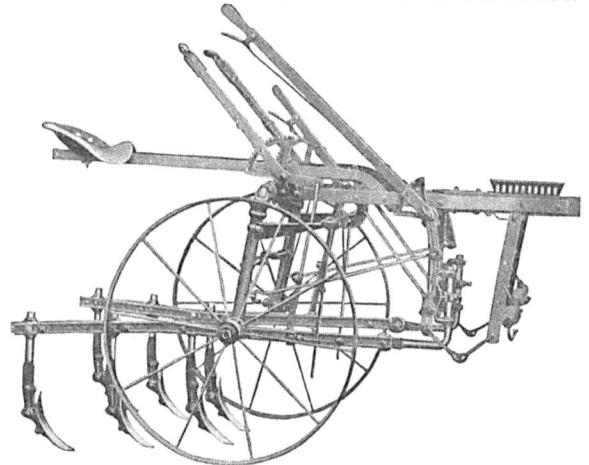

ROCK ISLAND NO. 86 SEAT SHIFT CULTIVATOR

Parallel Gang
Movement

ROCK ISLAND NO. 85 LONE STAR CULTIVATOR

For Southern Farmers

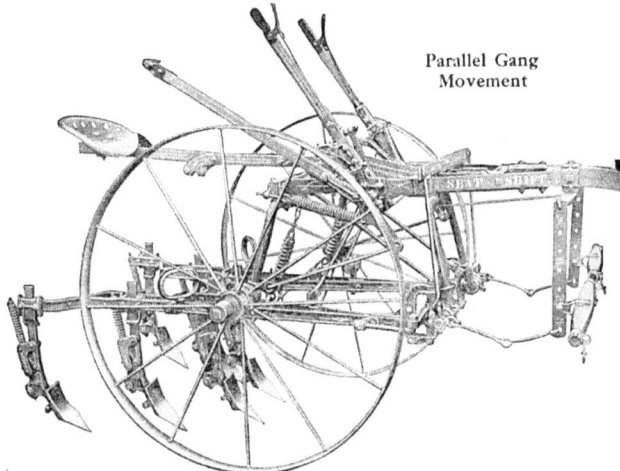

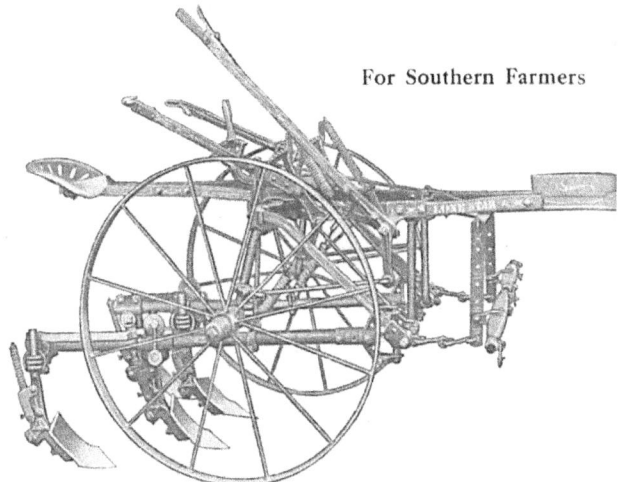

ROCK ISLAND NO. 80 BULLY BOY CULTIVATOR

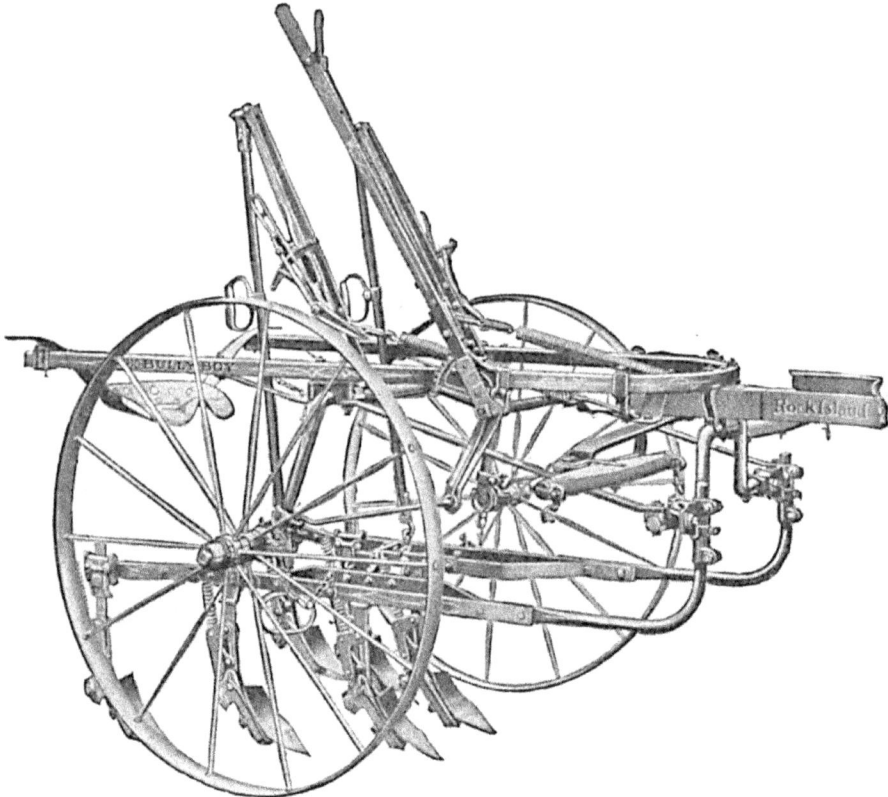

Adjustable
Beams
and Axles

Pivot or Rigid Pole

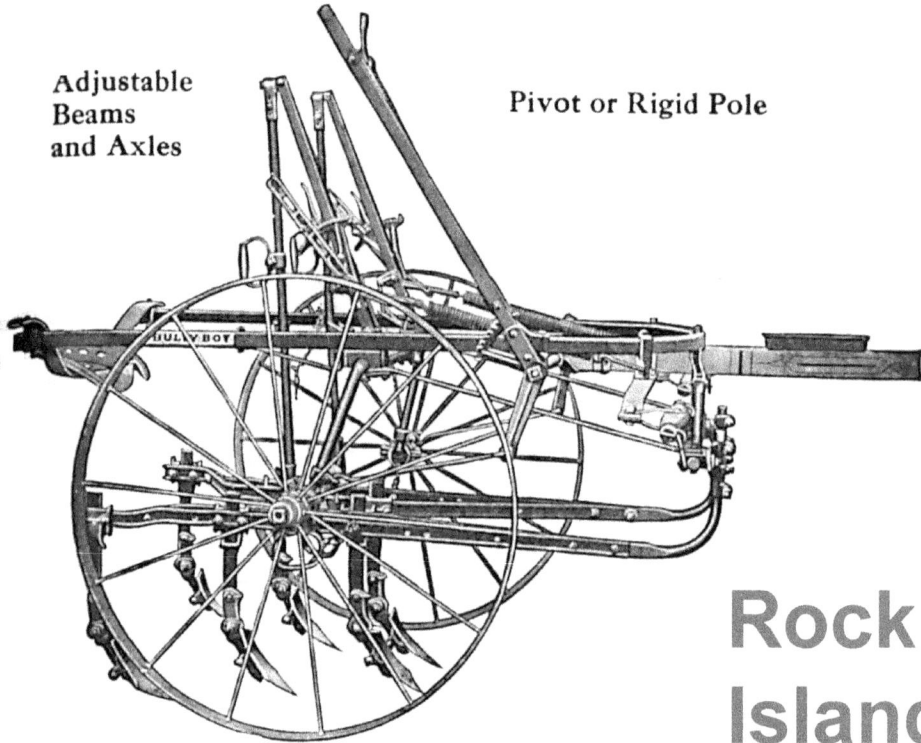

Rock
Island

Equipped with R-18 Gangs

ROCK ISLAND NO. 80 BULLY BOY CULTIVATOR

Equipped with R-18X (Southern) Gangs

To demonstrate the perfect balance of the Rock Island No. 70 the neckyoke was unattached and removed. It may be difficult to see in this old photo but the tongue, or pole, is hanging in mid air with the driver in the seat.

Rock Island

ROCK ISLAND NO. 86 SEAT SHIFT CULTIVATOR

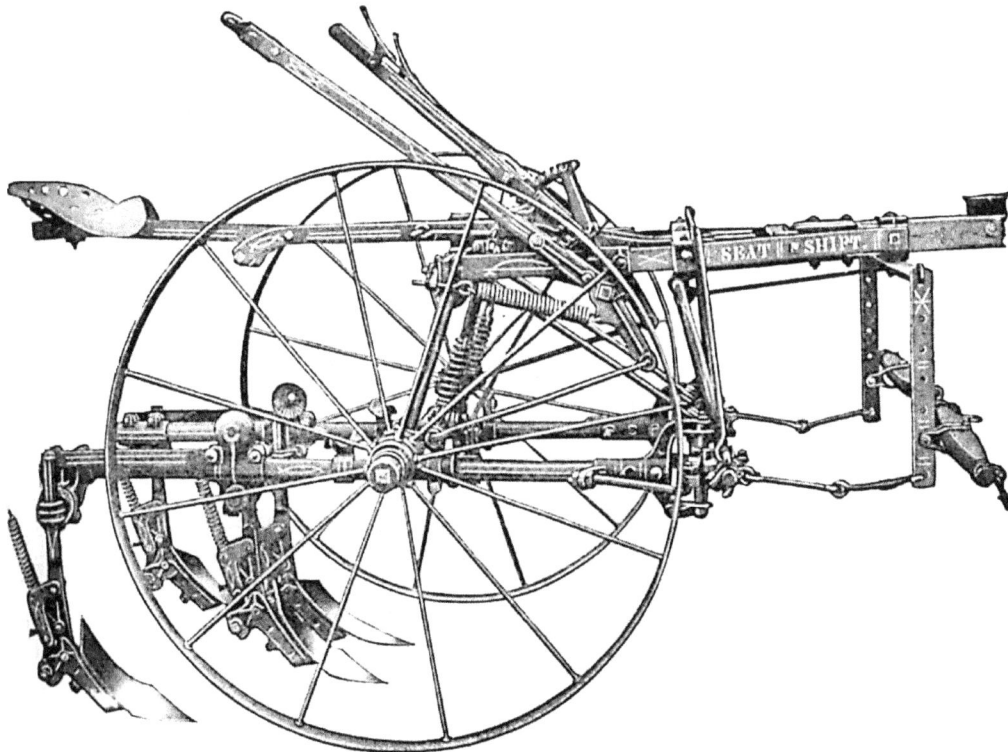

ROCK ISLAND NO. 86 SEAT SHIFT CULTIVATOR

No. 86 Equipped with No. 41 Disc Gangs

Rock
Island

ROCK ISLAND NO. 95 SURFACE CULTIVATOR

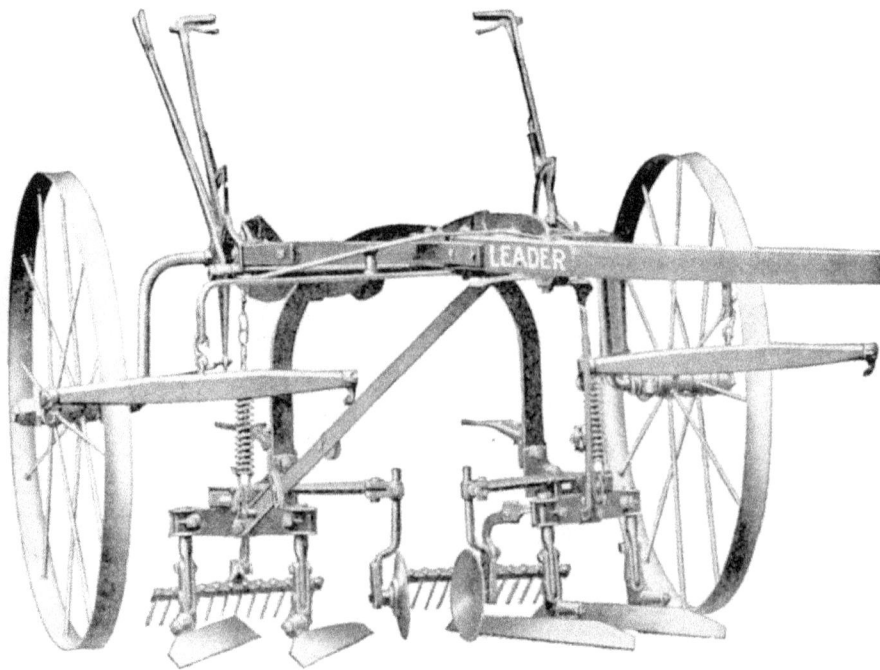

ROCK ISLAND NO. 100 DISC CULTIVATOR

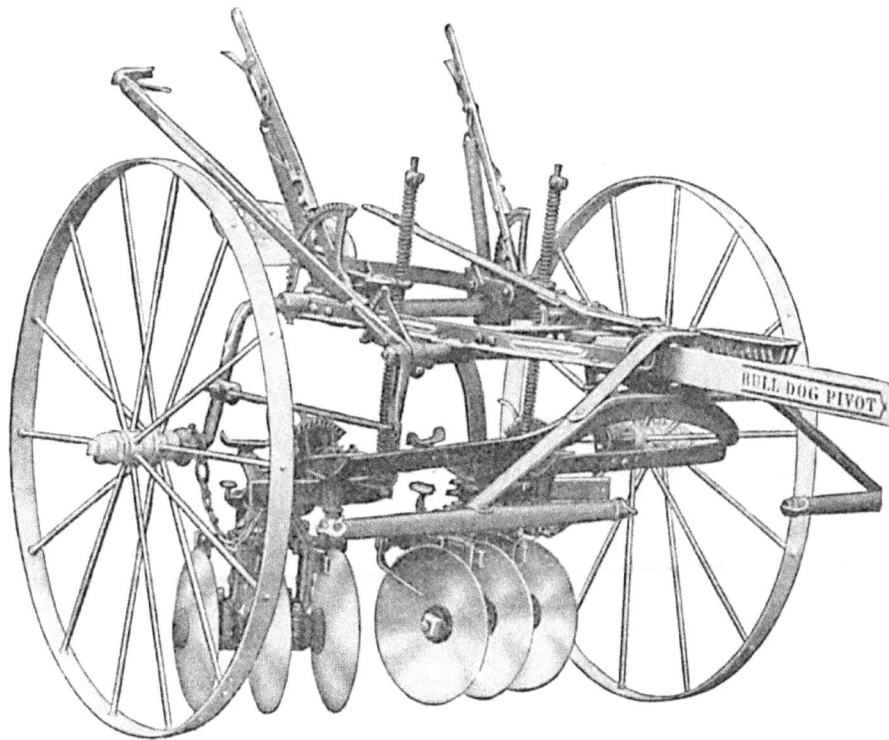

Rock Island

Showing front view—Pivot pole, levelers detached

ROCK ISLAND NO. 105 DISC CULTIVATOR

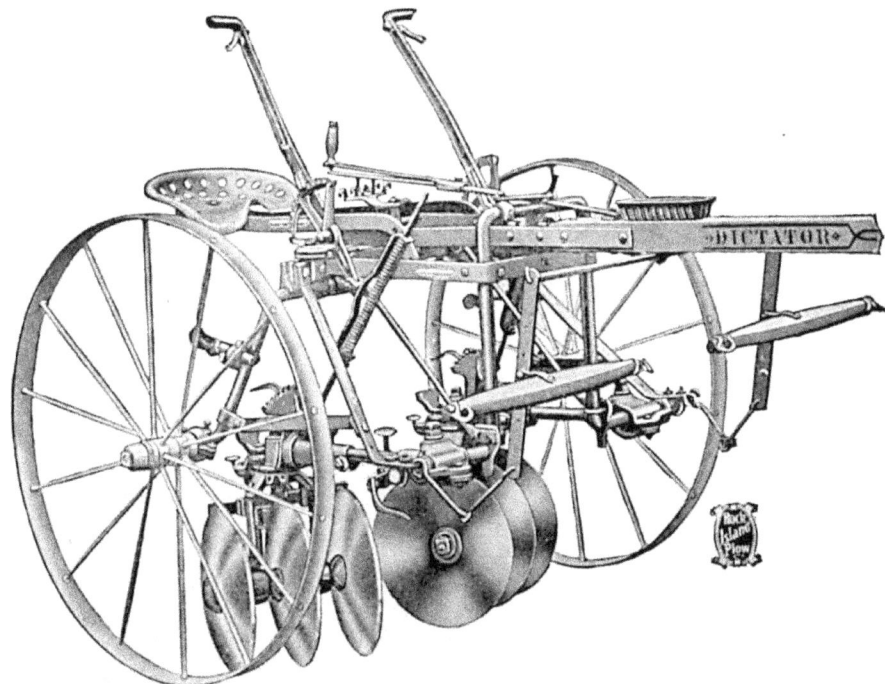

Front View of Rock Island No. 105

The Rock Island No. 105 is steered entirely by the center lever and the feet —either one or both. The steering lever is in the center of the frame, by which means the cultivator is steered, assisted by the feet on the foot treadles.

ROCK ISLAND NO. 105 DISC CULTIVATOR

Rock
Island

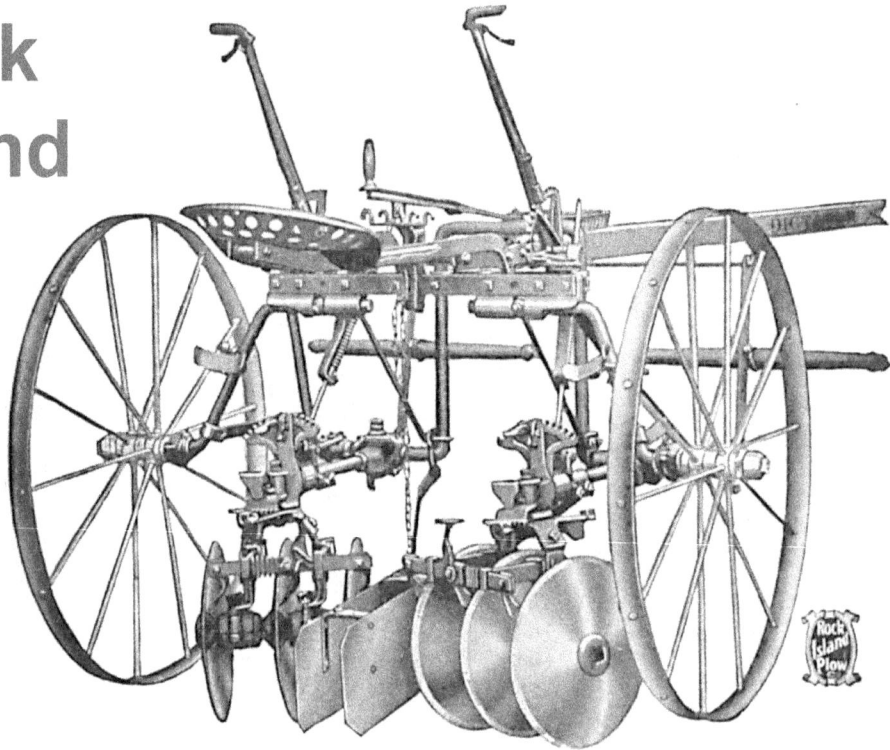

Rear View of Rock Island No. 105

The main frame is adjustable in width and can be widened or narrowed, as desired, from 48 to 58 inches between the wheels.

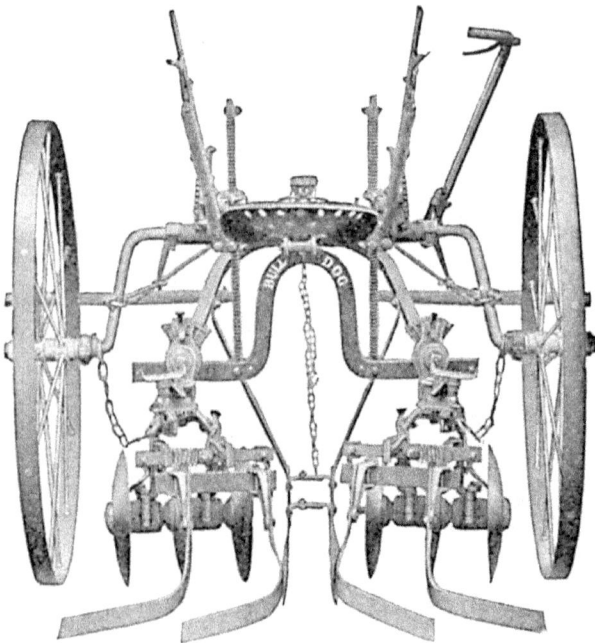

Discs set for in-throw. Levelers work dirt away

Rock Island

Rock Island No. 86 Seat Shift Cultivator
The seat yields instantly with each movement of the body. It can be operated and guided all day with but little exertion. A slight swing of the body shifts the frame and rigs at the same time, carrying the machine wherever desired, but always maintaining the rigs parallel. This helps while working on a hillside as the weight of the operator is thrown toward the lower side of the cultivator. That immediately leads the rigs uphill. On the flat, when the operator shifts the gangs he also changes the entire line of draft, accomplishing twice the action for dodging crooked corn that he has on the ordinary cultivator.

The Vulcan No. 5 Walking Cultivator

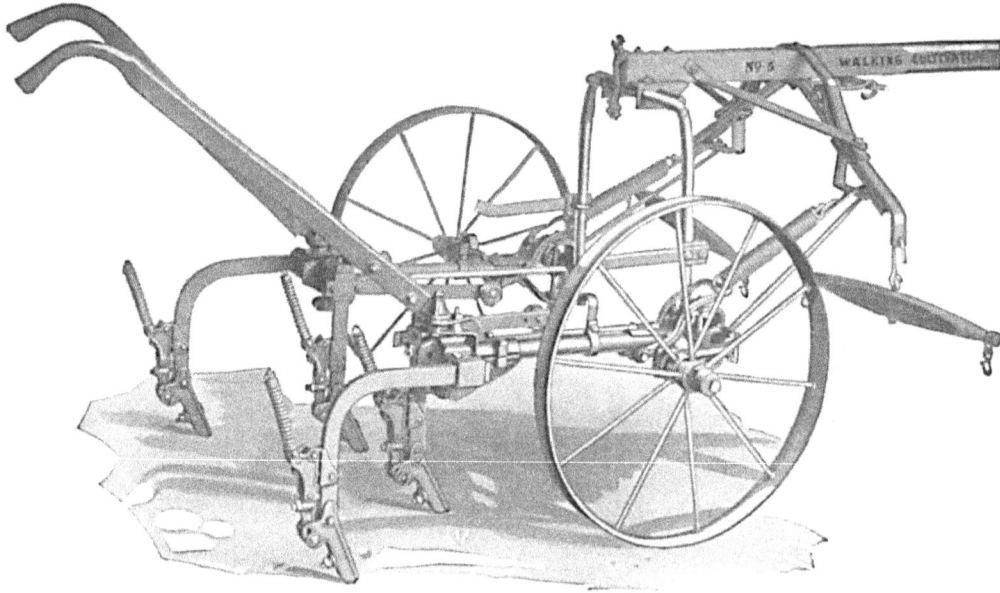

Vulcan

No. 7 Pivot Axle Cultivator

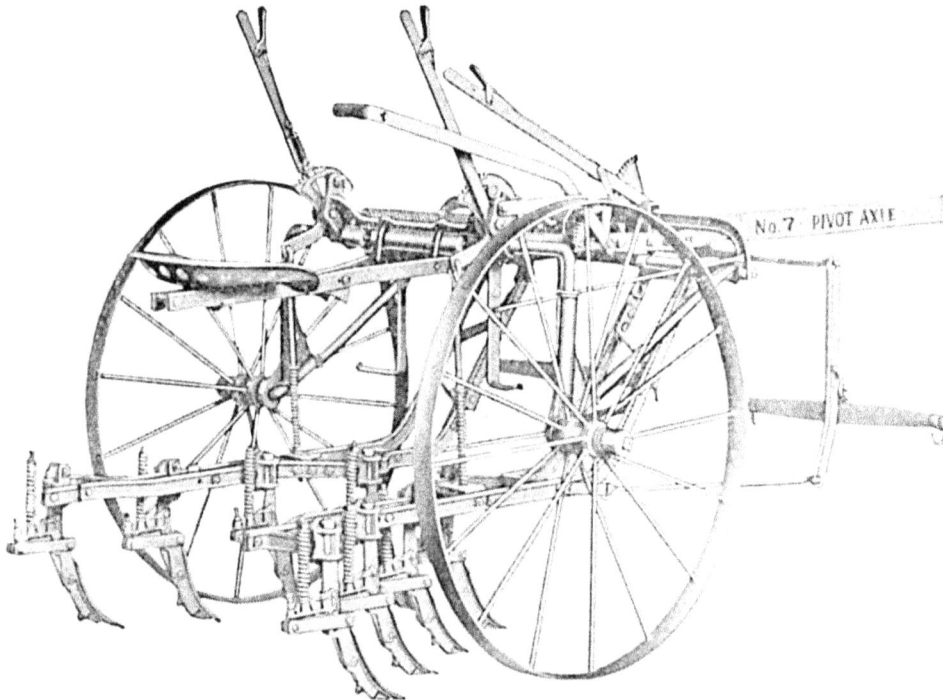

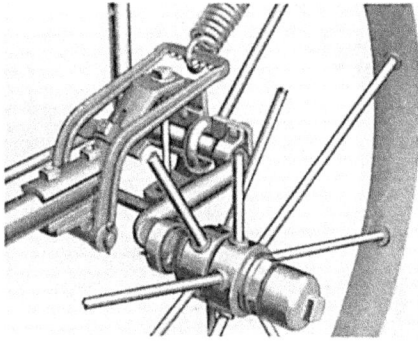

**ALL STEEL FRONT
GANG COUPLING**

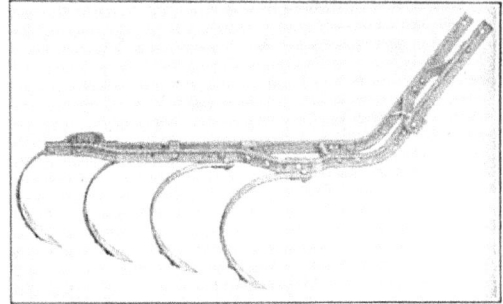

No. 7 Pivot Axle Cultivator Gang
Equipped with Spring Teeth.

Vulcan

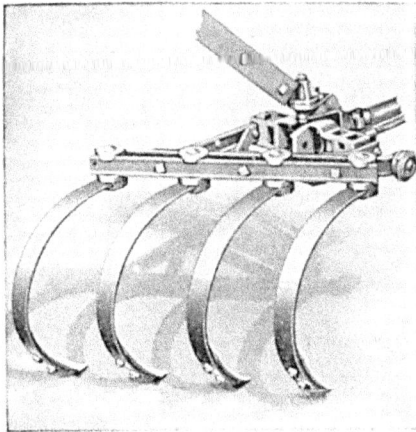

**SPRING TOOTH GANG
QUICKLY ATTACHED**

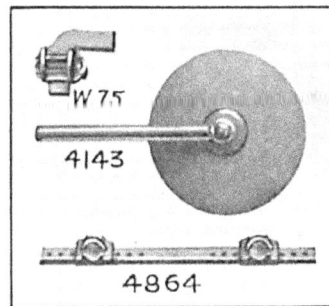

W 75
4143
4864

No. 7 Pivot Axle Cultivator

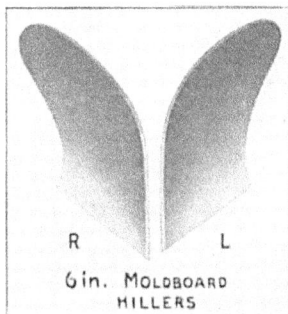

R   L

6 in. MOLDBOARD
HILLERS

Made of best quality plow
steel, with riveted back to fit
the regular No. 5 Open Boot.

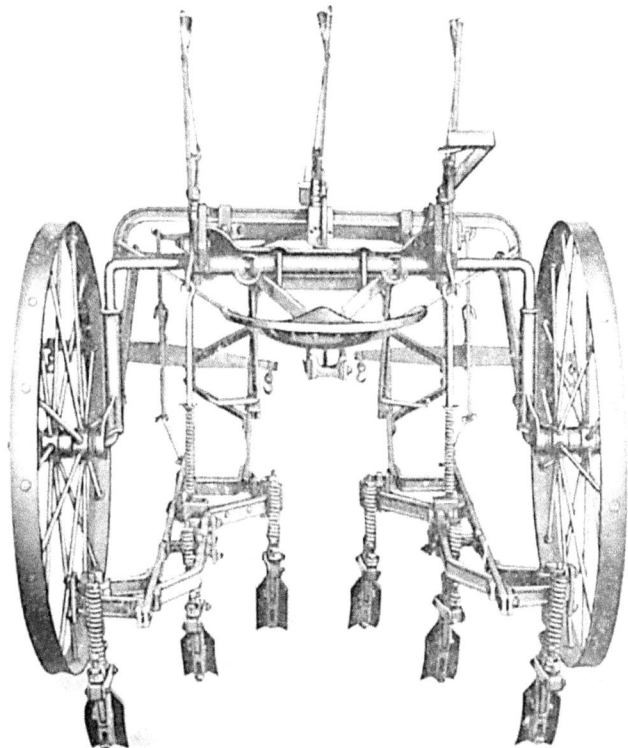

*Two Horse
Cultivators*

A-59 CULTIVATOR

Oliver

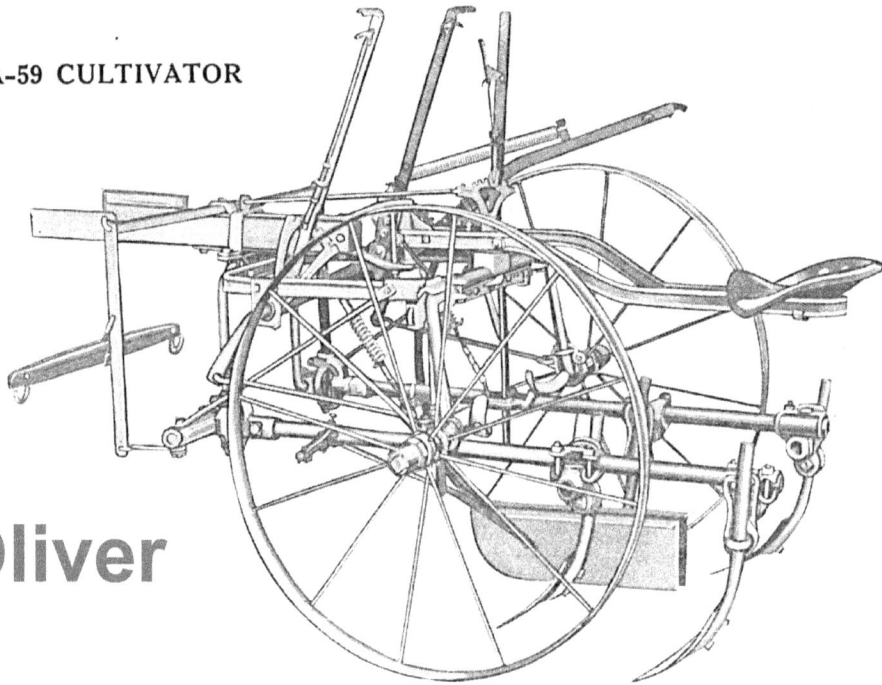

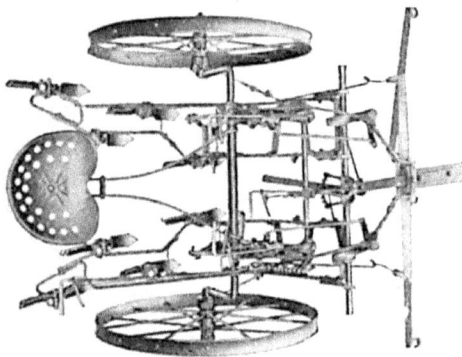

A top view shows the wide range of shift possible with the Oliver No. 35.

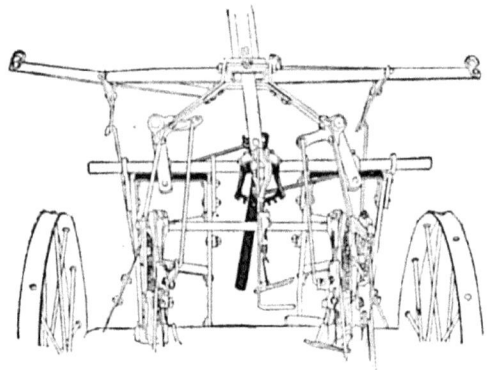

Right in front of the operator within easy reach of his hand is an expansion lever which spreads or closes the gangs. It is a great advantage especially where conditions are such as to produce plants of varying size in the same row.

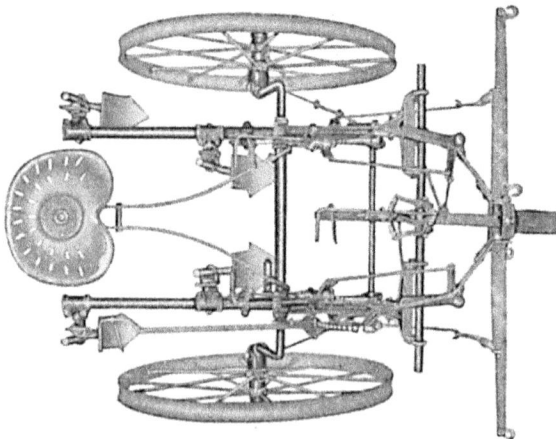

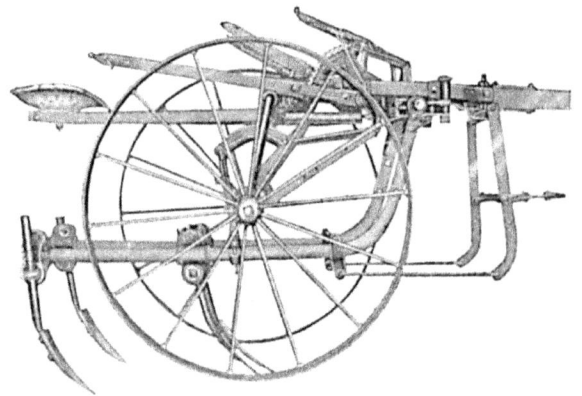

There are also friction trip gangs for the No. 35 Southern.

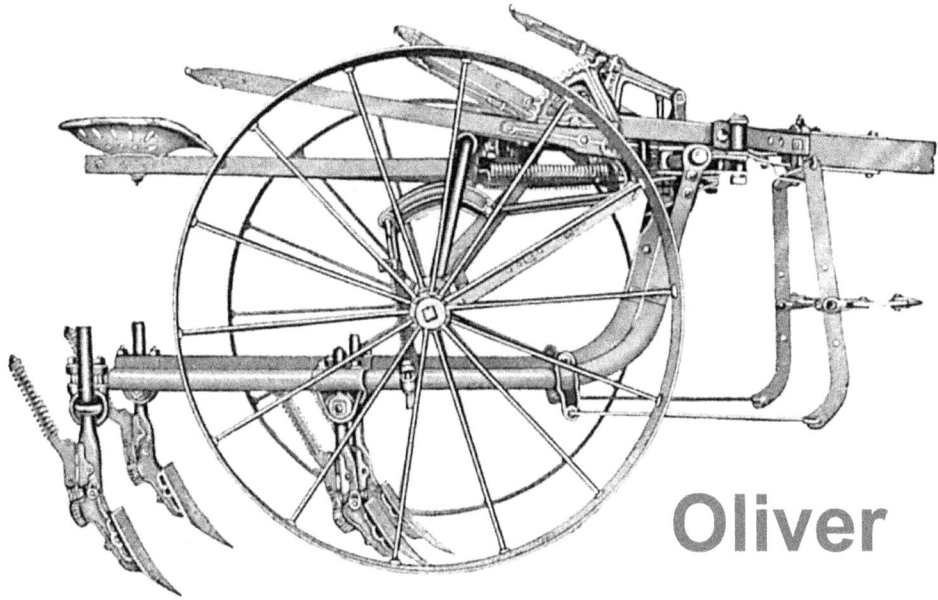

Oliver

The No. 35 Southern cultivator with pipe beams is built especially for southern conditions. Here are the four shovel spring trip gangs.

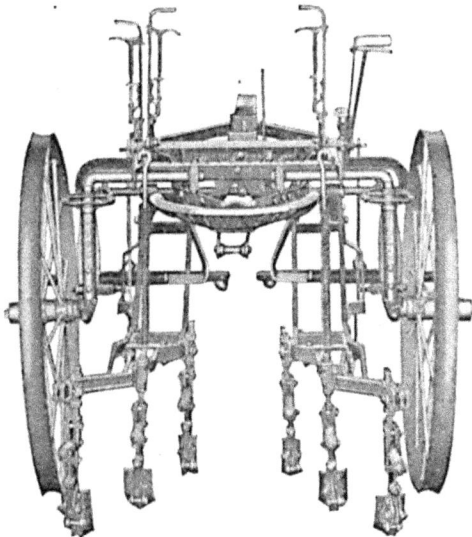

Rear view of the 957. Simplicity, a quick, wide dodge and excellent performance make the 950 series the farmers' favorite.

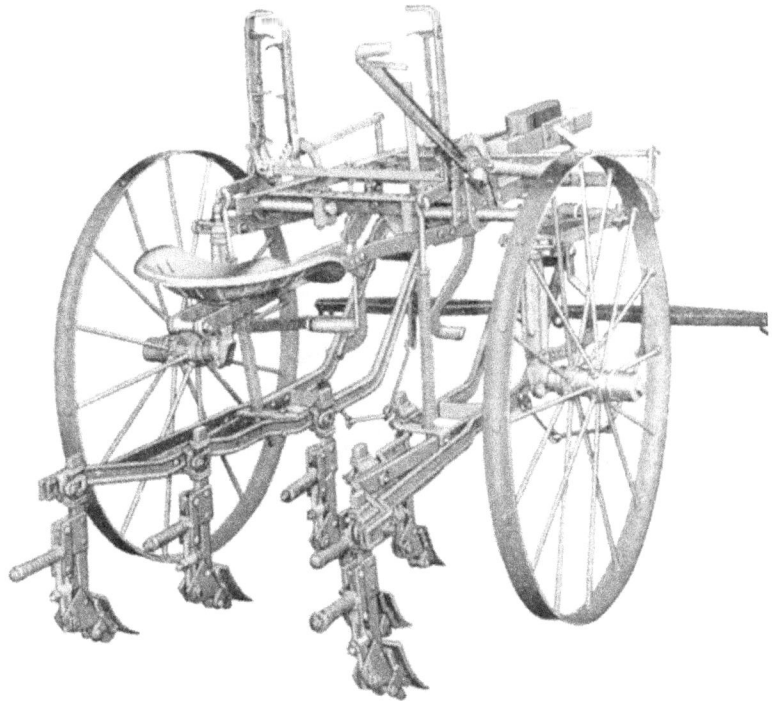

The Oliver Superior 957 one-row cultivator is popular wherever row crops are grown.

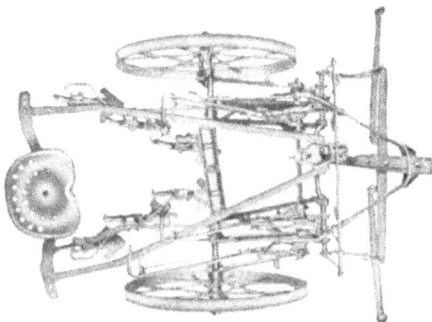

Top view of the No. 1 Southern showing how the seat guide works. As the seat is swung to either right or left by the operator, the gangs are immediately shifted. The main frame permits the use of either a hammock or handle seat. Here it is shown equipped with a hammock seat.

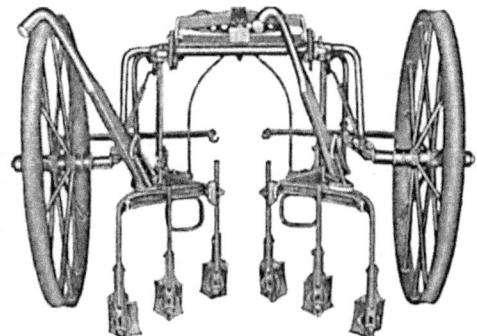

Rear view of the 33. Simplicity and ease of handling makes this the ideal cultivator for use by unskilled help.

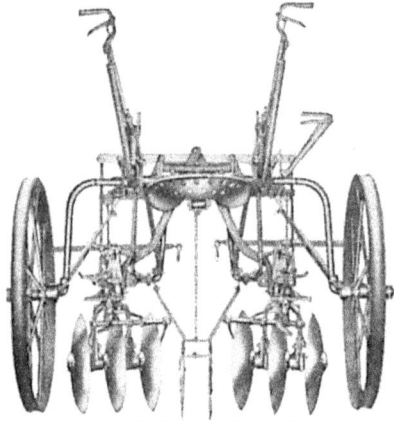

Oliver No. 25 with discs set for in-throw.

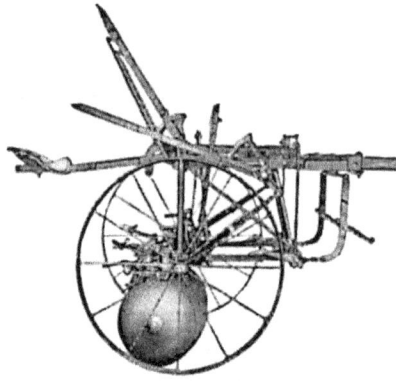

No. 25 from the side

Oliver

No. 25 with discs set for out-throw.

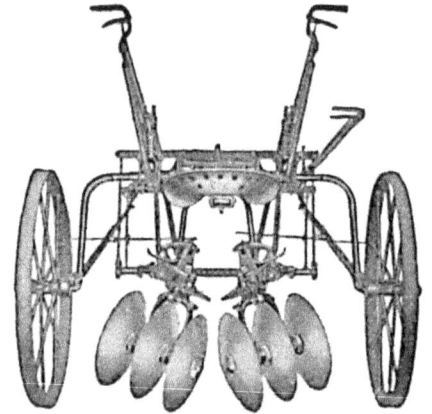

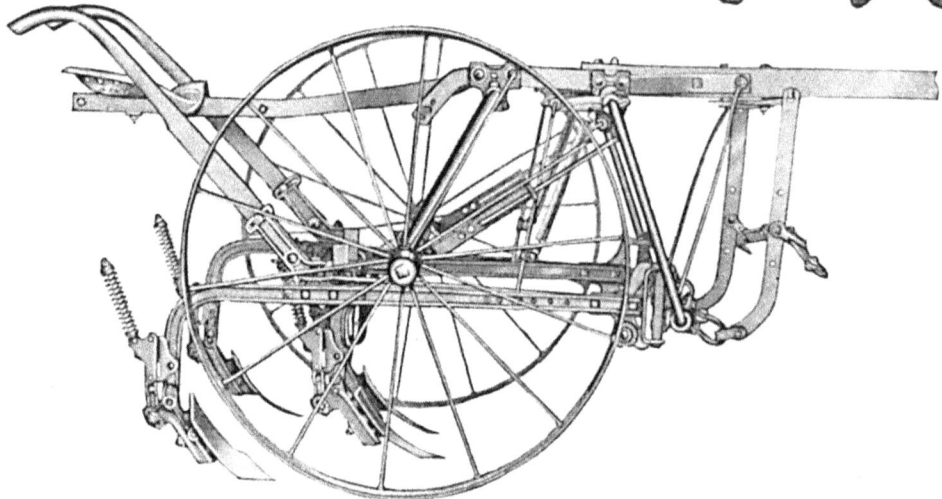

The 33 equipped with B.ASO gangs. This combination riding and walking cultivator has light draft—is easy on horses.

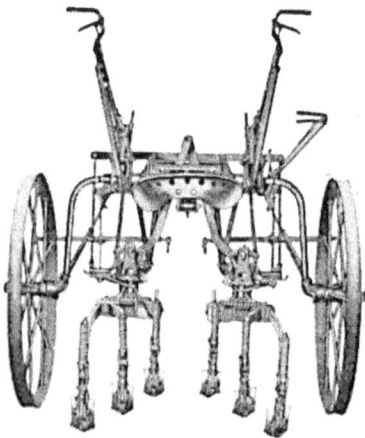

No. 25 equipped with six shovel spring trip gangs.

No. 25 with spring tooth attachment.

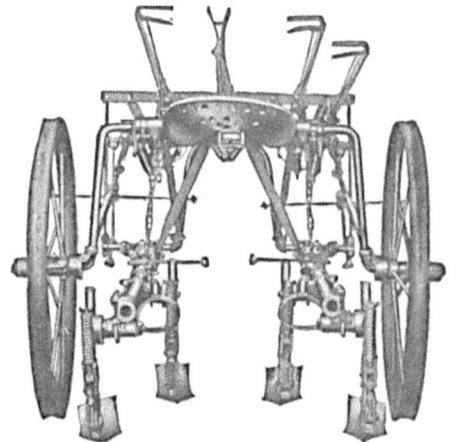

Rear view of the Oliver No. 1 Southern. The gangs and wheels are adjustable for working different widths.

Oliver

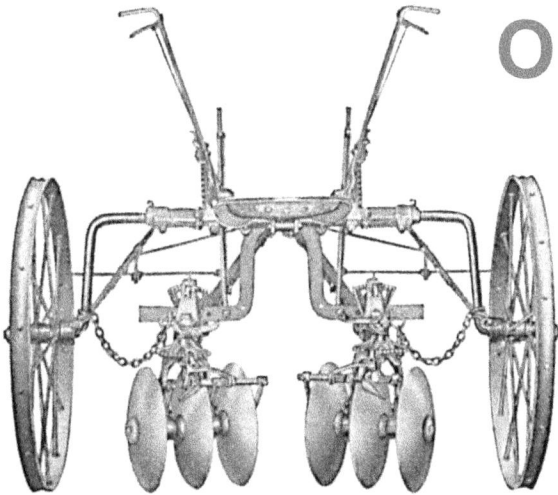

With discs set for in-throw. The 45 is a clean cut simple machine. The gangs are easily set in or out on the jockey arch.

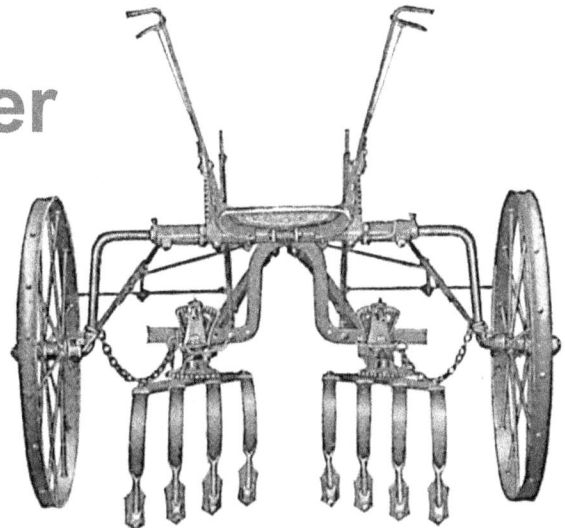

No. 45 with spring tooth gang equipment.

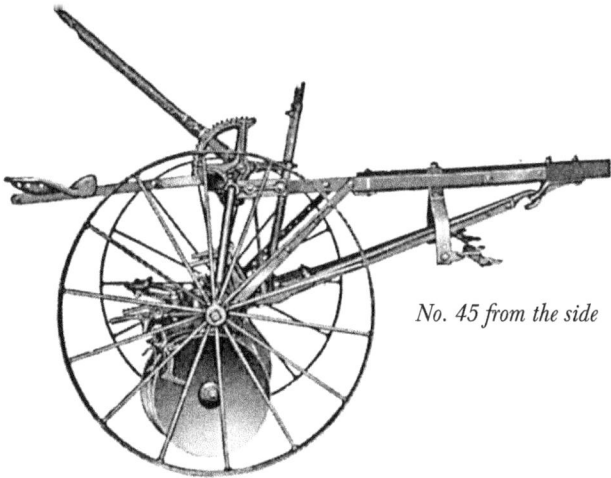

No. 45 from the side

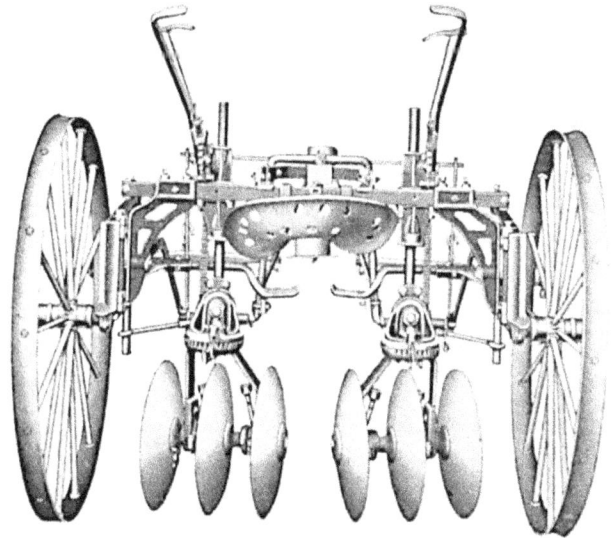

The Oliver Superior No. 60 disc cultivator has single beam heads.

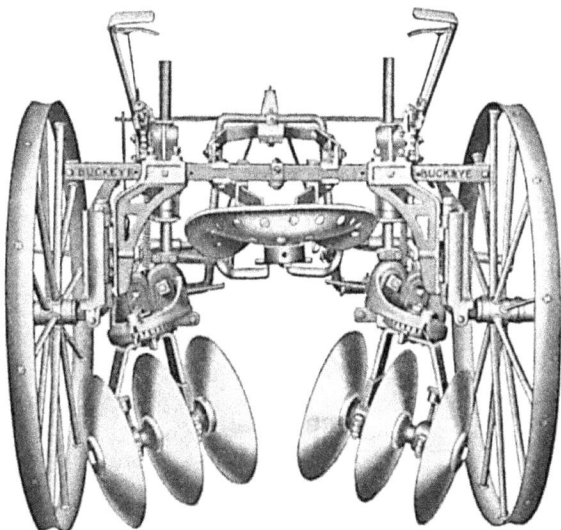

The Oliver Superior No. 61 disc cultivator is similar to the No. 60 except that it has double beam heads.

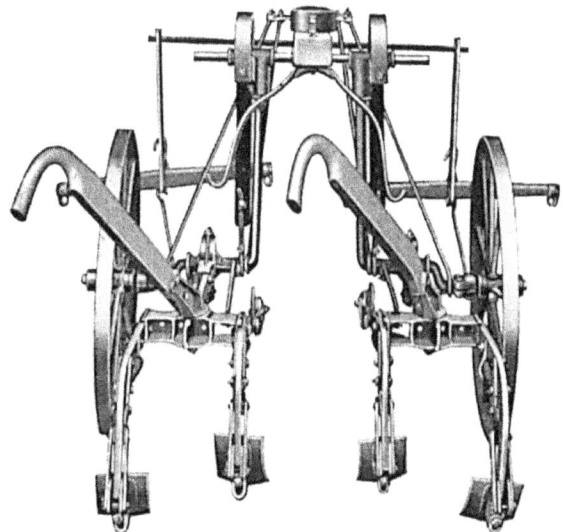

Rear view of the No. 123.

Two Horse Cultivators

239

The Oliver No. 2 Improved is one of the most popular walking cultivators ever built. It is a quality cultivator from handles to pole.

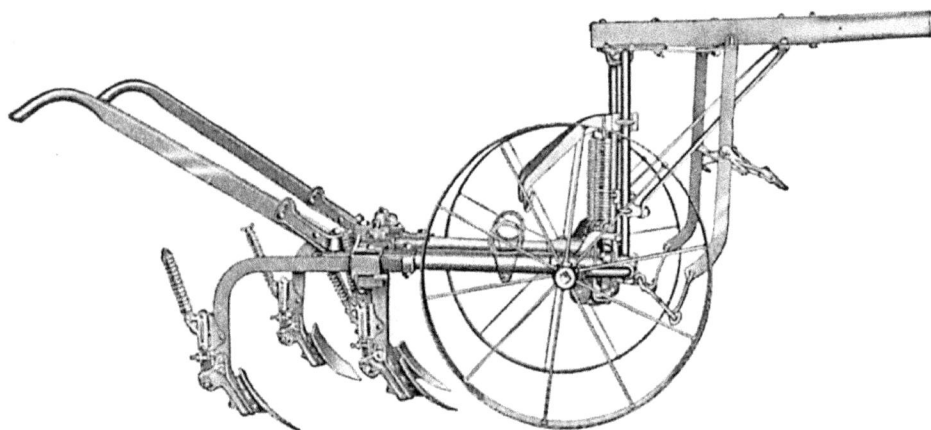

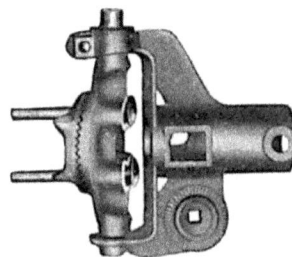

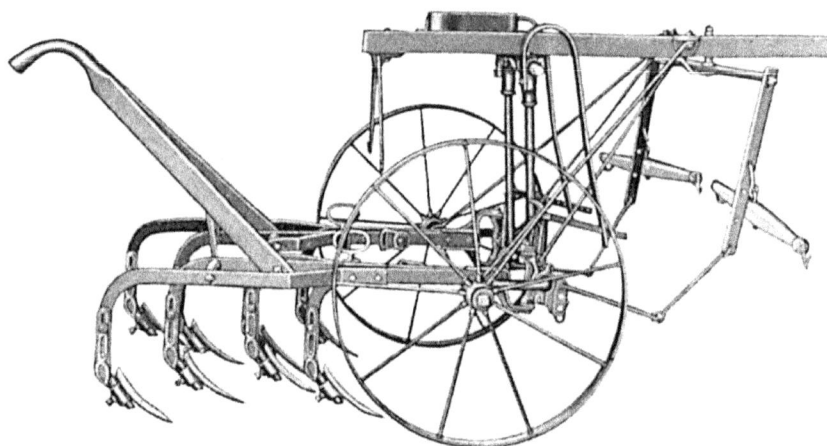

Each gang has an adjustable cone coupling to take up wear on the gang coupling.

The No. 127 Oliver Superior walking cultivator.

Oliver

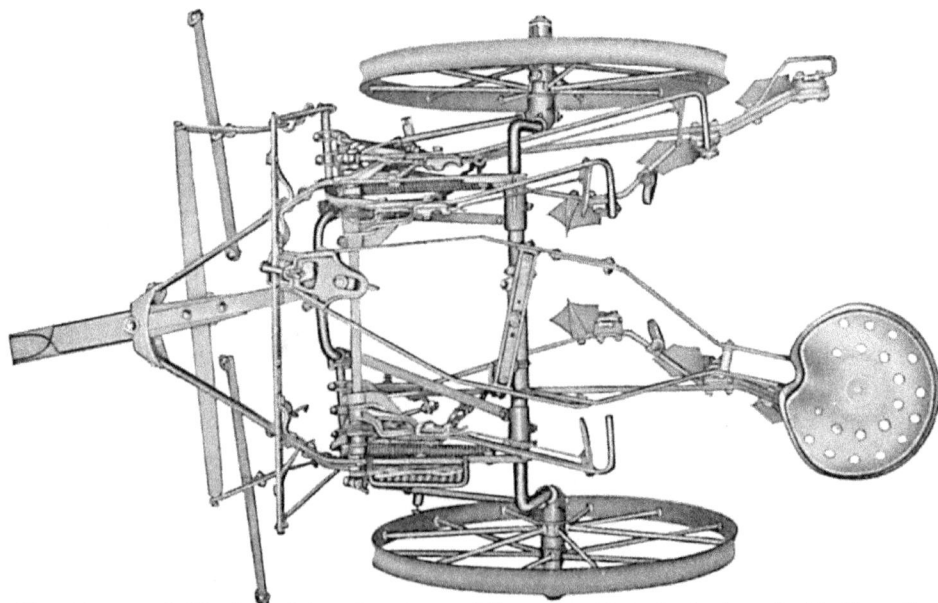

The wide range of shift obtained through the famous Oliver seat guide is clearly shown in this illustration. Note that the gangs are parallel when in shifted position.

Oliver Nos. 1 and 4 Imp. Cultivators

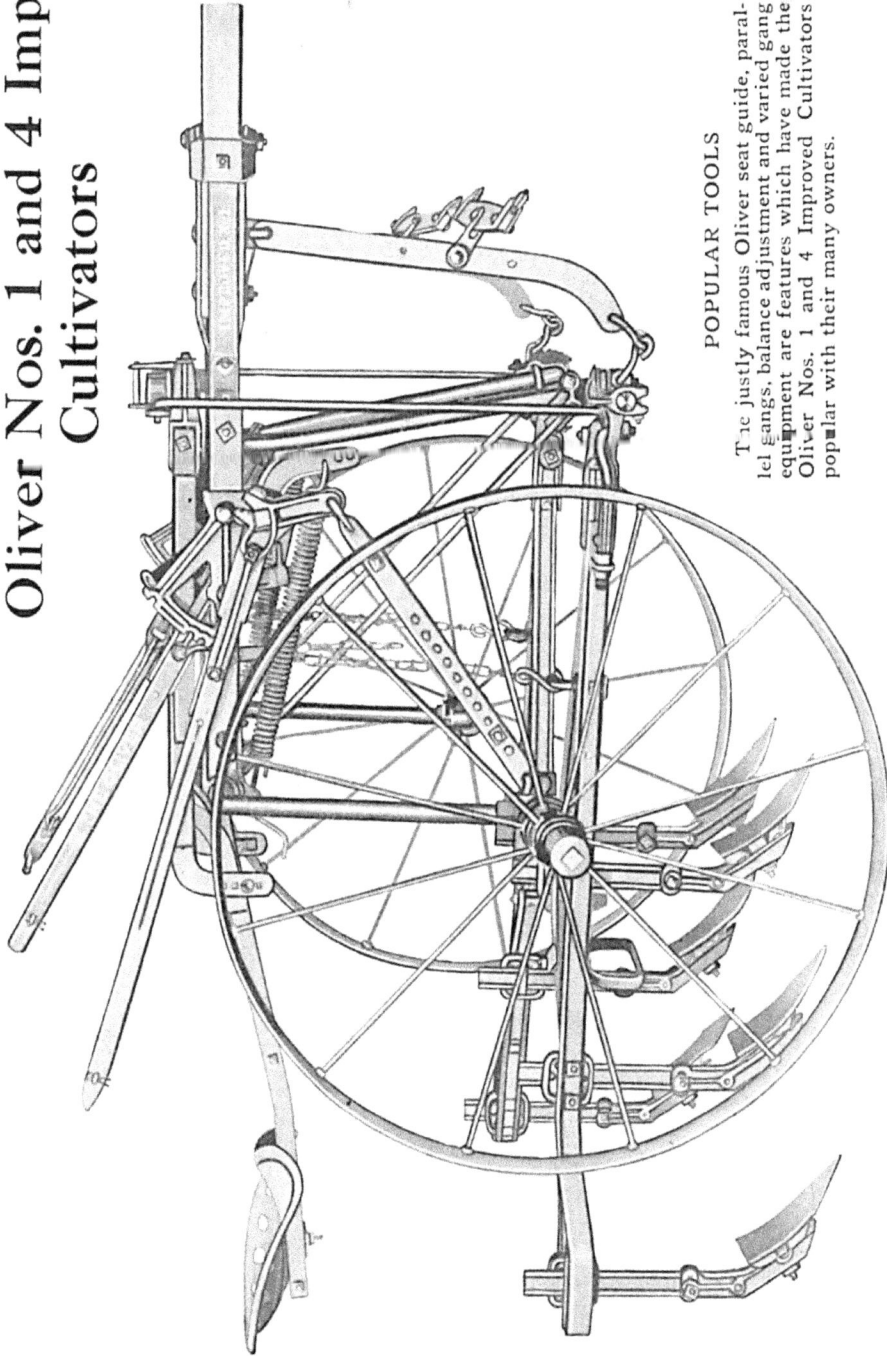

POPULAR TOOLS

The justly famous Oliver seat guide, parallel gangs, balance adjustment and varied gang equipment are features which have made the Oliver Nos. 1 and 4 Improved Cultivators popular with their many owners.

Two Horse Cultivators

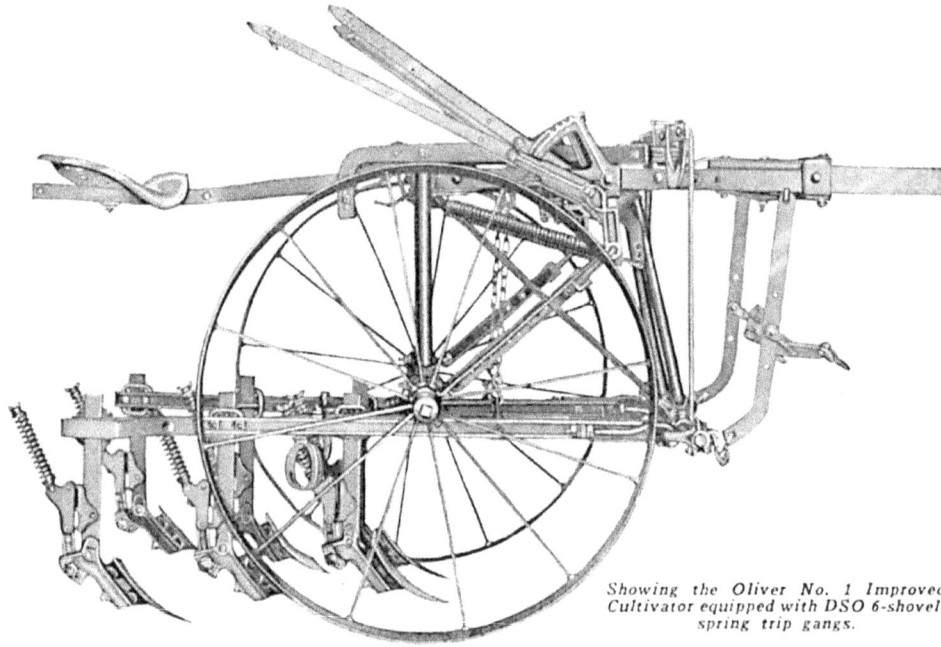

Showing the Oliver No. 1 Improved Cultivator equipped with DSO 6-shovel, spring trip gangs.

Oliver

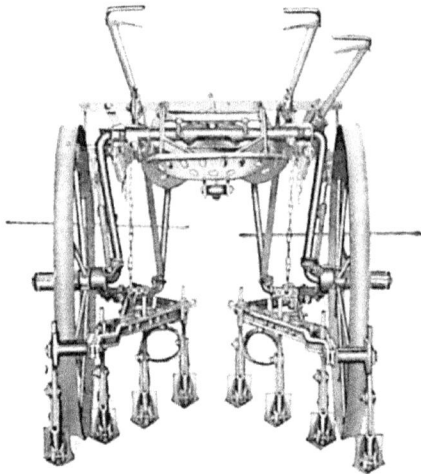

Rear view showing the No. 4 which is identical with the No. 1 except that it has a set-in axle providing a narrower track for narrow row crops such as beans, peas, and peanuts.

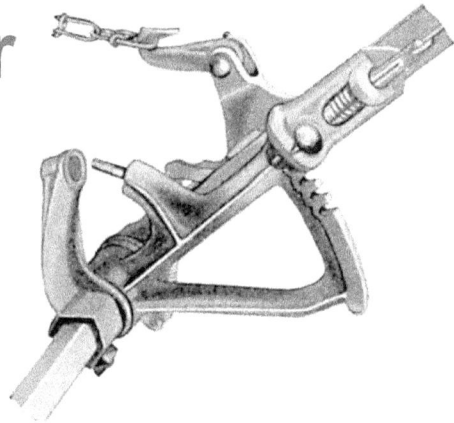

Showing how the gangs can be made rigid or floating. Loosening the bolt as shown in the illustration above causes the gangs to float.

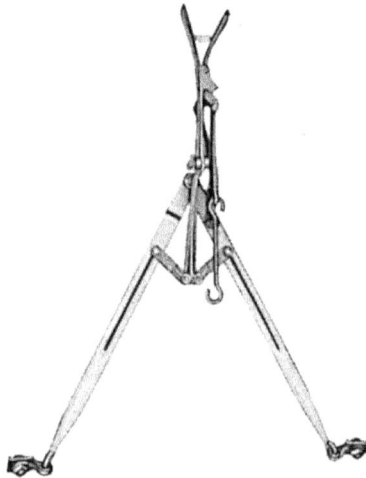

Adjustable jockey arch for use on the No. 1 Improved Cultivator.

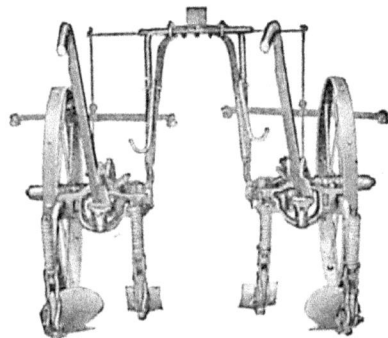

Showing the Oliver No. 2 Improved Cultivator equipped with shovels and hillers.

OLIVER

OPERATION MADE EASY

A distinct advantage of this cultivator is the ease with which it is controlled. This ease of control is produced by springs which counterbalance the weight of the gangs against the weight of the tongue, eliminating neck weight and making it easy to lift the gangs or swing them to or from the row.

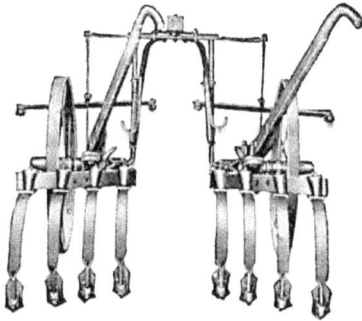

Showing the Oliver No. 2 Improved Cultivator equipped with shovels and hillers.

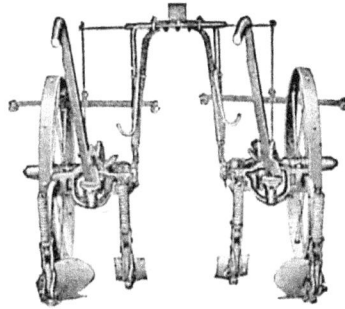

Oliver No. 2 Improved Walking Cultivator equipped with No. 21 spring tooth attachment.

PROPER SUCK

Gangs have proper suck for various soil conditions. This makes it unnecessary for the operator to bear down on the handles to secure penetration.

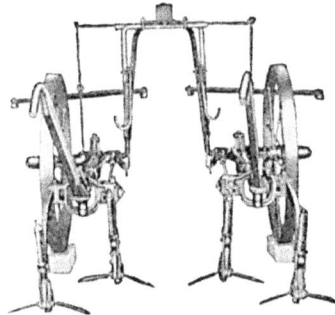

Here the Oliver No. 2 Improved Cultivator is equipped with sweeps.

PERFECT BALANCE

The weight of the Oliver No. 2 Improved Cultivator is not thrown on the horses' necks when the gangs are raised because the wheels are placed far enough back of the arch to balance the weight of the gangs. This affords perfect balance of the cultivator whether the gangs are raised or lowered.

THOROUGHT CULTIVATION

Since it requires only slight effort on the part of the operator to shift the gangs to or from the row it is possible to practically hoe the row with the No. 2 Improved Cultivator.

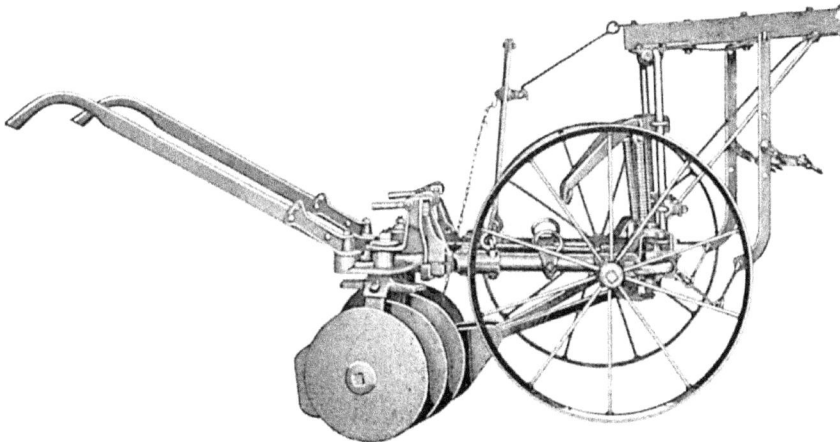

Oliver No. 2 Improved Cultivator equipped with No. 2-M Disc Gangs. Shields are furnished as an extra.

ADJUSTABLE ARCH AND HANDLES

The arch is adjustable to cultivate wide or narrow rows. Handles can be moved up or down for different heights of operators or they can be turned either to right or left to enable the operator to walk between the rows.

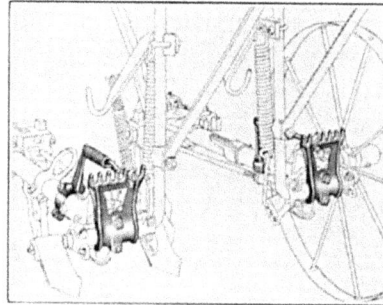

Showing guide spring attachment for all No. 2 Improved Cultivator pipe beam gangs except XO.

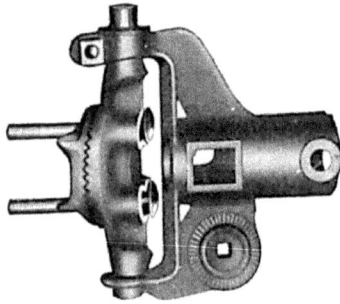

Adjustable cone gang coupling.

ADJUSTABLE CONE COUPLINGS

The gangs on the Oliver No. 2 Improved Cultivator are equipped with adjustable cone couplings. This enables the operator to take up the wear on the gang coupling to prevent the gang tipping sidewise and running untrue. On some gangs the coupling is swiveled to allow tilting the gangs to cultivate on the ridge or in the furrow.

HITCH—WHEELS

Close, direct hitch lightens the draft by putting the pull directly on the ends of the gangs. Singletrees are adjustable up or down for different heights of teams. Wheels are fitted with strong spokes and dust-proof bearings.

Oliver

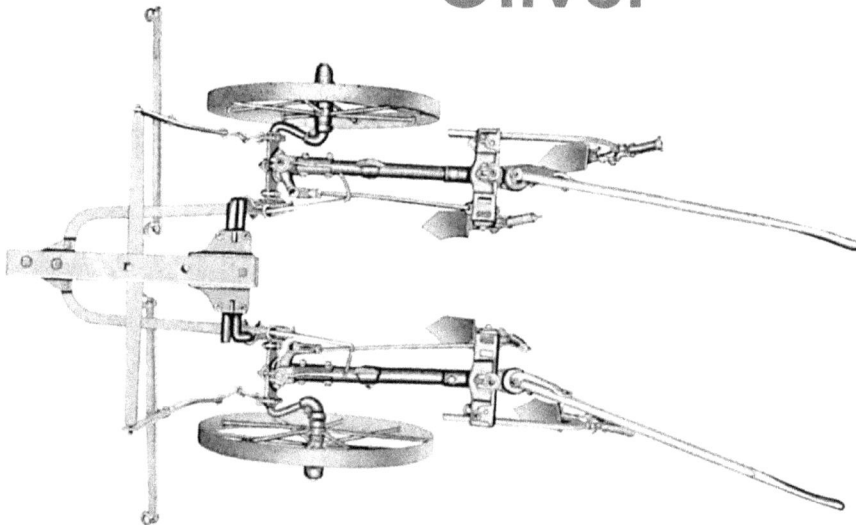

The arch is adjustable to cultivate wide or narrow rows. The handles can be set to right or left so that the operator can walk between the rows.

Oliver

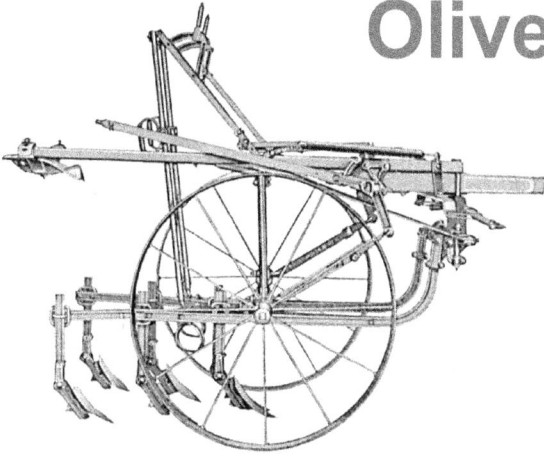

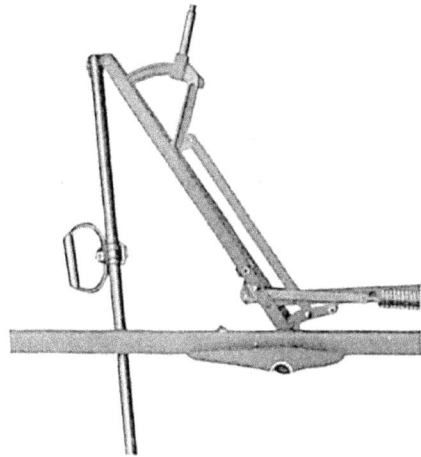

Balance springs, shown above, assist in raising the gangs, preventing the gangs from swinging into the wheels and act as depth regulators. The tension of the springs is easily controlled by a ratchet lever, convenient to the operator.

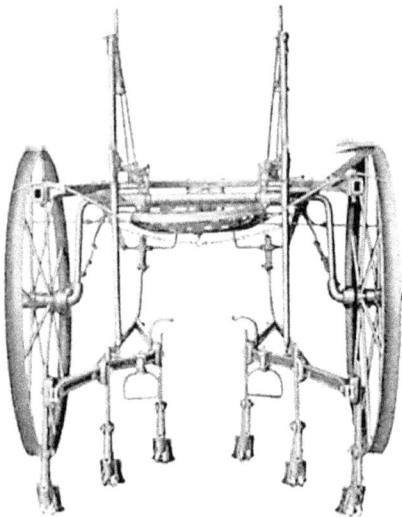

Farmers who have uneven ground, making it necessary to cultivate over knolls and into depressions, appreciate the extreme ease with which the gangs of the No. 12 are controlled. There is abundant clearance for cultivating tall corn.

A double adjustment, shown above, on the gang coupling keeps the gangs in proper position even after long use. All wear in the couplings can easily and quickly be taken up and the gangs restored to their original positions.

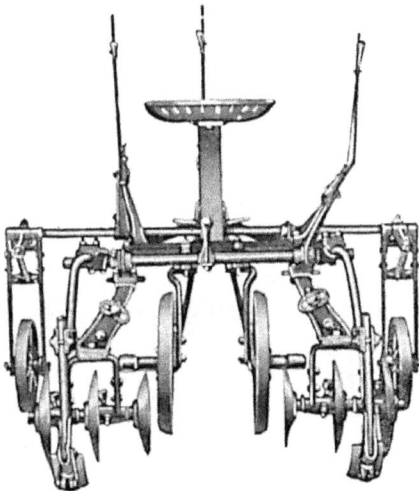

Here the No. 19-A has the shovels and discs set for the second cultivation or to throw the dirt to the plants.

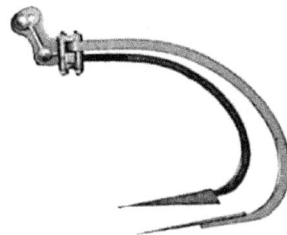

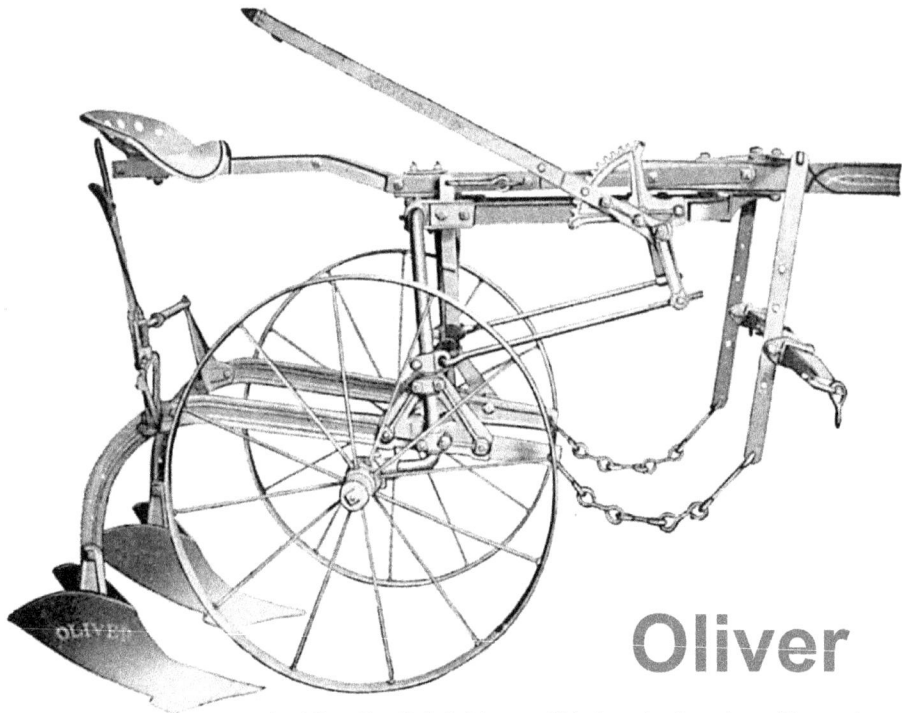

Oliver

Note the close, direct hitch on the Oliver No. 17-A Cultivator. This is a simple tool sturdily constructed.

EASY
ADJUSTMENT

EASY
ADJUSTMENT

BLOUNT LATEST WALKER CULTIVATORS

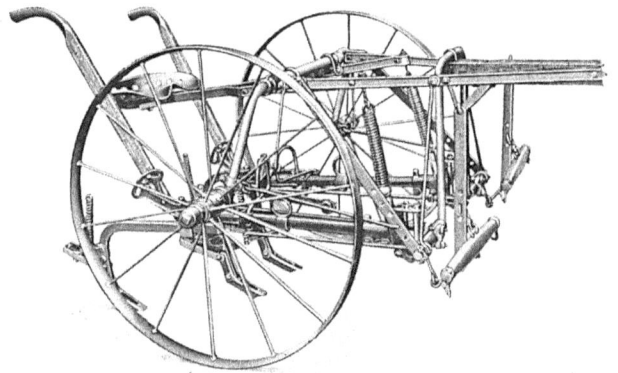

Blount

"TRUE BLUE" LEVERLESS RIDING OR
WALKING CULTIVATOR

BLOUNT PIVOT AXLE CULTIVATOR

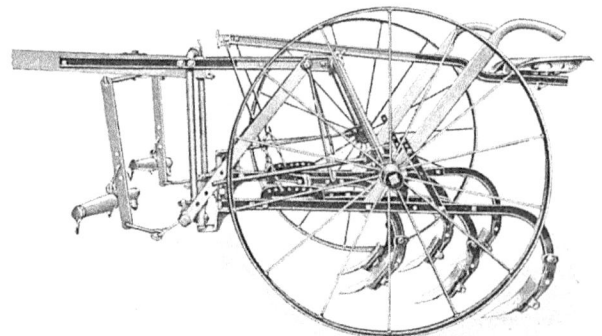

NEW AGE LEVERLESS RIDING OR
WALKING CULTIVATOR

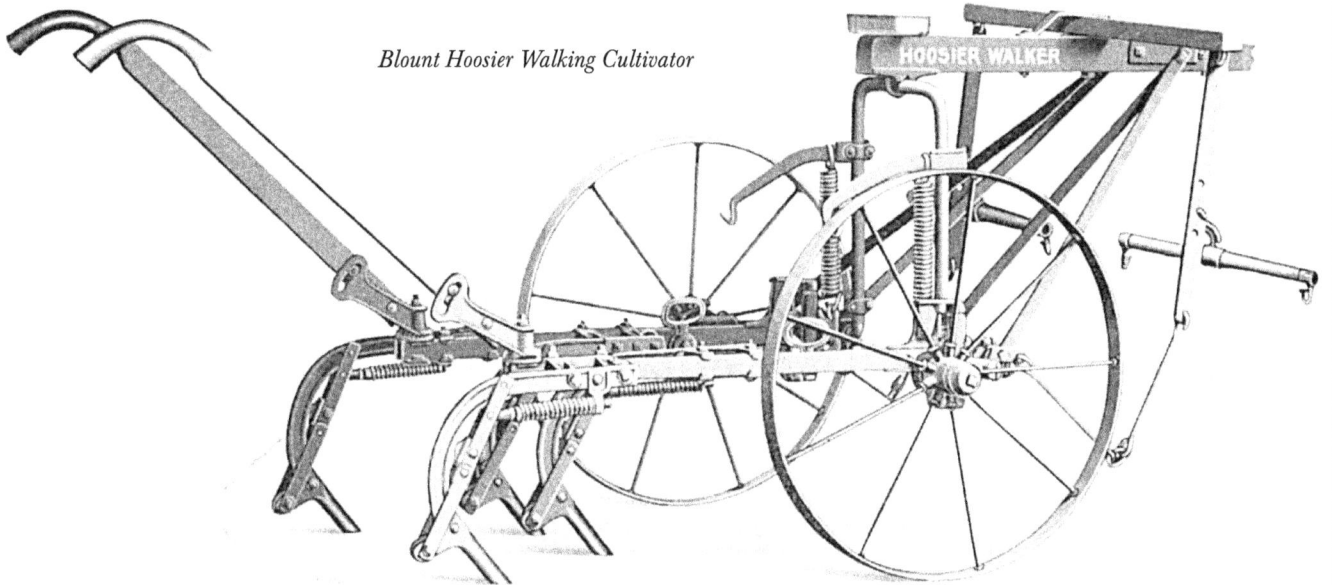

Blount Hoosier Walking Cultivator

P. & O. VOLUNTEER CULTIVATORS.

P & O

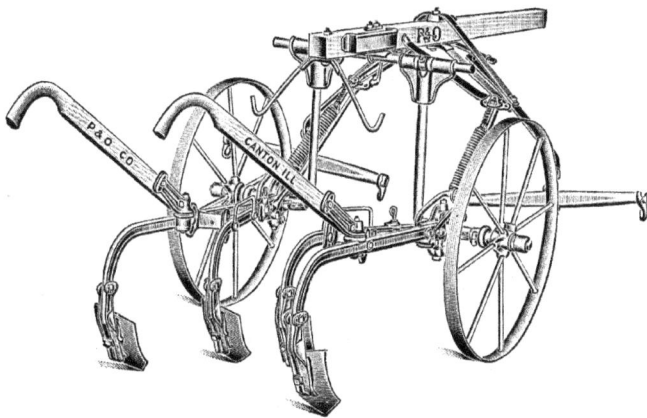

VOLUNTEER CULTIVATOR WITH NO. 1 GANGS.

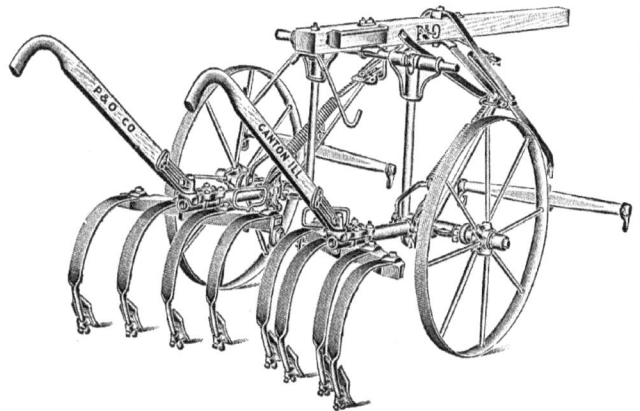

VOLUNTEER CULTIVATOR WITH NO. 76 GANGS

VOLUNTEER CULTIVATOR

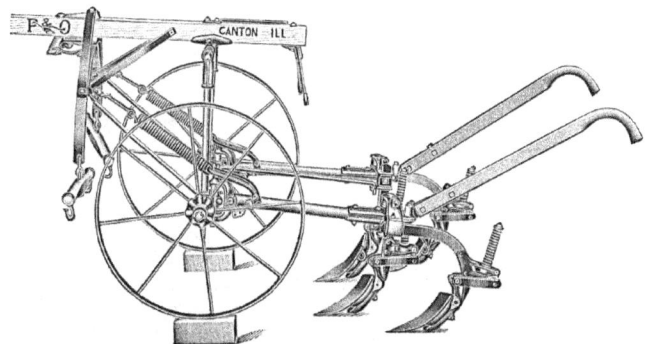

VOLUNTEER CULTIVATOR WITH NO. 13 GANGS.

Two Horse Cultivators

247

TONGUELESS ADJUSTABLE CULTIVATOR

P & O

CANTON TONGUELESS CULTIVATOR

QUEEN TONGUELESS CULTIVATOR

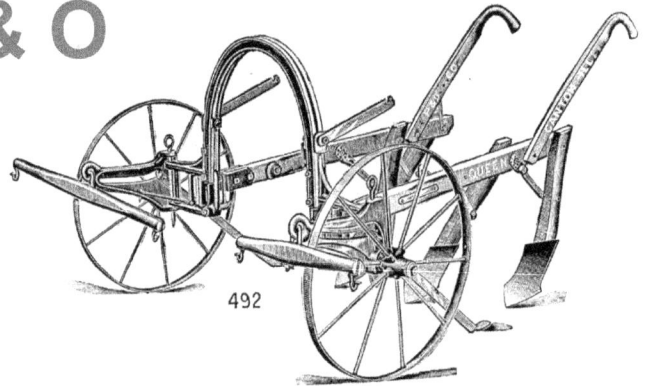

QUEEN CULTIVATOR WITH NO. 2 GANGS.

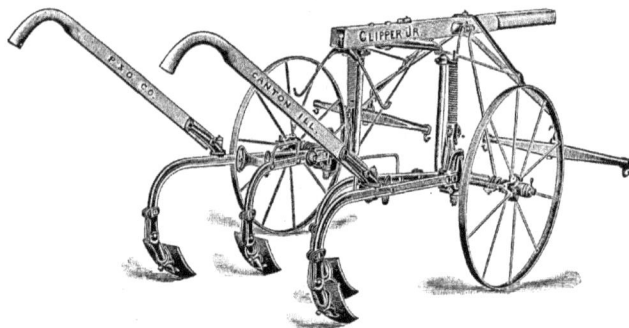

CLIPPER JUNIOR CULTIVATOR

CLIPPER CULTIVATOR WITH NO. 1 GANGS.

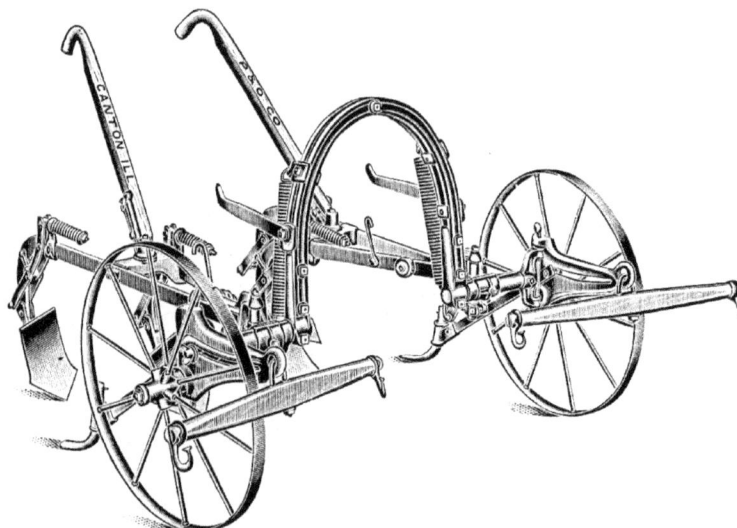

SPRING LIFT CULTIVATOR WITH NO. 11 GANGS.

P & O

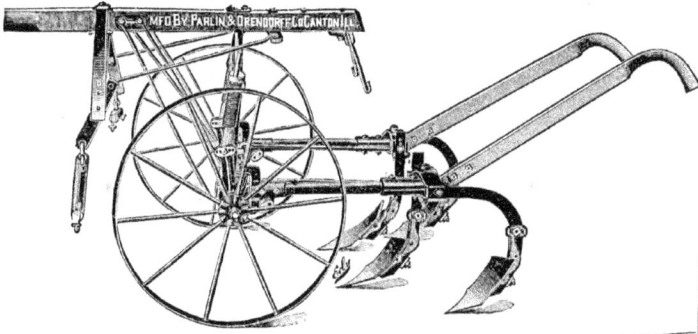

CLIPPER SENIOR CULTIVATOR

P. & O. KING CULTIVATORS.

(Style H.)

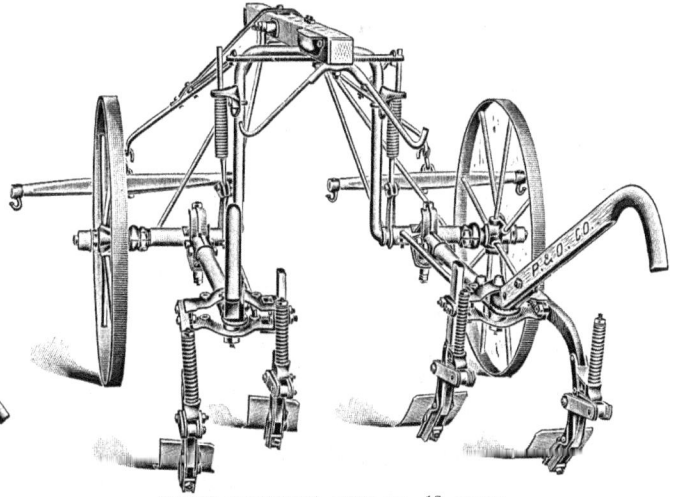

KING CULTIVATOR WITH NO. 21 GANGS.

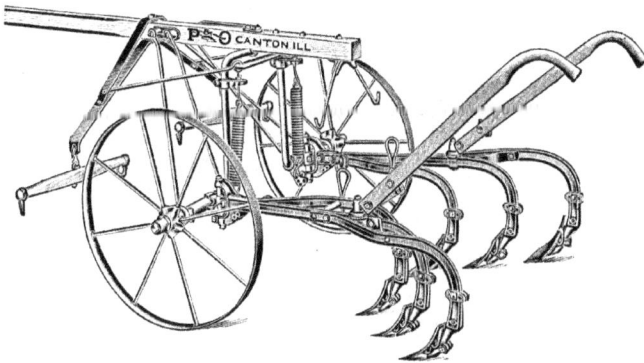

CLIPPER CULTIVATOR WITH NO. 13 GANGS.

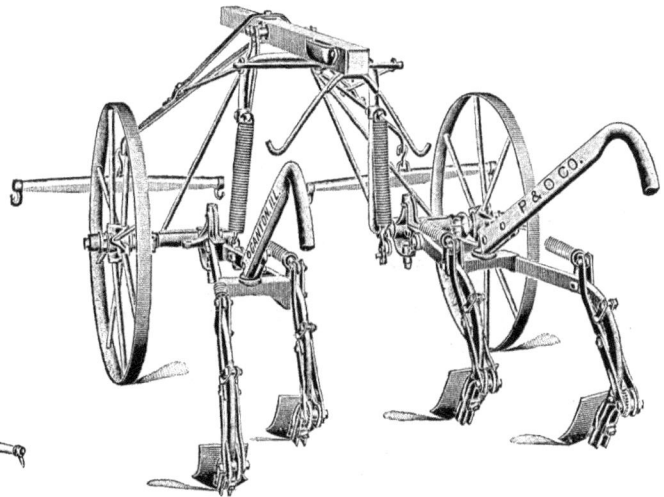

KING CULTIVATOR WITH NO. 11 GANGS.

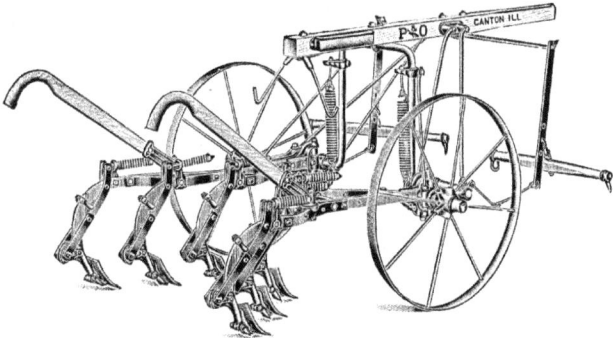

KING BALANCE FRAME CULTIVATOR WITH NO. 31 GANGS.

KING BALANCE FRAME CULTIVATOR WITH NO. 11 GANGS.

KING BALANCE FRAME CULTIVATOR WITH NO. 13 GANGS.

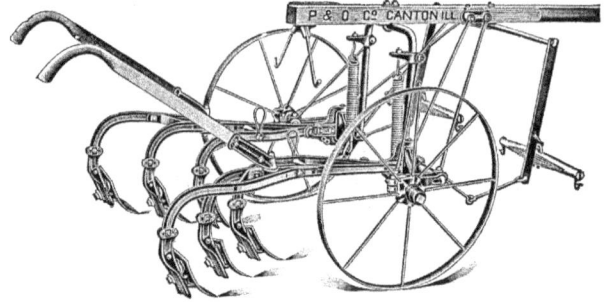

KING BALANCE FRAME CULTIVATOR WITH NO. 21 GANGS.

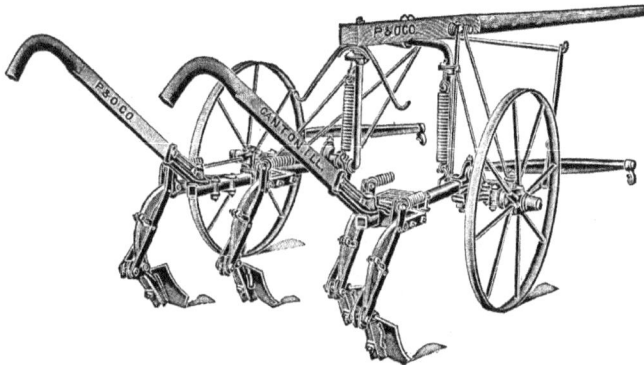

KING BALANCE FRAME CULTIVATOR WITH NO. 34 GANGS.

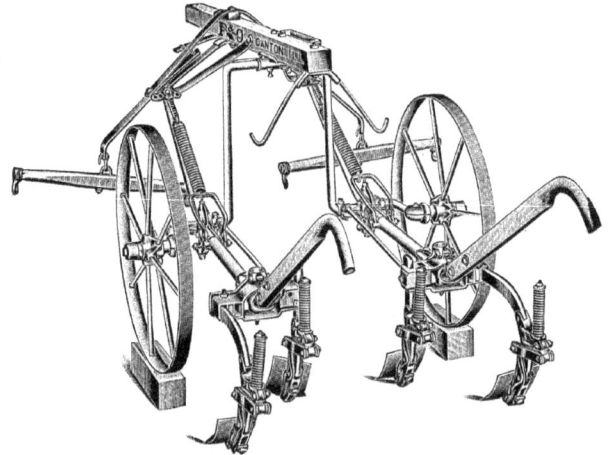

VOLUNTEER BALANCE FRAME CULTIVATOR WITH NO. 13 GANGS.

PARLIN CULTIVATOR

P & O

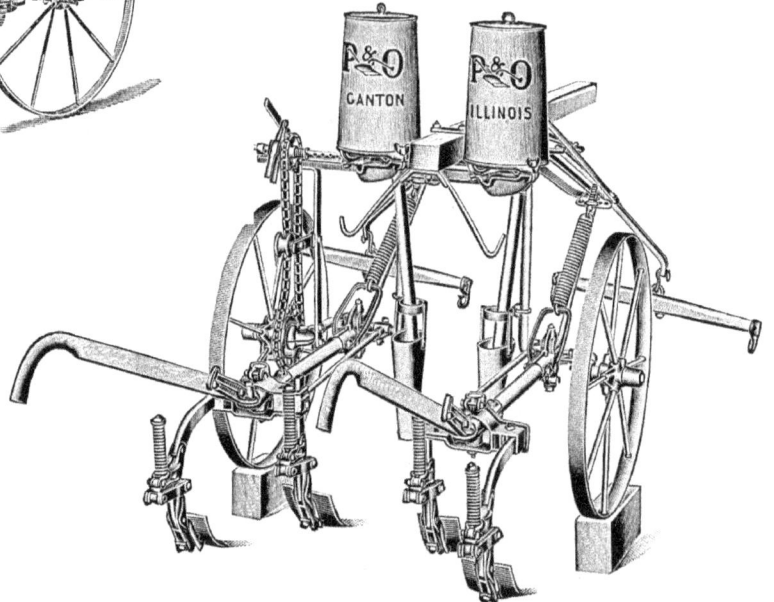

VOLUNTEER BALANCE FRAME CULTIVATOR WITH FERTILIZER ATTACHMENT AND NO. 13 GANGS.

P. & O. TEXAS VICTOR CULTIVATORS.
(Style Y)—Continued.

P. & O. TEXAS VICTOR CULTIVATORS.
(Style Y.)

P & O

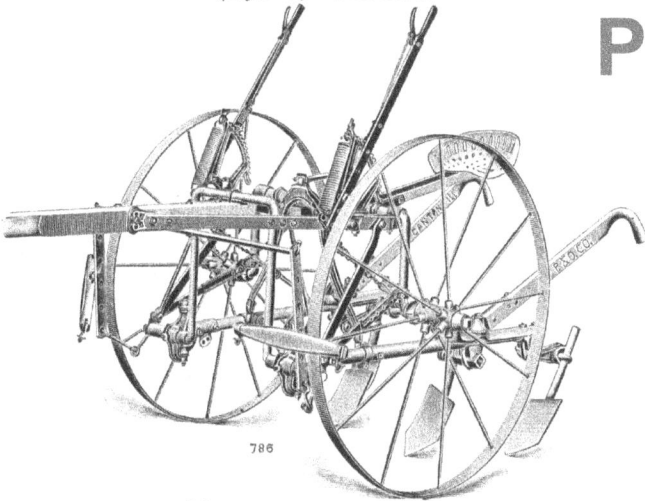

"Y" CULTIVATOR WITH NO. 4 GANGS.

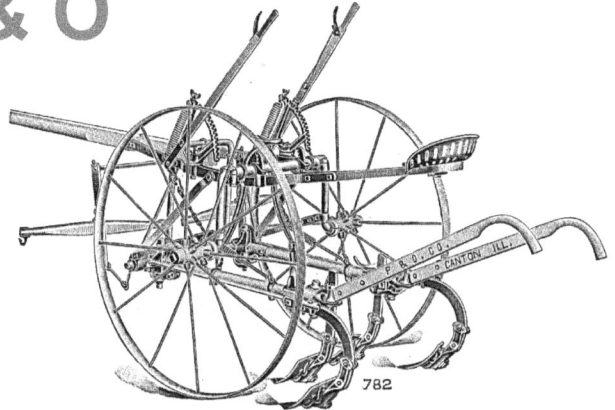

"Y" CULTIVATOR WITH NO. 3 GANGS.

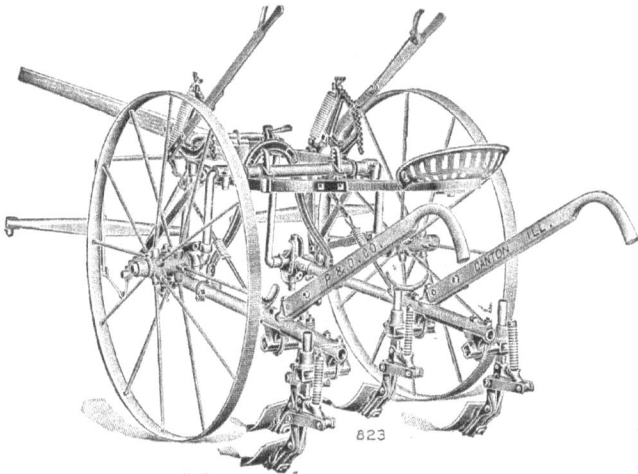

"Y" CULTIVATOR WITH NO. 20 GANGS.

"YC" CULTIVATOR WITH NO. 4 GANGS.

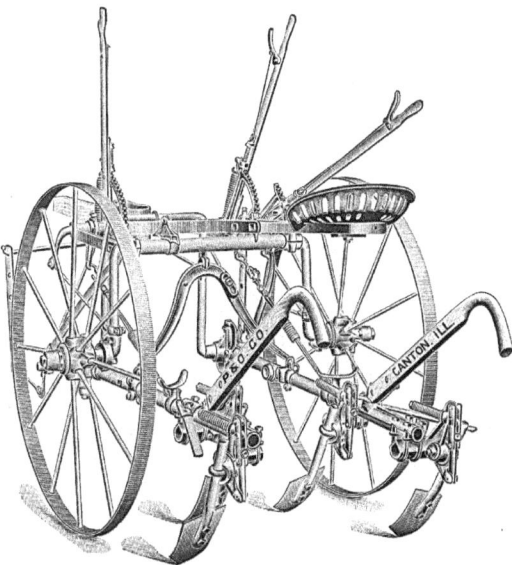

"Y" CULTIVATOR WITH
NO. 13 GANGS.

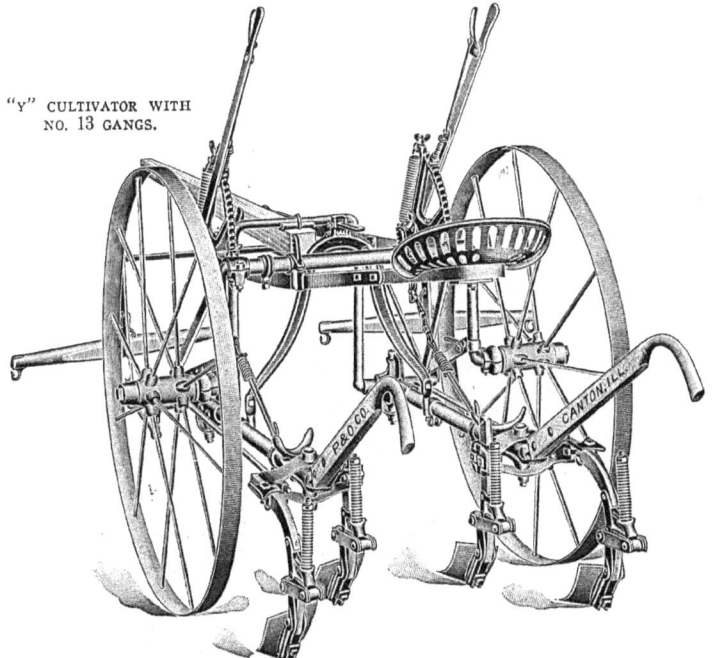

"YC" CULTIVATOR WITH NO. 14 GANGS.

FURNISHED WITH NECKYOKE, SHIELDS, NO. 8 SPREADER AND CONCAVE TIRE WHEELS.

PARLIN & ORENDORFF COMPANY, CANTON, ILLINOIS

Repair List, P. & O. Co. Canton Line

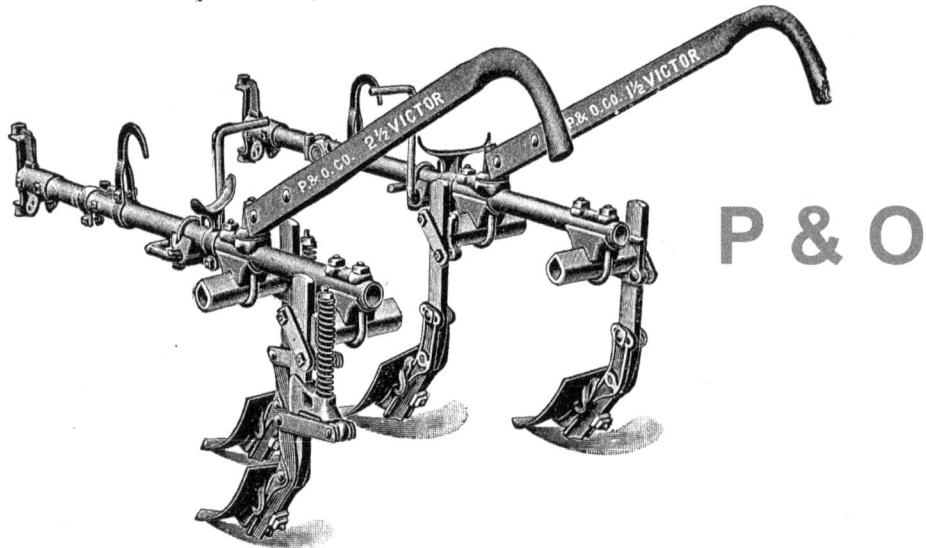

P & O

No 2 Spring Trip Breakoff as used on jointed beam cultivator gangs.

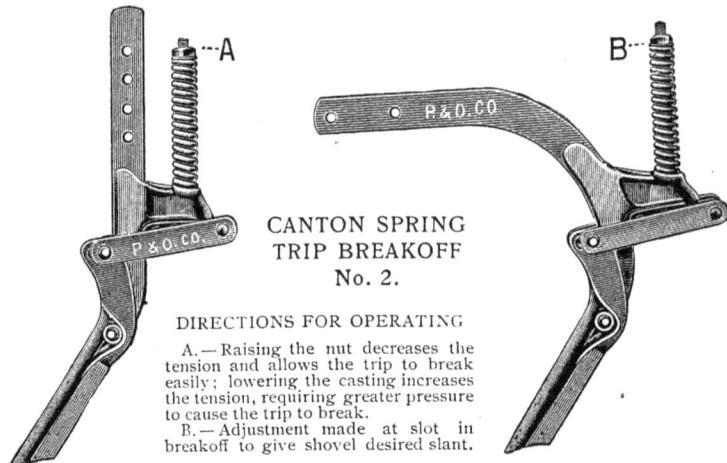

CANTON SPRING TRIP BREAKOFF No. 2.

DIRECTIONS FOR OPERATING

A.— Raising the nut decreases the tension and allows the trip to break easily; lowering the casting increases the tension, requiring greater pressure to cause the trip to break.

B.— Adjustment made at slot in breakoff to give shovel desired slant.

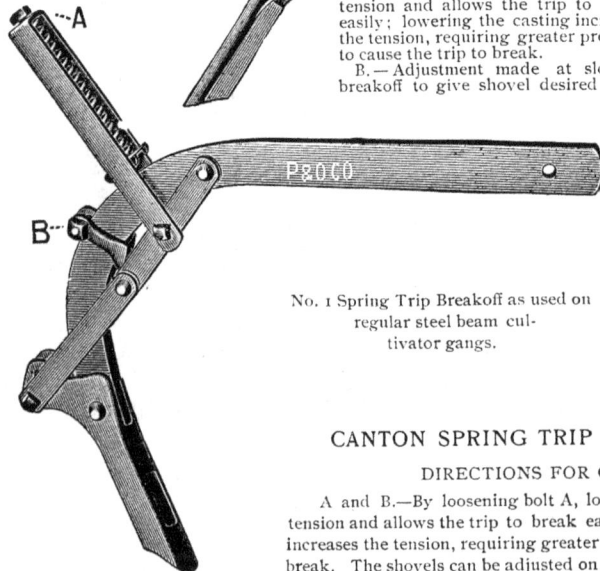

No. 1 Spring Trip Breakoff as used on regular steel beam cultivator gangs.

CANTON SPRING TRIP BREAKOFF No. 1

DIRECTIONS FOR OPERATING

A and B.—By loosening bolt A, lowering A and B, decreases the tension and allows the trip to break easily, and raising the castings increases the tension, requiring greater pressure to cause the trip to break. The shovels can be adjusted on shank to any desired slant.

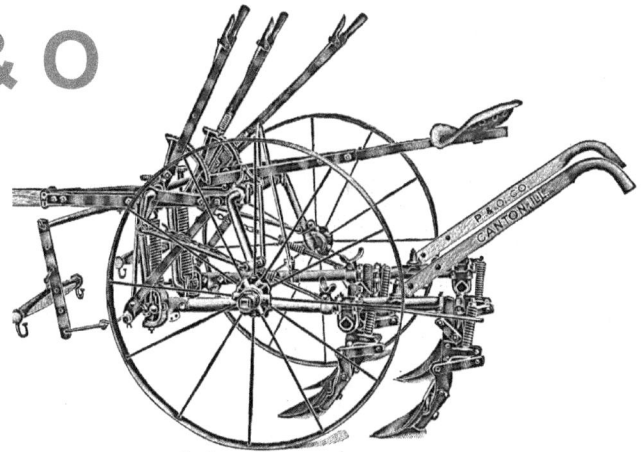
"YB" CULTIVATOR WITH NO. 7 GANGS.

(Style YB)

P & O

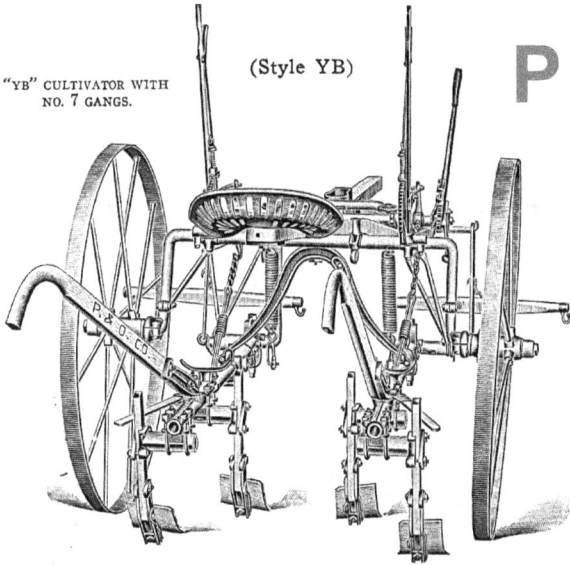
"YD" CULTIVATOR WITH NO. 17 GANGS.

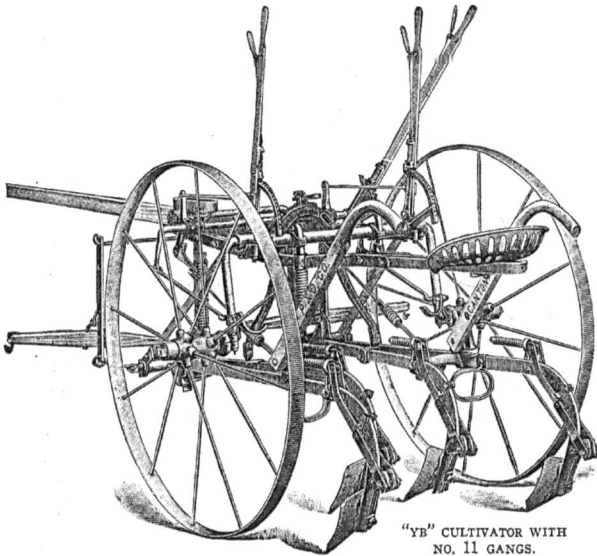
"YB" CULTIVATOR WITH NO. 11 GANGS.

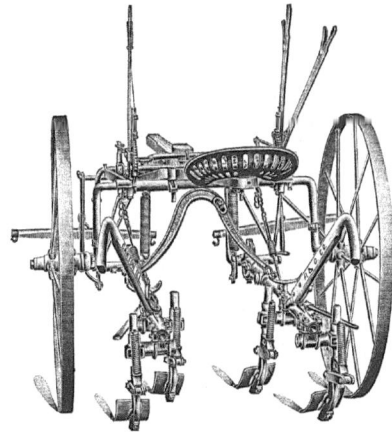
"YB" CULTIVATOR WITH NO. 20 GANGS.

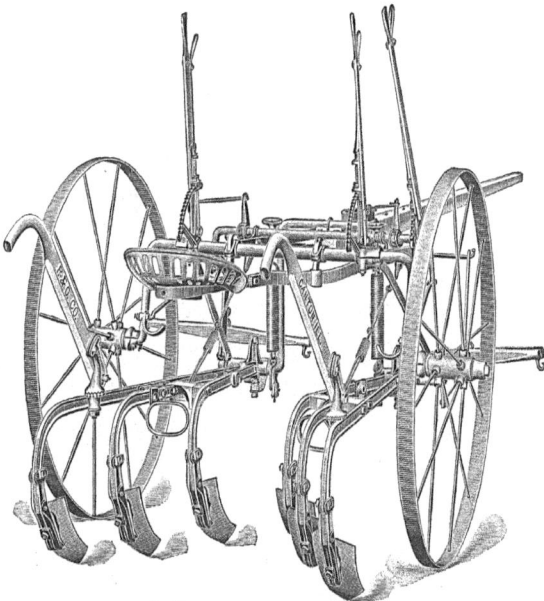
"YB" CULTIVATOR WITH NO. 21 GANGS.

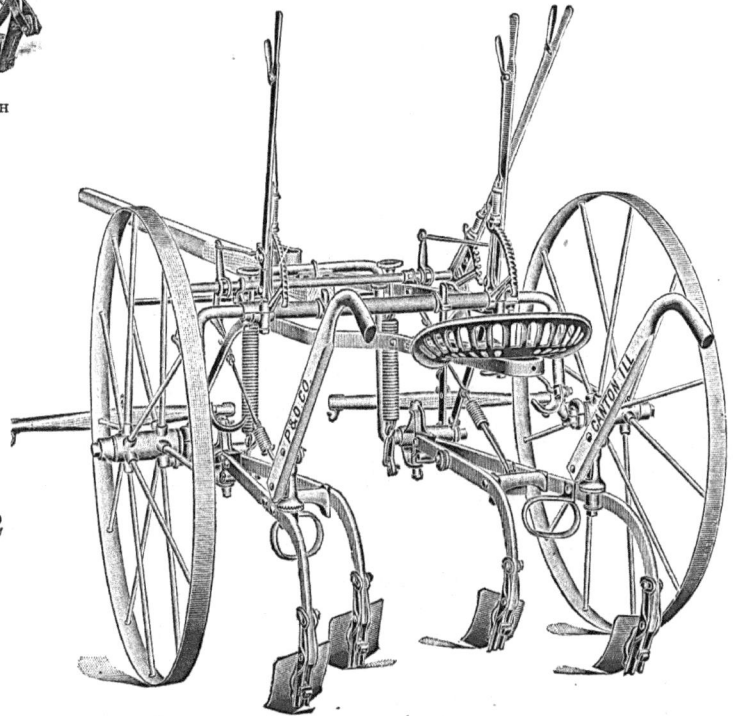
"YB" CULTIVATOR WITH NO. 1 GANGS.

"YB" CULTIVATOR WITH
NO. 13 GANGS.

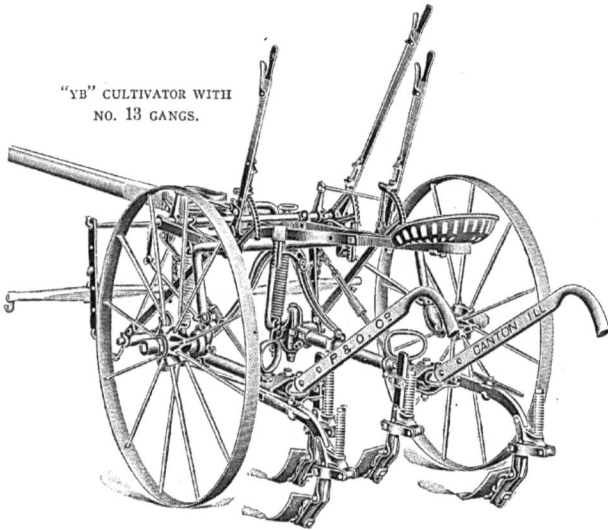

PARLIN CULTIVATOR, WITH NO. 1 GANG AND GOPHER BLADES

"YB" CULTIVATOR WITH
CONCAVE TIRES AND NO. 31
JEWEL HAMMOCK TYPE
GANGS.

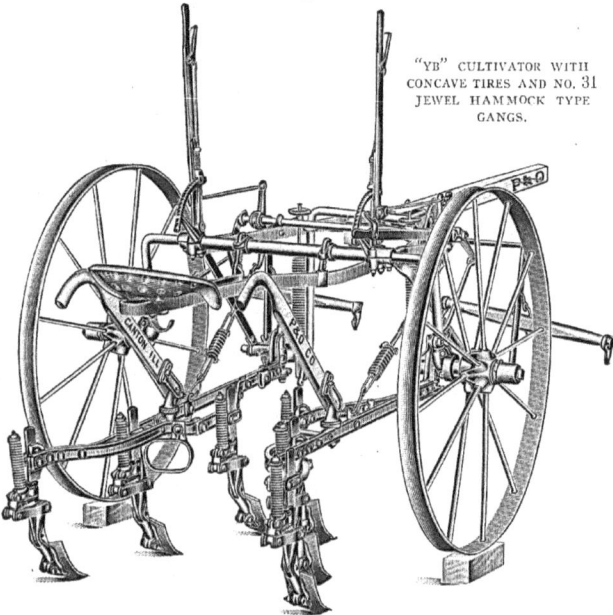

PARLIN CULTIVATOR, WITH NO. 11 GANG

The Parlin Cultivator as equipped above with Spring Trip Gangs. This favorite cultivator works equally well with any kind of a gang.

COMBINED BALANCE FRAME CULTIVATOR WITH NO. 21 GANG

P & O

"YF" CULTIVATOR WITH NO. 1 GANGS.

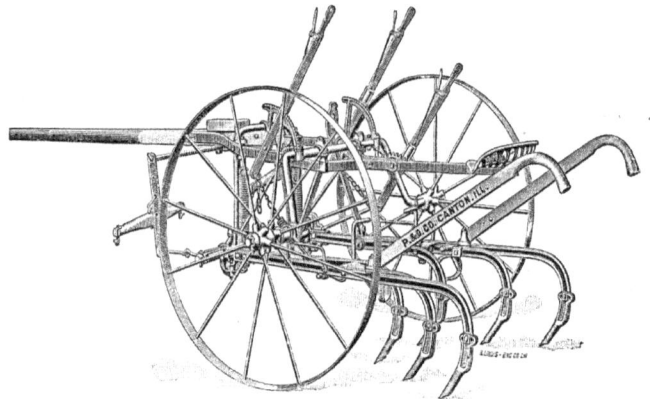

P & O

Parlin Cultivators

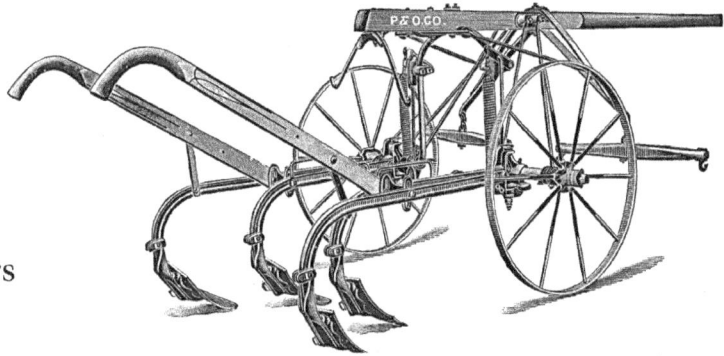

Invincible Cultivators

INVINCIBLE CULTIVATOR, WITH NO. 1 GANG

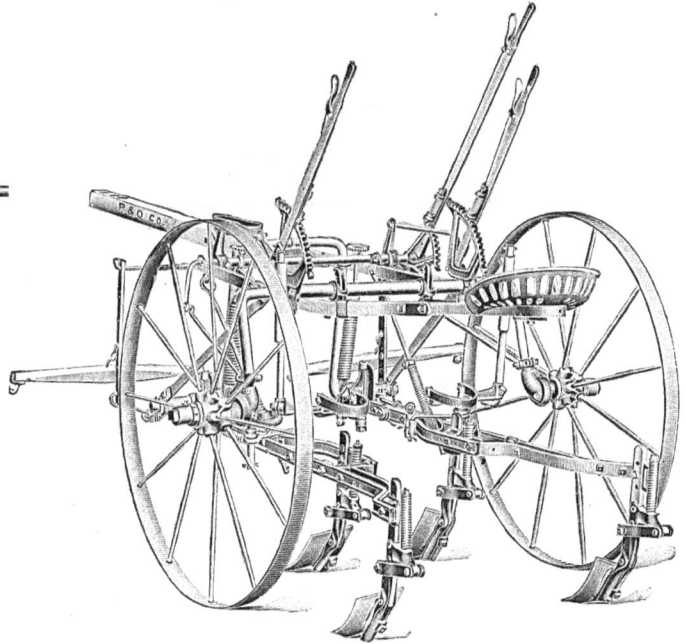

"YBX" CULTIVATOR WITH NO. 11 GANGS.

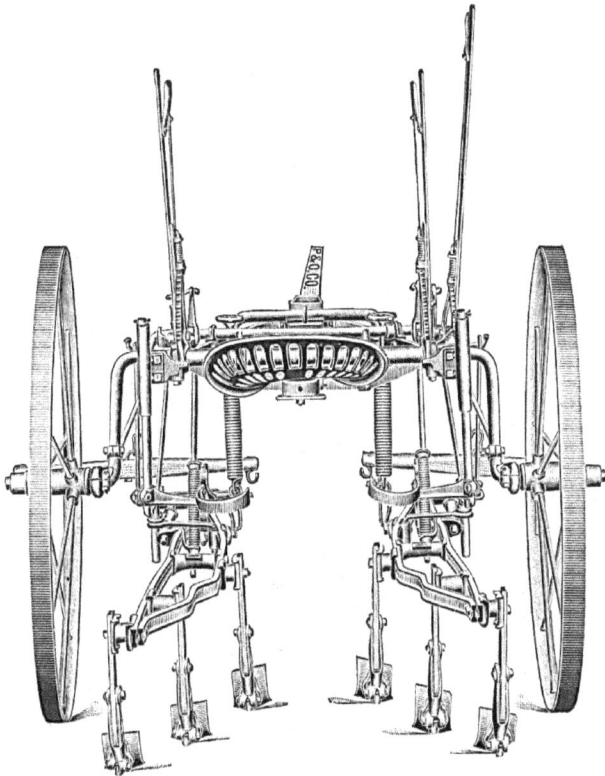

"YBX" CULTIVATOR WITH NO. 21 GANGS.

VICTOR SR. PIVOT TONGUE CULTIVATOR

Two Horse Cultivators

Combined Balance Frame Cultivators

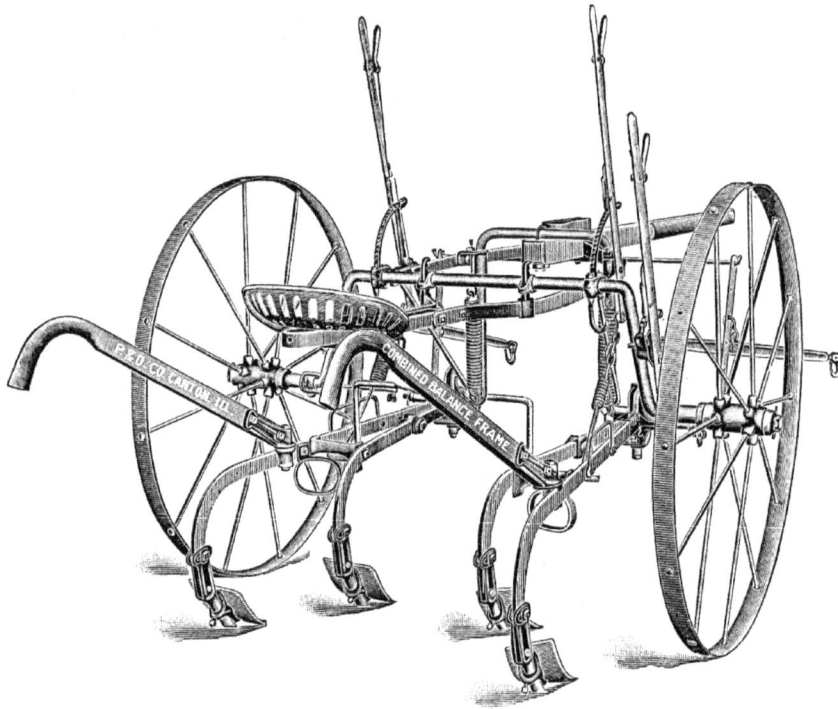

COMBINED BALANCE FRAME CULTIVATOR, WITH NO. 1 GANG

P & O

Detail of Balancing Lever.

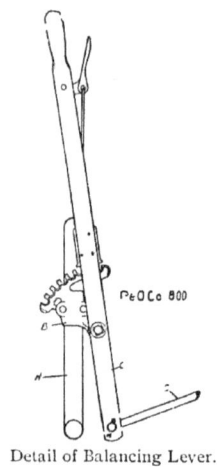

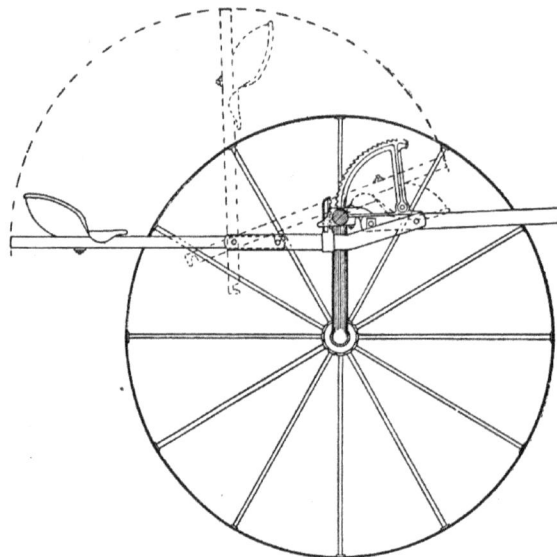

VIEW SHOWING SEAT ADJUSTMENT.
P. & O. CO. # 750.
When used as a Walking Cultivator the seat can be thrown forward, leaving the driver a clear view of his work.

Combined Balance Frame Pivot Tongue Cultivators

P & O

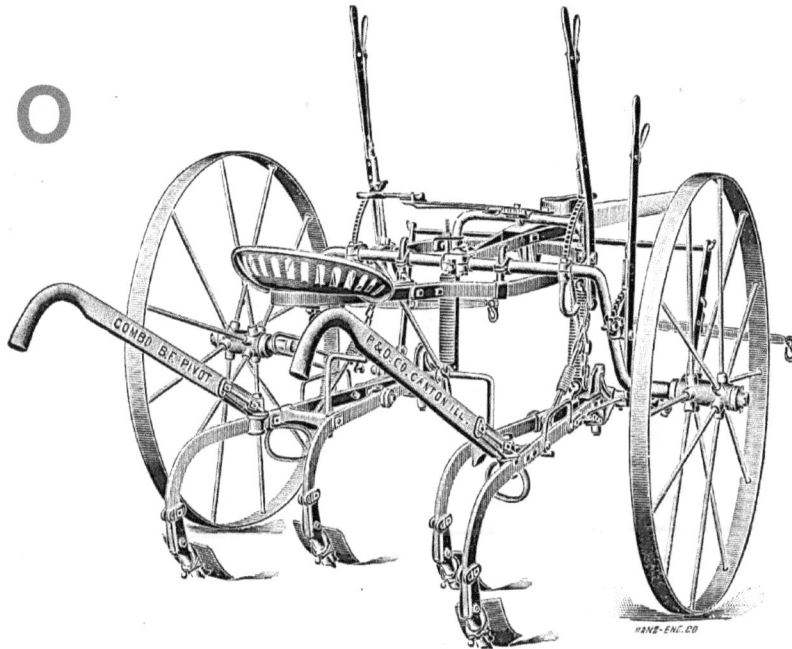

COMBINED BALANCE FRAME PIVOT TONGUE CULTIVATOR, WITH NO. 1 GANG

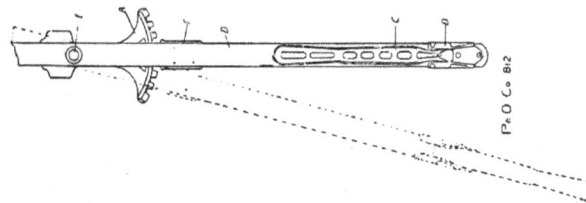

Detail of Steering Lever. The lever can be folded over when used as a rigid tongue lever.
The dotted lines indicate its powerful leverage when used as a steering lever.

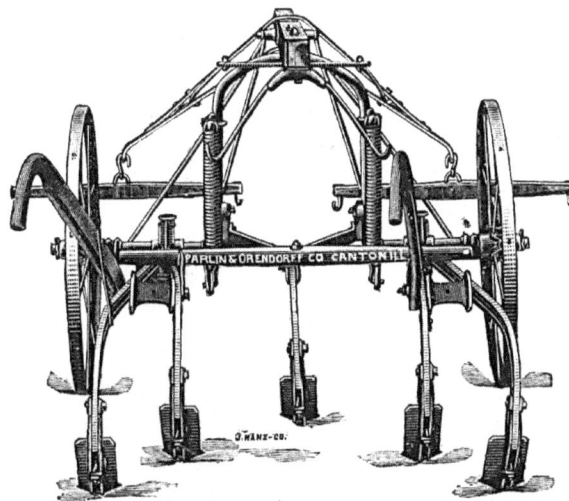

PARLIN CULTIVATOR, WITH NO. 1 GANG AND FIFTH SHOVEL ATTACHMENT

The above cut illustrates the Parlin Planter with a four shovel gang and fifth shovel attached. A cultivator equipped in this manner is a first-class implement for preparing the ground for putting in oats and plowing summer fallow.

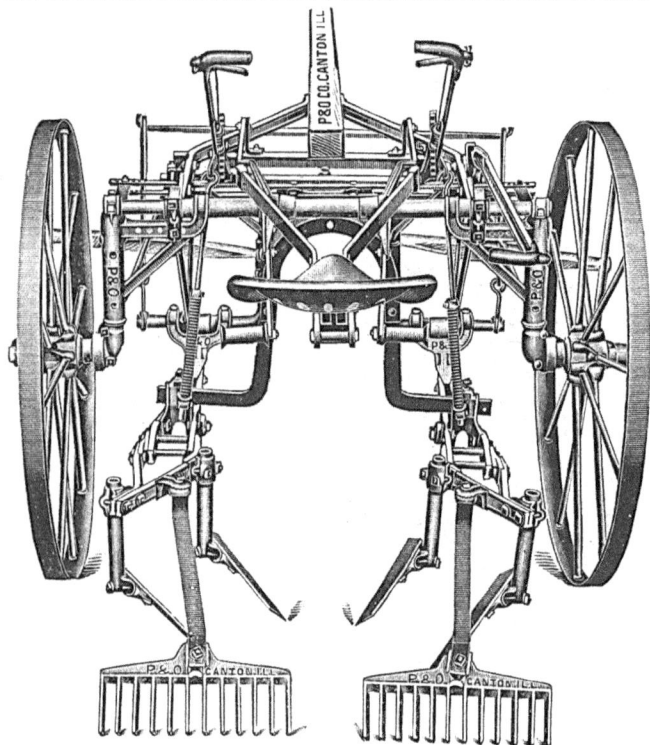

CANTONIAN CULTIVATOR WITH SURFACE ATTACHMENT.

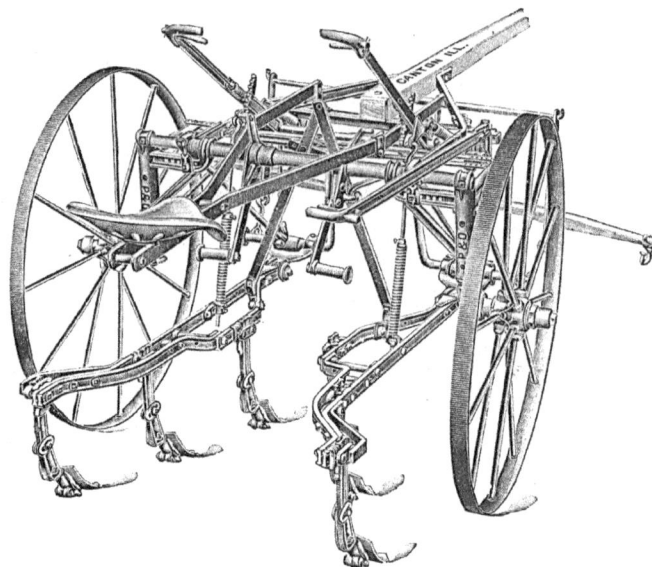

CANTONIAN CULTIVATOR WITH NO. 21 GANGS.

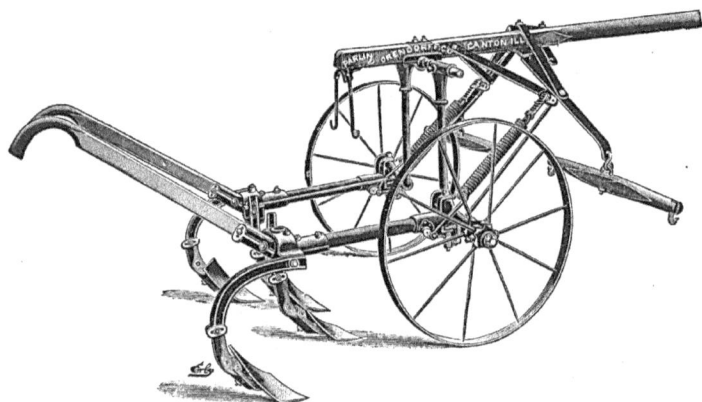

VOLUNTEER CULTIVATOR, WITH NO. 3 GANG

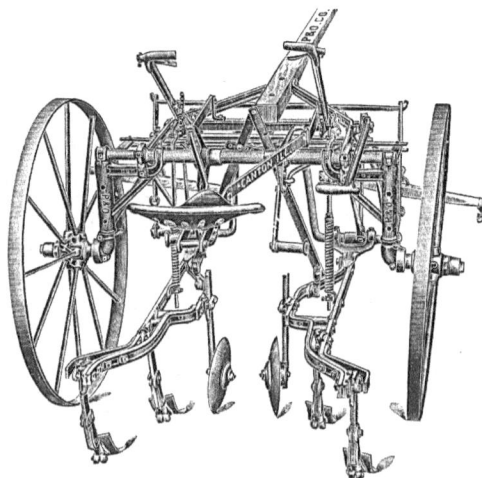

CANTONIAN CULTIVATOR WITH NO. 21 GANGS AND NO. 1 VINE CUTTER ATTACHMENT.

P & O

Potato Cultivator

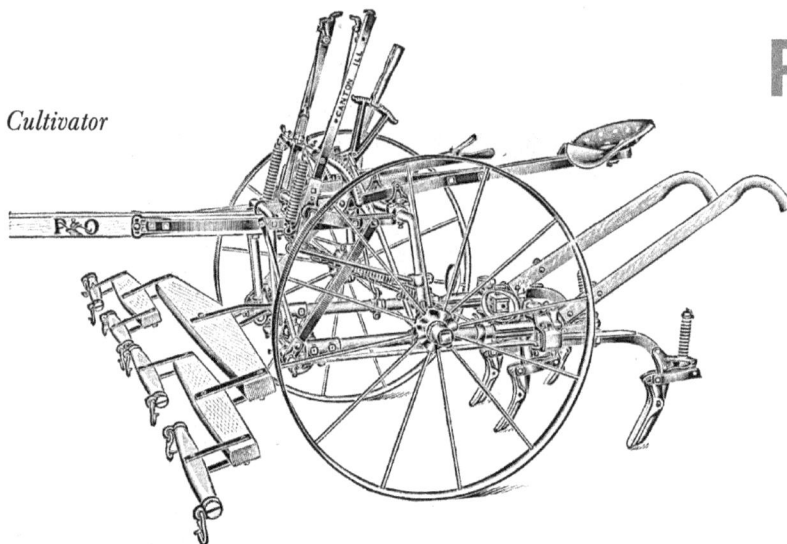

(Style TA.)

P & O

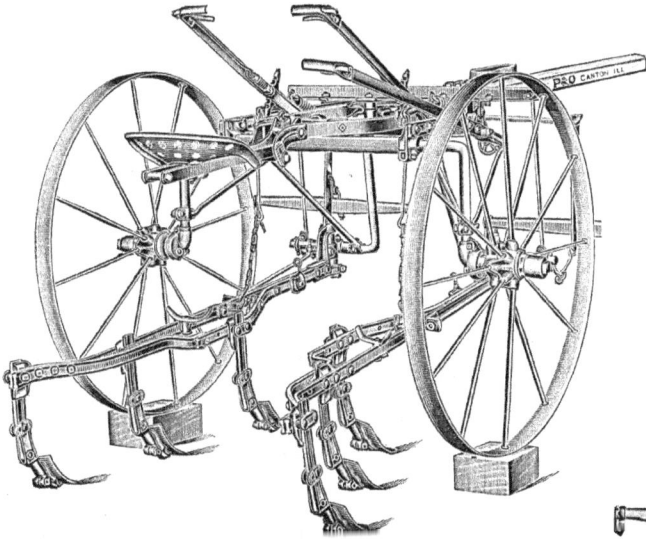

WIGGLETAIL CULTIVATOR WITH NO. 21 GANGS.

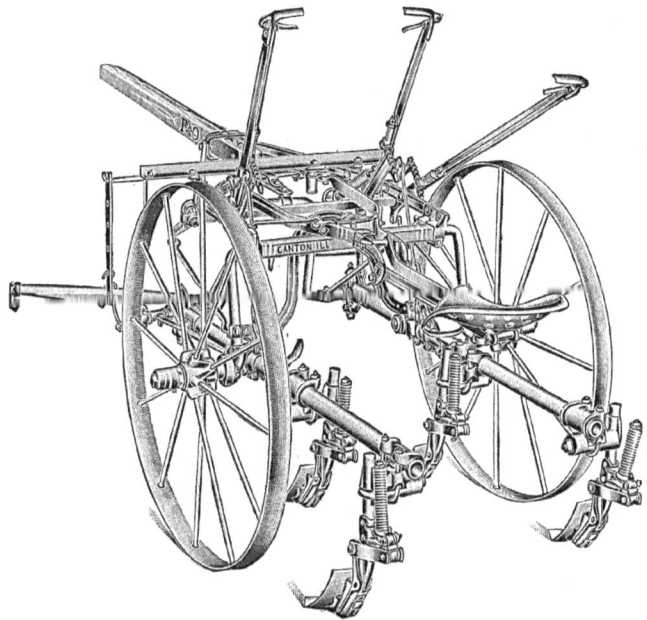

WIGGLETAIL CULTIVATOR WITH NO. 20 GANGS.

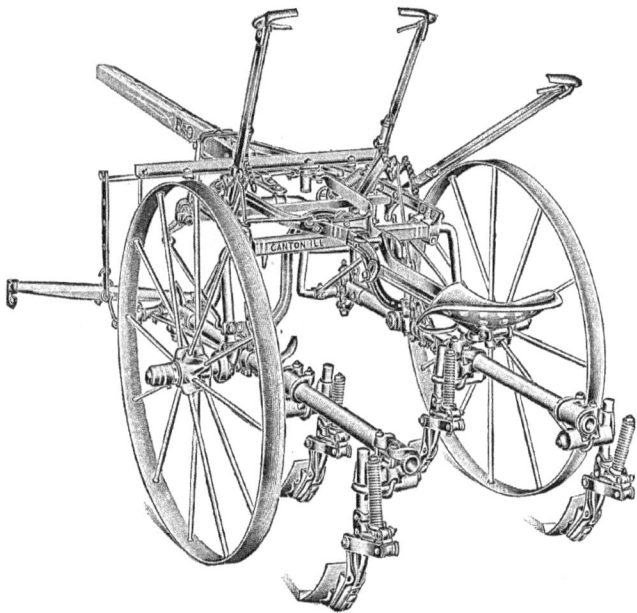

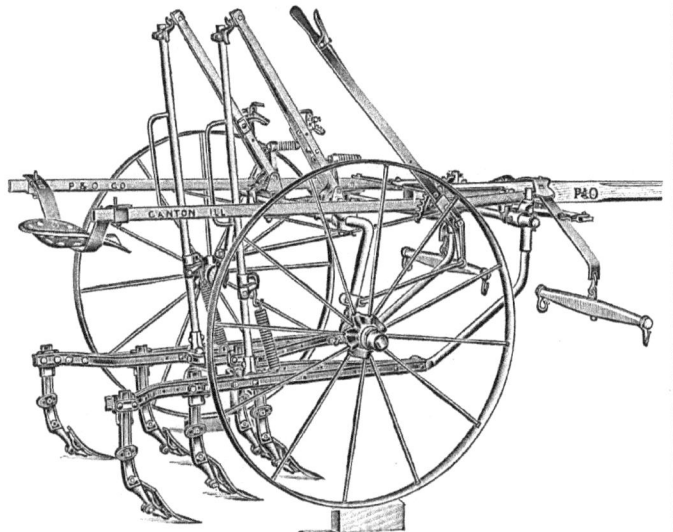

WIGGLETAIL CULTIVATOR WITH NO. 20 GANGS.

JEWEL HAMMOCK WITH NO. 21 GANGS.

JEWEL HAMMOCK CULTIVATOR WITH NO. 11 GANGS.

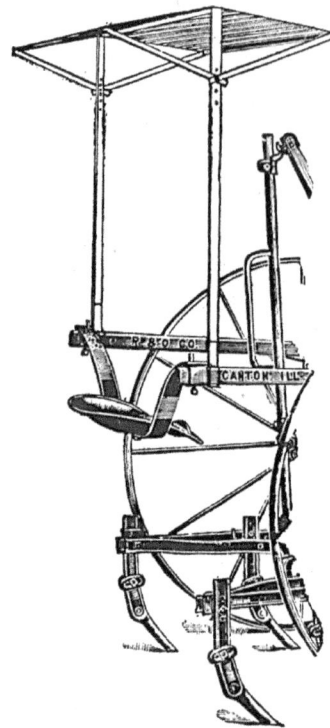

CANOPY FOR JEWEL HAMMOCK CULTIVATOR.

JEWEL HAMMOCK CULTIVATOR WITH NO. 31 GANGS.

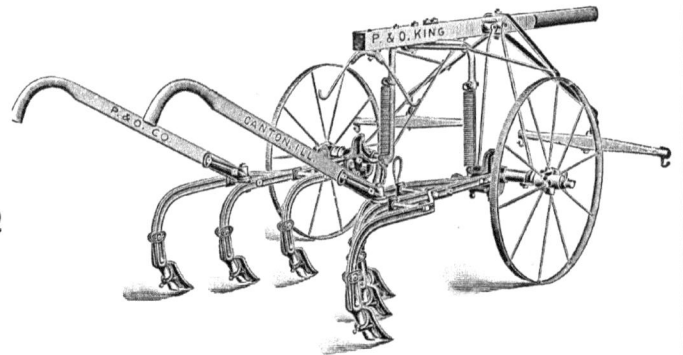

P & O

KING CULTIVATOR, WITH NO. 21 GANG

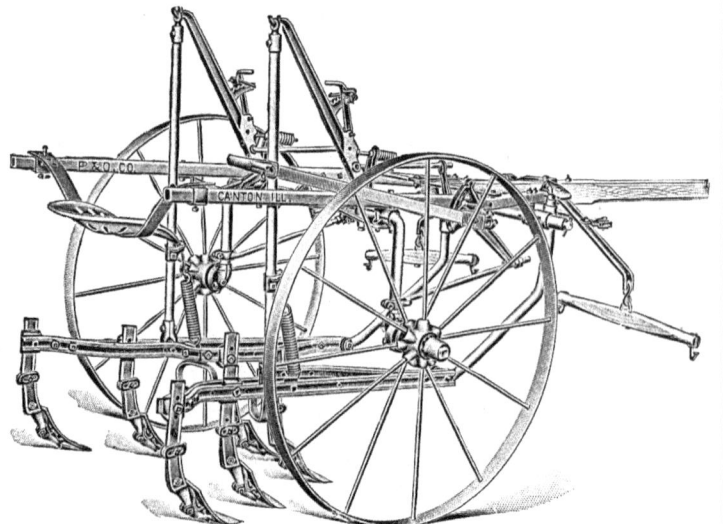

HAMMOCK HORSE LIFT CULTIVATOR WITH NO. 21 GANGS.

NO. 5 JEWEL SURFACE CULTIVATOR.

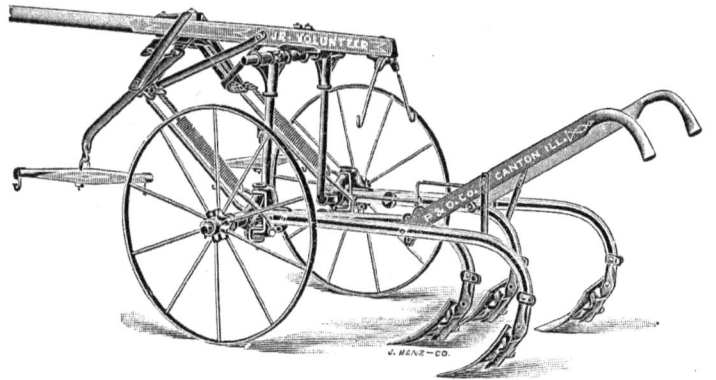

VOLUNTEER CULTIVATOR, WITH NO. 1 GANG

P & O

VOLUNTEER CULTIVATOR, WITH NO. 13 GANG

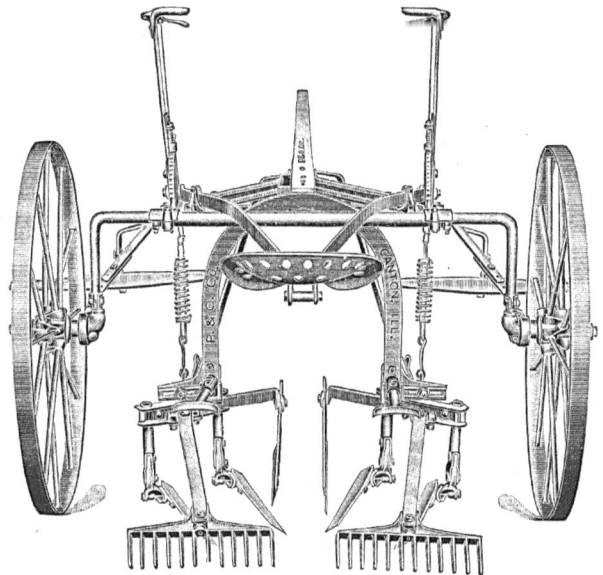

NO. 15 SURFACE CULTIVATOR.

VOLUNTEER CULTIVATOR, WITH NO. 76 GANG

CLIPPER SENIOR CULTIVATOR, WITH NO. 1 GANG

Two Horse Cultivators

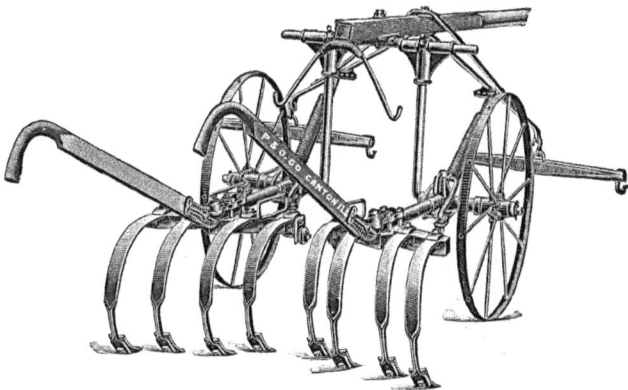

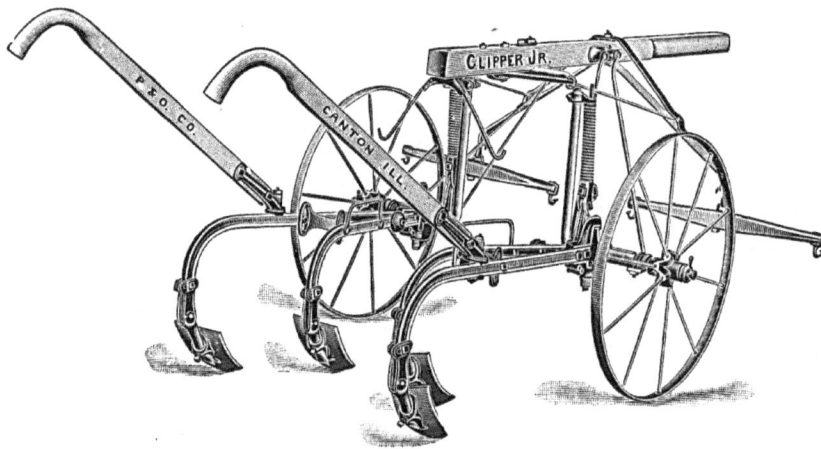

CLIPPER JUNIOR CULTIVATOR, WITH NO. 1 GANG

P & O

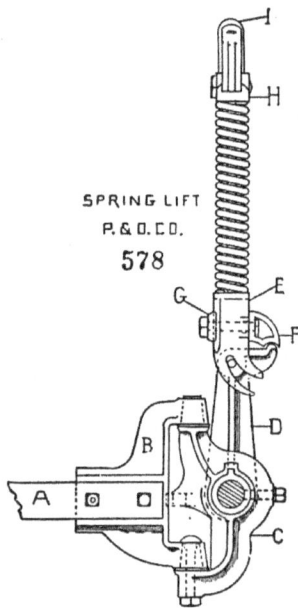

SPRING LIFT
P. & O. CO.
578

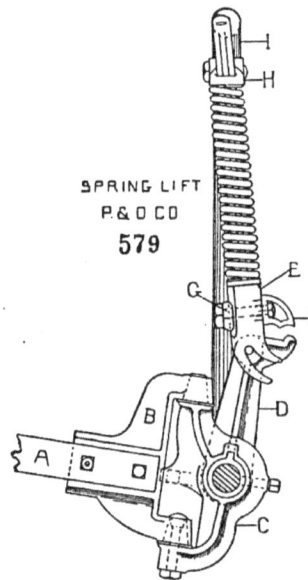

SPRING LIFT
P. & O. CO
579

shows gangs in working position with sleeve arm D in contact with adjustable block F. When in this position the gangs are perfectly balanced. The gangs can be raised or lowered by moving adjustable block F.

shows the beam partly raised, with the spring exerting a downward pressure on sleeve arm D, thus requiring no apparent effort to raise the gangs.

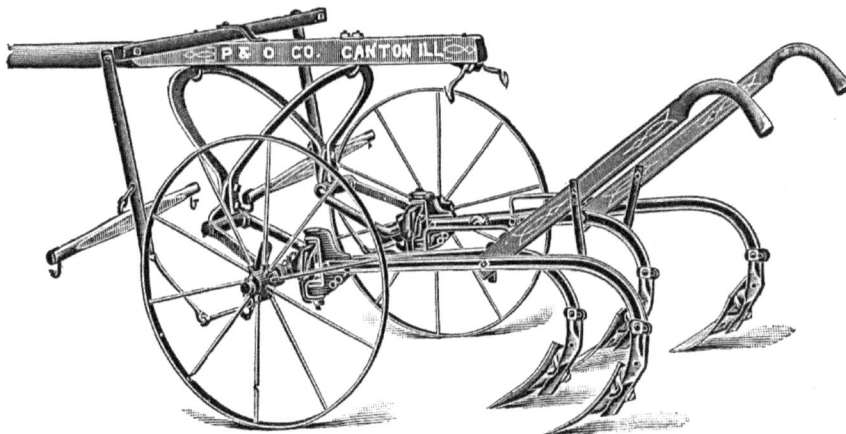

BALANCE FRAME CULTIVATOR, WITH NO. 1 GANG

P & O

Queen Tongueless Cultivator, with No. 2 Gang

Tongueless Adjustable Cultivator, with No. 2 Gang

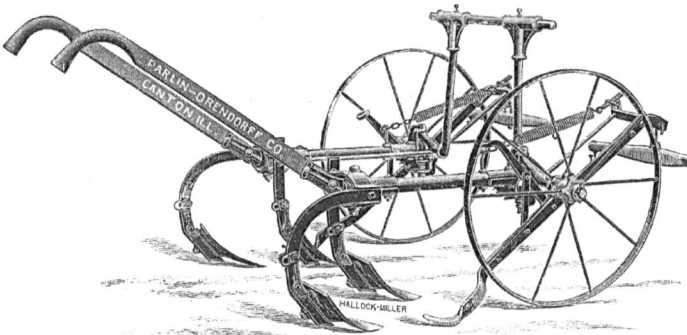

Tongueless Adjustable Cultivator, with No. 3 Gang

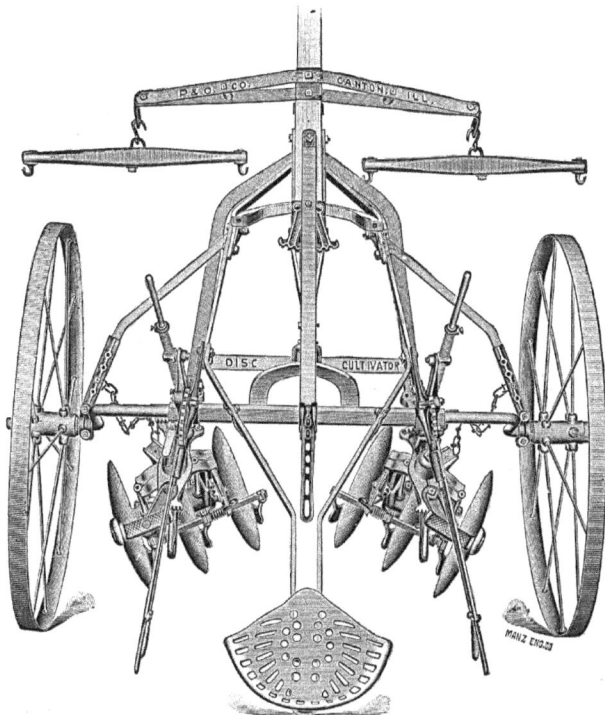

Canton Tongueless Cultivator, with No. 2 Gang

Canton Riding Disc Cultivator, Pivot Tongue

Rigid Tongue

The Steering Lever can be folded when used as a rigid tongue cultivator. Dotted lines show the lever in an operative position.

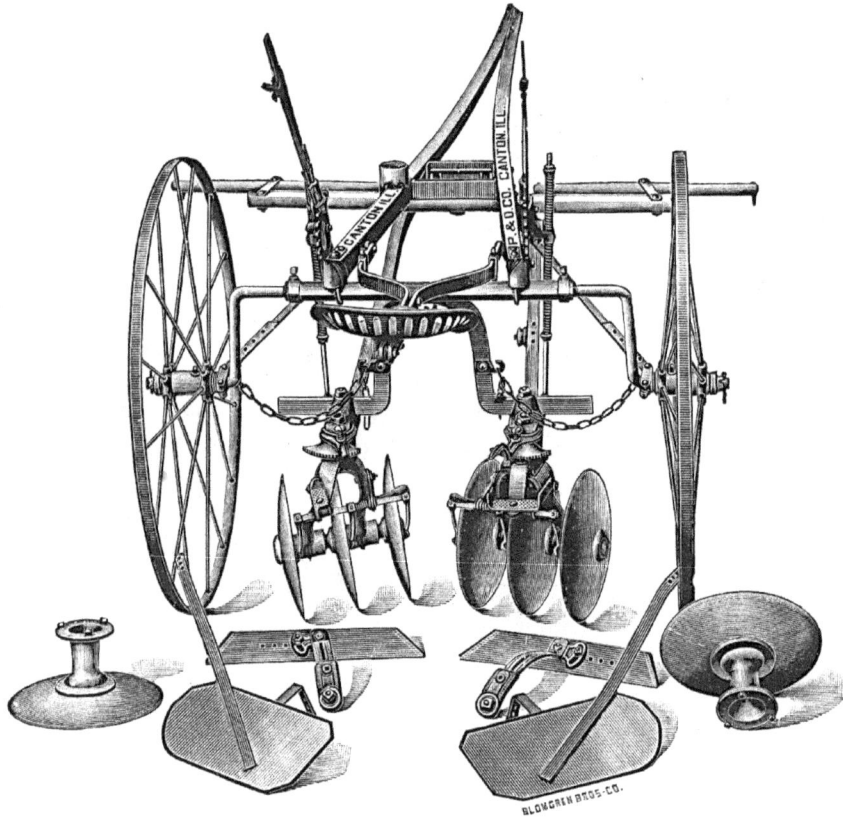

CANTON DISC CULTIVATOR, WITH LEVELERS AND HARROW ATTACHMENTS

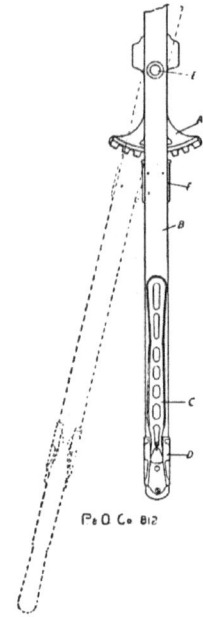

P & O

Detail of Disc Bearing Boxes.

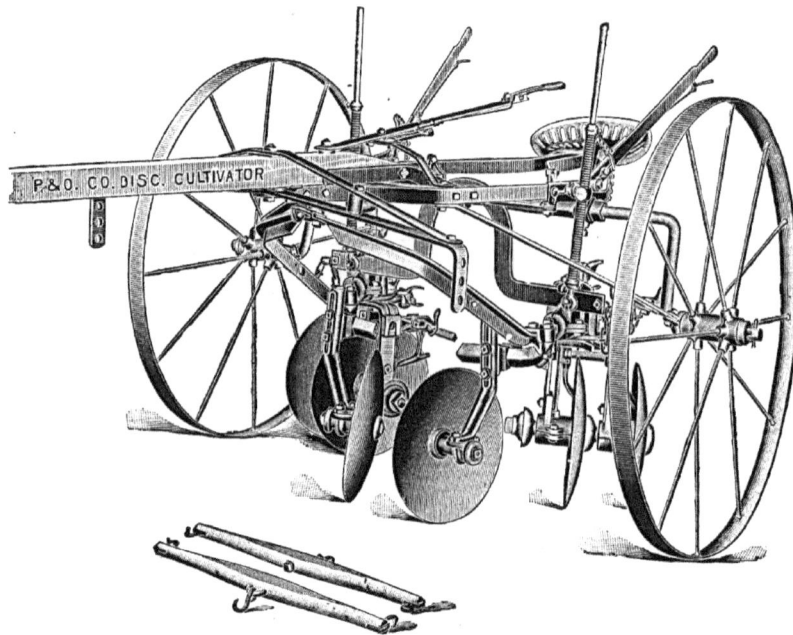

CANTON DISC CULTIVATOR, SHOWING THE MANNER IN WHICH THE FRONT DISCS ARE ATTACHED TO THE FRONT ARMS

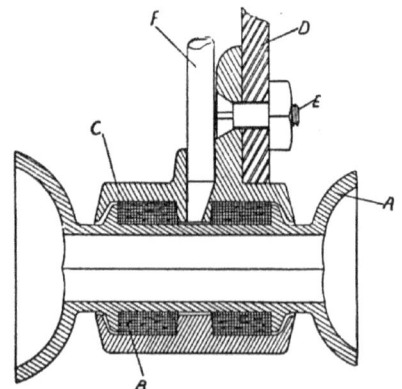

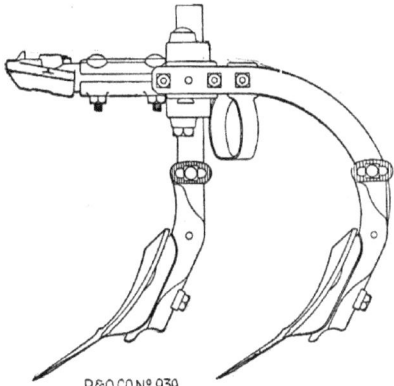

P&O.CO.Nº 939.
The Shovel Attachment.

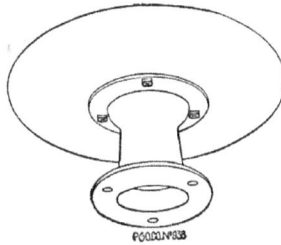

Detail of Harrow Attachment.

P & O

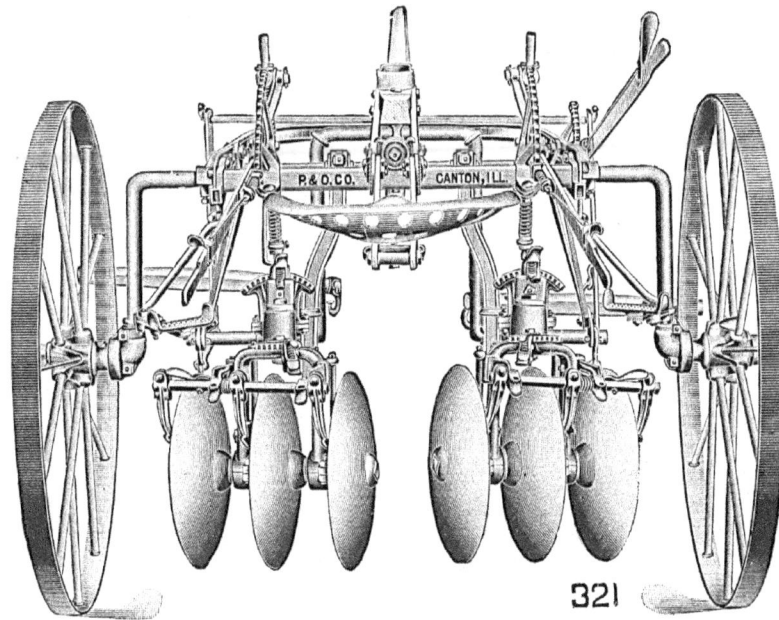

NO. 96 RIVAL DISC CULTIVATOR.

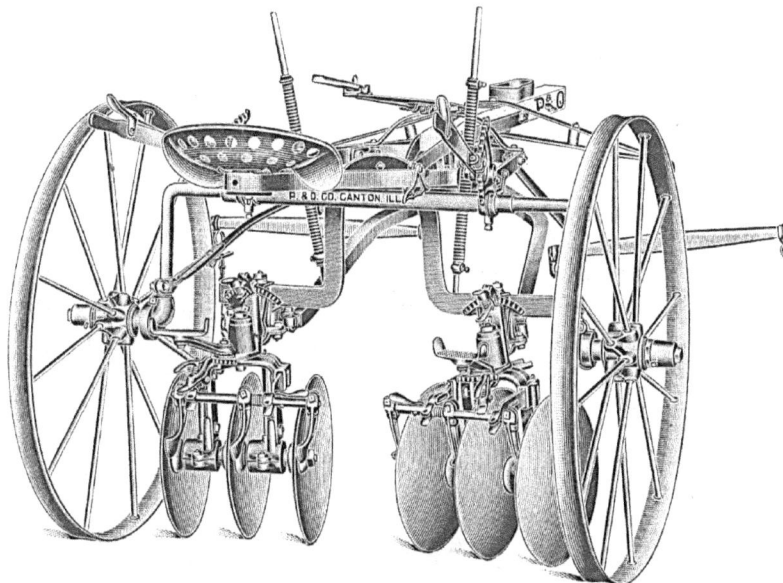

NO. 92 DISC CULTIVATOR.

P & O

P. & O. RIVAL DISC CULTIVATORS.

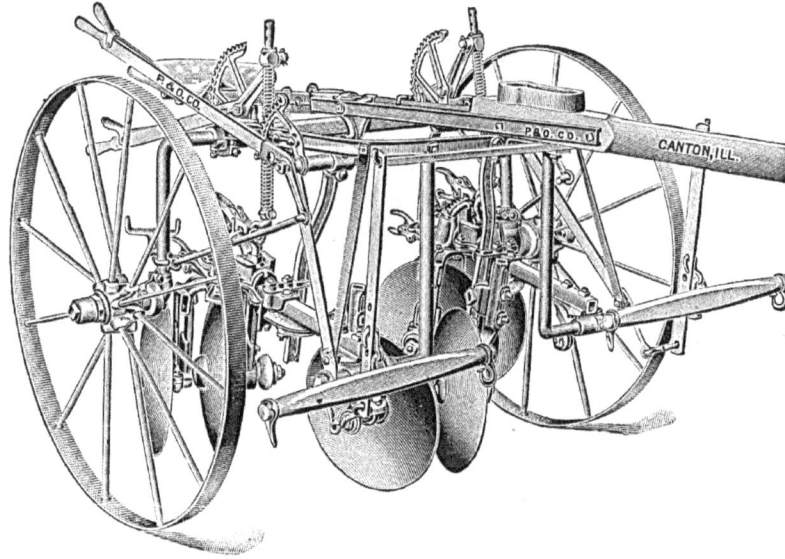

NO. 96 RIVAL DISC CULTIVATOR WITH FRONT ARM ATTACHMENT.

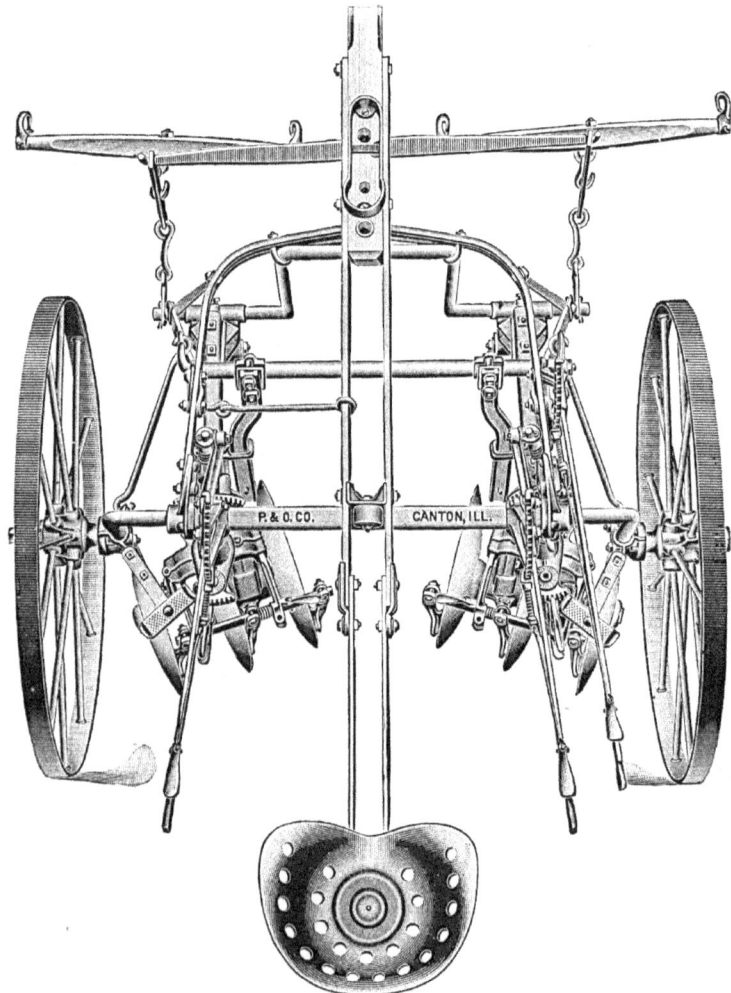

Top View—Showing Pivot Feature.

Nos. 98 and 99.

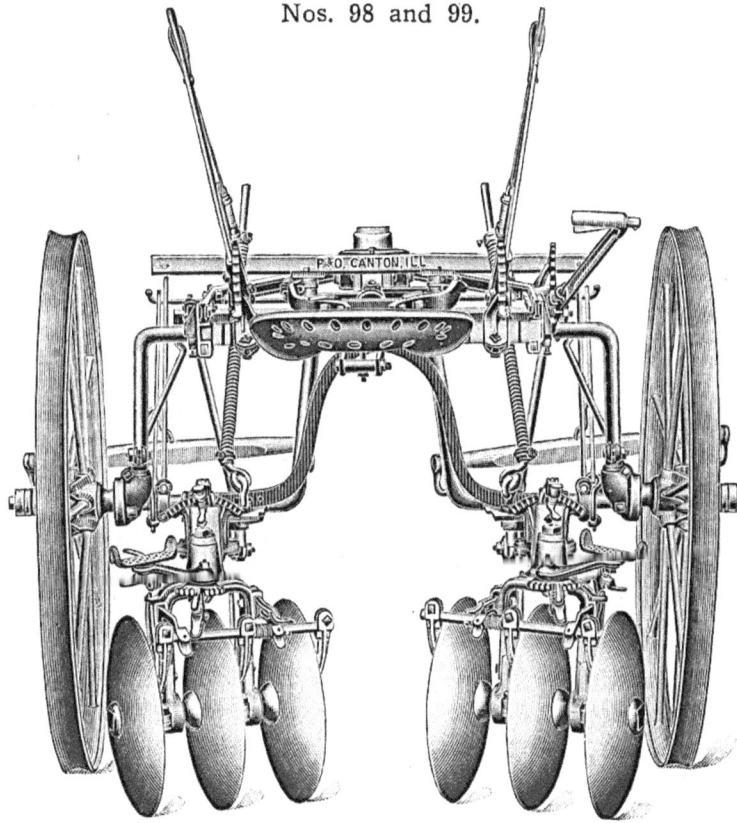

NO. 98 WIGGLETAIL DISC CULTIVATOR.

P & O

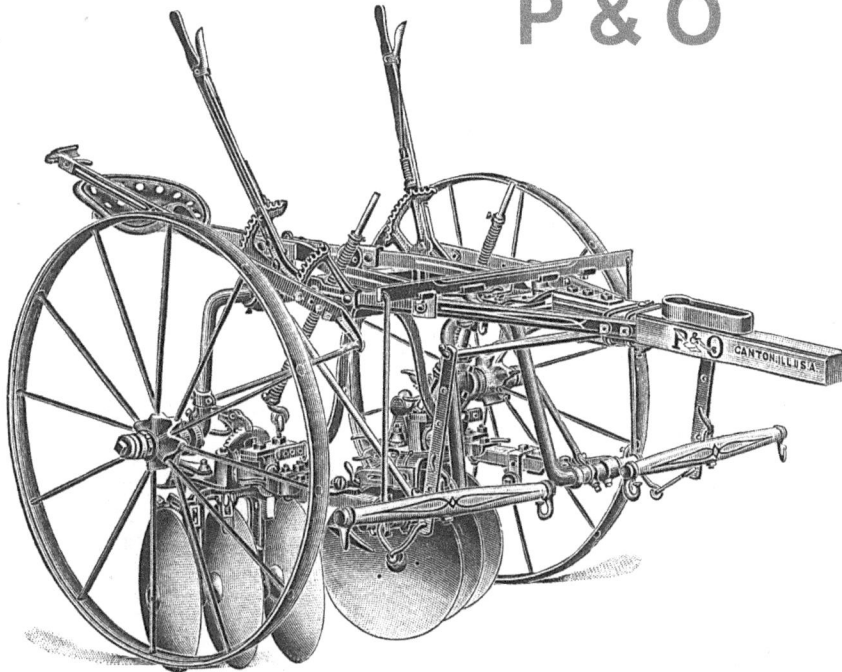

NO. 98 WIGGLETAIL DISC CULTIVATOR.

Company literature

MOLINE BALANCE-FRAME WALKING CULTIVATOR

This adjustable-arch, balance-frame walking cultivator is especially designed to balance the heavier type gangs used in the South.

Regular Equipment

Pole, neckyoke, evener, handles and solid shields, 2-1/2 inch concave or flat tire wheels, as ordered. Four and six shovel gangs with shovels.

Extra Equipment

Spring-tooth side harrow, adjustable rear shank coupling and center shovel attachments for parallel gangs. Rotary Shields. Special shovel equipment.

Lifting Spring Control

Good penetration, as well as ease of handling, has resulted in the method of application of gang lifting spring pressure. The lifting spring is attached to a simple spring loop casting, pivoted on the gang

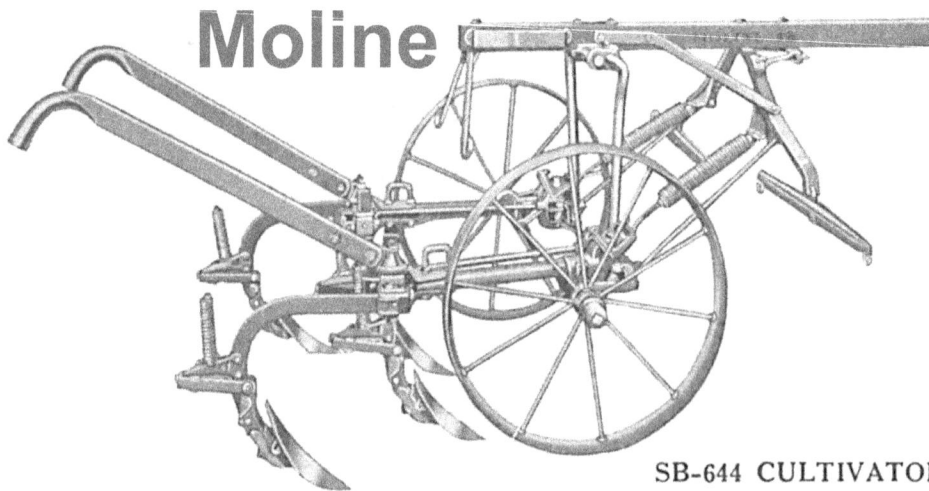

SB-644 CULTIVATOR

and shaped to extend around and over the gang coupling. As the gang is lowered the loop passes over center and the spring automatically assists in holding the gang to its work; yet as the gang is lifted the leverage of spring increases making the gangs easily handled. Tension is adjustable.

Depth Regulator

A very simple wing screw adjustment regulates depth of penetration. The spring loop comes in contact with the wing screw as the gang reaches the proper depth. This screw can be regulated without aid of wrench, creating a cushion spring depth stop upon which the spring loop rests. The efficiency of the gangs in irregular ground is not interfered with by the stop, for pressure may be applied upon the handles to force the shovels into depressions or dead furrows.

Parallel Gangs Quickly Tilted

The pipe beam parallel gangs for Moline Walking Cultivators are adjustable for tilting to conform to ridges. Merely loosen one bolt on pipe coupling, twist

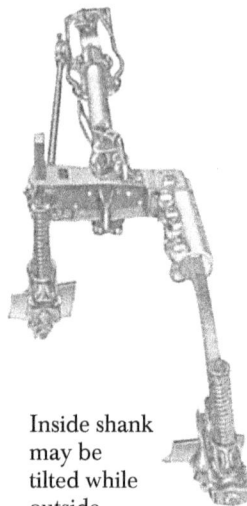

Inside shank may be tilted while outside shank is held verticle

gang to desired tilt, reset corrugated washer and tighten bolt.

Adjustable Rear Shank Coupling

Adjustable rear shank couplings are furnished as an attachment for parallel pipe beam gangs. With these couplings the outside shovel shanks may be adjusted to a vertical position while the gangs are tilted. An important feature desired by many farmers.

Gangs Hug Ridges

When cultivating ridges or hilling, the lifting spring can be regulated to hold the gangs to the side of ridge, or to or away from the row by simply adjusting spring to correct position in spring loop. A series of notches are provided and change is quickly made. This device makes possible thorough ridge and flat cultivation with little effort by operator.

Cone-Bearing Gang Couplings

The improved cone-bearing type gang couplings assure dependable performance of the gangs after long service. These bearings are adjustable and can be kept tight after wear to avoid play; therefore the gangs are held rigidly in correct alignment and pressure on the handles is effective.

Free Gang Swing

The two-piece arch provides a wide range of wheel adjustment to conform to row spacings. It gives the maximum range of gang swing making possible the most careful cultivation. Moline shovels are designed to penetrate. They have ample clearance behind the cutting edge so that they do not ride on their backs.

Load Is Equalized

The evener is correctly designed to allow team ample fore and aft equalization. One horse can be as much as 12 inches ahead or behind the other without affecting load on either horse. Moline cultivators pull light. The gangs are of proper length and connected to arch at correct height, improving draft. The pull is direct to arch.

Adjustments Are Easily Made

Moline Walking Cultivators handle easily. The efficient application of lifting spring pressure gives a free easy gang swing. All adjustments can be made without effort; none are complicated. Depth regulation is just a matter of seconds, relieving the operator of controlling depth with handles. The gang may be tilted quickly to hug the ridges without any effort of operator. Another feature of remarkable simplicity is the adjustment of shovel sleeves on spring-trip gangs; just adjust one bolt to set shovel as desired.

Handles are adjustable for height and may be set to allow operator to walk comfortably in middle of row.

Moline

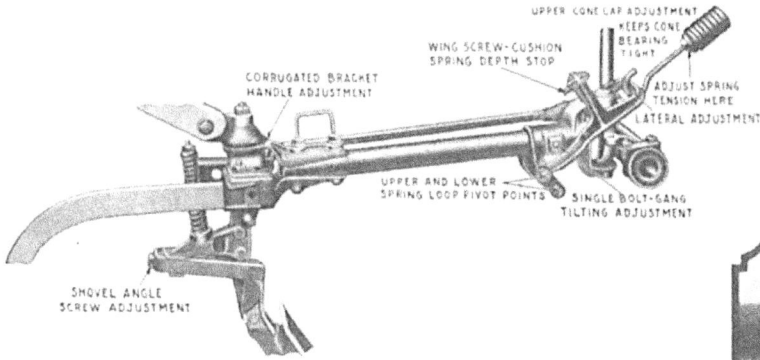

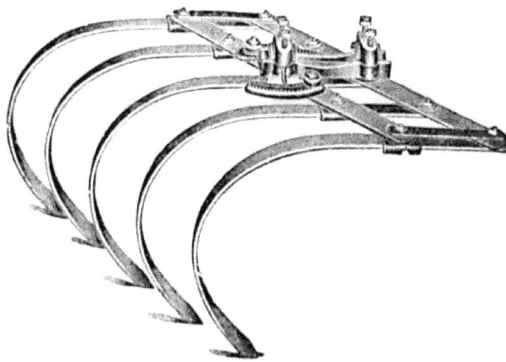

SB-644 at work

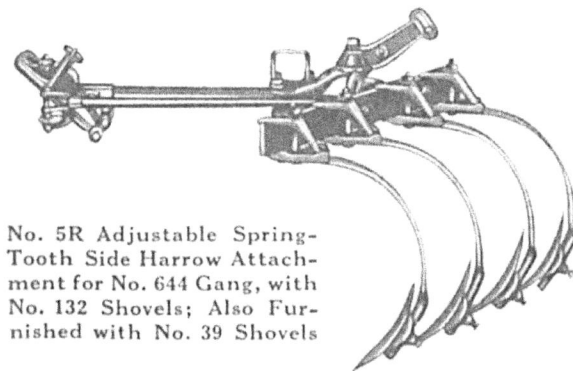

No. 5R Adjustable Spring–Tooth Side Harrow Attachment for No. 644 Gang, with No. 132 Shovels; Also Furnished with No. 39 Shovels

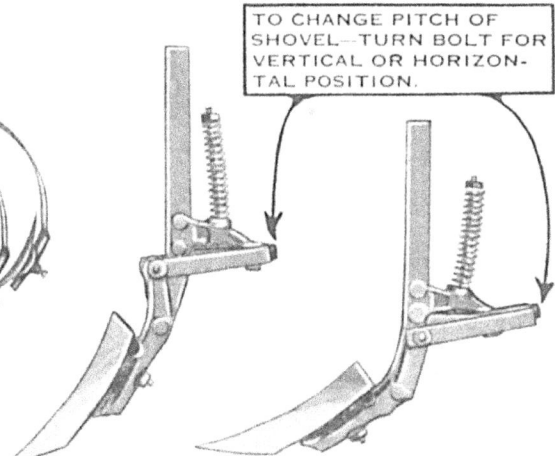

TO CHANGE PITCH OF SHOVEL—TURN BOLT FOR VERTICAL OR HORIZONTAL POSITION.

Moline Spring-Trip Shank—Quickly Adjustable for Pitch of Shovel

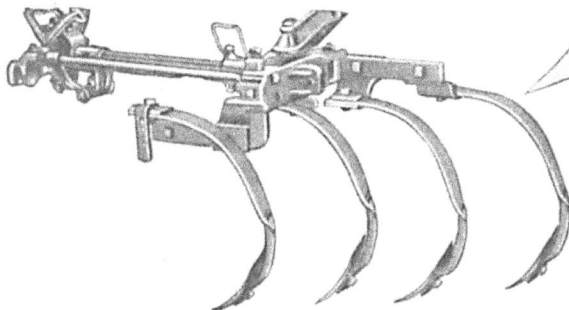

No. 2F Spring-Tooth Side Harrow Attachment for No. 644 Gang, with No. 39 Shovels; Also Furnished with No. 132 Shovels

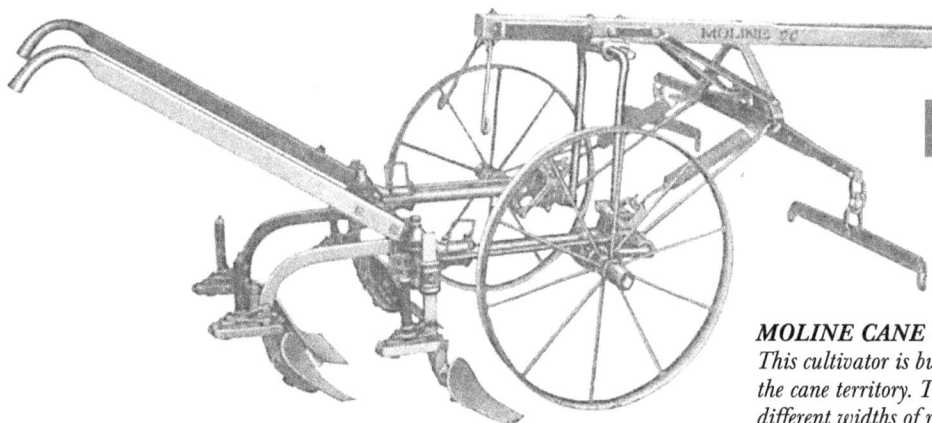

Moline

SC-64 CULTIVATOR

MOLINE CANE CULTIVATOR
This cultivator is built to meet the particular need of the cane territory. The evener is adjustable to the different widths of rows. The gang is particularly adapted to covering the seed cane, for first and second cultivation and for throwing out middles.

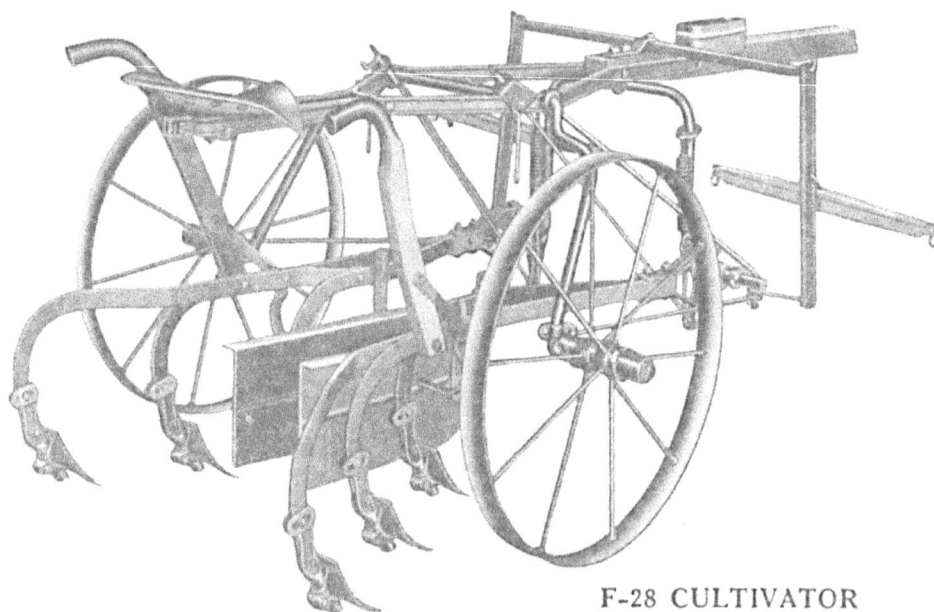

F-28 CULTIVATOR

MOLINE CORN BELT CULTIVATOR
This cultivator, as the name indicates, is built especially for corn belt farmers. It is stripped of all unnecessary parts required on cultivators sold in other territories, yet its simple construction provides all essential adjustments that any cultivator should have for work in the Corn Belt.

Wide Gang Swing
It is easy to dodge misplaced hills with the Moline Corn Belt Cultivator and to cultivate with the same thoroughness as with a walking cultivator, for the low wheels and specially shaped gangs provide an unusually wide gang swing. The wheels are spaced for rows planted 40 inches to 44 inches apart. Clearance under the arch is greater than on many high wheel cultivators and ample to lay by corn; it is wide to avoid danger of breaking down corn planted in crooked rows. The gangs may be set to or away from the row by a simple set screw adjustment on the coupling.

Properly Balanced
Since the low wheels set farther to the rear on this cultivator, the operator's weight has less effect on pole. It is balanced for operators

up to 200 pounds; light operators do not create excessive neck weight. No balance adjustment for various weights of gangs is necessary. The singletrees are connected to vertical pendulum bars at proper height to give ample fore and aft equalization and yet not high enough to produce excessive neck weight.

Non-Clog Shields
The shield blade cuts through clods instead of riding over them and is controlled by a spring on the shield bolt which allows it to move away from shovel upon striking an obstruction; the obstacle passes through without danger of covering small corn. Shields are adjustable up or down on beam.

No Dirt Thrown in Operator's Face
The low wheels do not throw dirt high enough to be blown in the face of operator. Another feature is the natural vision of row and team, for there are no levers and other devices to obstruct the view. The gangs have an easy vertical and lateral movement, made possible by the efficient gang spring application and the cone bearing couplings, therefore, good clean cultivation is possible with little effort.

Moline

is adjustable to vary the wheel tread from 42 to 56 in.; while the gangs may be set closer to or away from row by moving coupling on the arch. Different weights of operators and gangs are accommodated by an adjustment on the rocker shaft.

Swinging Gangs

The swinging gang construction permits thorough cultivation. The operator has absolute control of the shovels. He may cultivate as close to the row as is desirable, dodge in between hills to cut out weeds, leave a thoroughly pulverized mulch between the rows and dodge misplaced hills or follow crooked rows easily. Briefly it affords all of the advantages of walking cultivator thoroughness with the present day necessity of a riding cultivator.

Automatic Gang Control

The seat bars are pivoted under arch pipe and extend forward to a pivoted cross yoke. Links extending from the ends of the yoke down to steel strap on gangs cause the gangs to be lifted by weight of operator when his feet are removed from the stirrups. The seat can be moved forward or backward on seat bars to suit operator and adjustment made in front link connection to beam to balance gangs.

The gangs can be operated independently because of the pivoted cross yoke. This allows them to be handled efficiently and also to vary the depth of gangs. Pressure on the foot stirrups regulates depth; one gang may be run deeper than the other to account for variations in field.

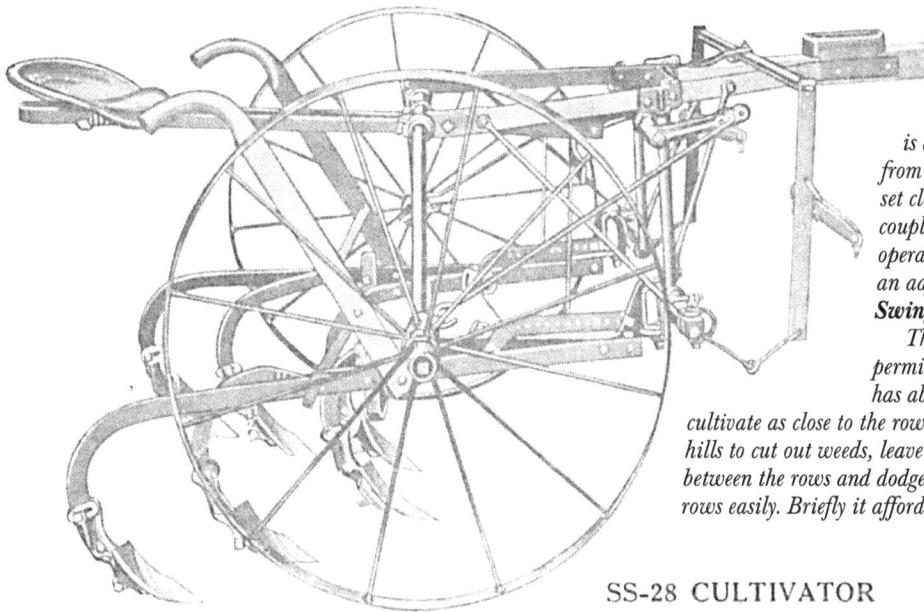

SS-28 CULTIVATOR

Two Horse Cultivators

MOLINE HIGH WHEEL LEVERLESS CULTIVATOR

A single-row, leverless, balance frame cultivator with adjustable axle, remarkable for its simplicity, perfect balance and quality of field performance.

There are many farmers who prefer the leverless type of cultivator because of its automatic gang action, eliminating the necessity for raising levers and making other adjustments. Full attention may be given to driving the team and guiding the shovels assuring at all times a good job of work. Particularly, this type of cultivator is better adapted to work in fields heavy with vines and weeds for the gang can be raised automatically and more quickly to shed trash. The axle

MOLINE BALANCE-FRAME HIGH WHEEL CULTIVATOR

Being a combined walking or riding cultivator, it has an unusual degree of free gang swing. The axles are adjustable for spacing wheels from 42 to 52 inches apart; suitable for cultivating various width of rows. The balance of cultivator is adjustable for various weights of operators and gangs, or for walking instead of riding.

For Riding or Walking

To change from a riding to a walking cultivator, it is simply necessary to loosen a bolt in the seat bar and fold the seat forward on top of the axle out of the way of operator. The frame may then be balanced by lever.

Convenient Foot Pedals

When turning at the end of rows, or to clean shovels of trash, it is only necessary for operator to place his feet on the foot cranks to raise the gangs. This is done without touching the levers. On completing the turn, the feet are lifted and the gangs return to their proper depth. Thus the operator has at all times perfect control of the team.

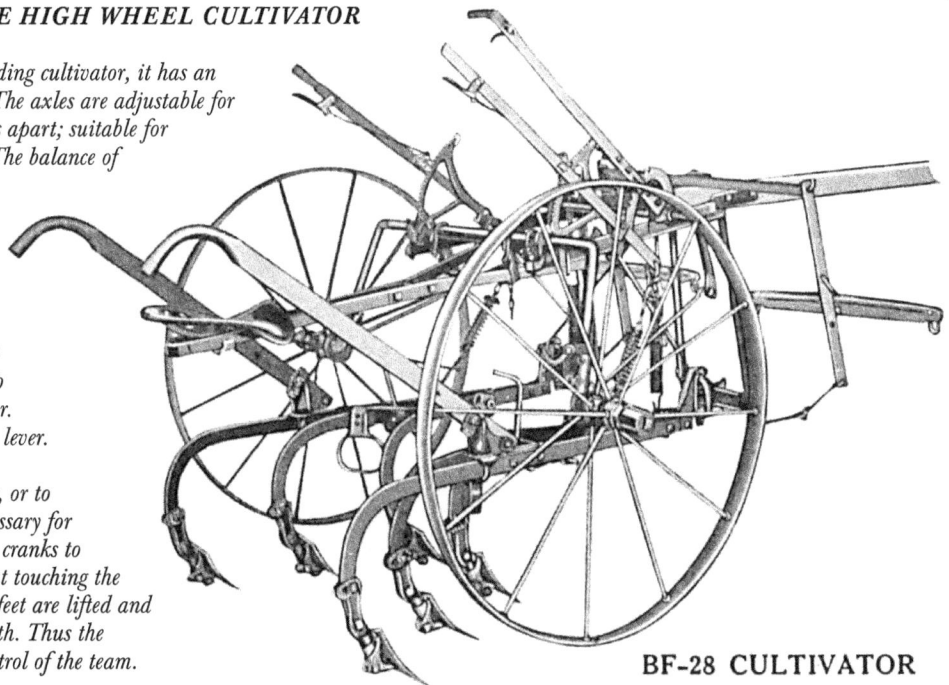

BF-28 CULTIVATOR

Moline

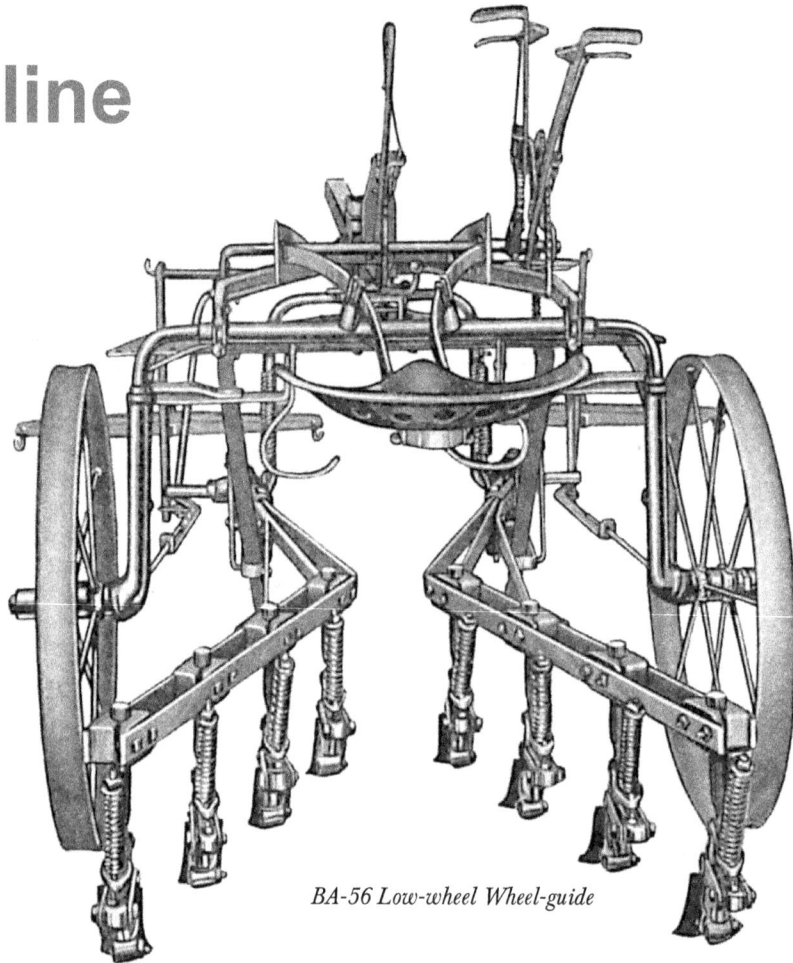

BA-56 Low-wheel Wheel-guide

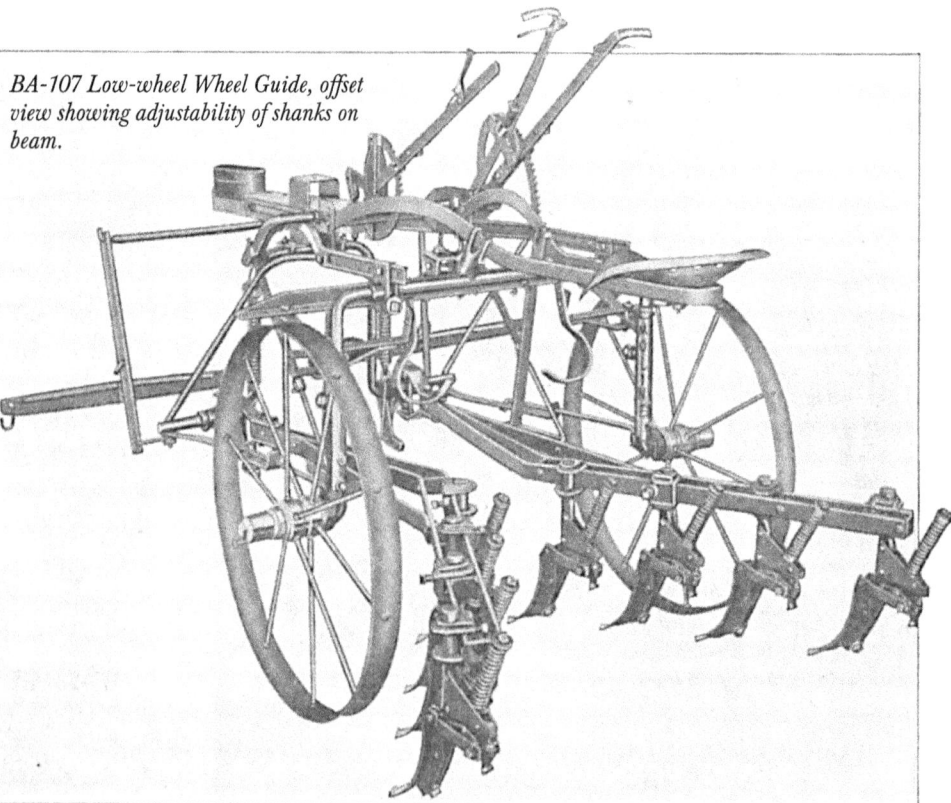

BA-107 Low-wheel Wheel Guide, offset view showing adjustability of shanks on beam.

*BH-131 High-wheel
Wheel guide*

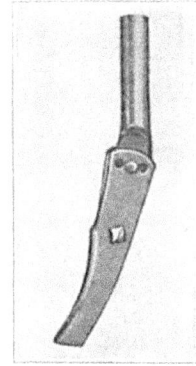

Pin-break (shear-pin) shank

*Moline No. 2 Disc Hilling
Attachment*

Moline

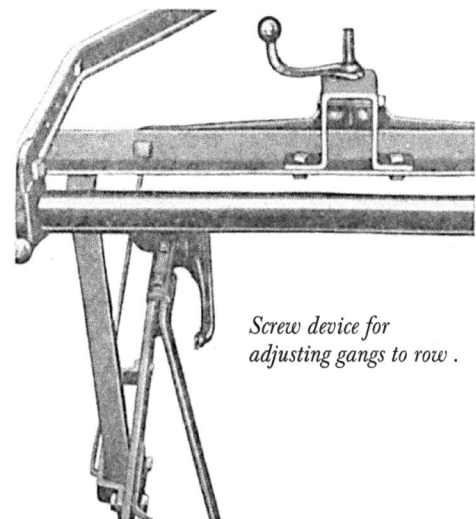

*Screw device for
adjusting gangs to row .*

Overhead view of the Moline High-wheel Wheel Guide cultivator

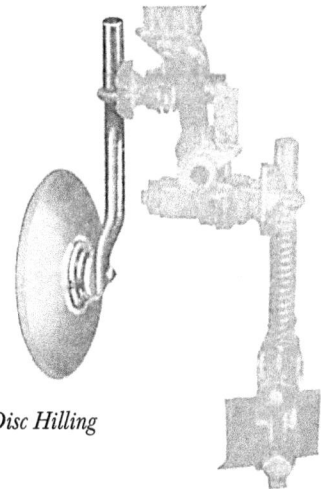

KA-28 CULTIVATOR

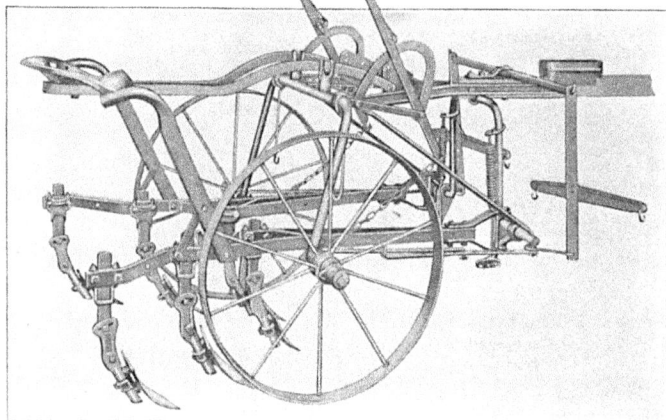

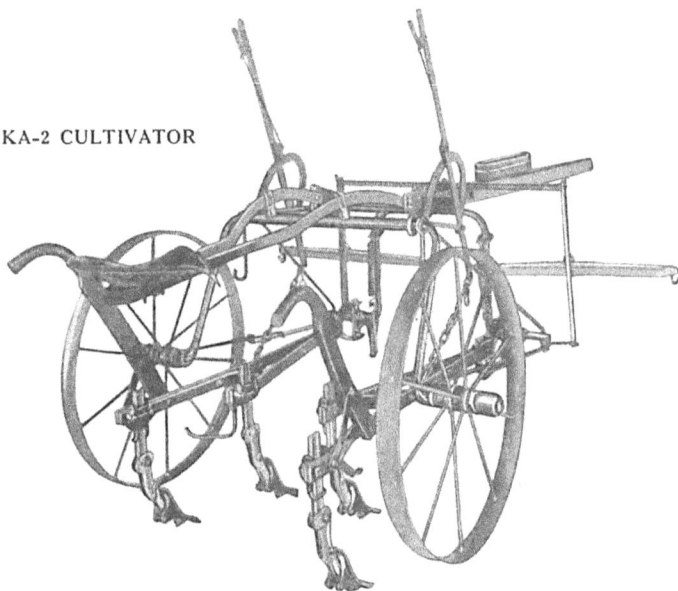

A-24 CULTIVATOR

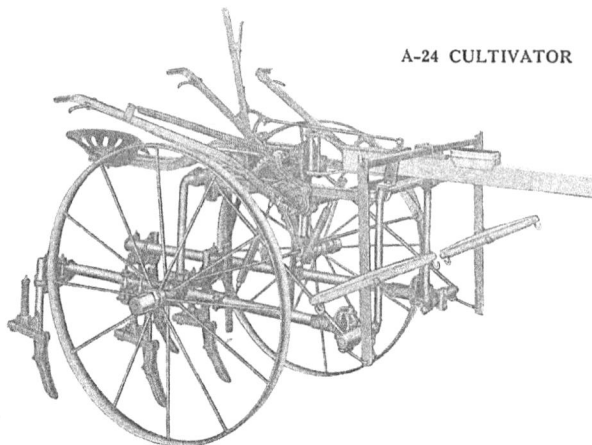

KA-2 CULTIVATOR

S-28 CULTIVATOR

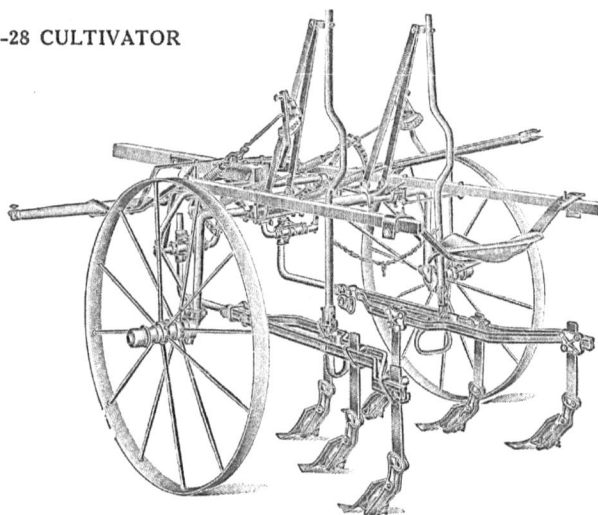

Moline

A-24 Seat guide cultivator

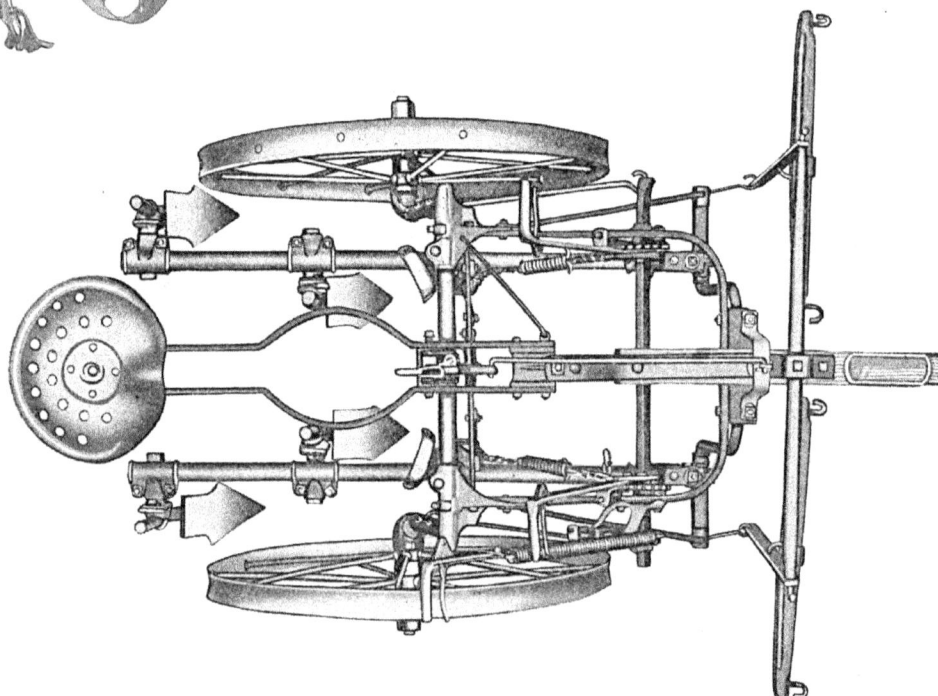

TEXAS DANDY No. 2 CULTIVATORS

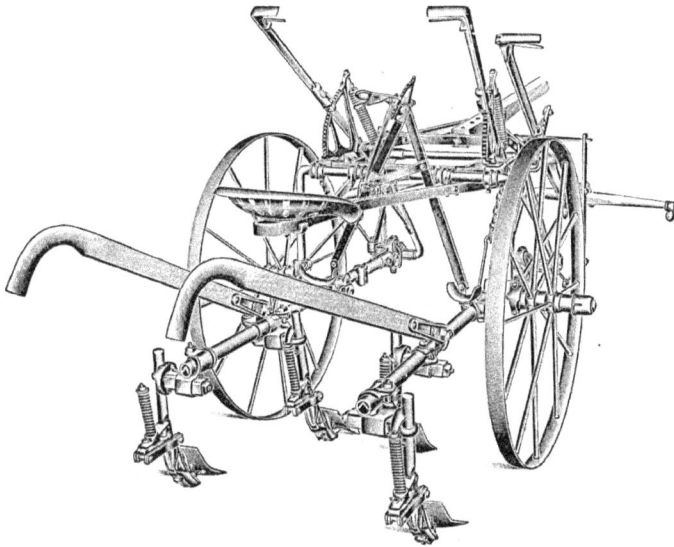

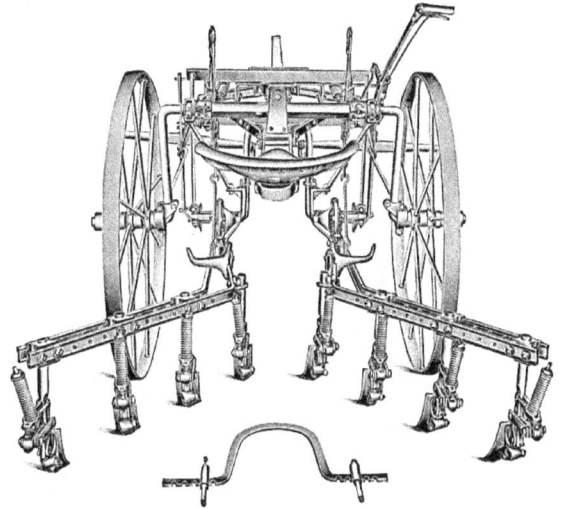

Dutch Tango No. 2 with 8-Shovel Diagonal, Round Shank, Spring Trip Gangs.

SOUTHERN DANDY CULTIVATORS

Moline

TEXAS DANDY No. 1 COMBINED CULTIVATORS

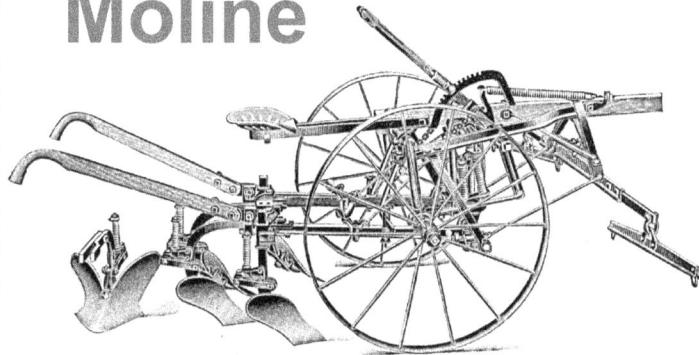

Southern Dandy Cultivator Set for Throwing Out Middles.

BALANCE FRAME DANDY CULTIVATORS

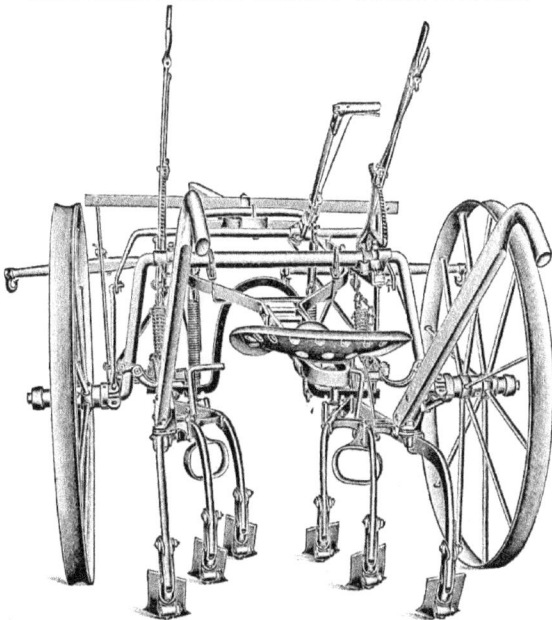

Dutch Tango No. 2.
Equipped with 6-shovel Round Sleeve Steel Beam Gangs.

Two Horse Cultivators

DUTCH UNCLE RIDING CULTIVATORS

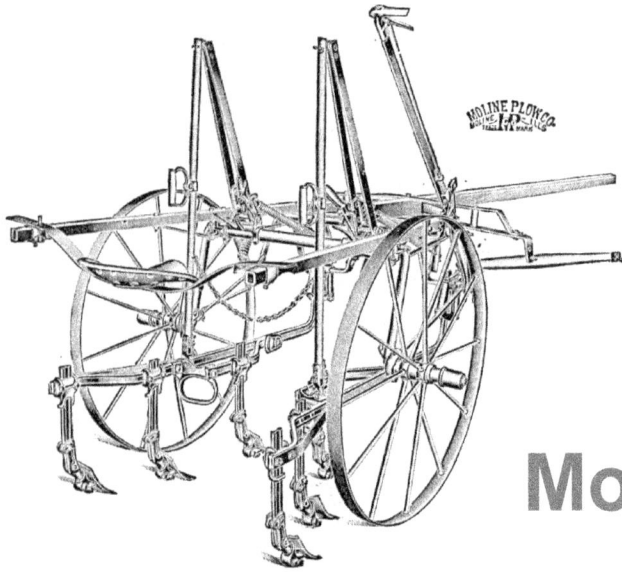

DUTCH TANGO SEAT GUIDE CULTIVATORS

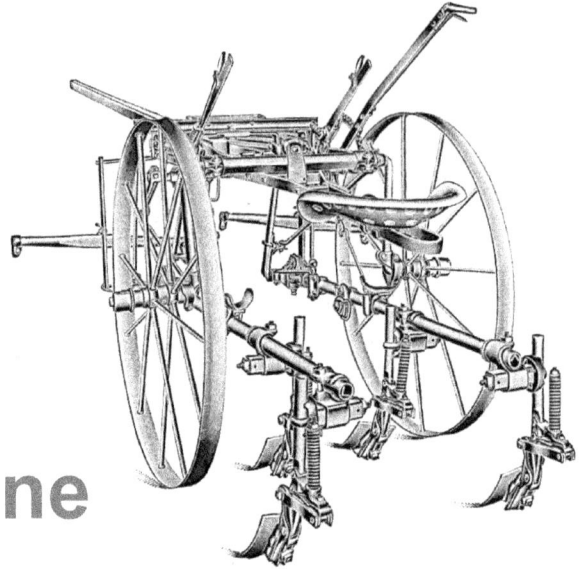

Moline

Dutch Uncle No. 4.

Dutch Tango No. 1.

Sun Shade for Dutch Uncle Cultivator

PIVOTAL FRAME DUTCH UNCLE CULTIVATORS

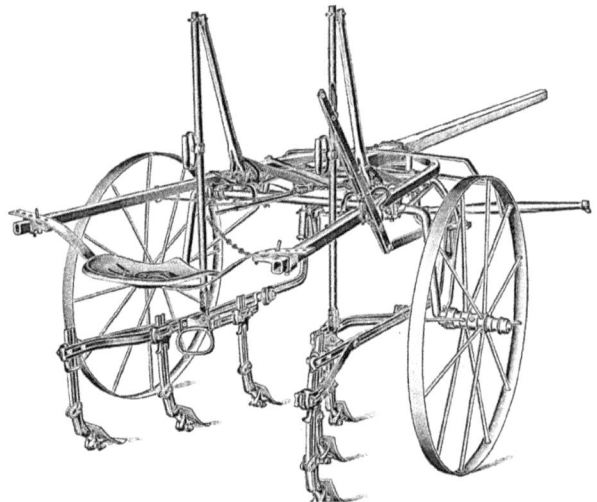

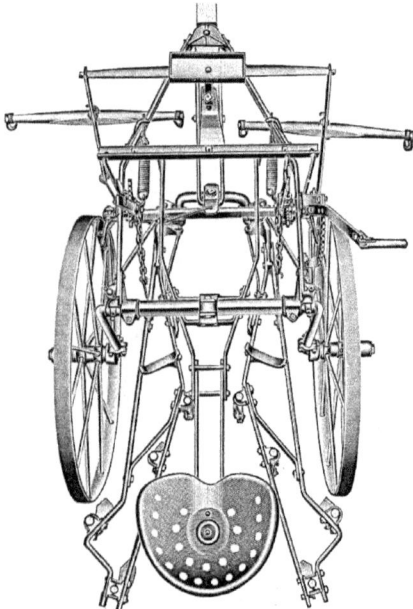

Top View Showing Unobstructed View of the Operator.

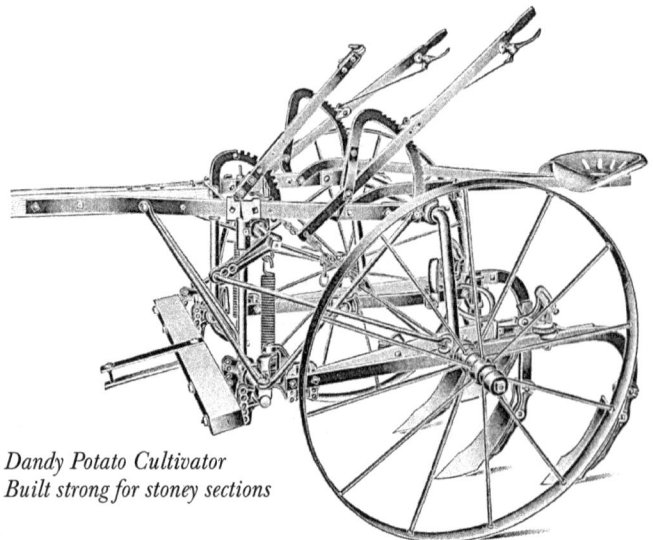

Dandy Potato Cultivator
Built strong for stoney sections

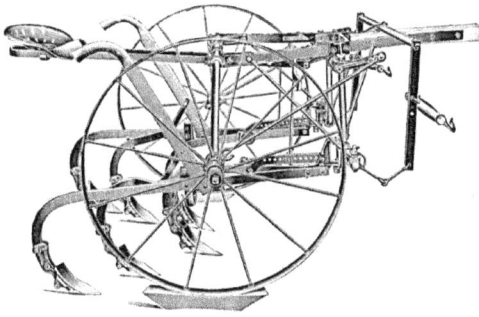

See Saw Cultivator
No levers, no springs. The weight of the operator's body does all the work, and it responds readily to the slightest movement. All that is necessary to secure perfect balance is to move the seat forward or backward on the seat bar until the operator's weight balances the cultivator.

Moline

Flying Dutchman No. 1 Pivot Axle Cultivator
Built expecially for corn, potatoes, tobacco, beans, peas and other crops grown in narrow rows. The wheels can be set as close as 32 inches apart, which is considerably closer than on other makes and gives an 8 inch wheel clearance in 24 inch rows. the maximum spacing is 44 inches.

A foot guide cultivator with levers which pivot the wheels and guide the machine in the desired direction. The foot levers slide in and out on the axles and are within easy reach of the operator

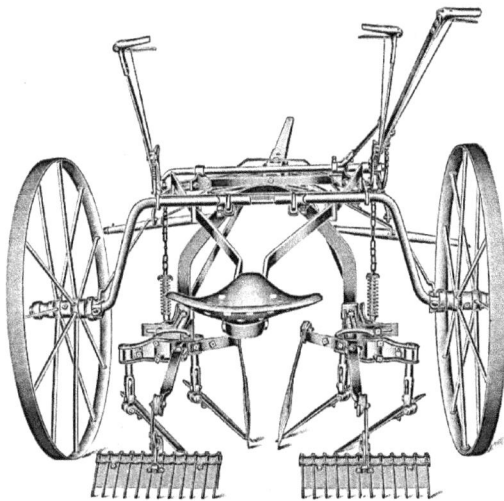

Moline No. 3 Surface Cultivator
It may be used for surface or deep cultivation.

PIVOTAL DANDY CULTIVATORS

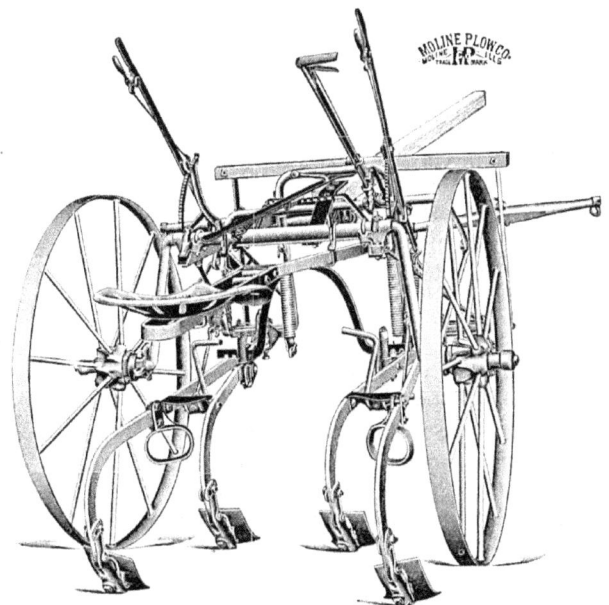

Note the Perfect Pulverization of the soil by the Moline Surface Cultivator.

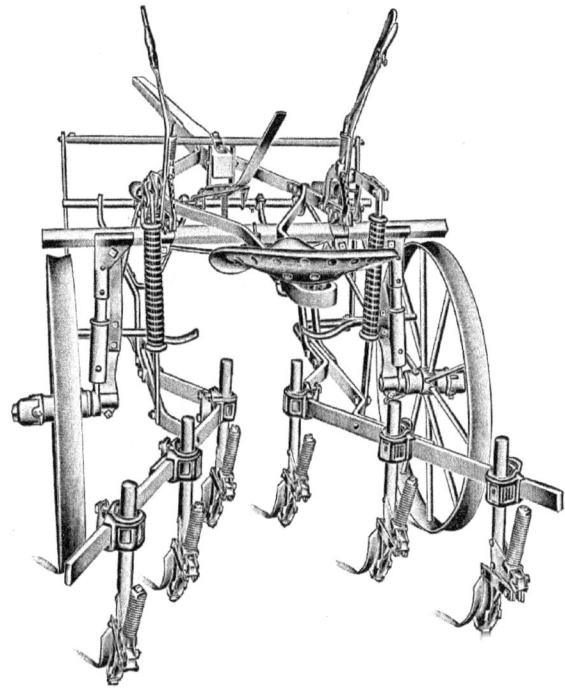

CAPTAIN KIDD DISC CULTIVATORS

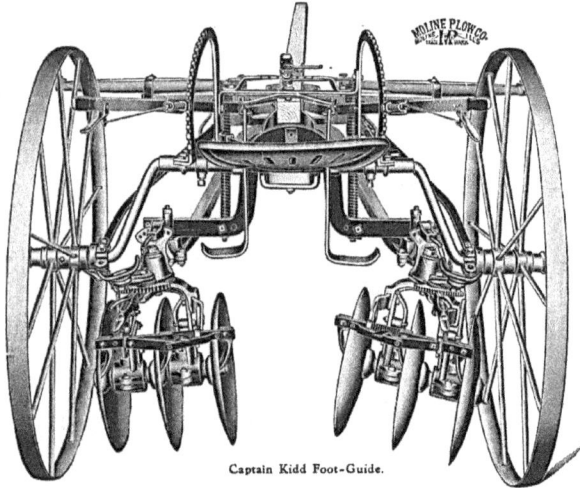

Captain Kidd Foot-Guide.

Moline

The Captain Kidd Disc Cultivator is a great favorite, owing to its ability to do a great variety of work and do it well with the least effort on the part of the operator and team.

Foot Guide.

The frame is attached to the pole on a pivot, enabling the operator, by means of convenient foot levers, to swing the cultivator to the right or left, to cultivate hills not planted in a direct line with the team meanwhile traveling straight. Hills out of the row are easily cultivated without cutting or destroying any of the stalks. The frame is made rigid at will.

Gang Adjustment.

A very desirable feature of this machine is the easy and quick adjustment of the gangs.

The gangs are adjustable and may be set to and from the rows and at any angle for hilling up or laying by corn, or for cultivating listed corn. They are held in the ground by the weight of the frame and the driver, assisted by pressure springs which can be adjusted to give any weight on the gangs desired, thus adapting the Captain Kidd to all conditions of soil, and permitting the cultivation of corn when the ground is hard and dry, pulverizing and turning the earth much better than can be done with any other class of cultivators.

Scrapers.

The scrapers are held to the discs by spring pressure. When not in use, they may be thrown from the discs by a foot-lever and are automatically locked in that position, and may be applied or thrown from the discs as often as desired without stopping the team.

Shields.

The shields are long and adjustable in width.

Leveling Blades.

These blades may be attached and are useful in many ways. They effectually crush the clods and leave the ground smooth and even, thereby better holding the moisture. The arms are made of spring steel, are adjustable and may be placed at almost any angle. Furnished as extras.

Front Arm Attachment.

The front arm attachment consists of arms fastened to and extending at right angles with the beams and just forward of the discs. To these arms are attached adjustable standards. In cultivating small corn the inside discs are removed from the gangs and are attached to these standards so as to throw the dirt from the corn, the rear discs meanwhile throwing the dirt toward the corn. This device not only cultivates the corn more thoroughly than can be done in any other way, but serves as a balance, for while the forward discs are bearing one way, the rear discs are bearing the other; thus one counteracts the pull of the other. The standards are corrugated and cannot slip, and are adjustable up and down and also at any angle. All discs may be used to throw the dirt either to or from the corn. This attachment is furnished at a small additional cost and is readily and easily attached.

Front Arm Attachment.

Front Arm Attachment in Use.

Harrow Attachment.

The harrow attachment consists of two extra discs, two spools and two scrapers, which are easily attached. By the use of these an eight-disc harrow is obtained, making an excellent pulverizer and an invaluable tool for fallowing in small grain, etc.

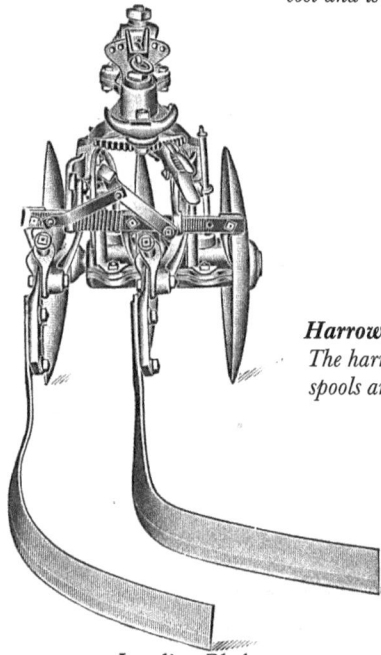

Leveling Blades.

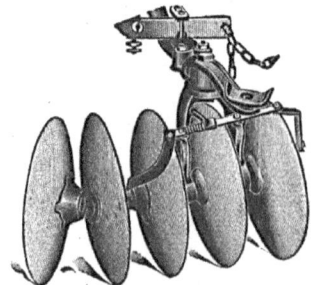

Harrow Attachment.

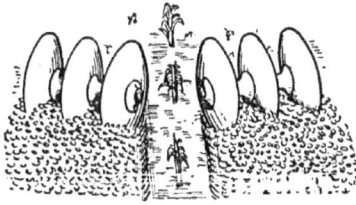

Gangs set for throwing out. It is best to use the leveling blades as they will throw some of the soil back.

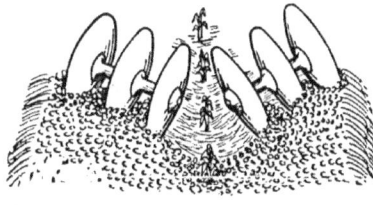

Gangs tilted for cultivating listed corn. They may also be reversed to throw in when desired.

Gangs set for laying by or throwing the ground in.

Bull Tongue.

When so desired, the Captain Kidd may be supplied with bull tongues, to be used instead of front arms. The small shovels may be set to cultivate close to the corn. They may be used with the discs on each side, and the entire space between rows will be cultivated. The shovels are adjustable up or down, or sidewise, to throw more or less soil to the corn.

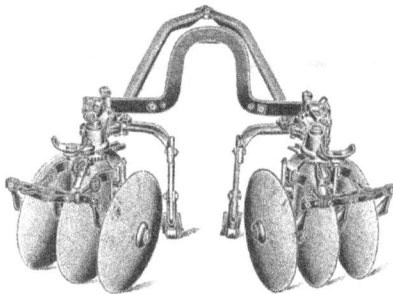

Shovel Attachments.

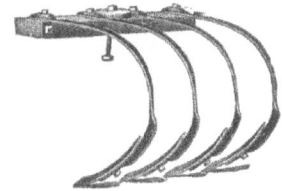

When shovel attachments are used on the Captain Kidd, it is practically a shovel cultivator. They are easily attached and adjusted. With these and the rest of the attachments shown, the Captain Kidd becomes a universal cultivator, and one which can be used for all classes of work.

The Captain Kidd may also be equipped with spring tooth attachments.

Shovel Attachment.

Spring Tooth Attachment.

Bull Tongue Attachment.

Attachment for Potatoes.

The attachment for cultivating potatoes consists of two bent rods attached to the forward end of shield straps. These rods moving in front of the shield, lift the vines and guide them between the shields, the discs cut the weeds and hill up the potatoes. Furnished at a small extra cost.

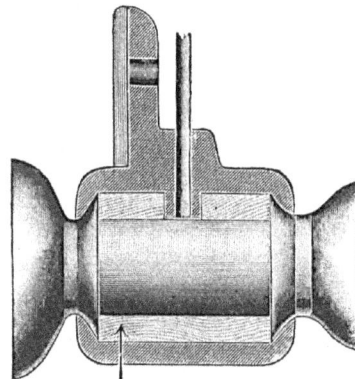

HARD WOOD.
Bearing and Wood Bushings.

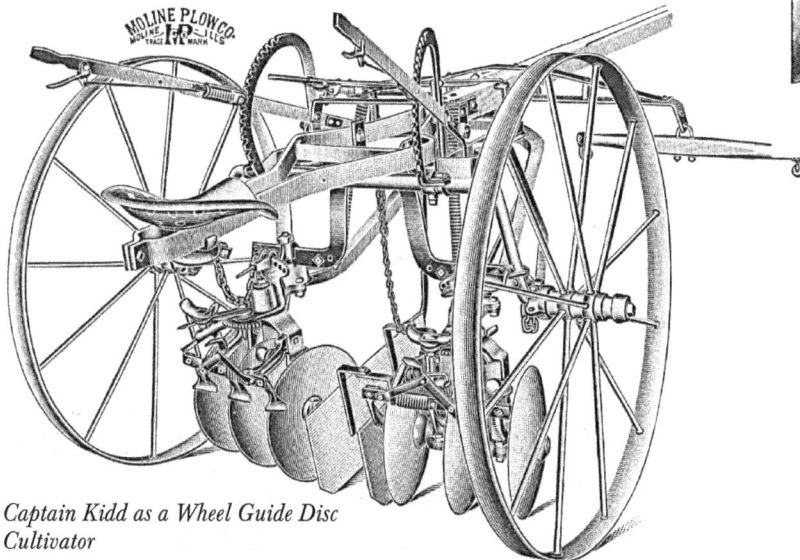

Captain Kidd as a Wheel Guide Disc Cultivator

Moline

SOUTHERN KIDD DISC CULTIVATORS

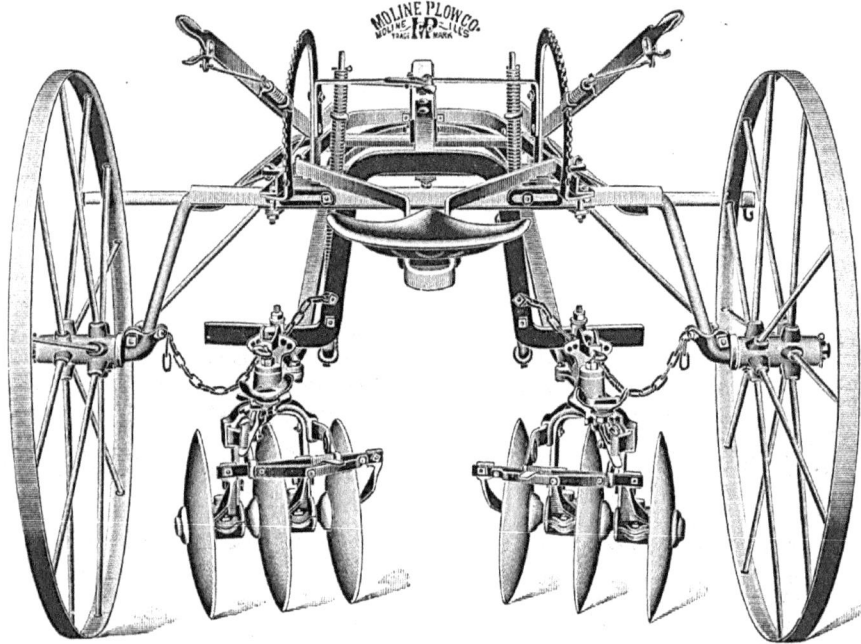

The Southern was designed as a mid-grade disc cultivator

SHANGHAI KIDD No. 2 DISC CULTIVATORS

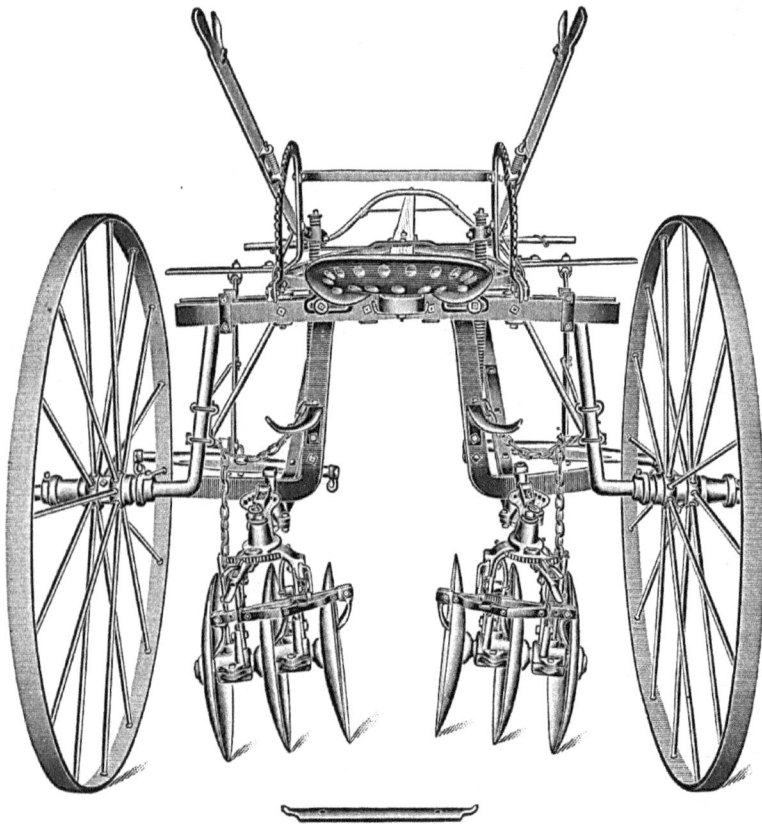

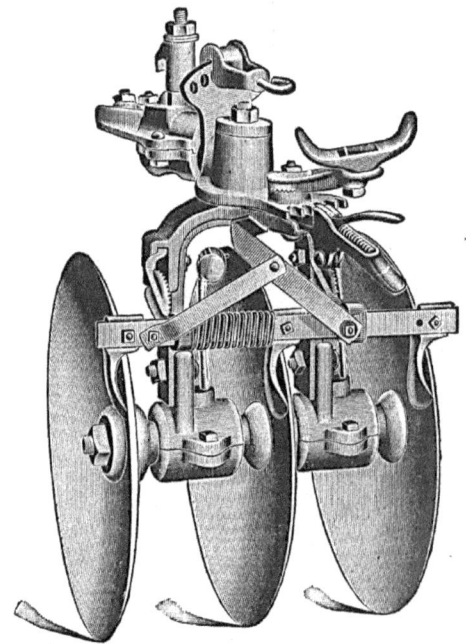

Showing the disc gang and scraper assembly for the Kidd Cultivators

COW PEA ATTACHMENT FOR SHANGHAI KIDD CULTIVATORS.

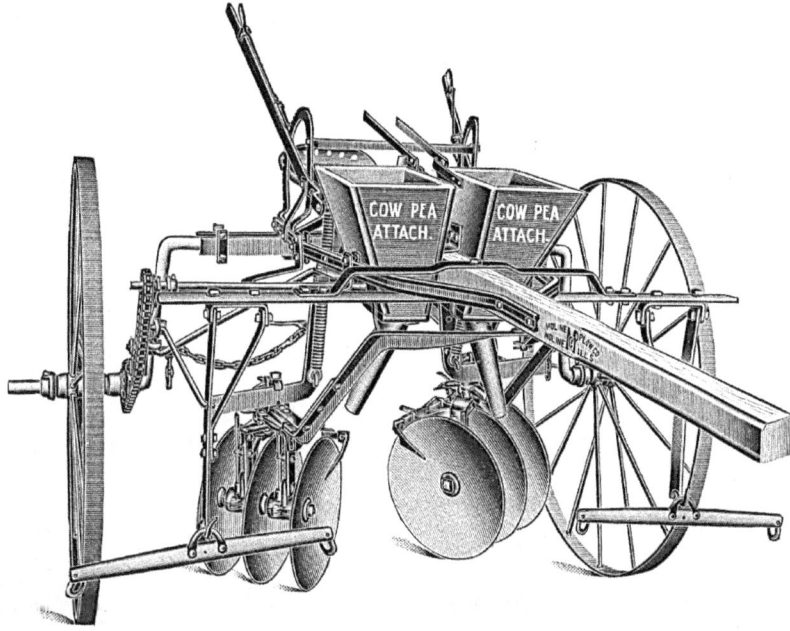

Moline

BW-64 a Pivot Axle Cultivator

A father and son go to the field in 1920 with two Moline cultivators

281

Emerson Brantingham

EB Walking Cultivtor

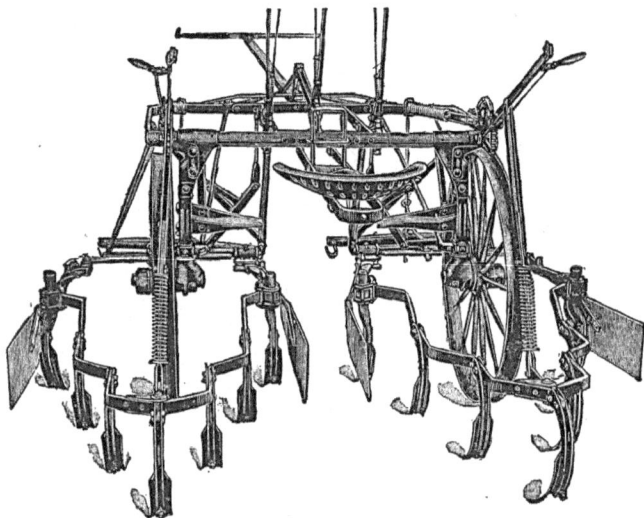

Two-row cultivator which straddles one row cultivating two middles.

EB Riding Cultivator

P & O

NO. 2 CANE CULTIVATOR, WITH DISCS SET FOR OUT-THROW.

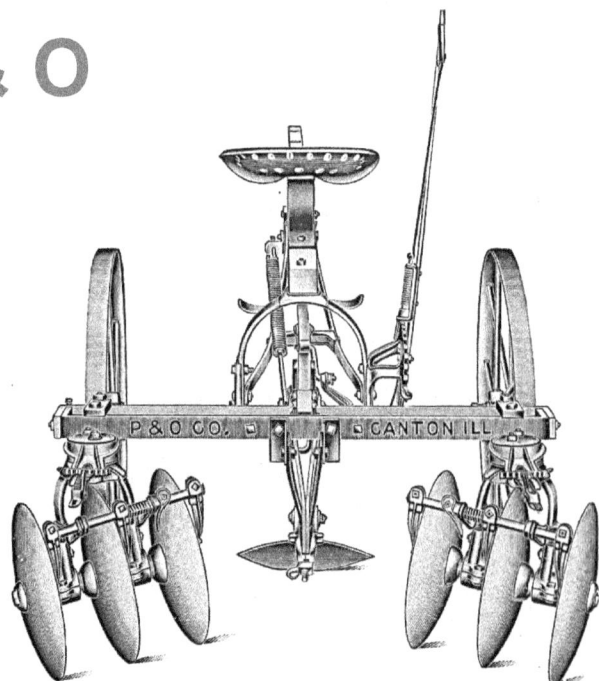

NO. 2 CANE CULTIVATOR WITH SWEEP ATTACHMENT. DISCS SET FOR IN-THROW.

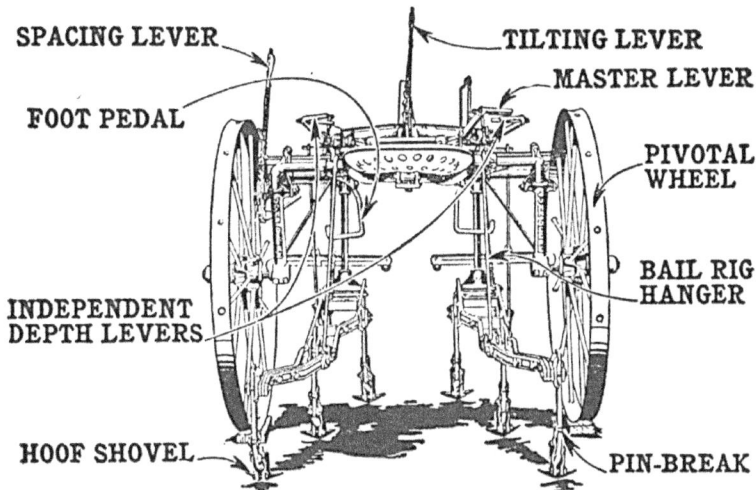

SPACING LEVER
FOOT PEDAL
TILTING LEVER
MASTER LEVER
PIVOTAL WHEEL
BAIL RIG HANGER
INDEPENDENT DEPTH LEVERS
HOOF SHOVEL
PIN-BREAK

One-row riding cultivator of the parallel rig, lever shift type.

Care and Maintenance of Cultivators

In order to have the cultivator operate economically, efficiently and effectively it must be properly lubricated, adjusted and kept in good repair. Before using it, all parts should be checked and any badly worn parts replaced. It should then be lubricated and properly adjusted. Occasionally it should be thoroughly cleaned and painted.

Some of the parts of a cultivator to check are:

1. Tongue and yoke--tighten yoke coupling and replace broken or decayed tongue on horse-drawn
2. Wheels--replace worn boxings.
3. Toe-in of wheels.
4. Evener and singletrees on horse-drawn.
5. Steering assembly.
6. Pressure springs.
7. Lifting mechanism.
8. Shovels--sharpen or repair worn shovels and adjust properly.
9. All bolts.

Lubricate all working parts at regular intervals. The shovels should be covered with a thin coat of oil at the close of the day and a thick coat of grease before being stored to prevent rusting. Remember, preventive maintenance may save many costly repair bills.

Two Horse Cultivators

(Above) Cultivator placed on saw horses ready for inspection. The wheel is being examined for play, indicating whether or not the boxing is worn. If so, it should be replaced with a new one. Give the cultivator the proper care and few repairs will be necessary.

(Right) Cleaning the axles with kerosene. After the wheel is put in place, clean hub cap, pack with grease and screw on, forcing grease completely through the hub.

(Below) To set the shovels, block the wheels off the floor to a height equal to the depth the cultivator will work in the field. Adjust the levers so as to lower the gangs to cultivating position. Place a carpenter's level on the gang beam. Then adjust the tilting level until the gangs are level. Adjust the depth of the shovels so that the points touch the floor and tighten shank standard securely. Adjust the angle of the shovels, using a T-bevel set at an angle of 135 degrees as illustrated.

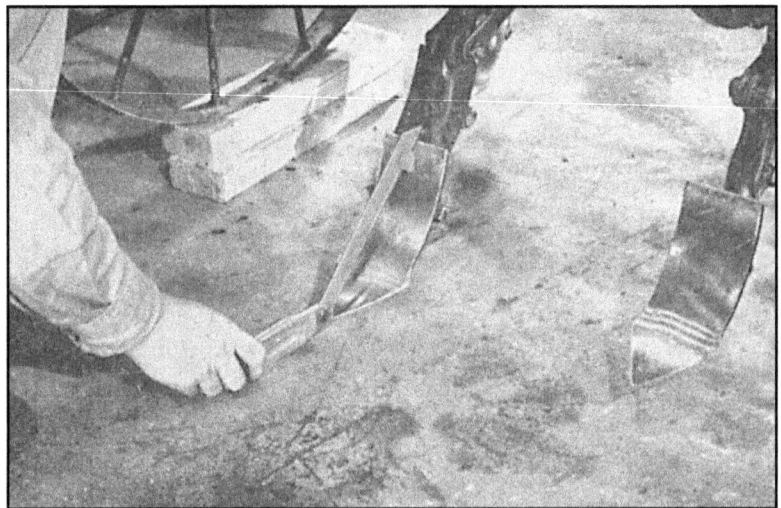

Measuring the distance between the wheels in determining wheel toe-in. In determining toe-in first measure the distance between the wheels at the rear. This cultivator measures 42" between wheels at this point. Then measure the distance between the wheels at the front. This distance should be ½" to 1" less than at the rear, making a toe-in of ½" to 1."

*This information was taken from **Farm Mechanics Text and Handbook** published in 1951 by Interstate Printers*

A worn drawlink may break and cause a runaway with a horsedrawn cultivator. Replace it.

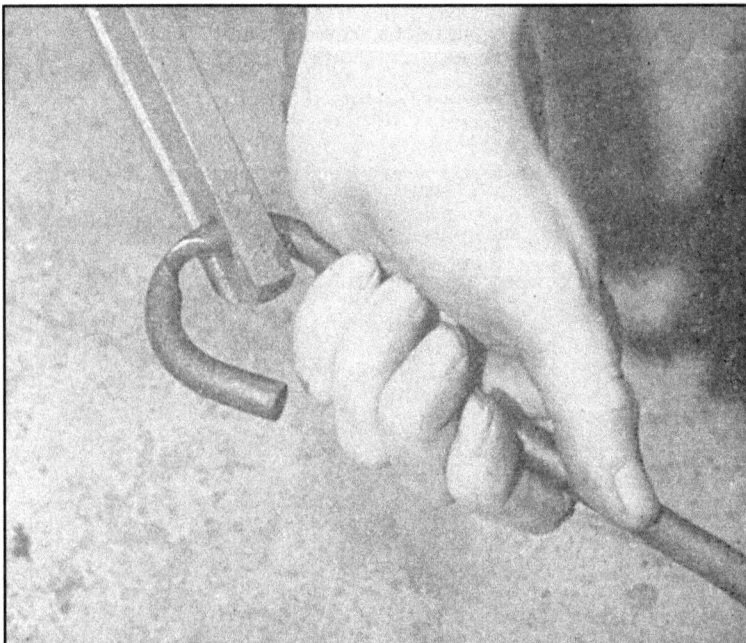

Restoration Notes: Many people enjoy restoring older farm implements as a hobby or for profitable resale. This author was closely involved with the auction sale of HD implements for half of a century and would like to encourage that these implements be taken completely apart, every nut loosened, and all parts cleaned to bare metal. Then the ideal next step, for resale purposes, would be to liberally coat all pieces, wood and metal, with a mixture of 25% turpentine and 75% boiled linseed oil applied warm. And DON'T paint it. For the serious collector/ horsefarmer, paint serves to hide ills and, in many minds reduces value.

I & J
Mfg

This brand new one row two horse cultivator with the "comfortable" seat is being manufactured in Pennsylvania by folks who actually farm with horses. It comes with four different shovels, optional disc hillers, additional tines for conversion to field status and adjustable row shields. For more information see the back of this book.

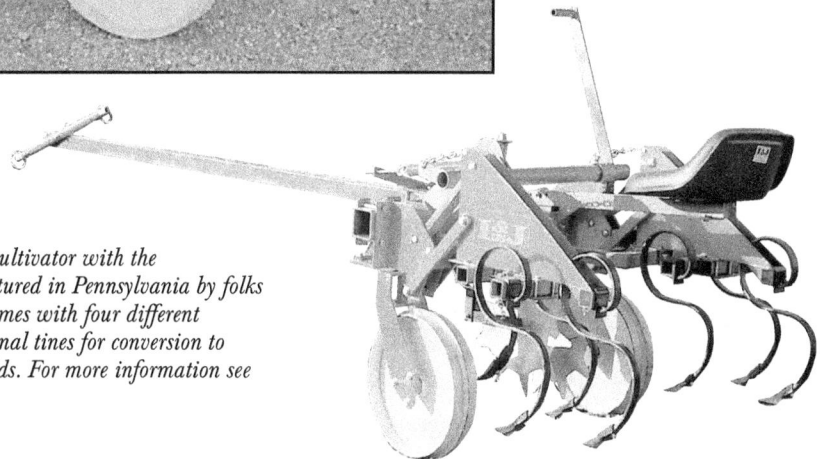

Avery Jack Rabbit Cultivators
Twin-Jack

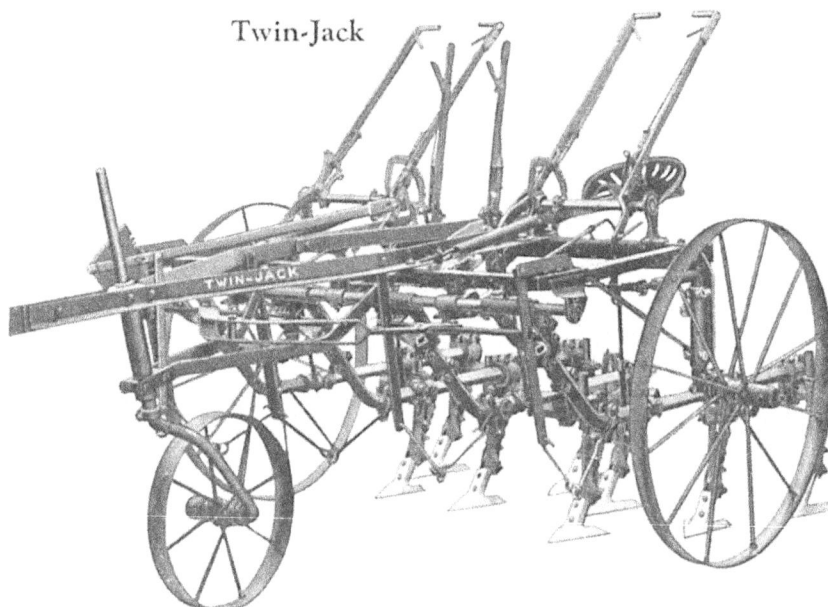

Eight Shovel. Equipped with No. 572 Pipe Beam Gangs, Round Standard, Spring Trip Open Foot

Chapter Fifteen

Two and Three Row Cultivators

I & J two row cultivator with fertilizer tank is pulled by three Percherons at Ohio Horse Progress Days '98

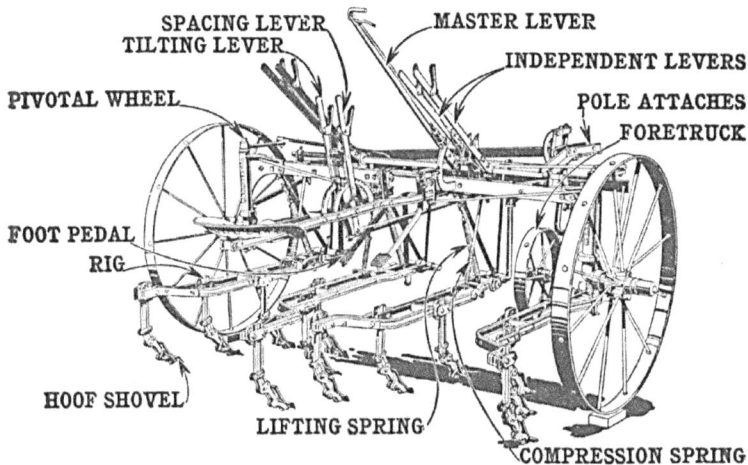

SPACING LEVER — MASTER LEVER
TILTING LEVER — INDEPENDENT LEVERS
PIVOTAL WHEEL — POLE ATTACHES
FORETRUCK
FOOT PEDAL RIG
HOOF SHOVEL — LIFTING SPRING
COMPRESSION SPRING

John Deere

John Deere Two Row Cultivator

How to Cultivate with a Two-Row

If the crop was planted with a two-row planter, follow the "twin-rows" with your cultivator. This is important because, regardless of how careful a driver you may have been when planting corn, there will be some variation between the pairs of rows.

If you have taken the precautions that you should in the adjustment of the shovels and shields, your entire time can be devoted to following the rows properly, guiding the cultivator, and observing that no hills are covered or dug out.

See that the two pairs of rigs are spaced so that the front shovels straddle both rows equally. There is an adjustment on all two-row cultivators to accomplish this. On most of them it is accomplished by means of levers, so the adjustment can be made instantly. With this adjustment once made, it will not be necessary to change it while you are cultivating the first time through the straight way of planting. Keep this in mind, so you will understand clearly, then, that if one pair of front shovels is straddling one row, the other pair of front shovels must straddle the other row in exactly the same fashion.

So, then, until you become accustomed to the general operation of the cultivator, watch one row

A new Canadian-built three row cultivator.

287

most of the time, and by guiding the cultivator and keeping the shovels properly spaced on that row, the other one will take care of itself. After a little experience, it will be easy to look about and observe both rows generally, so as to make sure that a clod or old corn stalk does not cover a hill of corn.

Now, we must consider the second cultivation when, in checked corn, we are cultivating crosswise. If you have been careful when planting corn, and your hills are checked straight, then you have nothing more to consider than in cultivating the straight way.

A three row cultivator at work

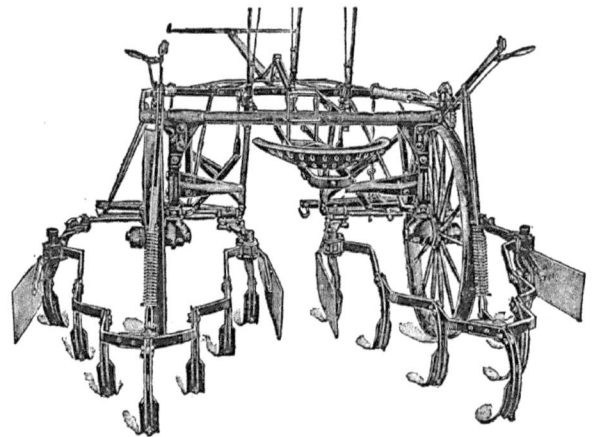

Two-row cultivator which straddles one row cultivating two middles.

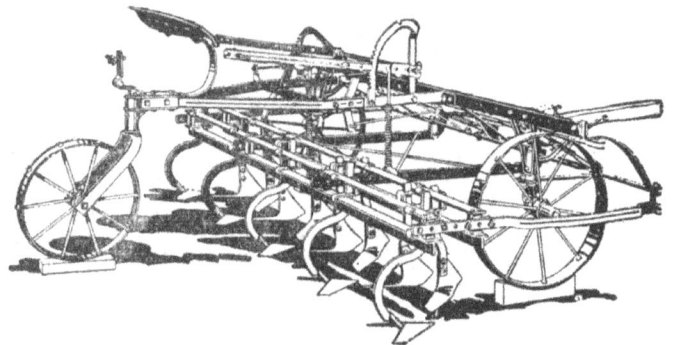

Beet and bean cultivator.

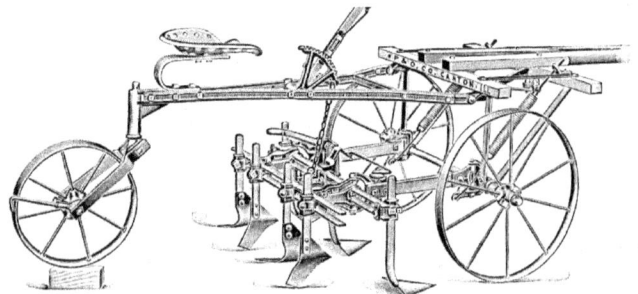

Two views of a Rock Island No. 112 Two-row cultivator at work.

Rock Island

NO. 1 TWO-ROW BEET CULTIVATOR WITH RIDING ATTACHMENT.

Easily Follows the Rows

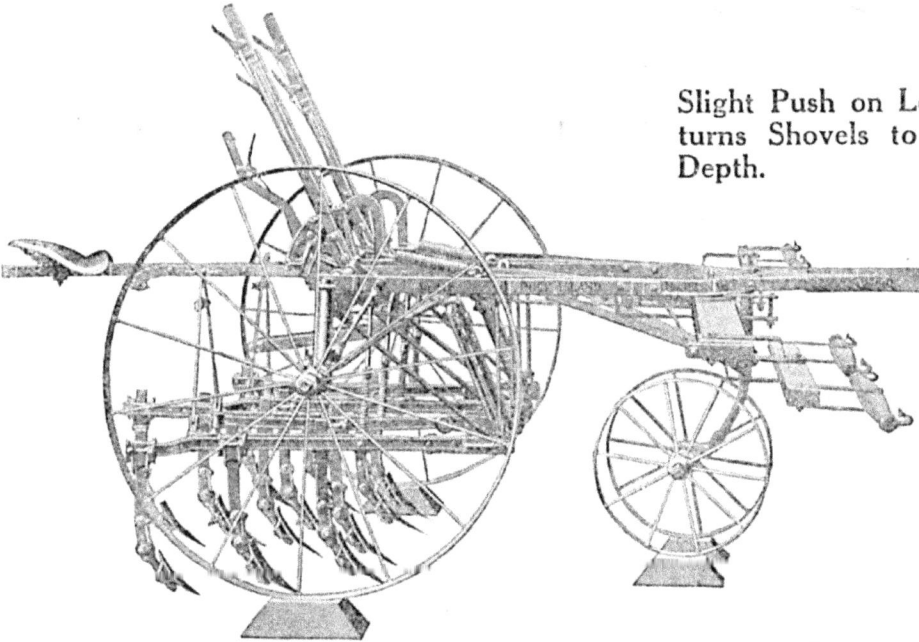

Slight Push on Levers Returns Shovels to Original Depth.

Rock Island No. 112 Two-Row Cultivator

P. & O. No. 4 Four-Row Beet Cultivator.

Rock Island Fore-Carriage

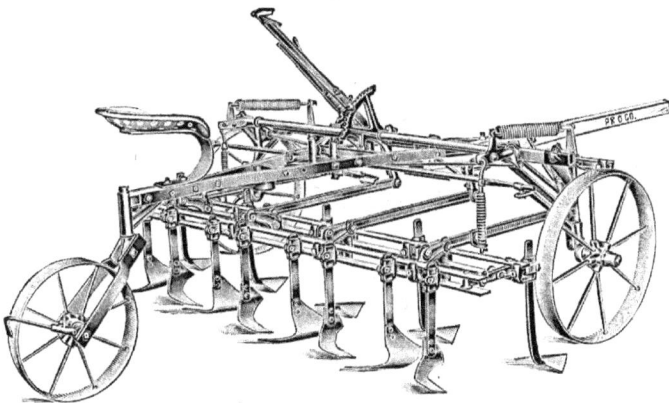

NO. 4 FOUR-ROW BEET CULTIVATOR.

Avery Jack Rabbit Cultivator
Triple-Jack

Avery Jack Rabbit Cultivators
Twin-Jack

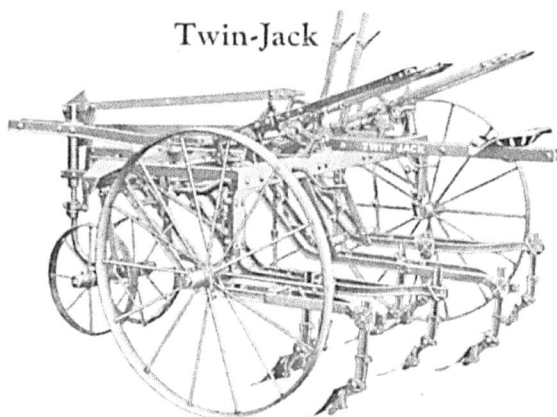

Moline

Dirt and Weather Proof Wheel Pivot—Equipped with Alemite Oiler

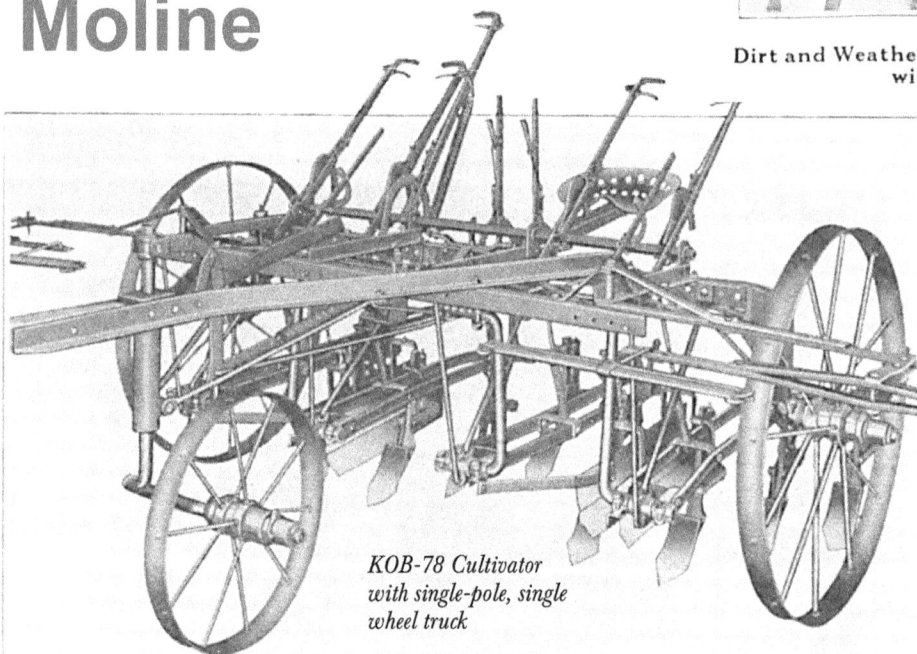

KOB-78 Cultivator with single-pole, single wheel truck

This cultivator may be used as either a combined wheel pivot and gang shift or as a wheel guide only. Wheels and gangs may be locked for transporting.

The seat is adjustable up or down, as well as from front to rear. Therefore the operator can get off and on quickly.

Passes through twelve foot gates. The eveners can be moved toward the center of the machine to let horses and cultivator fit inside twelve foot width.

As a Cross Arch cultivator. The gang shifting lever, which is placed immediately in front of the operator, controls the arches. By moving this lever fore and aft, the right hand gangs of each pair are moved opposite to the left hand gangs, thus varying the spacing between each right and left hand gang. This arch arrangement permits close or wide cultivation of the rows and is used on fields planted with a two-row planter.

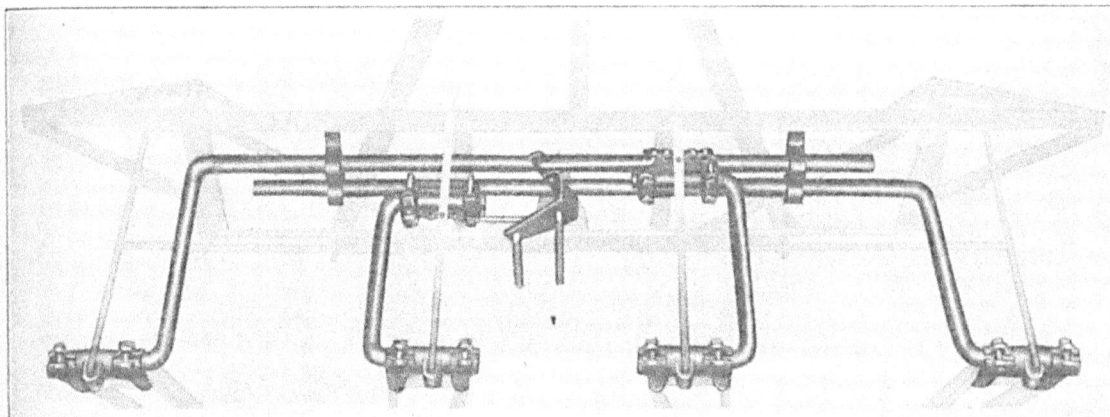

Moline

Features of the two
row KOB-78

Moline Single-Wheel Truck

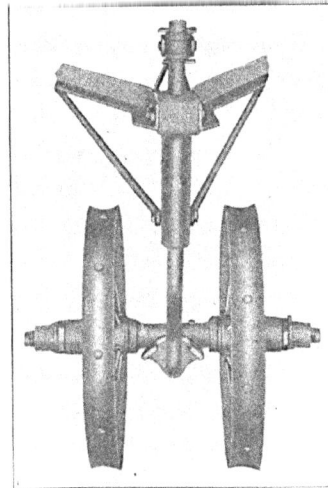

Moline Lever Adjustable Shield—Permits
Quick, Close Adjustment from Cultivator Seat

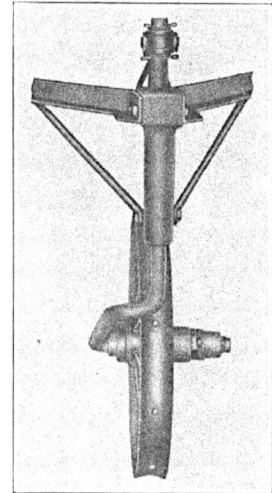

*Moline Two-wheel Truck
with wheels set for narrow
tread. Width is 13 inches*

Moline Two-wheel truck
with wheels reversed for
wide tread - 17 inches.

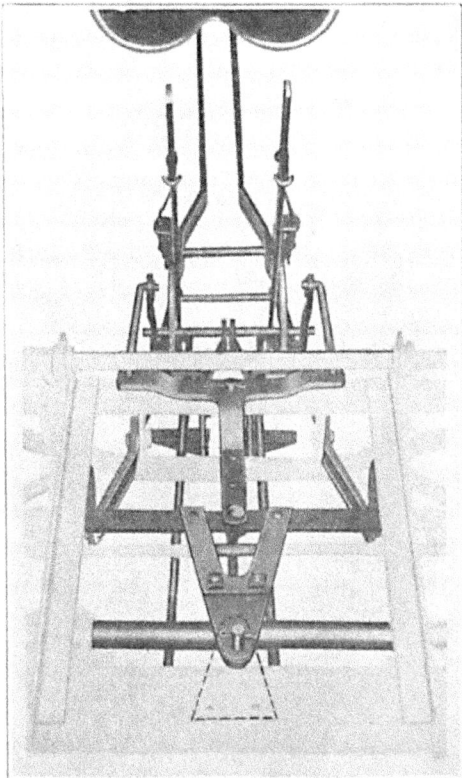

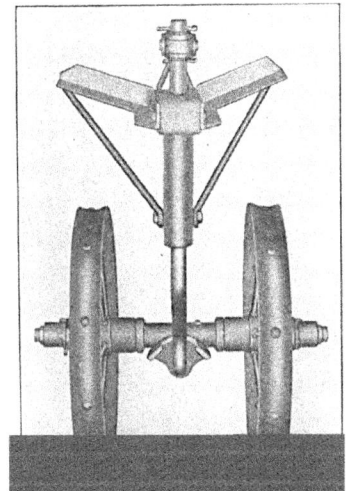

*Two & Three
Row Cultivators*

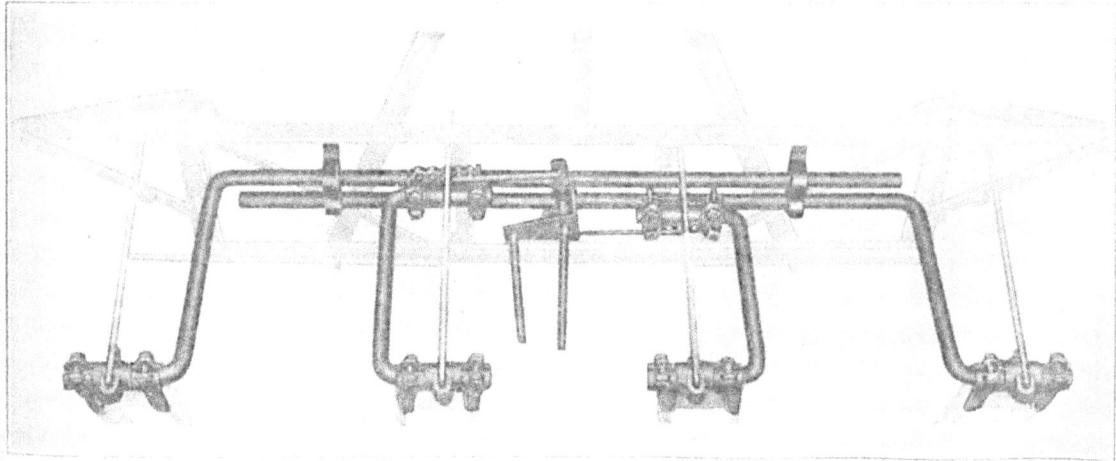

KOB-78 as a Straight Arch cultivator. By reversing two clamps its converted from cross arch to straight arch. The arch for each pair of gangs is then separate and, as the gang shifting lever is operated, the arches are moved toward or away from each other. For use with crops planted by a single-row planter and possible non-parallel.

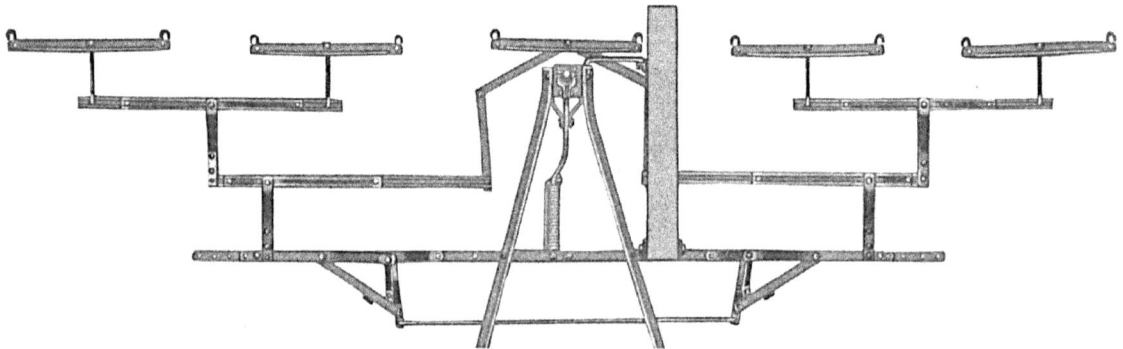

Moline Five-Horse All-Steel Hitch with Drop Pole—Observe Its Strong, Sturdy Construction— No Wood Parts to Rot or Break

KOB-116 with single pole, single wheel truck.

Moline

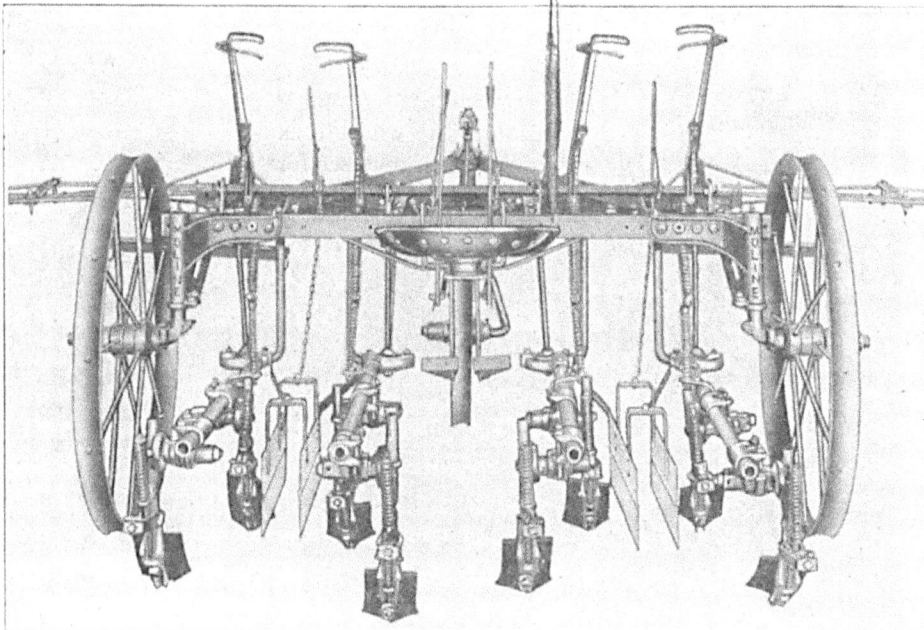

Rear view of Moline 4 row beet and bean cultivator

Moline

Equipped for three row
cultivation

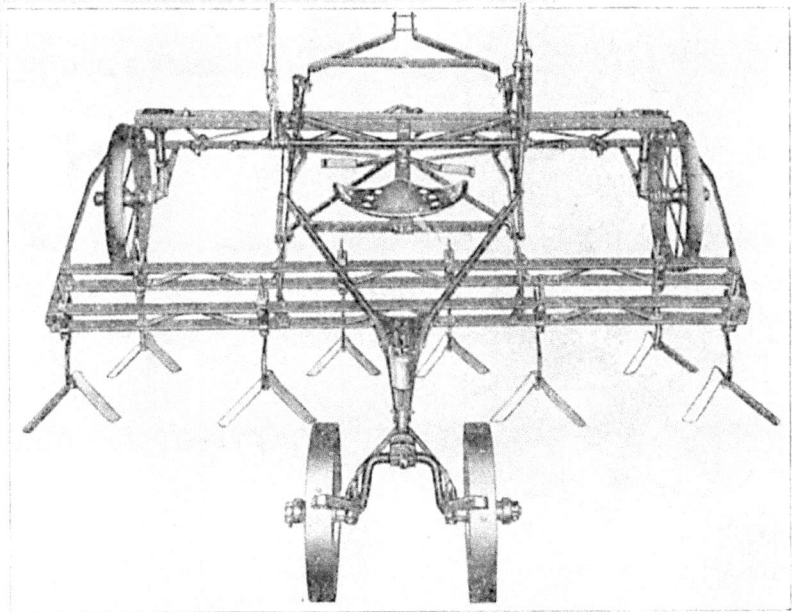

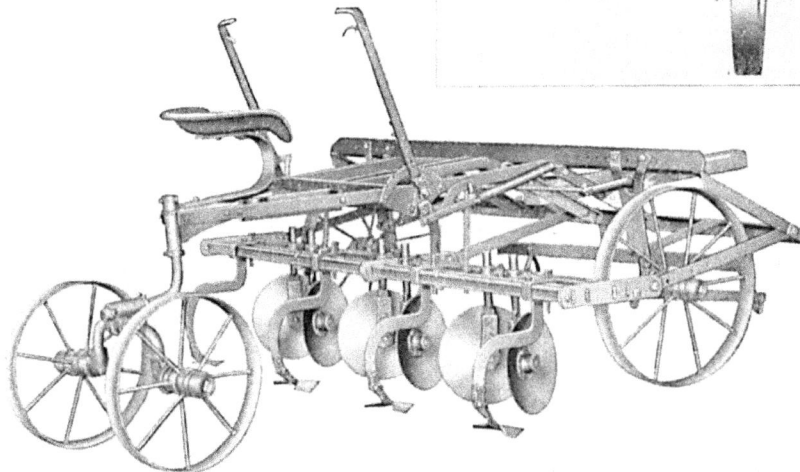

Oliver

The No. 30 three row can be converted to a four row machine.

4-ROW CULTIVATOR

Moline

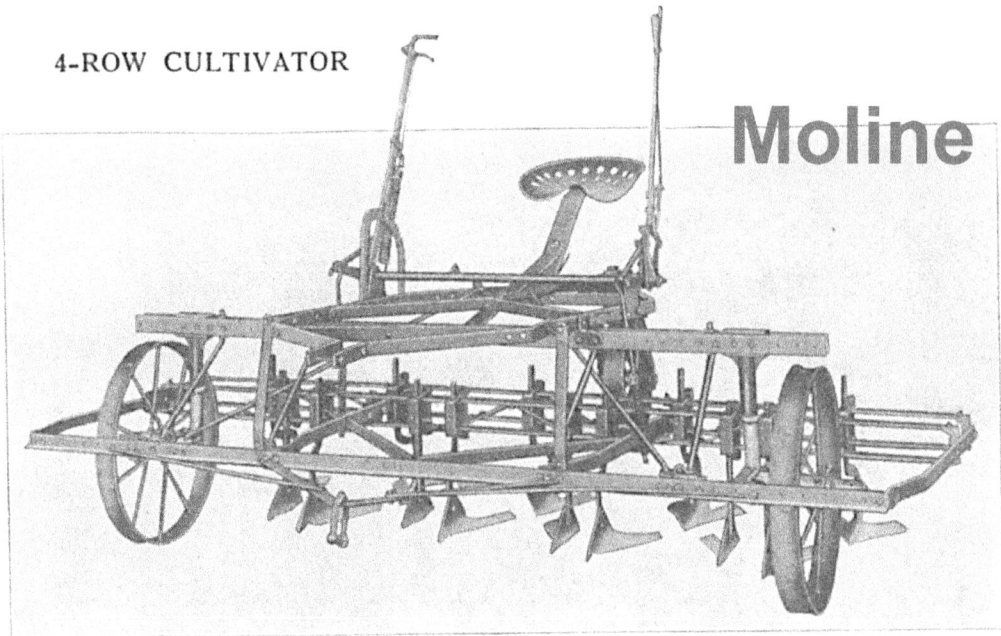

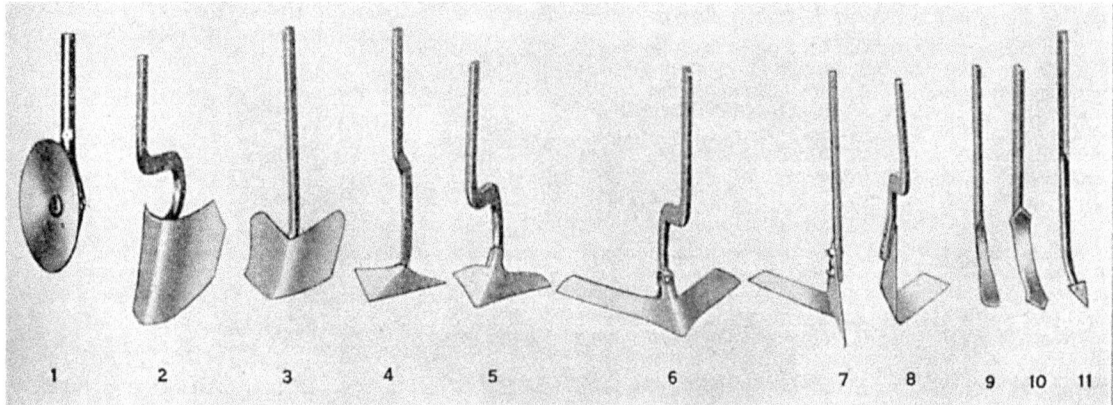

ROCK ISLAND NO. 112 TWO-ROW CULTIVATOR

Rock
Island

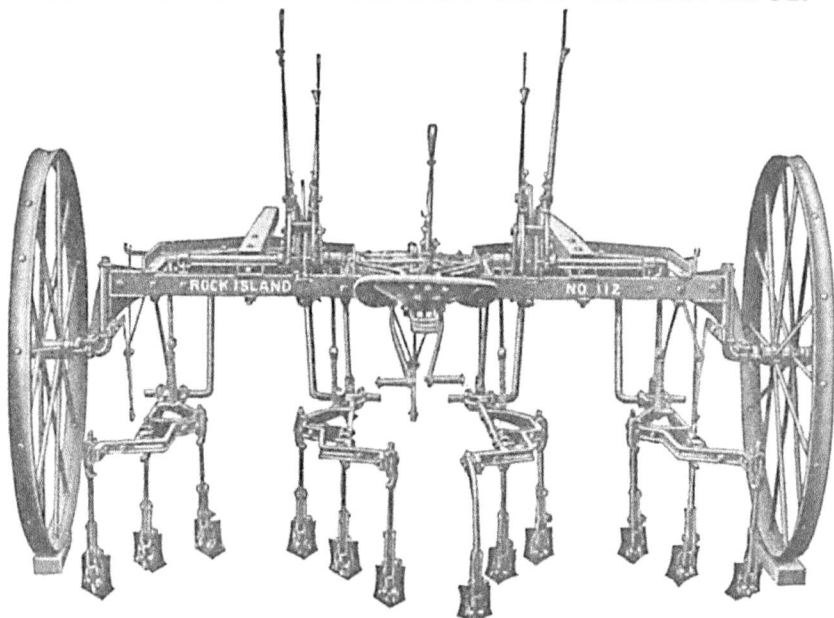

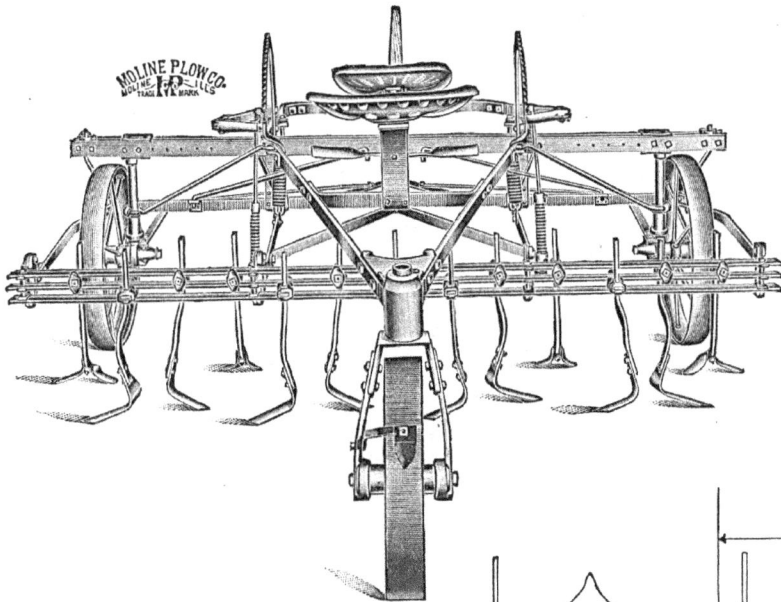

Beet Cultivator No. 3

Handles four rows at a time. Controlled entirely by foot steering. This cultivator is adjustable for rows of either 16, 18 or 20 inches in width.

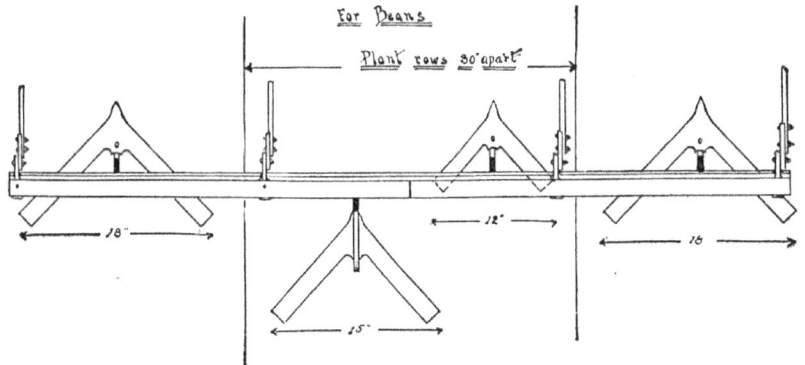

Moline

Moline Two-row Cultivator with 12 shovel, friction break, round sleeve gangs.

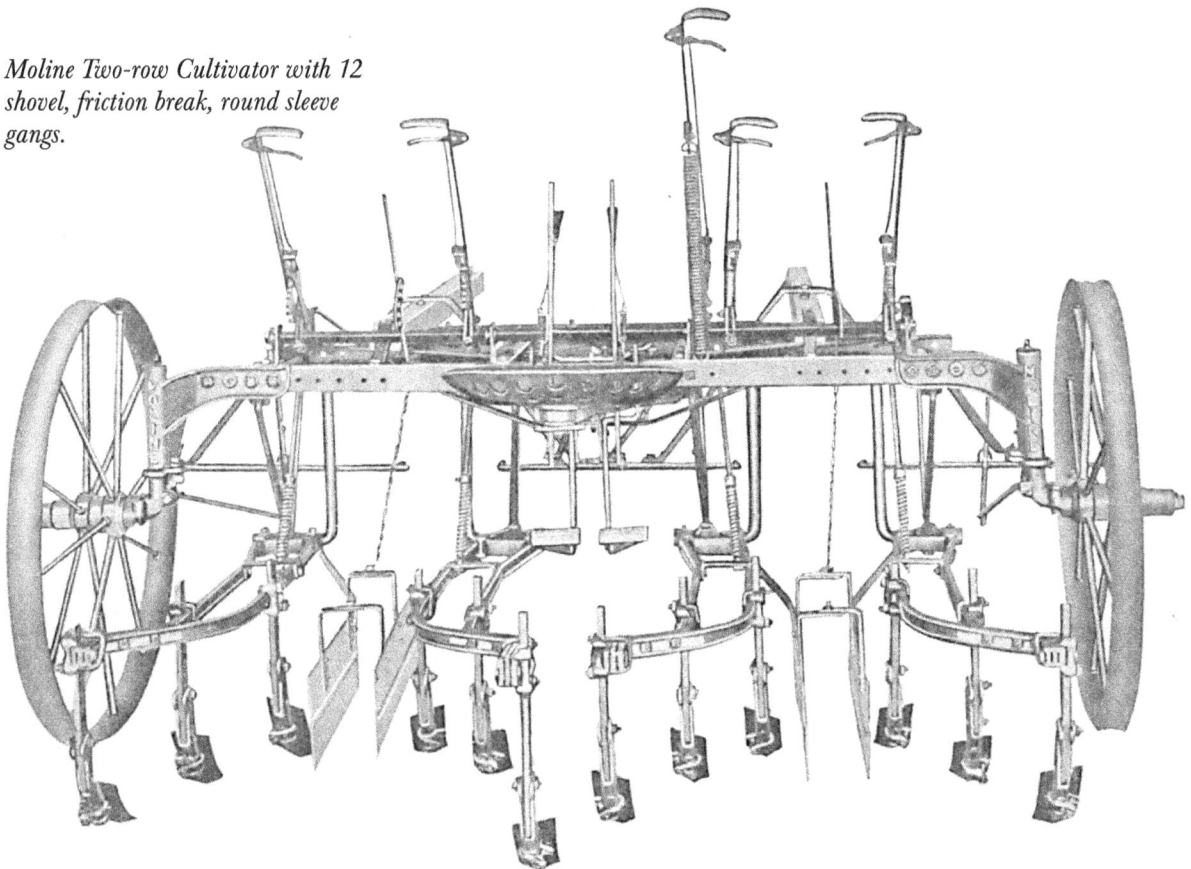

Moline Two-Row Cultivator with 12-Shovel, Friction-Break, Round-Sleeve Gangs

Two & Three Row Cultivators

ROCK ISLAND NO. 112 TWO-ROW CULTIVATOR

Rock
Island

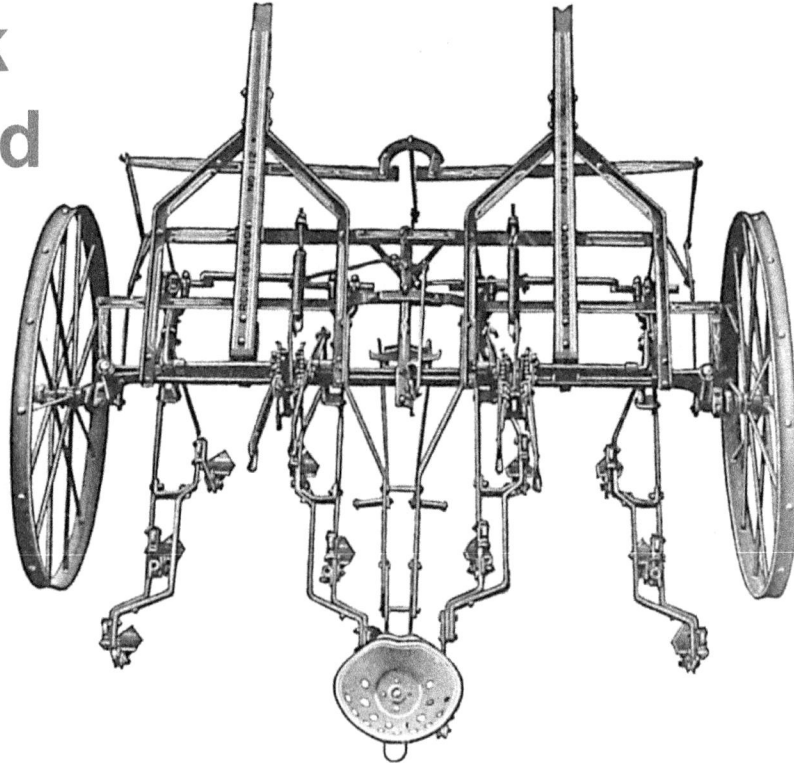

No. 37 three row from above and at work.

Oliver

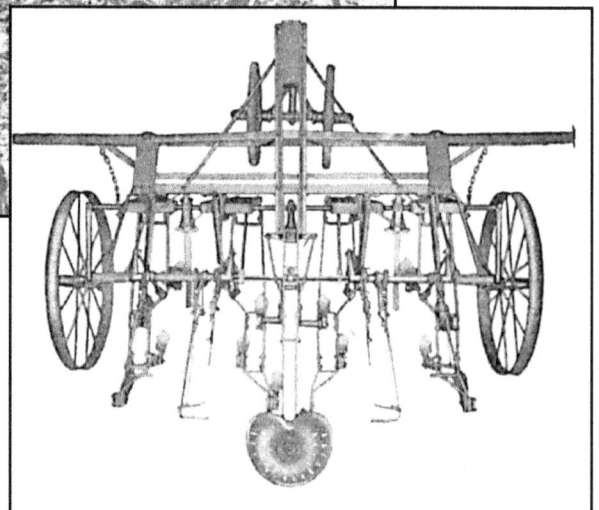

Oliver

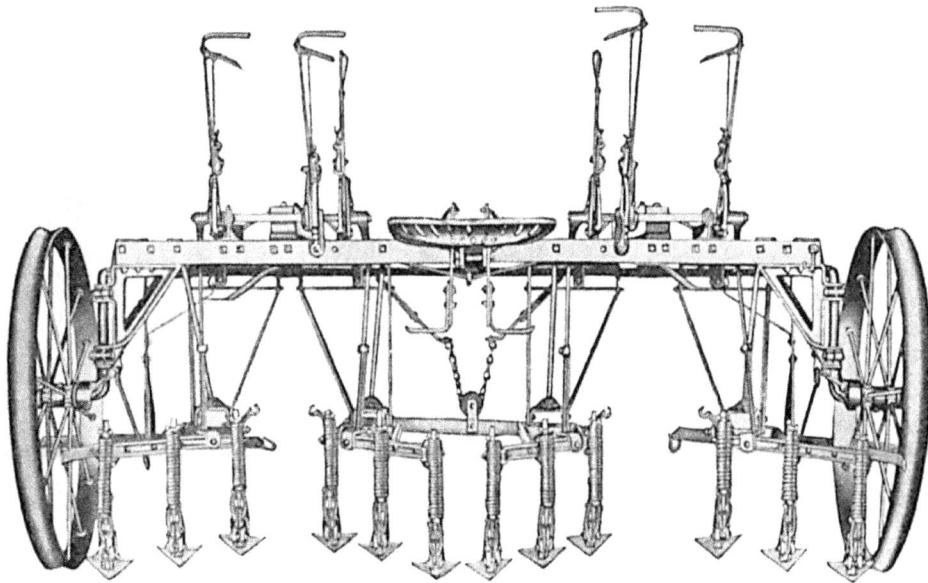

The No. 37 may be equipped with a double pole three horse hitch as shown here.

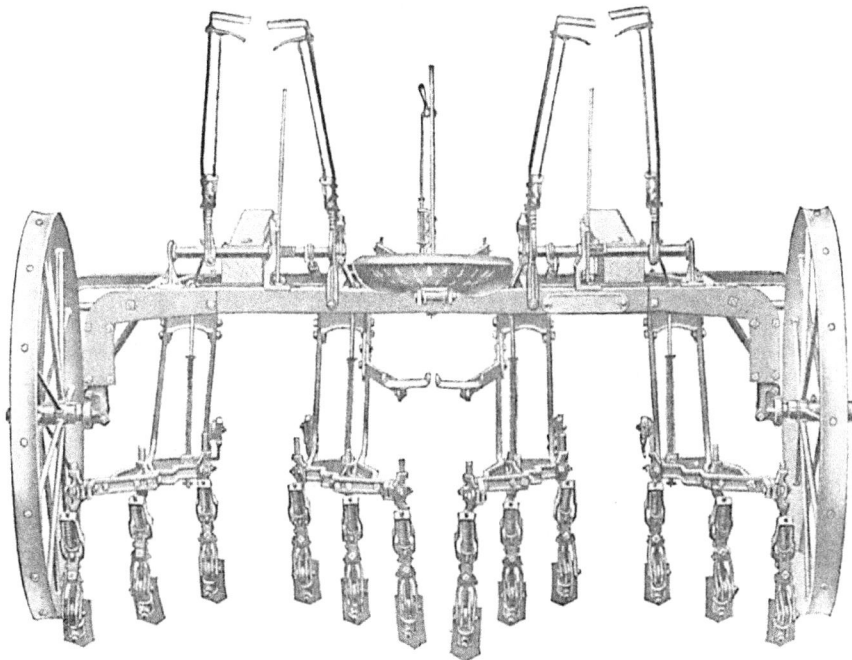

Rear view of the 557 Oliver Superior two-row cultivator.

Rock Island

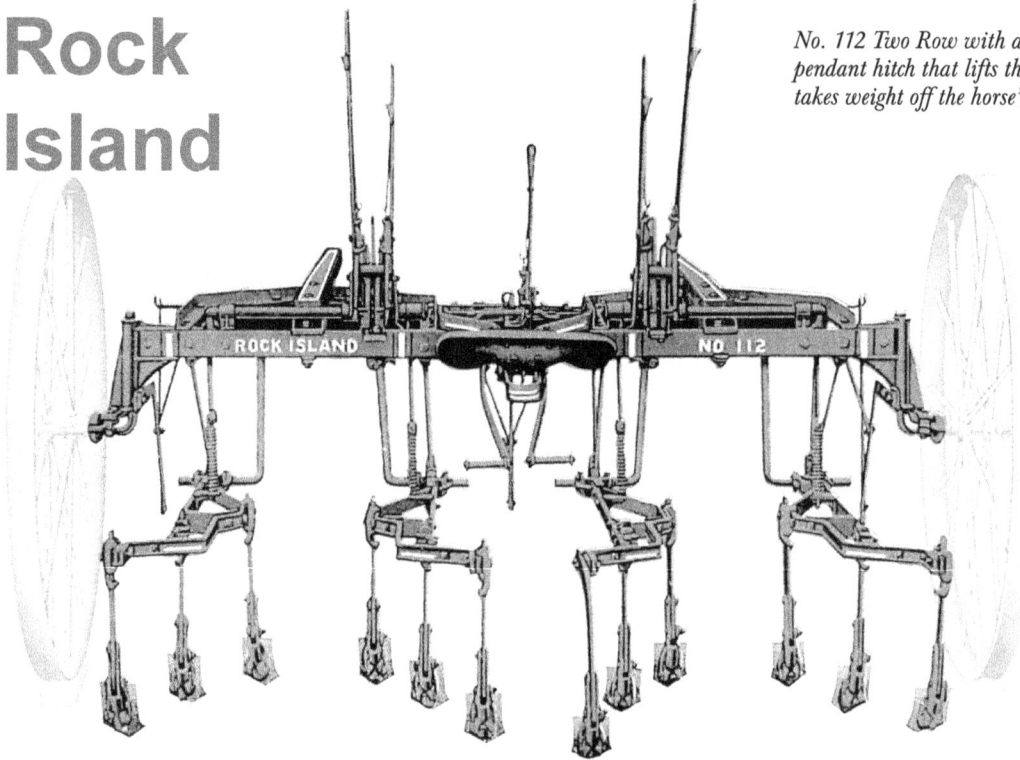

No. 112 Two Row with a three horse pendant hitch that lifts the pole and takes weight off the horse's necks.

Oliver

Oliver No. 23
Two-Row Cultivator

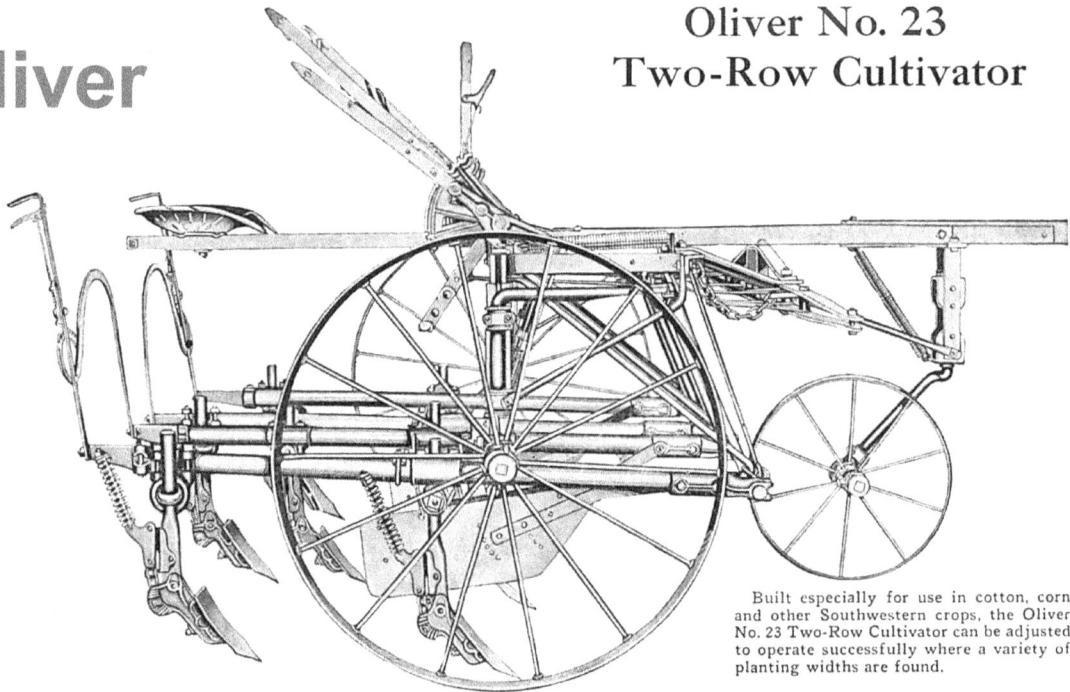

Built especially for use in cotton, corn and other Southwestern crops, the Oliver No. 23 Two-Row Cultivator can be adjusted to operate successfully where a variety of planting widths are found.

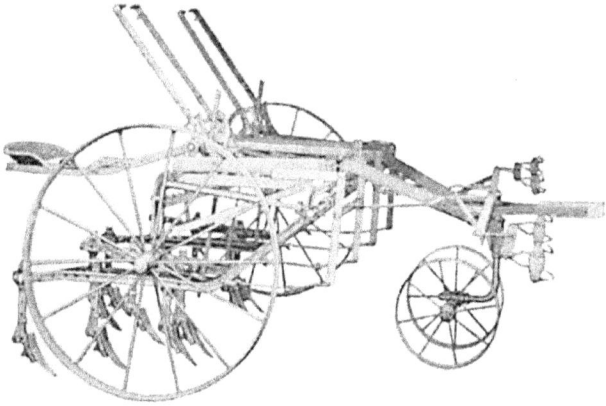

The 4-horse hitch and front truck is special equipment.

Disc beam for the No. 562 two row cultivator. Discs are quickly and easily adjustable to any position.

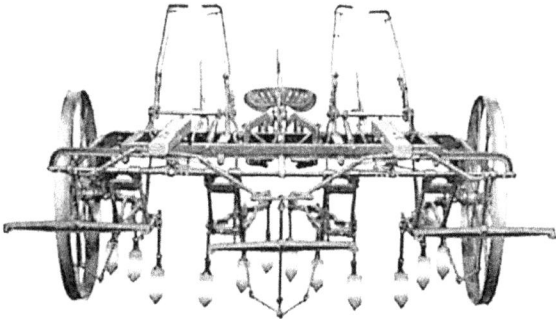

Front view with two poles and three horse hitch which is standard equipment.

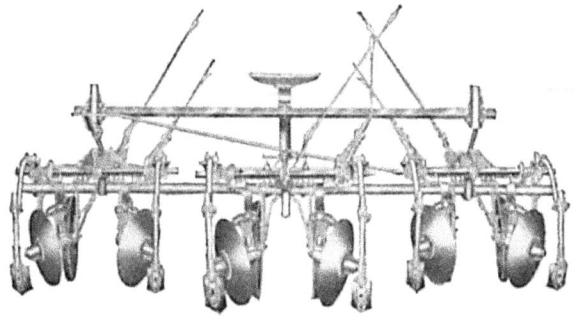

The 49-A three-row lister cultivator.

Oliver

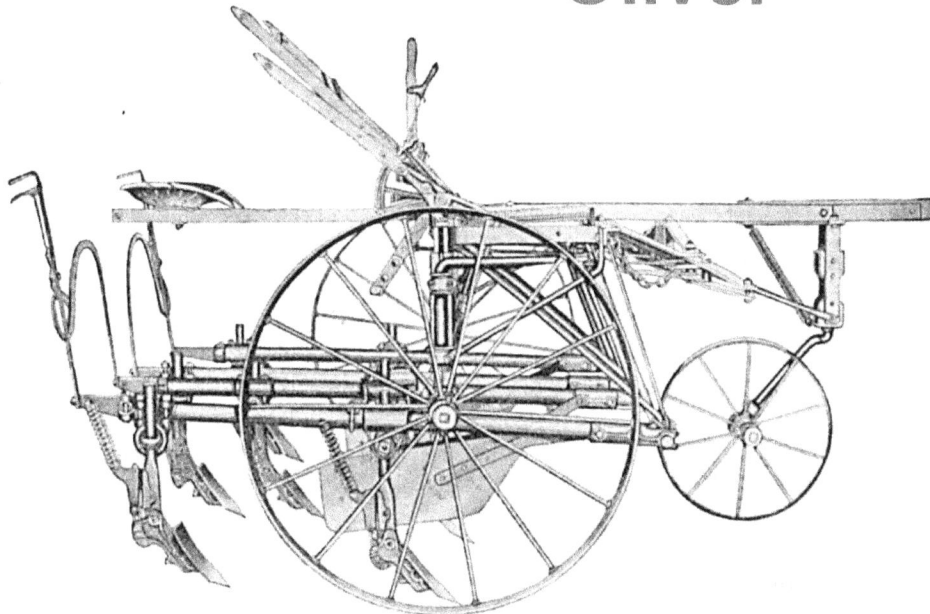

The Oliver No. 23 is an ideal machine wherever there is a need of a two-row that is adjustable to many row widths.

Two & Three Row Cultivators

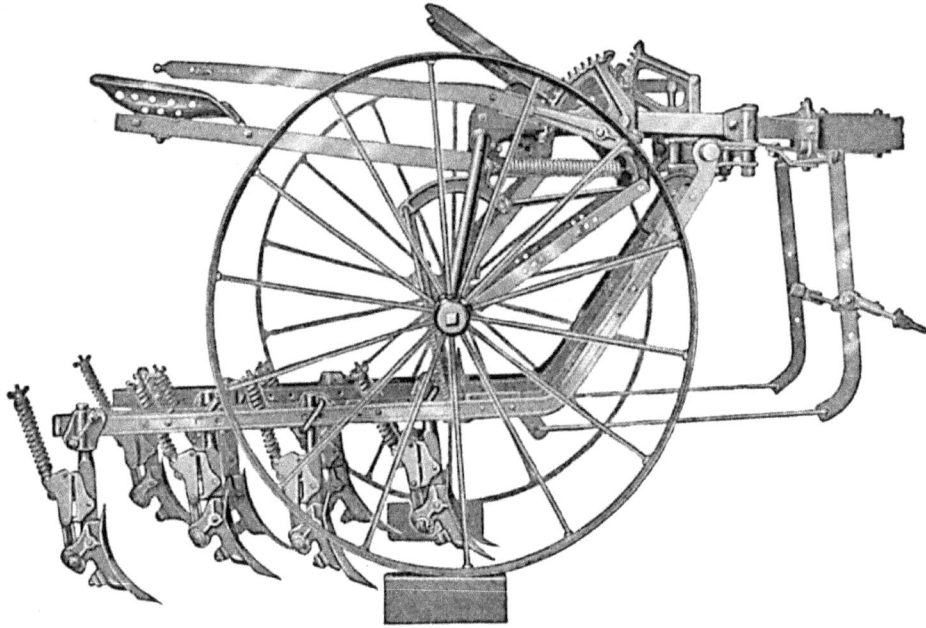

The Oliver No. 35 is reputed to be the best one-row cultivator ever built. Here it is with 8 shovel spring trip gangs.

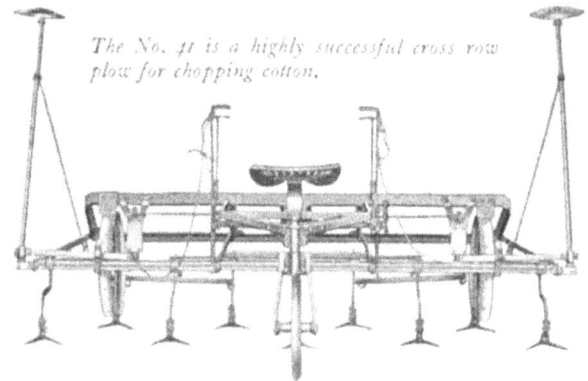

Oliver

The No. 41 is a highly successful cross row plow for chopping cotton.

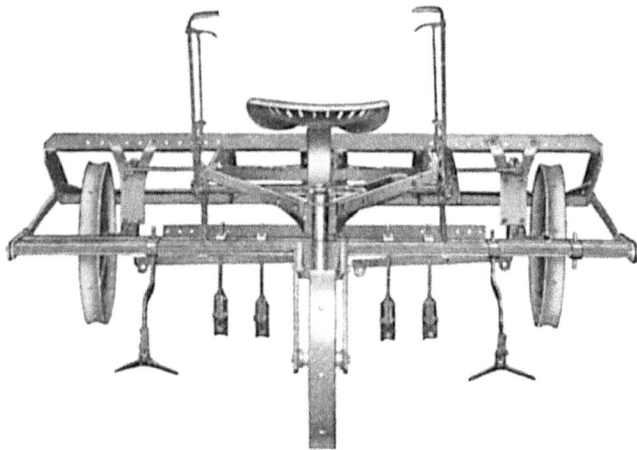

The No. 20 two row can be converted to a three row machine.

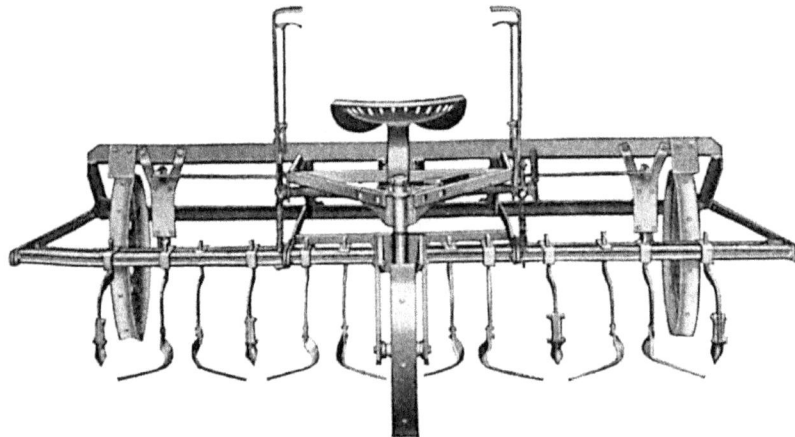

No. 40 four-row. The No. 50 is similar to this machine except its tool bars are 16 inches longer.

P. & O. TWO-ROW PIVOT WHEEL CULTIVATORS.

(Style UA.)

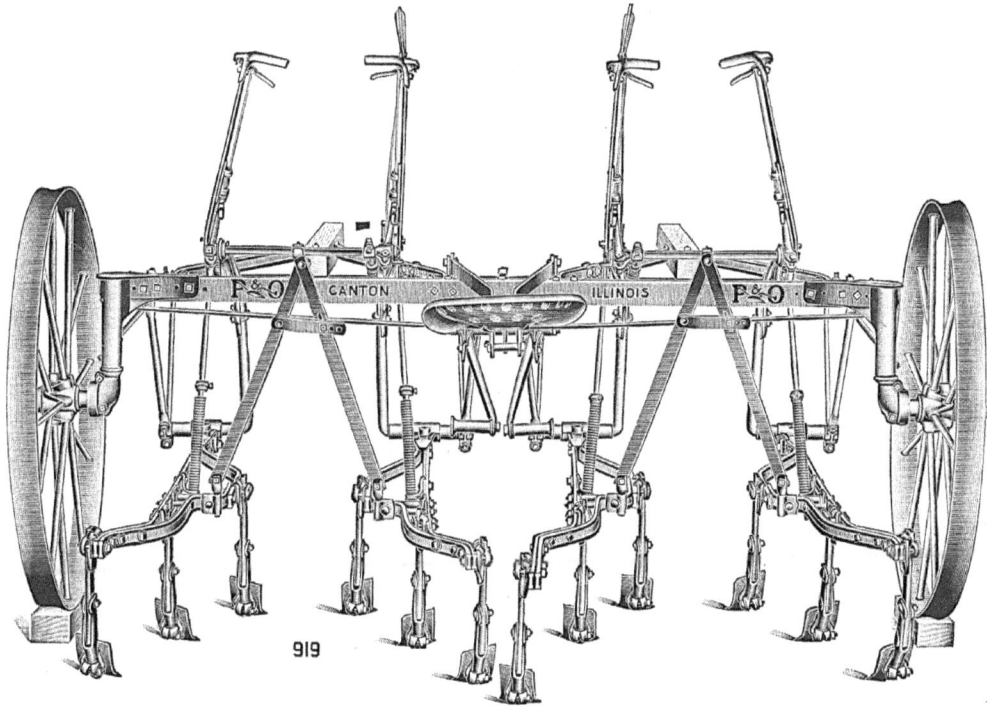

TWO-ROW PIVOT WHEEL CULTIVATOR WITH NO. 21 GANGS.

P & O

P. & O. TWO-ROW PIVOT WHEEL CULTIVATORS.

(Style UA.)—Continued.

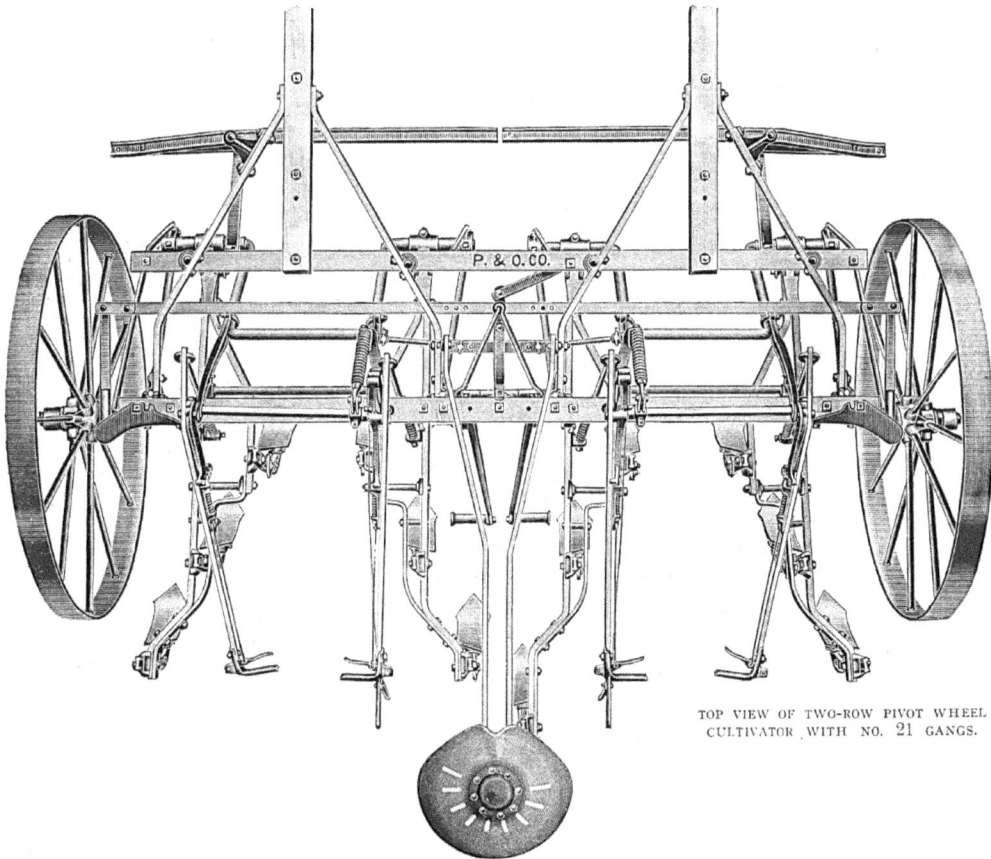

TOP VIEW OF TWO-ROW PIVOT WHEEL
CULTIVATOR WITH NO. 21 GANGS.

*Two & Three
Row Cultivators*

Oliver

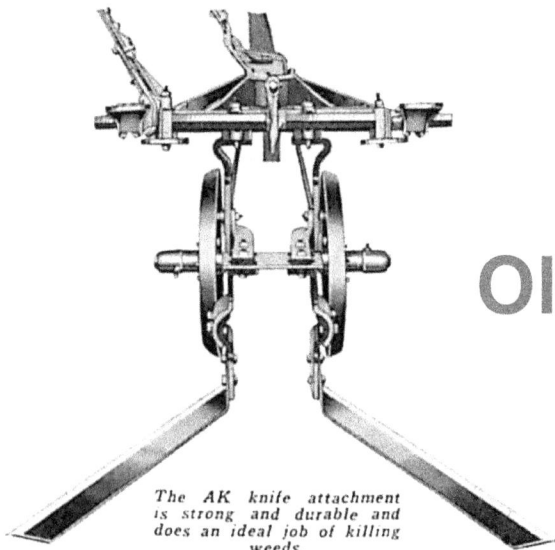

The AK knife attachment is strong and durable and does an ideal job of killing weeds.

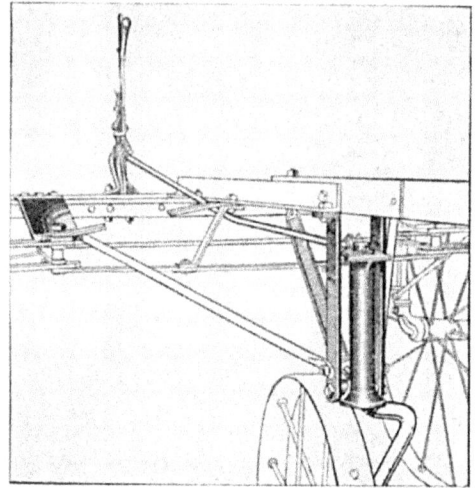

This tilting lever can be had with the four-horse hitch with single or double wheel truck. It provides for instant regulation of "set and suck" of the shovels.

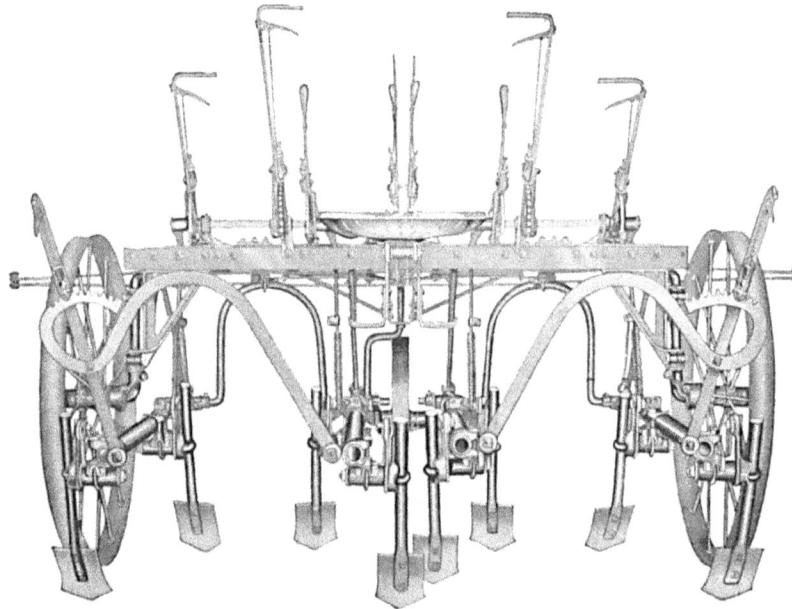

The unusually wide range of adjustment provided for on the Oliver No. 23 Two-Row Cultivator adapts it for a variety of conditions.

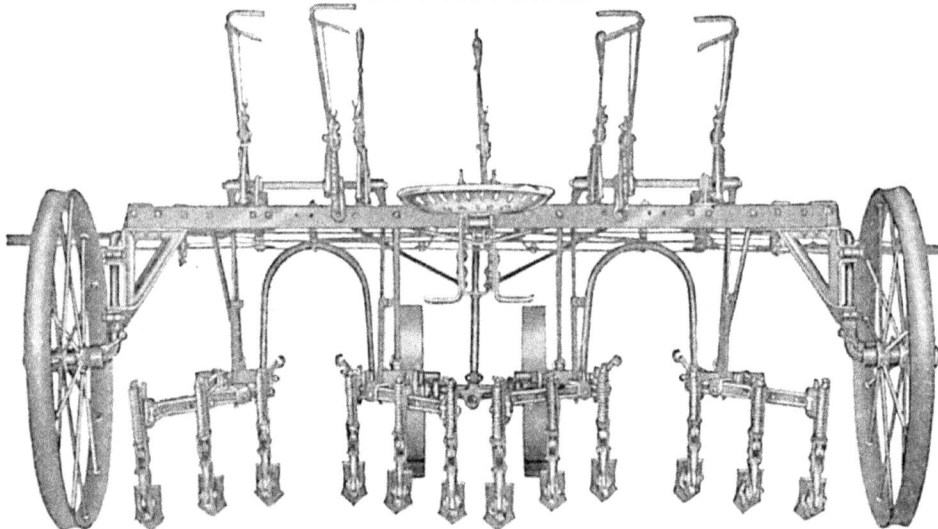

The Oliver No. 37 is a two-row pivot axle, pedal guide cultivator. It may be equipped with a tongue truck and four horse hitch as shown here.

Oliver

*Note the convenient location of the
foot pedals and the easily operated
levers. One master lever raises or
lowers all the gangs at once—no
stops at ends of rows.*

*A long view of your work is possible with the Oliver No. 23 Cultivator. You can see three hills at once
on the guide row. There is no digging out, lifting or covering of hills. This means a big saving each
year to the farmer.*

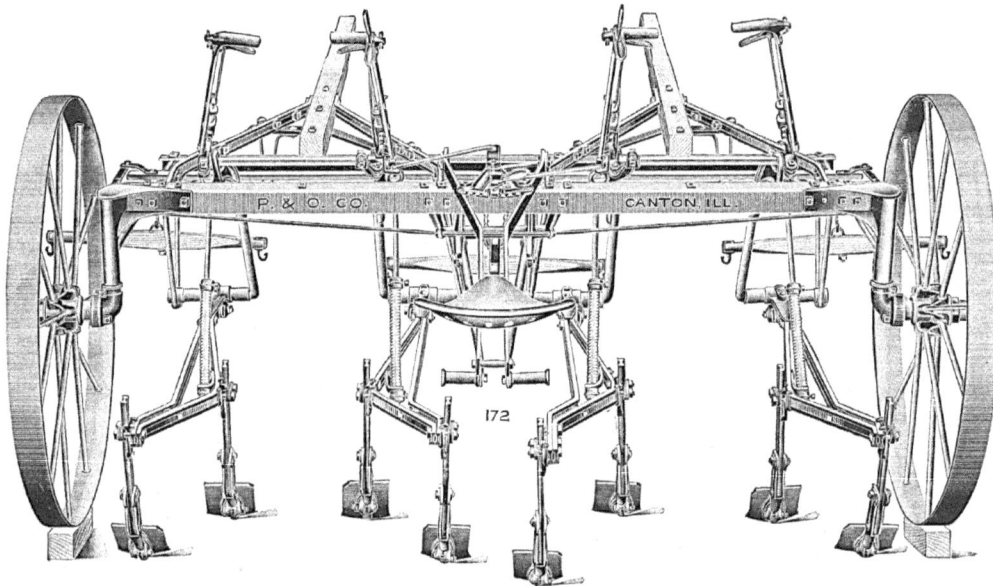

TWO-ROW PIVOT WHEEL CULTIVATOR WITH NO. 1 GANGS AND PIVOT LEVER ATTACHMENT.

FURNISHED WITH EITHER FLAT OR ROUND BREAK-OFFS.

P & O

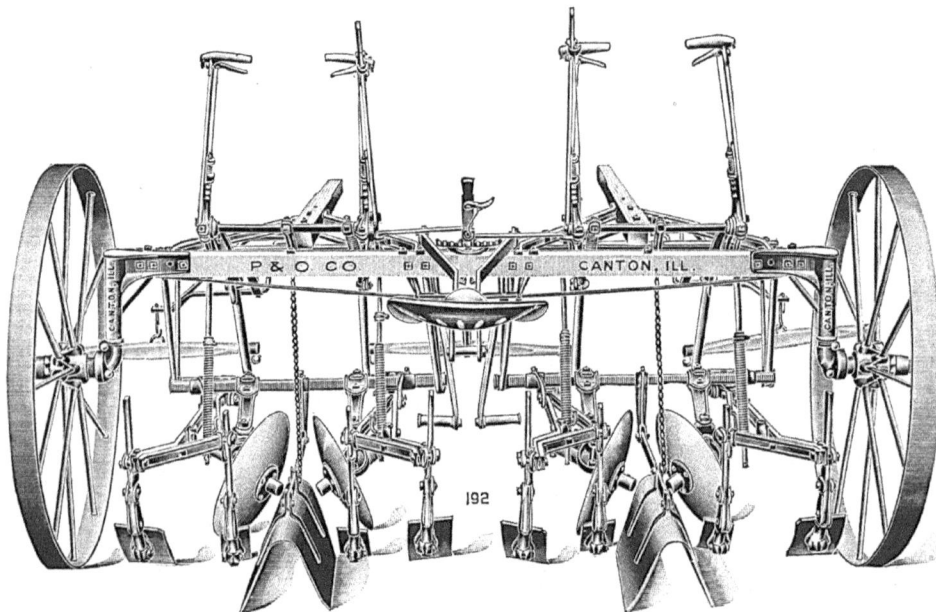

TWO-ROW PIVOT WHEEL CULTIVATOR WITH NO. 1 GANGS, PIVOT LEVER AND DISC LISTER ATTACHMENTS.

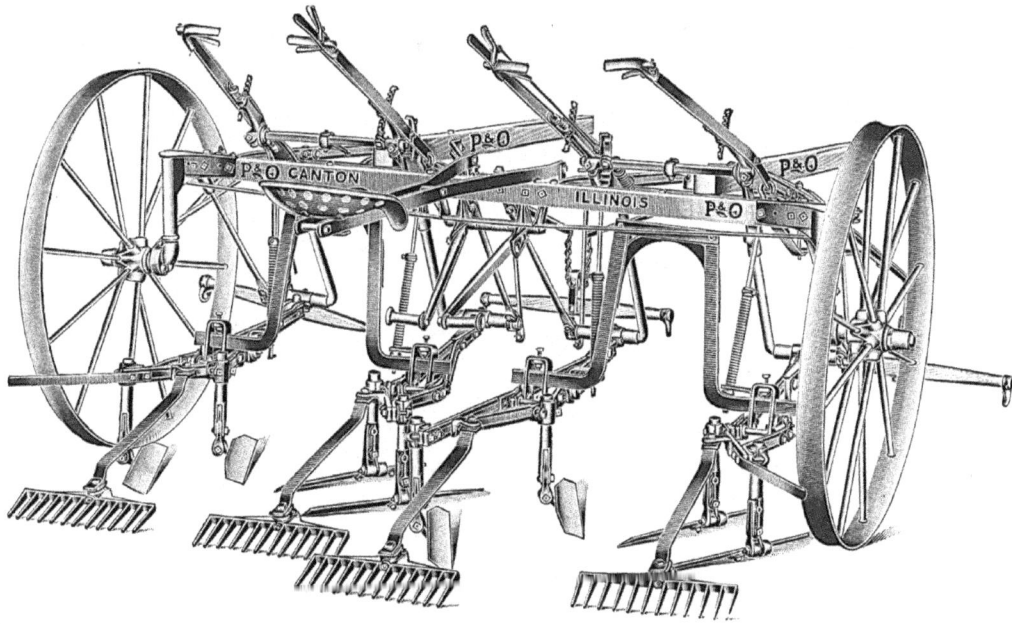

NO. 23 TWO-ROW SURFACE CULTIVATOR.

P & O

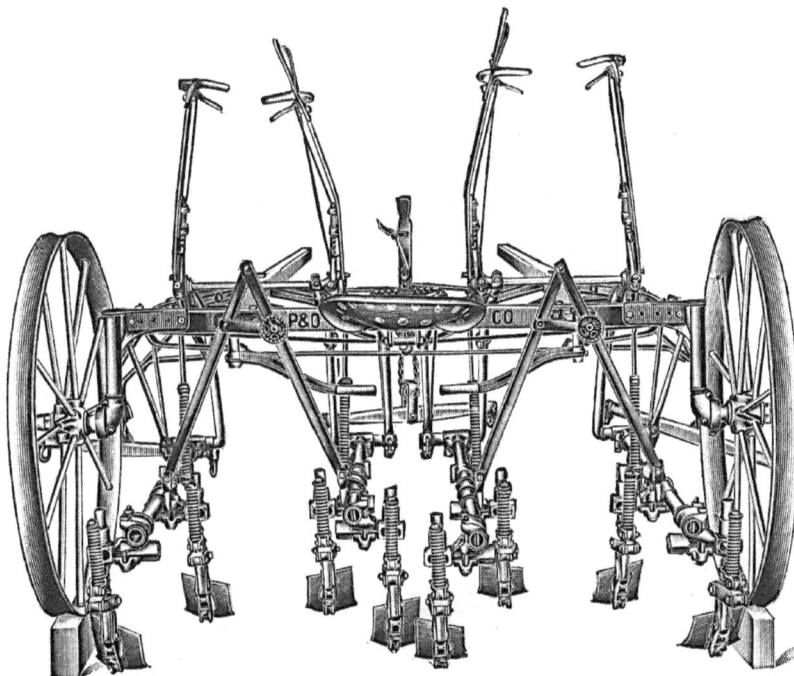

The Two-Row, Style UB, is a narrow frame cultivator built especially for use in cotton and in corn that is planted in rows narrower than the standard width. The general construction is the same as that of the Two-Row just described.

Two & Three Row Cultivators

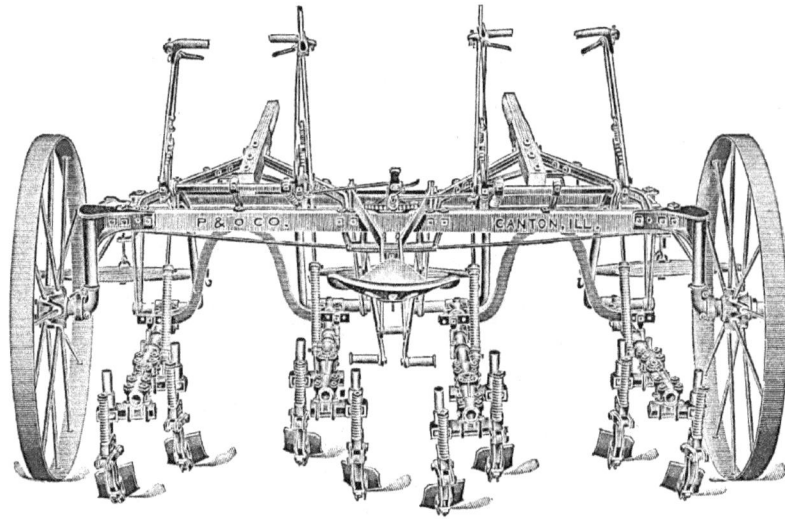

TWO-ROW PIVOT WHEEL CULTIVATOR WITH NO. 20 GANGS AND PIVOT LEVER ATTACHMENT.

P & O

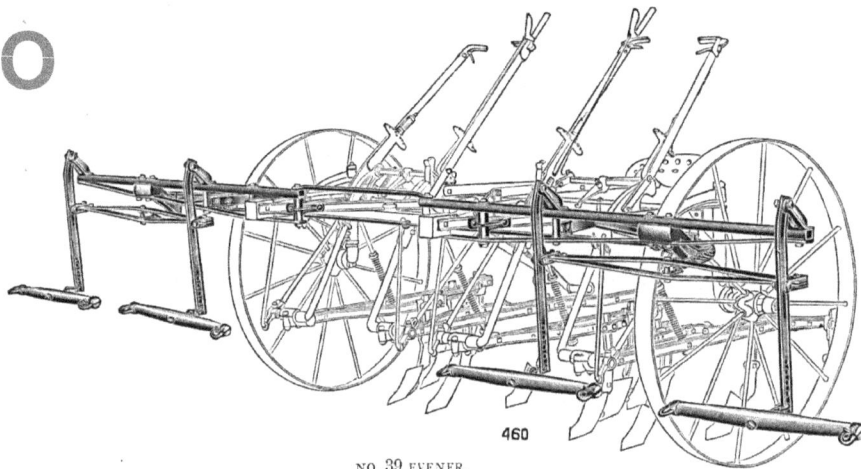

460

NO. 39 EVENER.

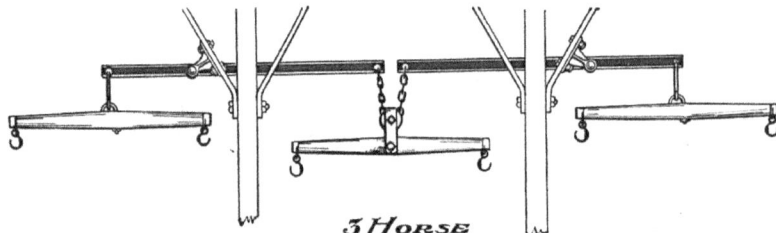

3 HORSE

NO. 35 EVENER.

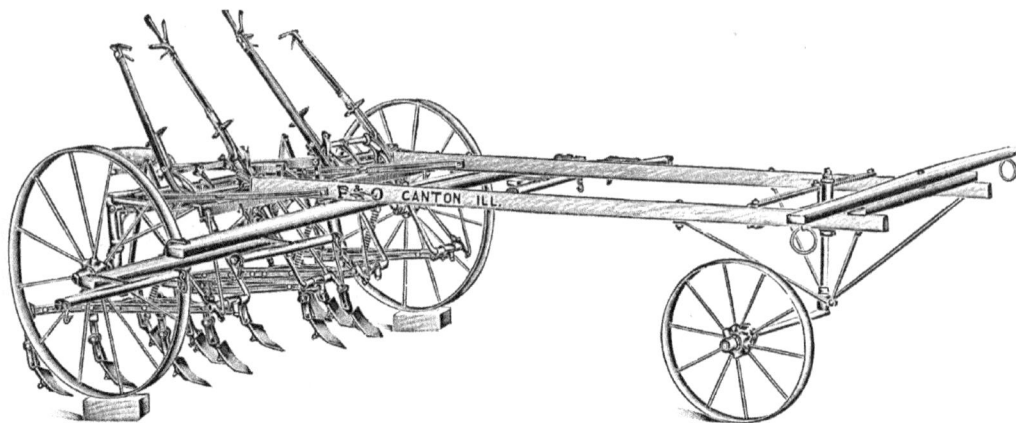

TWO-ROW CULTIVATOR WITH TONGUE TRUCK AND NO. 48 EVENER.

I & J Mfg

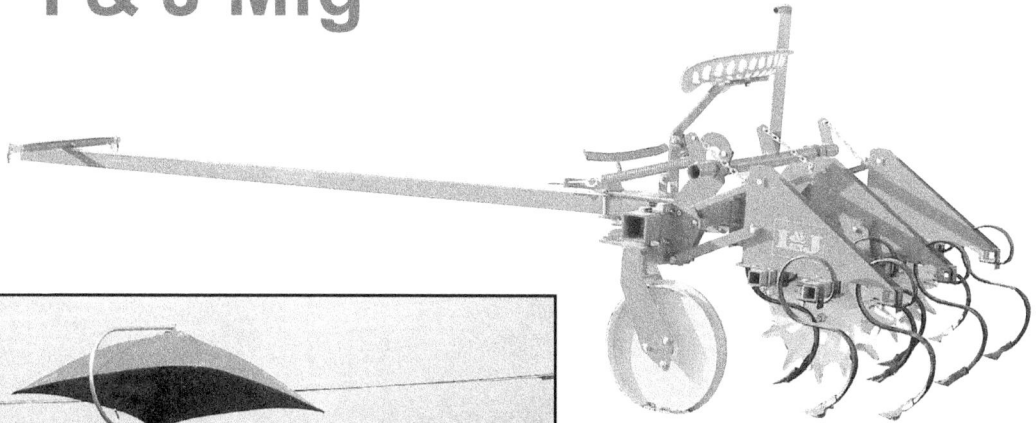

The I & J two row is a fully modern horsedrawn cultivator with many attachments available from the Gap, Pennsylvania manufacturer.

On the left is one setup with a gravity feed liquid side dresser allowing convenient appliction of fish emulsion or other liquid fertlizers.

This unit can be modified to use with plastic mulch.

See back of this book for contact information.

Two & Three Row Cultivators

Chapter Sixteen

Shovels, Points
& their gangs

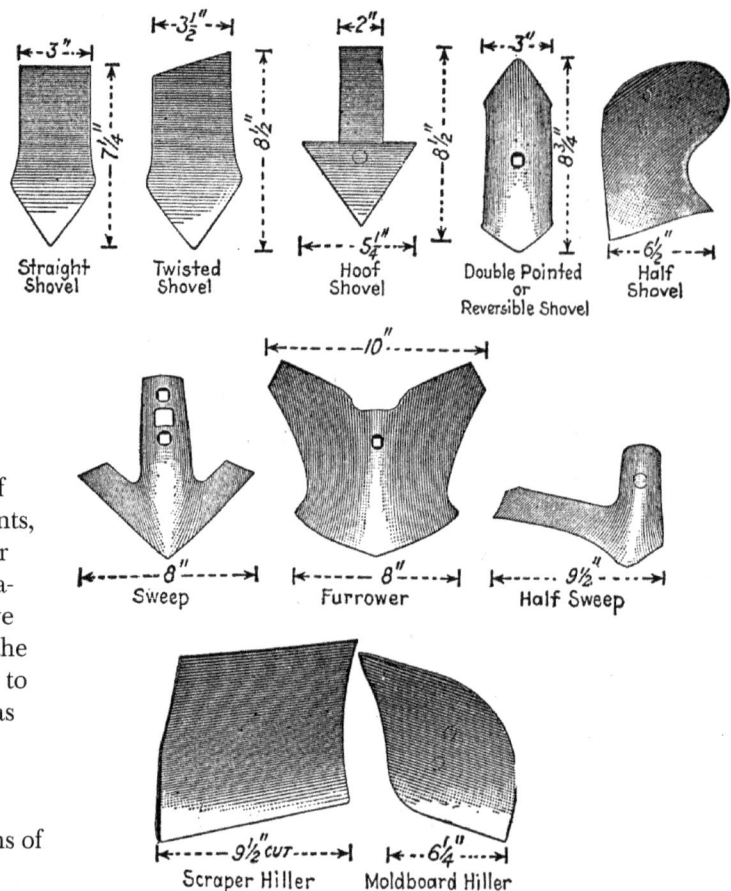

Straight Shovel — 3" — 7¼"
Twisted Shovel — 3½" — 8½"
Hoof Shovel — 2" — 8½" — 5¼"
Double Pointed or Reversible Shovel — 3" — 8¾"
Half Shovel — 6½"

Sweep — 8"
Furrower — 10" — 8"
Half Sweep — 9½"

Scraper Hiller — 9½" CUT
Moldboard Hiller — 6¼"

This chapter offers a broad smattering of pictorial information covering the shovels, points, furrowers, hillers, and disc styles which were or might be available for most horsedrawn cultivators. Though this may seem to be an exhaustive display in actuality it represents just the tip of the iceberg. Today, most farmers limit themselves to between four and six different shovel options as with tractors and brute hydraulic force, less concern remains for the subtleties and perfect pitch, balance and turn.

This chapter also shows off several dozens of shank assemblies and gang setups.

photo by John Nordell

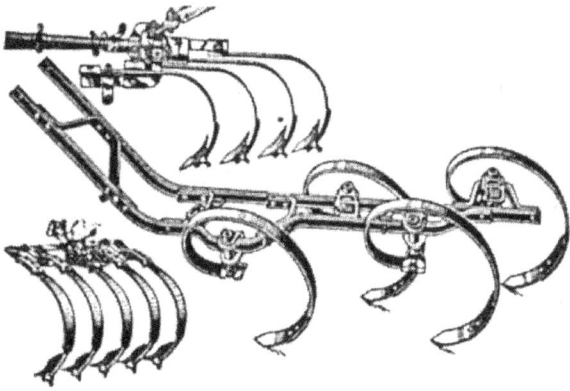

Spring-tooth attachments.

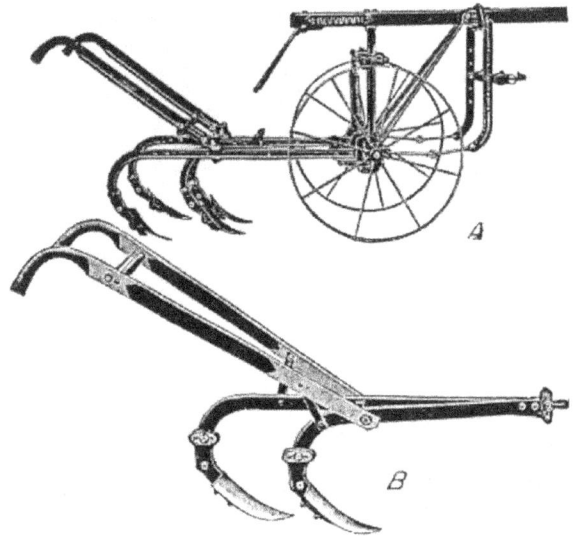

Two other types of walking cultivators. *A*, four-shovel; *B*, two-shovel cultivator, or double shovel plow. *B* is a cultivator commonly used in the South.

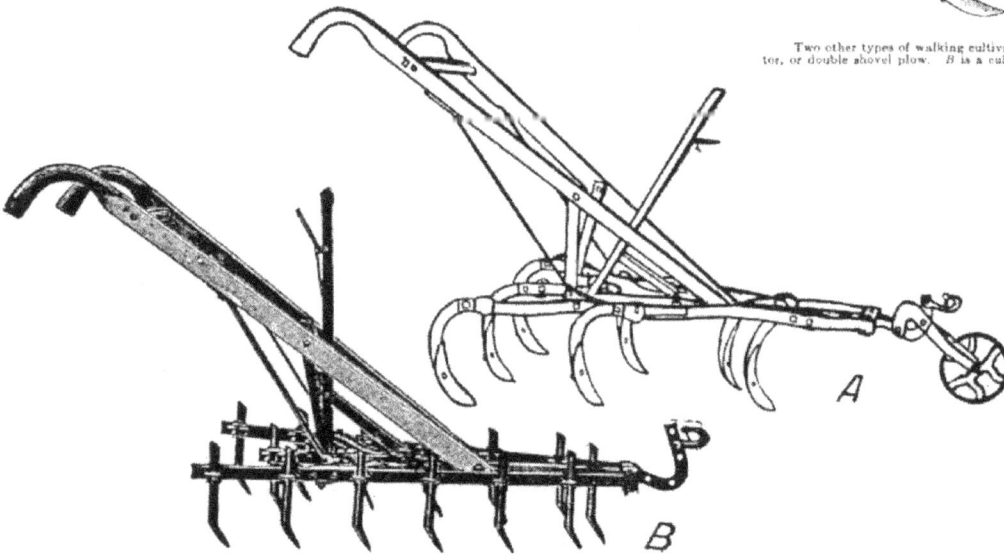

Two types of walking cultivators. *A*, seven-shovel one-horse cultivator; *B*, fourteen-tooth cultivator.

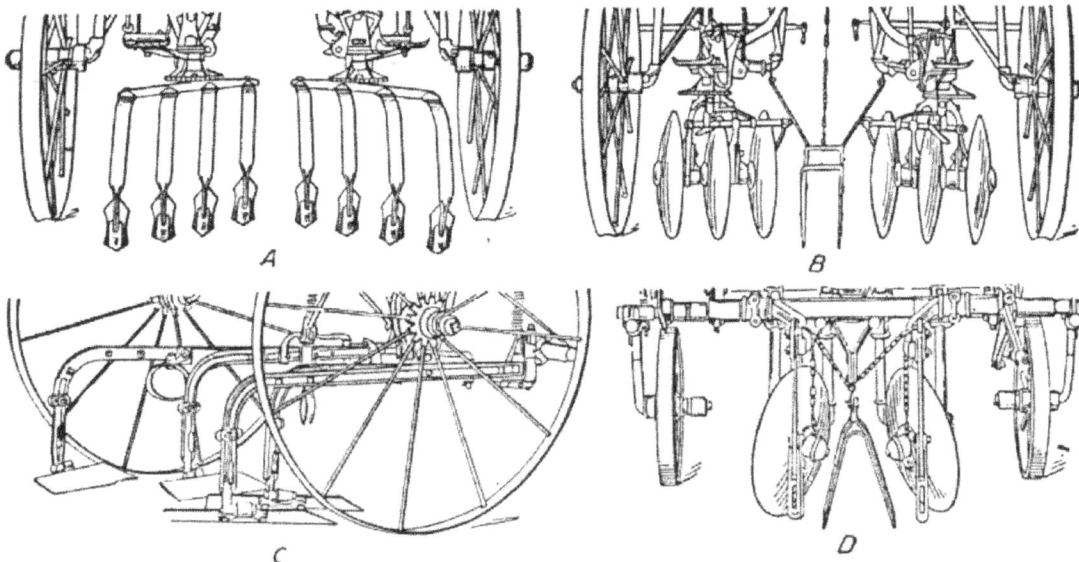

Other types of riding cultivators. *A*, four-shovel, spring-tooth gang; *B*, disk cultivator; *C*, surface cultivator; *D*, single row lister cultivator.

The B.F. Avery Plow Co of Kentucky was at the forefront of many horsedrawn implement innovations and was justifiably proud of their outstanding engineering and workmanship. No detail was too small for them to delve into forcefully and this included the perfection of their Blue Ribbon Sweeps. As they wrote:

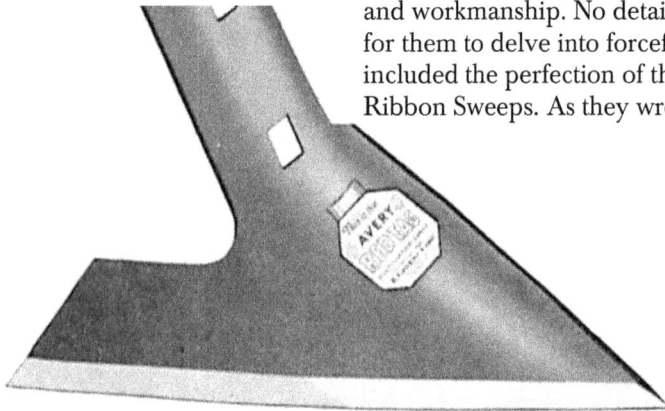

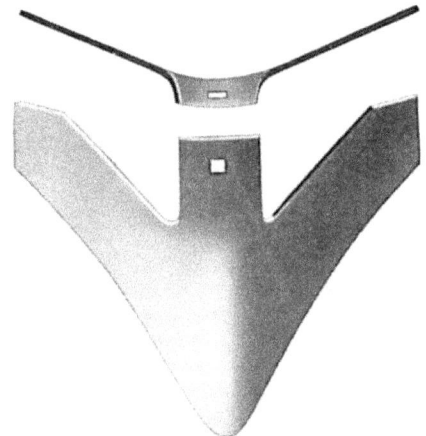

No. 4

"Our sweeps are so designed that they move smoothly and steadily through the earth, cutting and destroying the weeds and grass with their sharp edges and producing an ideal top mulch. They are made only of purest and finest steel,

therefore, taking and retaining a chisel-like edge. They scour evenly and perfectly and take on a smooth, bright land polish. They have a self-sharpening tendency, however, should extreme conditions dull their edges, they can easily be sharpened

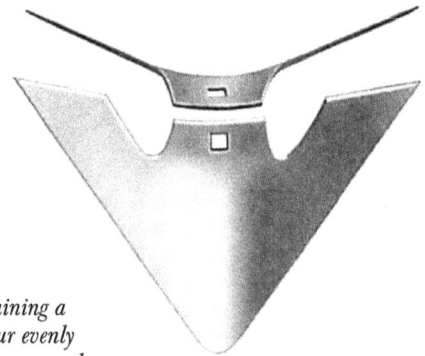

No. 5

and hardened without affecting the temper of the steel, or they can be sharpened with file or stone, without removing from the cultivator. These sweeps will fit any open foot cultivator.

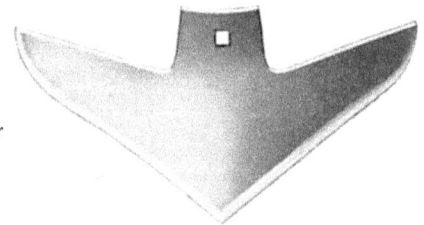

No. 51

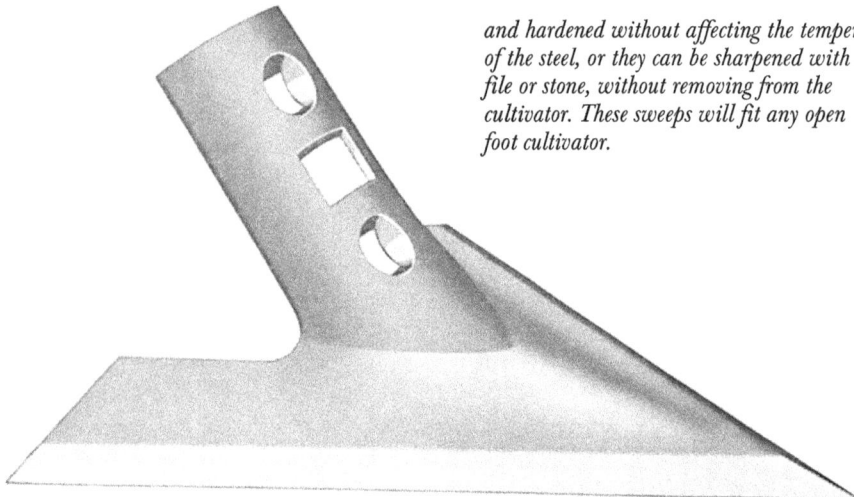

Avery
Sweeps

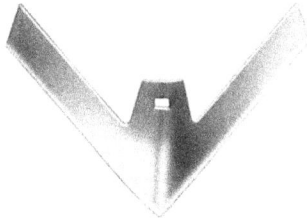

No. 235
Polished

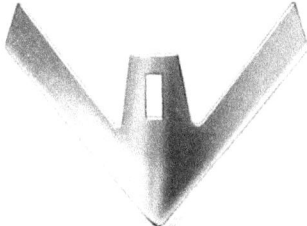

No. 236
Polished

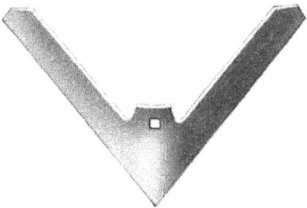

No. 212

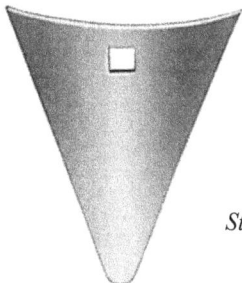

No. 9

Gang Equipment for Jack Rabbit Cultivators

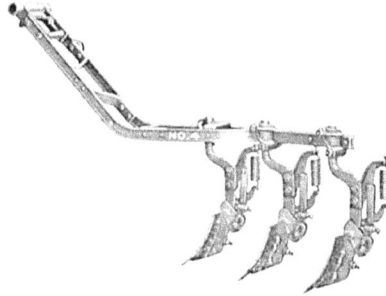

No. 4—Six-Shovel. Channel Beam, Round Standards, Spring Trip, Open Foot.

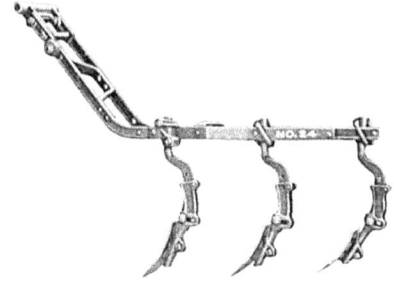

No. 24—Six-Shovel. Channel Beam Round Standards, Break Pin, Open Foot.

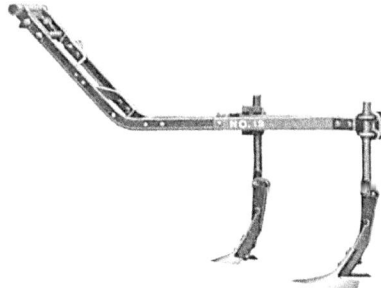

No. 12 — Four-Shovel. Channel Beam Round Standards, Break Pin, Open Foot.

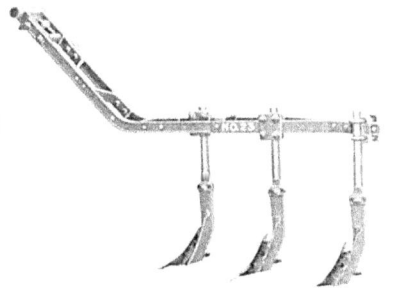

No. 23—Six-Shovel. Channel Beam Round Standards, Break Pin, Open Foot.

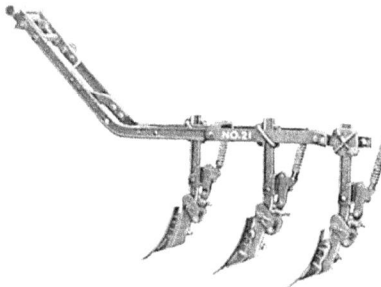

No. 21 — Six-Shovel. Channel Beam Spring Trip, Open Foot.

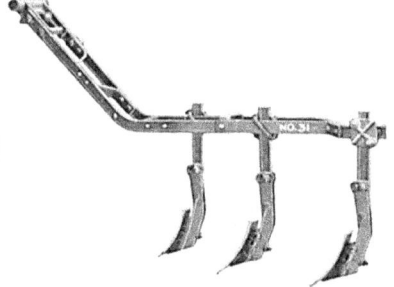

No. 31 — Six-Shovel. Channel Beam Break Pin, Open Foot.

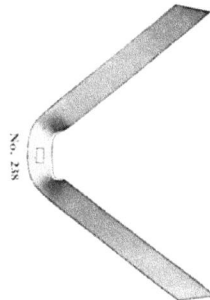

No. 218

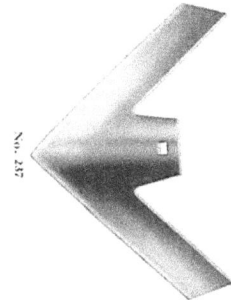

No. 237

Straight Shovels

No. 7

Avery

Shovels, Points, & Gangs

Gang Equipment for Jack Rabbit Cultivators

Gang Equipment for Jack Rabbit Cultivators

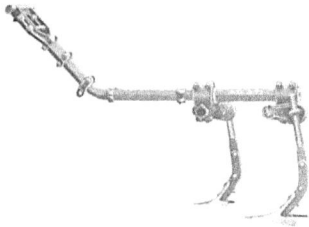

No. 371—Four-Shovel. Pipe Beam Friction Break Foot L J 81.

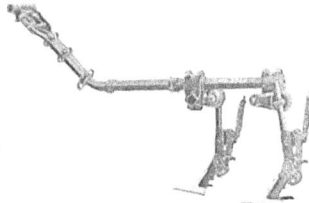

No. 371 X—Four-Shovel. Pipe Beam Friction Break, Spring Trip, Open Foot.

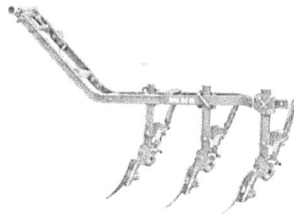

No. 21 B—Six-Shovel. Channel Beam Spring Trip, Round Foot.

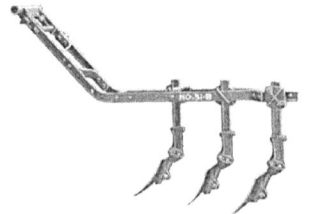

No. 31 B—Six-Shovel. Channel Beam Break Pin, Round Foot.

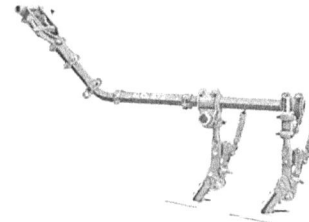

No. 372—Four-Shovel, Pipe Beam, Spring Trip, Open Foot.

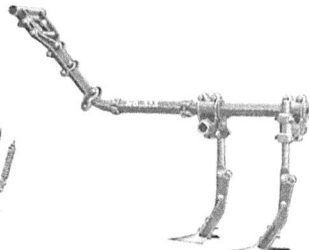

No. 373—Four-Shovel, Pipe Beam, Break Pin, Open Foot.

No. 41—Eight-Shovel. Channel Beam Round Standards, Spring Trip, Open Foot.

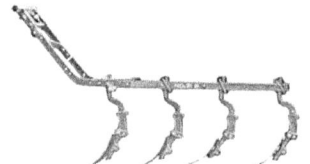

No. 51 B—Eight-Shovel. Channel Beam Round Standards, Break Pin, Round Foot.

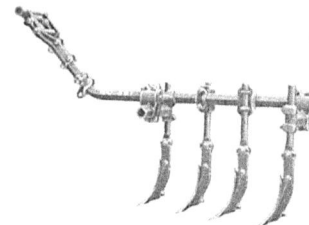

No. 381—Eight-Shovel, Pipe Beam, Break Pin, Open Foot.

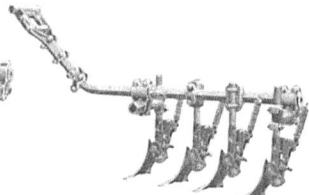

No. 382—Eight-Shovel, Pipe Beam, Spring Trip, Open Foot.

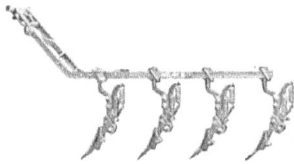

No. 41 B—Eight-Shovel. Channel Beam Round Standards, Spring Trip, Round Foot.

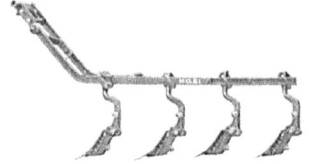

No. 51—Eight-Shovel. Channel Beam Round Standards, Break Pin, Open Foot.

Avery

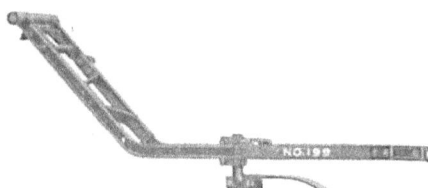

No. 199

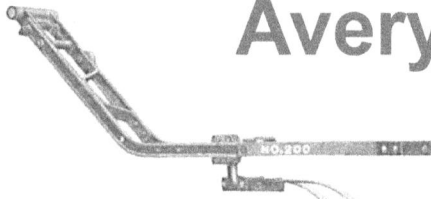

No. 200

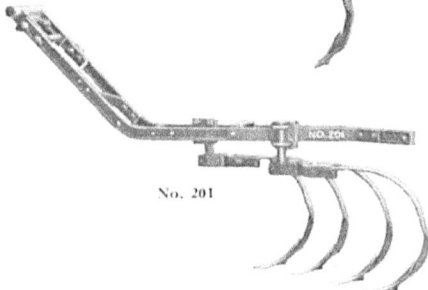

No. 201

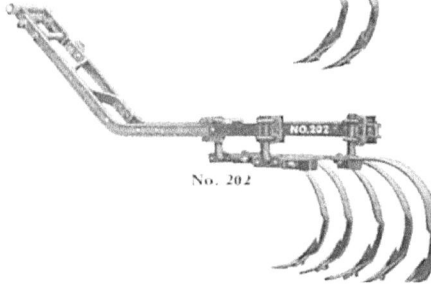

No. 202

Round shank attachments

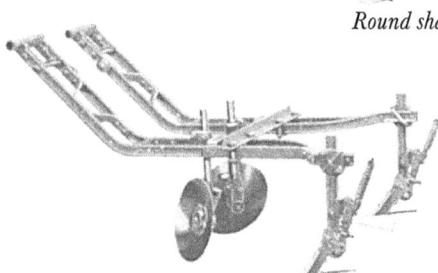

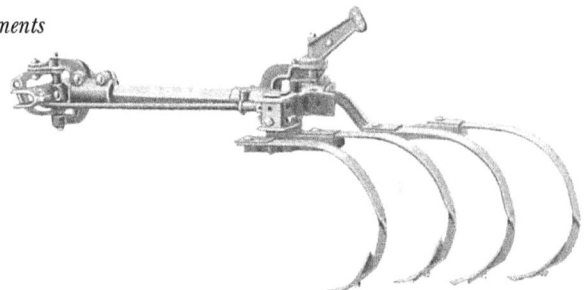

No. 99

No. 100

No. 101

No. 102

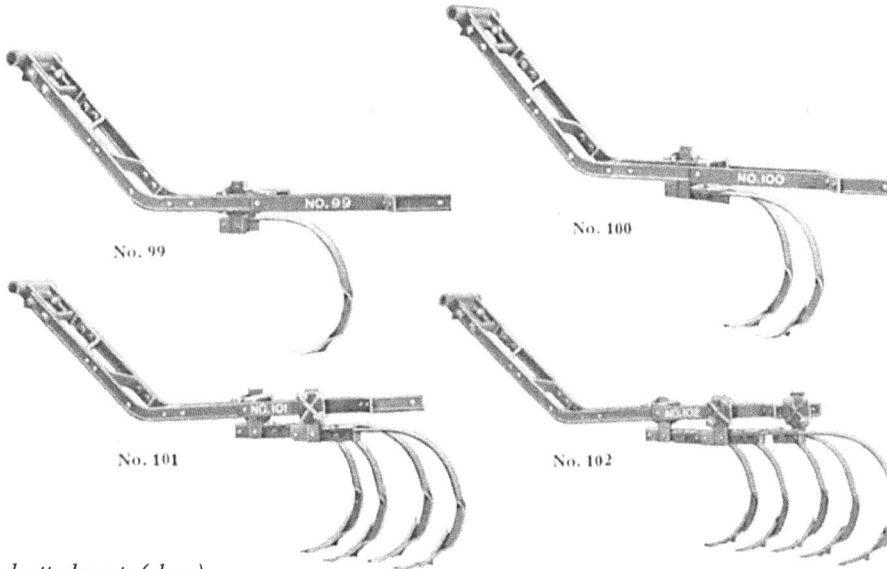

Flat shank attachments (above).

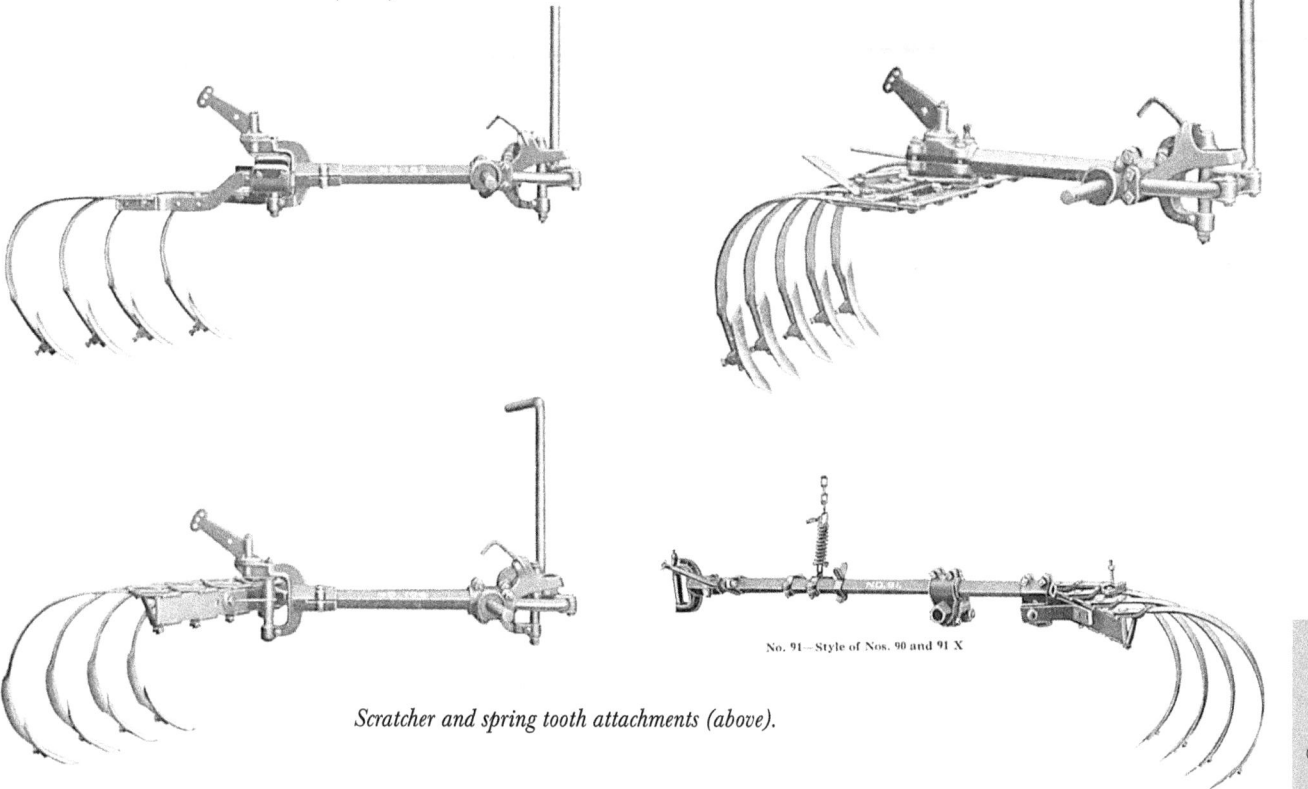

No. 91—Style of Nos. 90 and 91 X

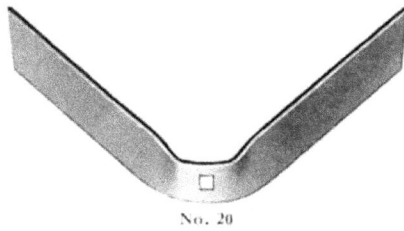

Scratcher and spring tooth attachments (above).

No. 20

Avery

Heel Sweep

Turning Shovels

Bull Tongues

No. 194

No. 14

No. 244

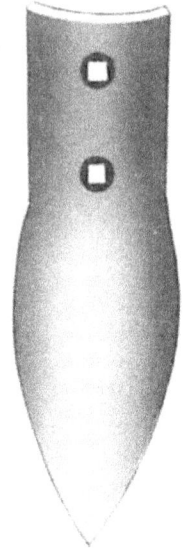

No. 219
Polished

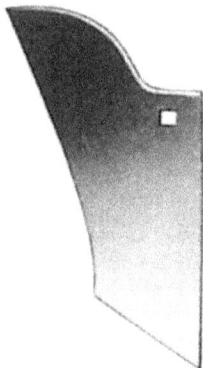

No. 246

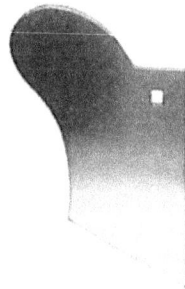

No. 147

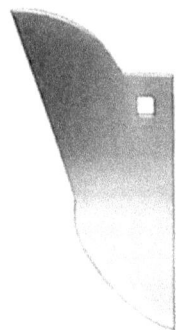

No. 213

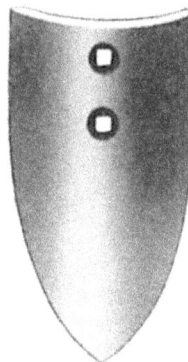

No. 120
Polished

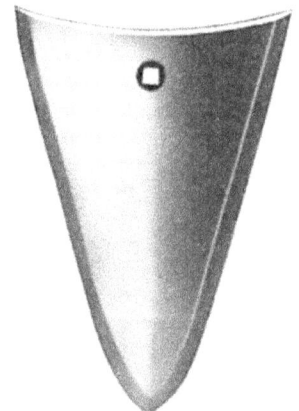

No. 133
Polished

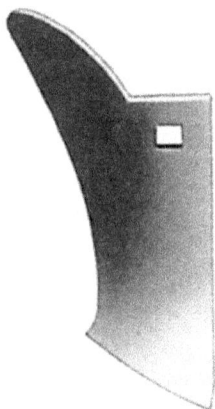

No. 16

No. 243

No. 41
Polished

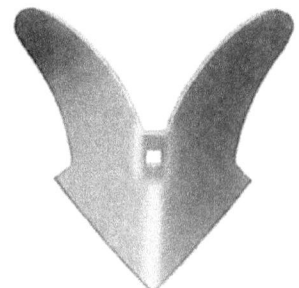

No. 134
Polished

Avery

314

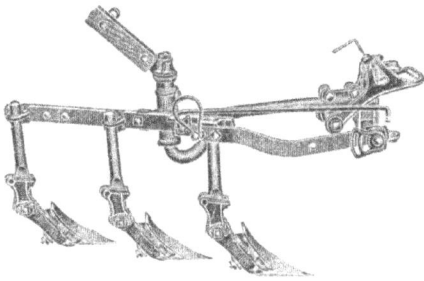

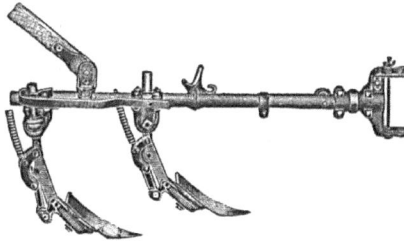

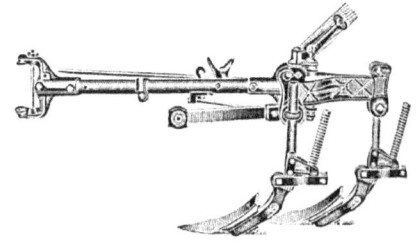

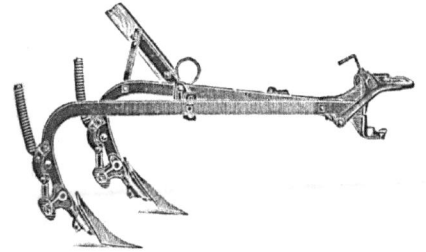

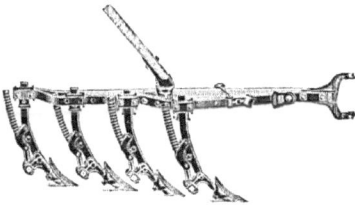

John Deere

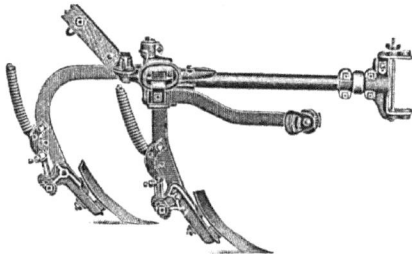

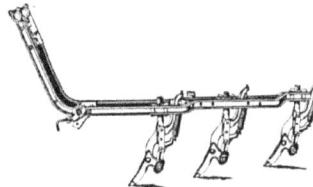

KL-198 — 6-shovel, spring-trip, round shank.

KL-196 — Same but 4-shovel.

KL-200 — 8-shovel, spring-trip, round shank.

KL-199—Same but with pin-break.

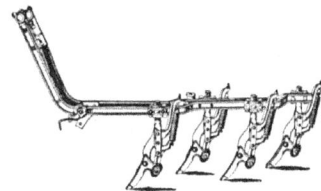

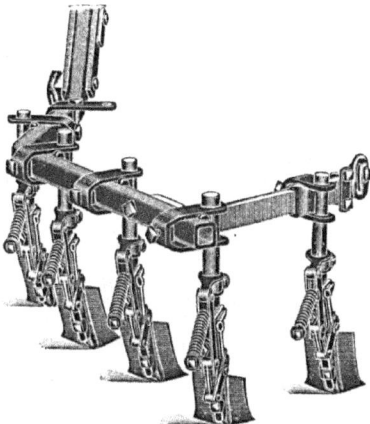

KL-201—8-shovel, spring-tooth, U-bolt clamp.

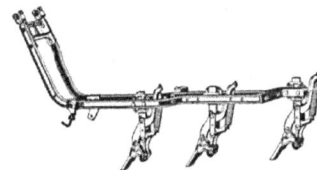

KL-206 — 6-shovel, spring-trip, round sleeve.

KL-216—Same as 206, but with 4 shovels.

KL-205 — Same but with pin-break.

KL-215—Same as 205, but with 4 shovels.

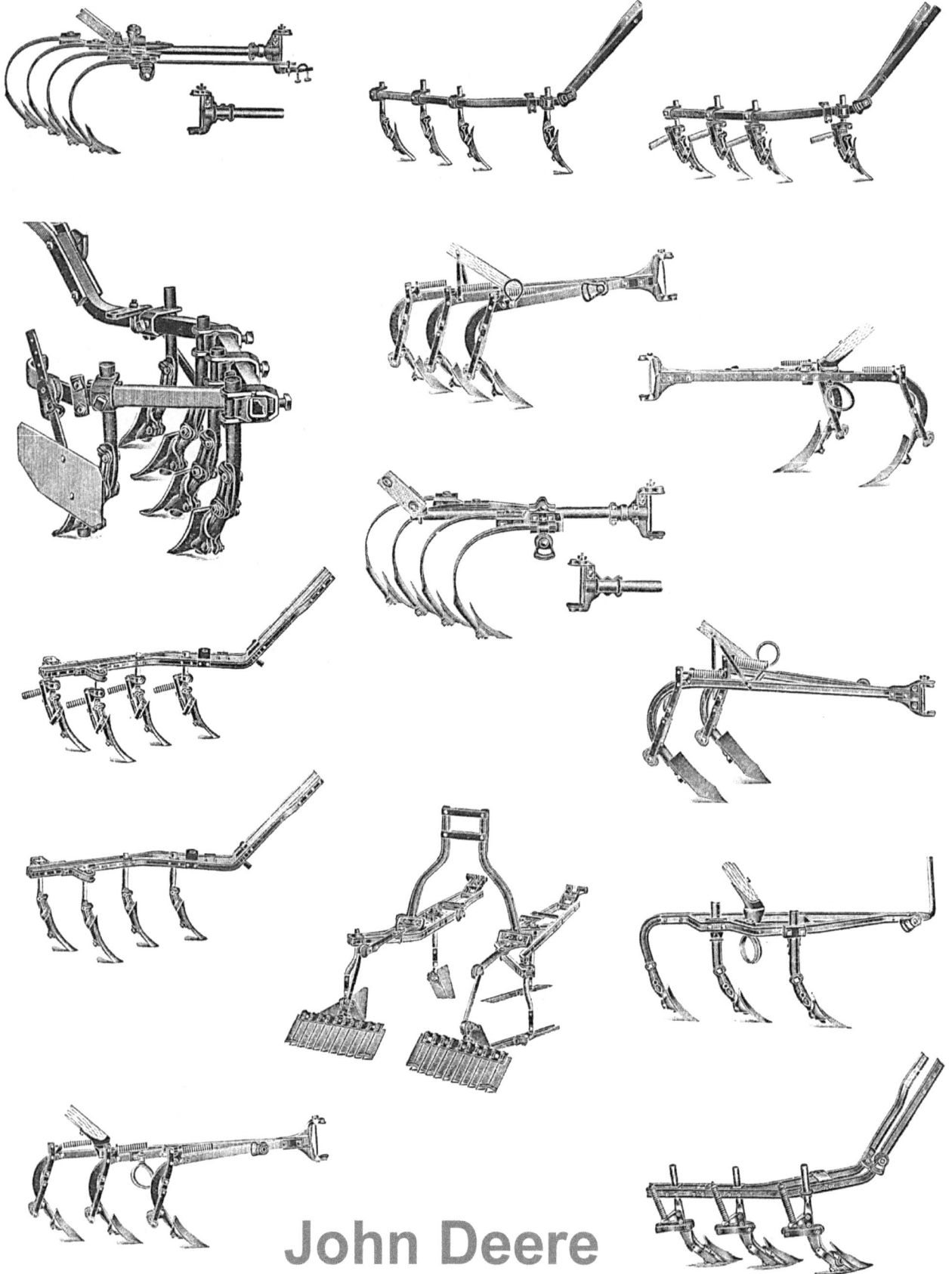

John Deere

John Deere

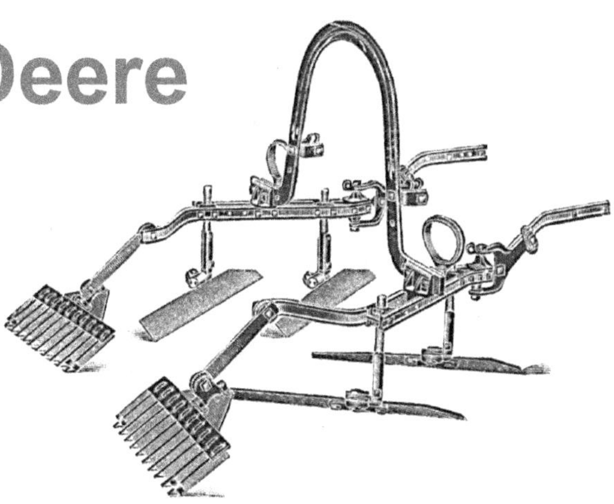

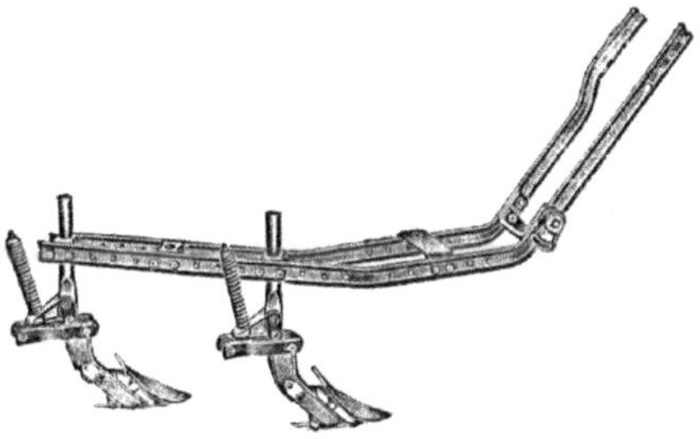

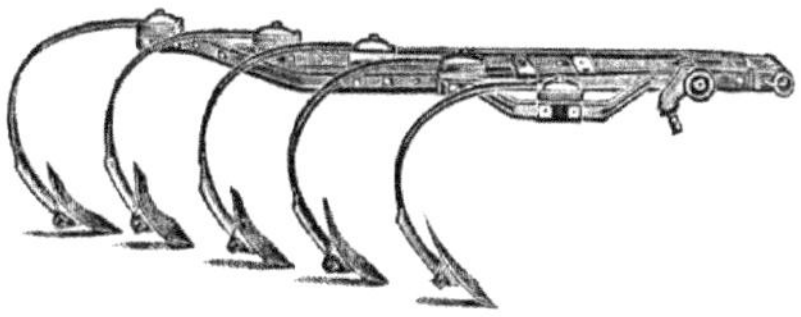

Six-shovel attachment illustrated at right, can be furnished to convert the DF 202 or 203 to a six-shovel cultivator.

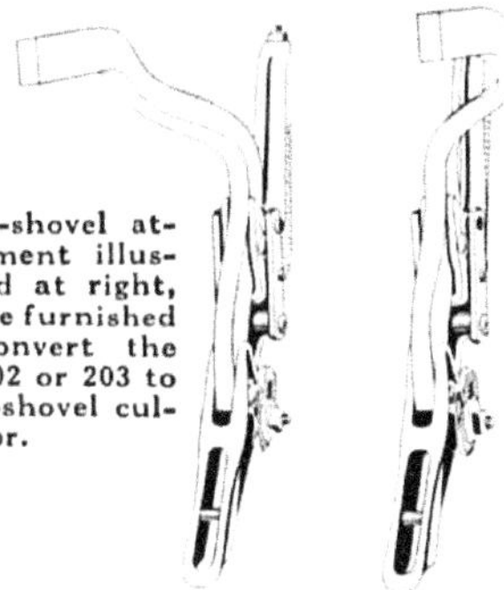

John Deere Spring Tooth Attachments

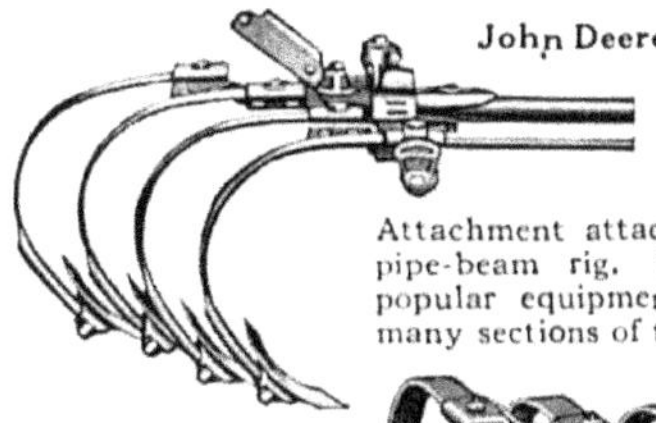

Showing John Deere No. 96 Spring - Tooth Attachment attached to DF-202 pipe-beam rig. This has been popular equipment for years in many sections of the South.

Showing front view of John Deere No. 96 Spring-Tooth Attachment. Be sure to see this attachment next time you are in your John Deere dealer's store.

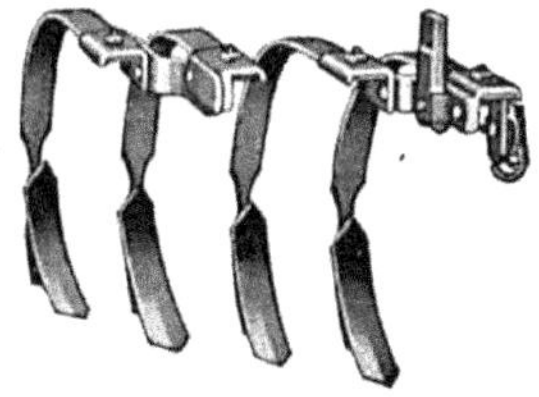

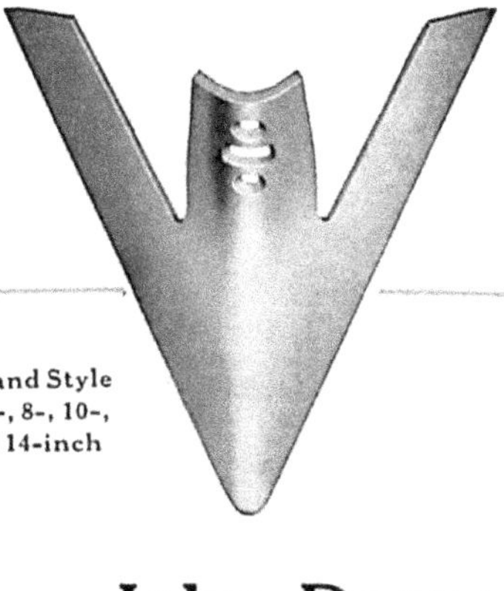

Black-Land Style
Sizes—6-, 8-, 10-, 12-, and 14-inch

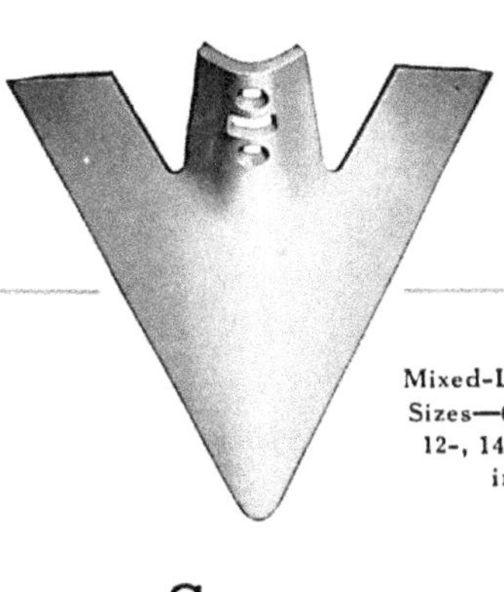

Mixed-Land Style
Sizes—6-, 8-, 10-, 12-, 14- and 16-inch

John Deere Southern Sweeps

No. 1—Steel Beam, Spring Trip, 4-Shovel
Used on AA1, E1, F1, R1, RP1, RA1, T1, TA1, WA1, XA1, YA1 Cultivators

No. 2—Steel Beam, Spring Trip, 6-Shovel
Used on AA2, D2, E2, F2, R2, RP2, RA2, T2, TA2, WA2, XA2 YA2 Cultivators

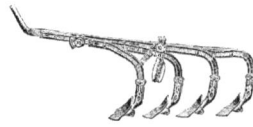

Steel Beam, Eagle Claw, 8-Shovel, Four Shovels Aligned

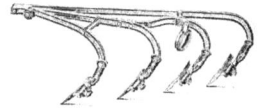

Steel Beam, 8-Shovel

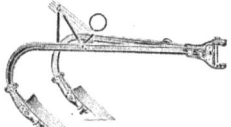

No. 3—Steel Beam, 4-Shovel
Used on AA3, D3, E3, F3, R3, RA3, T3, TA3, WA3, XA3 YA3 Cultivators

No. 4—Wood Beam, 4-Shovel
Used on AA4, E4, F4, WA4, XA4, YA4, Cultivators

Steel Frame, Spring Trip, 4-Shovel

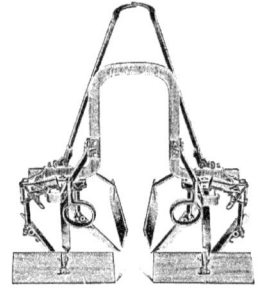

Surface Attachment

John Deere

No. 5—Wood Beam, Milton, 4-Shovel
Used on 5, A5, E5, EA5, F5, M5, R5, RA5, U5 Cultivators

No. 6—Wood Beam, Milton, 6-Shovel
Used on A6, E6, EA6, M6, R6, RA6, U6 Cultivators

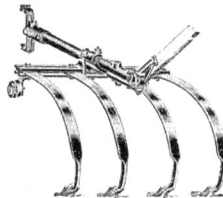

Attachment for all Pipe Beam Rig Cultivators Except No. 21 Rig

Pipe Beam, Parallel, with 2 V Spring Trip Shanks and Middle Attachment

No. 7—Pipe Beam, Parallel, 4-Shovel
Used on AA7, DB7, E7, HA7, HB7, HD7, HE7, HF7, HG7, S7, SP7, SA7, WA7, XA7, YA7 Cultivators

No. 8—Pipe Beam, Parallel, Spring Trip, 4-Shovel
Used on AA8, DB8, E8, HA8, HB8, S8, SP8, SA8, WA8, XA8 YA8 Cultivators

Steel Beam, Spring Trip, 4-Shovel
Used on AB, EB, HF, HG, RA, RC, RD, TA, WA, XA and YA Cultivators

Steel Beam, Spring Trip, 6-Shovel
Used on AB, EB, D, DB, HF, HG, RA, RC, RD, TA, WA XA and YA Cultivators

Steel Frame, Spring Trip, 8-Shovel

Pipe Beam Friction Break, 4-Shovel

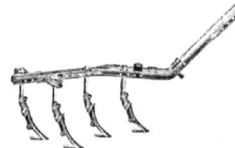

8-Shovel Rig, Breakpin
Used on KA Cultivator only
Furnished with Open or Round Sleeves as ordered

Steel Beam, Universal, Spring Trip, 4-Shovel
Used on AB, HF, HG, J, S, SA, TA, WA, XA and YA Cultivators

Steel Beam, Spring Trip, 4-Shovel

Steel Frame, Spring Trip, 4-Shovel

8-Shovel Rig, Spring Trip
Used on KA Cultivator only
Furnished with Open or Round Sleeves as ordered

Steel Beam, Universal, Spring Trip, 6-Shovels
HF, HG, J, RC, S, SA and FA Cultivators

Steel Beam, Spring Trip, 6-Shovel

Steel Frame, Spring Trip, 6-Shovel

Steel Frame, Spring Trip, 4-Shovel

Pipe Beam Oblique Rig, Spring

Surface Attachment, Breakpin
Used on NA Cultivator only

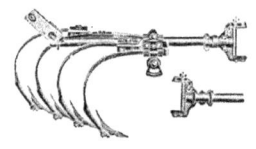

Pipe Beam, Adj. Head, 8-Shovel, Spring Tooth
Used on AB, D, DB, HF, HG, J, S and SA Cultivators

318

No. 70—Steel Frame, V Spring Trip 4-Shovel
Rig with Long Shanks for Listed Corn
Used on R, RA, RC, RD, S, SA Cultivators

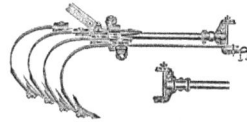

No. 71—Steel Frame, V Spring Trip 6-Shovel
with Long Shanks for Listed Corn
Used on R, RA, RC, RD, S, SA, Cultivators

Pipe Beam Parallel, 8-Shovel Spring Tooth

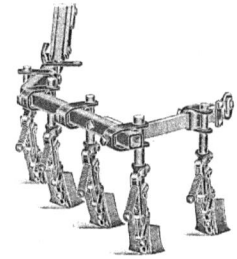

Pipe Beam, Spring Trip, 10-Shovel

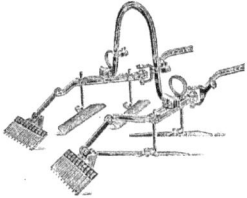

No. 72—Surface Attachment
Used on NN, R, RA, RC, Cultivators

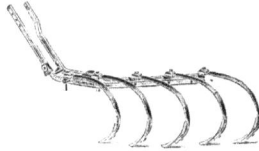

No. 73—10-Shovel Rig, Spring Tooth
Used on KA Cultivator Only
Furnished with Flat or Round Shanks as Ordered

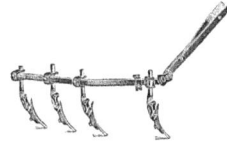

Pipe Beam Break Pin, 8-Shovel

Pipe Beam, Parallel Spring Trip, 4-Shovel

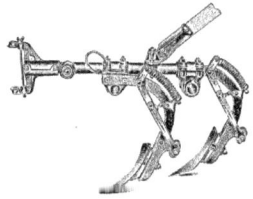

No. 74—Pipe Beam, Spring Trip, 4-Shovel Rig
Used on EB Cultivators

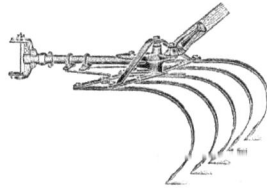

No. 75—Spring Tooth, Diverse, 10-Shovel Rig
Used on AA, SA, WA, XA, YA, D Cultivators

Pipe Beam Spring Trip, 8-Shovel

Steel Beam Spring Trip,

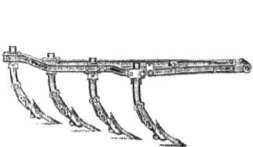

No. 76—Steel Frame, 8-Shovel Rig
Used on J, R, RA, RC, RD, S, SA, NA Cultivator

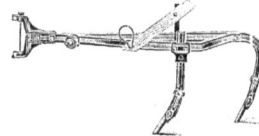

No. 77—Steel Frame, Adjustable Shank 4-Shovel Rig
Used on AA, WA, XA, YA, V, VA Cultivators

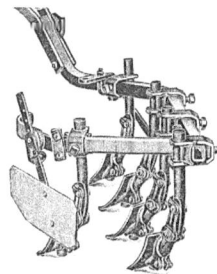

Pipe Beam Break Pin, 10-Shovel—Two-Row

Pipe Beam, Spring Trip,

Steel Beam, Spring Tooth, 10-Shovel

Steel Frame, 4-Shovel

Pipe Beam, Texas, 4-Shovel Rig

Pipe Beam, Texas, Spring Trip, 4-Shovel Rig

John Deere

Steel Frame, 6-Shovel

Steel Frame, 6-Shovel

Pipe Beam, Texas, 4-Shovel Rig with Friction Clutch

Pipe Beam, Texas, Spring Trip, 4-Shovel

Wood Beam, Milton, 4-Shovel

Wood Beam, Milton, 6-Shovel

Steel Frame, Spring Trip, 6-Shovel

Steel Beam, Universal, 6-Shovel

Steel Frame, 8-Shovel Rig

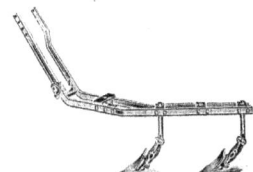

Four-Shovel Rig Break Pin

Shovels, Points, & Gangs

319

Rig Styles for John Deere Walking Cultivators

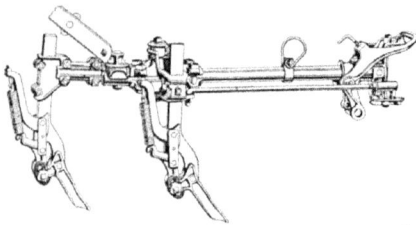

No. 203—4 Shovel, Pipe Beam, Spring Trip

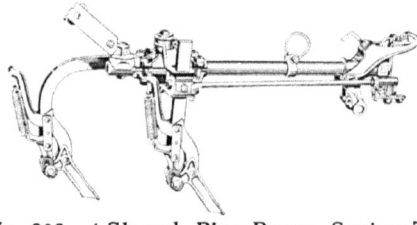

No. 202—4 Shovel, Pipe Beam, Spring Trip

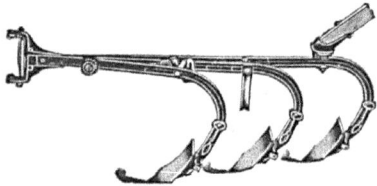

No. 38—Steel Beam, Universal, 6-Shovel

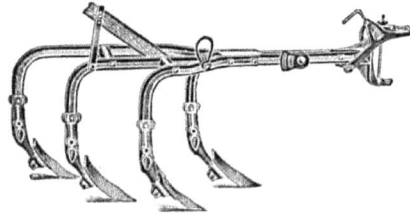

No. 14—Steel Beam, Eagle Claw, 8-Shovel, Three Shovels Aligned

No. 158—Solid Steel Beam, 6-Shovel Spring Trip

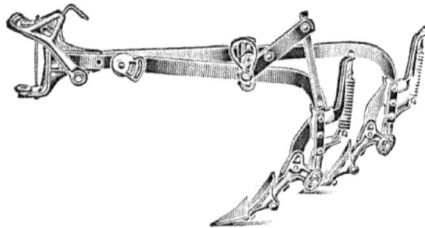

No. 157—Solid Steel Beam, 4-Shovel Spring Trip

No. 193 Adjustable Spring-Tooth Attachment Shown on No. 202 Rig

No. 162—Pipe Beam, Parallel, 6-Shovel Spring Trip. Peanut Rig No. 132 with Pin Break

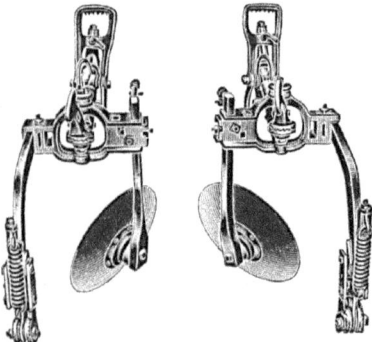

No. 202 DF Rig, with No. 1-A 12-inch Disk Hillers Attached

No. 194 Adjustable Spring-Tooth Attachment Shown on No. 202 Rig

No. 16—Spring Tooth, Wood Cross Bar, 8-Shovel

No. 17—Steel Beam, 6-Shovel

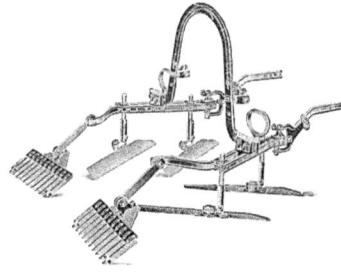

—Surface Attachment
Used on R, RA, RC and LB Cultivators

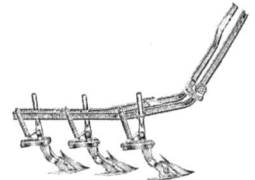

—6-Shovel, V Spring Trip
Used on KA Cultivator only

No. 19—Pipe Beam, Texas, 4-Shovel

No. 20—Pipe Beam, "X" Spring Trip,

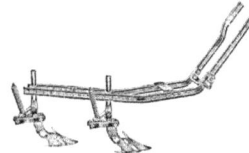

—4-Shovel, V Spring Trip
(Used on KA Cultivator only

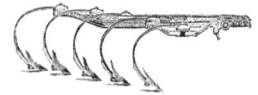

—Steel Frame, Spring-Tooth, 10-Shovel Rig
Used on NA Cultivator

No. 21—Pipe Beam, Oblique, 4-Shovel

No. 22—Steel Beam, Universal Spring 6-Shovel

Steel Frame, Adjustable Shanks, 6-Shovel Rig
Used on JV, VA Cultivators

Pipe Beam, Spring Trip, 4-Shovel Rig
Used on AA, D, WA, XA, YA Cultivators

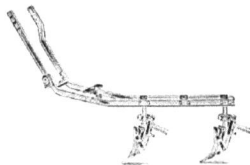

—Four-Shovel Rig, Spring Trip
Used on KA Cultivator only
Furnished with Open or Round Sleeve as Ordered

—Six-Shovel Rig, Break Pin
Used on KA Cultivator only
Furnished with Open or Round Sleeve as Ordered

—Pipe Beam, Parallel, Spring Trip, 4-Shovel Rig
Used on AB, D, HF, S, WA, XA, YA Cultivators

—Spring Tooth Attachment
For all Pipe Beam Rig Cultivators, Except No. 21 Rig

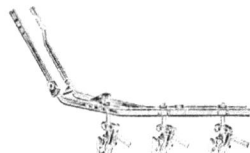

—Six-Shovel Rig, Spring Trip
Used on KA Cultivator only
Furnished with Open or Round Sleeve as Ordered

Eight-Shovel Rig, Break Pin
Used on KA Cultivator only
Furnished with Open or Round Sleeve as Ordered

9th, Shovel Attach. for AA, WA, XA, YA16.

John Deere

9th, Shovel Att. for S, SA, T, TA 14.

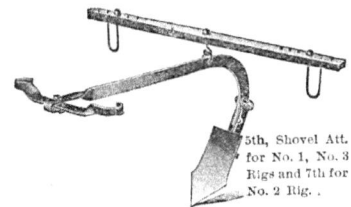

5th, Shovel Att. for No. 1, No. 3 Rigs and 7th for No. 2 Rig.

Eight-Shovel Rig, Spring Trip
Used on KA Cultivator only
Furnished with Open or Round Sleeve as Ordered

Eight-Shovel Rig, Spring Tooth
Used on KA Cultivator only
Furnished with Flat or Round Shanks as Ordered

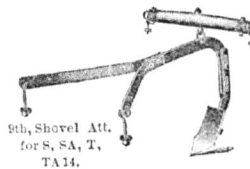

5th, Shovel Attach. for S 11, SA 11, T 3, TA 3.

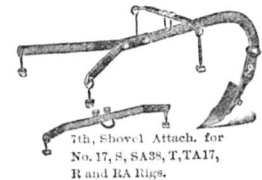

7th, Shovel Attach. for No. 17, S, SA 3s, T, TA 17, R and RA Rigs.

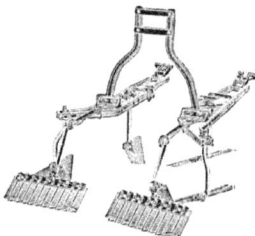

Surface Attachment
Used on NA Cultivator only

Pipe Beam, Oblique, Four-Shovel Rig
Used on DB, HA, HB, S and SA Cultivators

Center Att. for S, SA, T, TA 14 Rig.

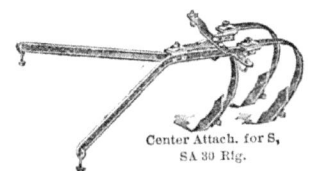

Center Attach. for S, SA 30 Rig.

Shovels, Points, & Gangs

Two-Horse Riding Cultivator Shovels
(with Bolts)

ONE-PIECE BACK
(with Bolts)

| Round Straight | Round R. H. Twisted | Round L. H. Twisted | Slotted Straight | Slotted R. H. Twisted | Slotted L. H. Twisted |

THREE-PIECE BACK

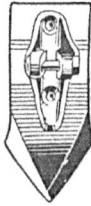  

| Round Straight | Round R. H. Twisted | Round L. H. Twisted |

SPEARHEAD SHOVELS

| One-Piece Back Round | One-Piece Back Slotted | Three-Piece Back Round |

TWO-PIECE SHOVELS AND SWEEPS

  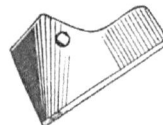 

| Shovel Point | Shovel | Sweep Right | Sweep Left | Full Sweep |

McCormick

Shovels and Sweeps

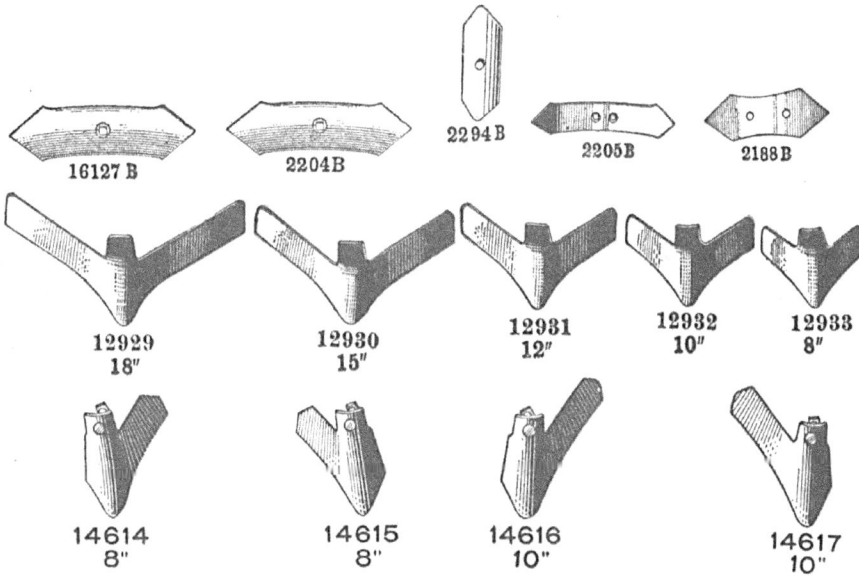

16127 B   2204 B   2294 B   2205 B   2188 B

12929
18"

12930
15"

12931
12"

12932
10"

12933
8"

14614
8"

14615
8"

14616
10"

14617
10"

FURROWER ATTACHMENT

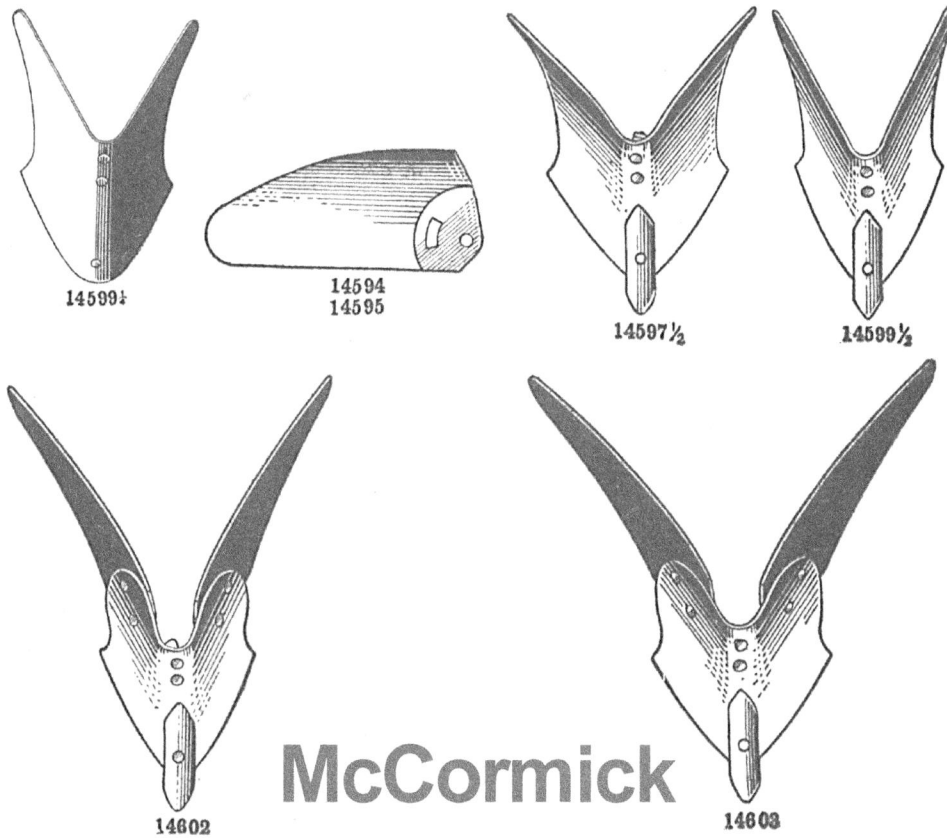

14599½

14594
14595

14597½

14599½

14602

14603

McCormick

McCormick

Two-Horse Riding Cultivator Shovels (Continued)

MOULDBOARD SHOVELS

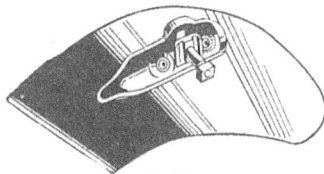 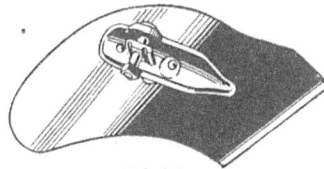

Left   Right

COTTON SCRAPERS

Left   Right

TURNING SHOVELS

 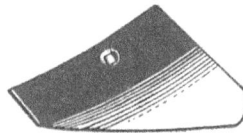

Right   Left

SWEEPS

  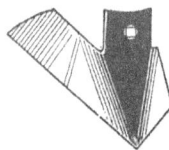 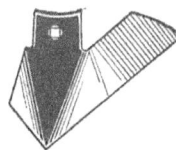 

Round   Slotted   Right Without Back   Left Without Back   Right With Back

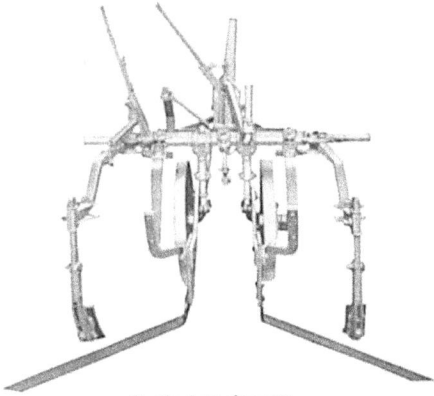

Knife Attachment

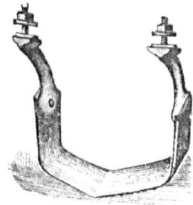

Onion harvesting
attachment

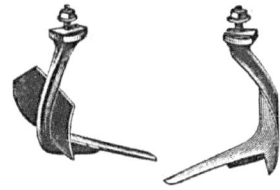

Peat knives

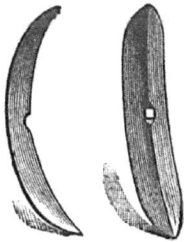

Alfalfa Steel.
Planet Jr.

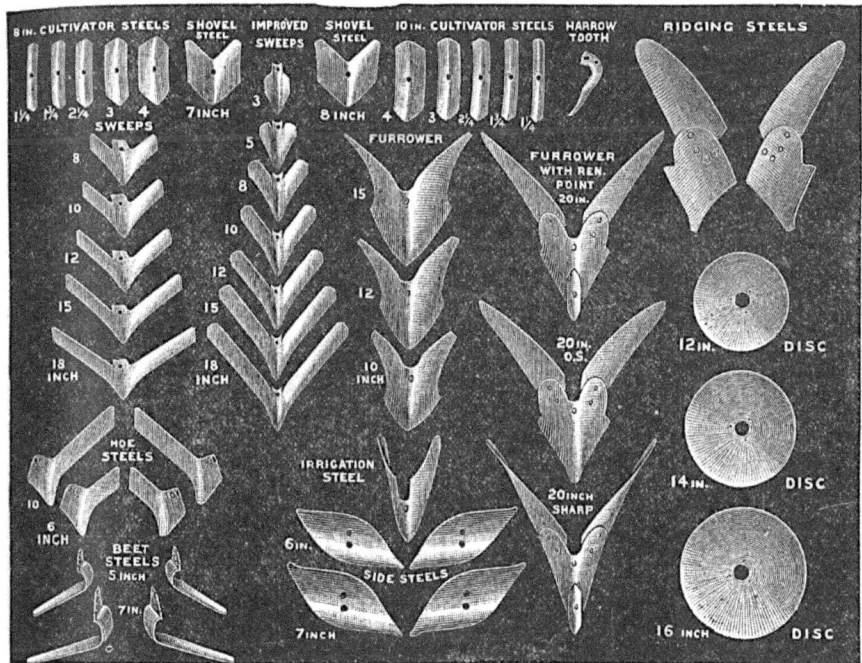

Planet Jr. Shovels

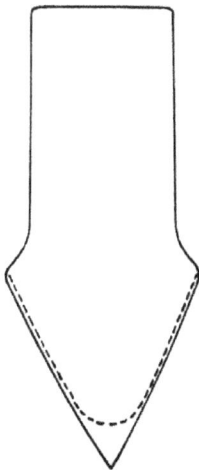

shovel with dotted lines
showing how point
looks when shovel
needs resharpening.
Obviously, a shovel in
this condition pene-
trates poorly and in-
creases draft.

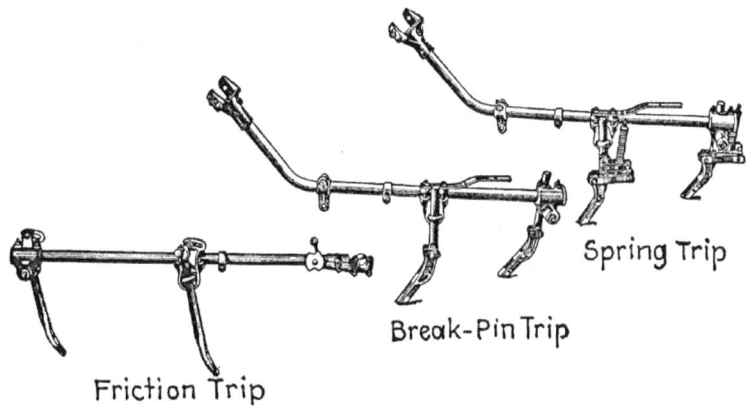

Friction Trip

Break-Pin Trip

Spring Trip

Types of cultivator trips.

1¾ in.   2½ in.   2½ in.   3½ in.   3½ in. R   3½ in. L   3½ in.

5 in.   5 in. R   5 in. L.   7 in.   6 in.

S.T. PT.   1¾ in.   2½ in.   3½ in.   4 in.

E 112 A   C 141   C 128   4143 HILLER

C 140   C 354 X   CN 60   CN 108 X   CN 109   CN 105   C 180

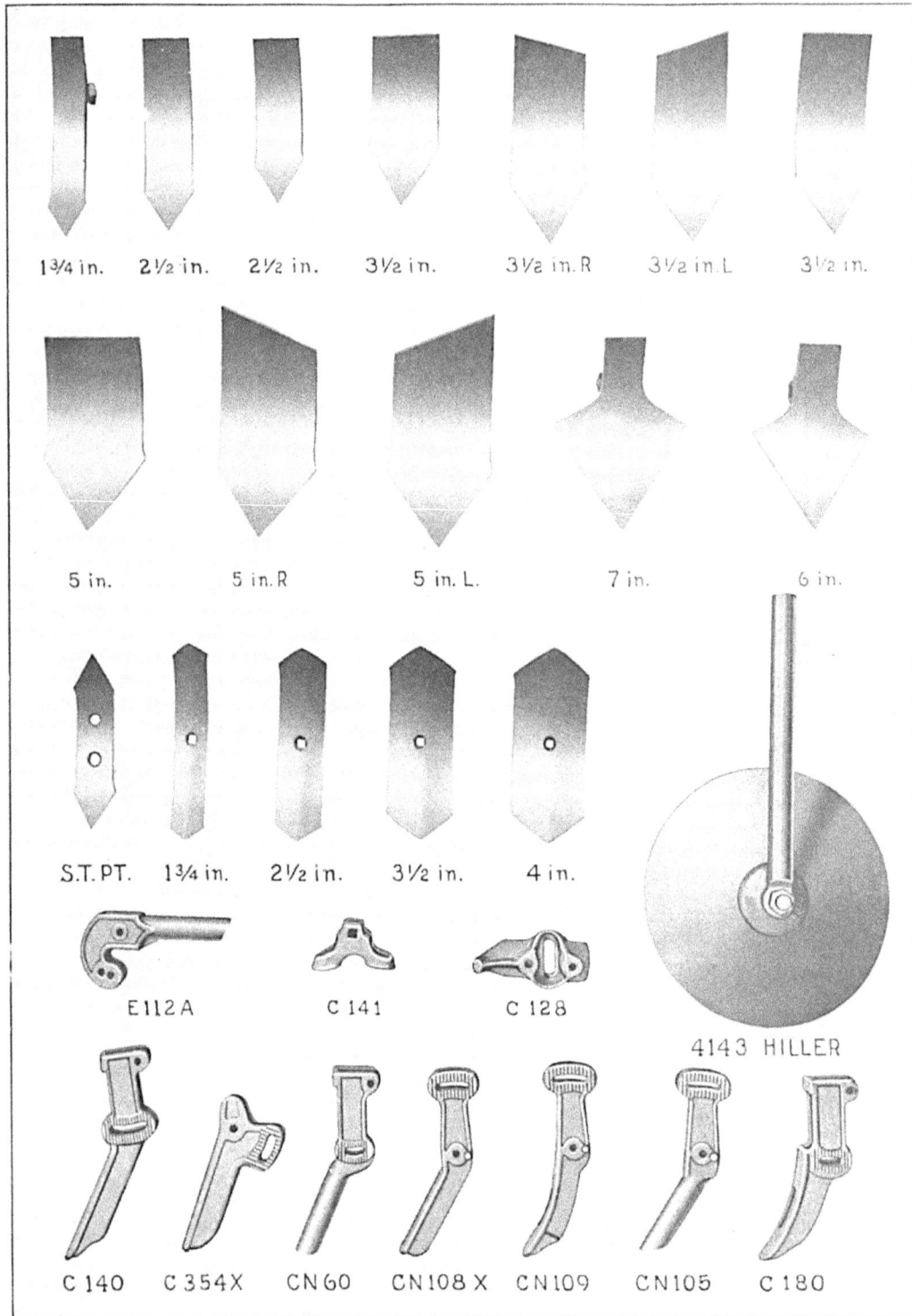

Roderick Lean
Shovels and Shanks

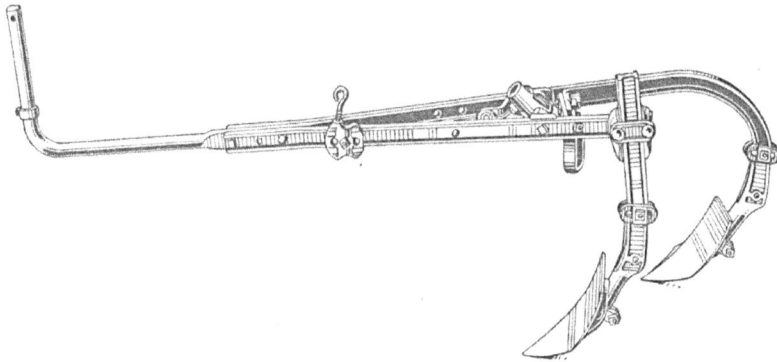

No. 5 Gang, 2-shovel, pin break—for "ZD" Corn Cultivator

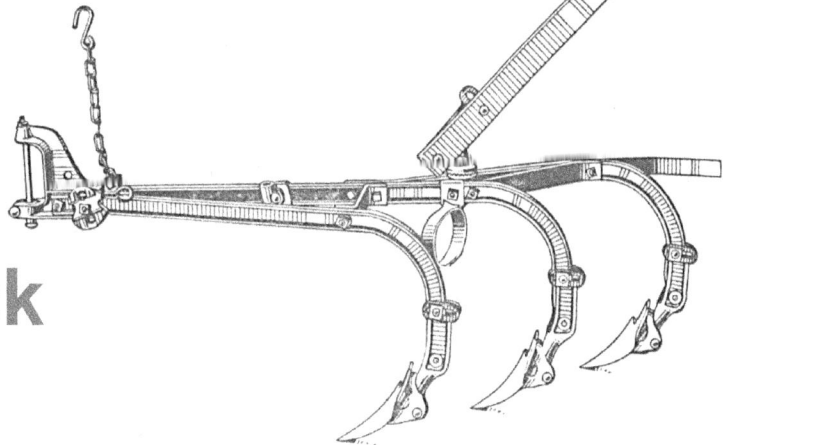

McCormick

No. 21 Gang, 3-shovel, pin break—for "SC" Cultivator

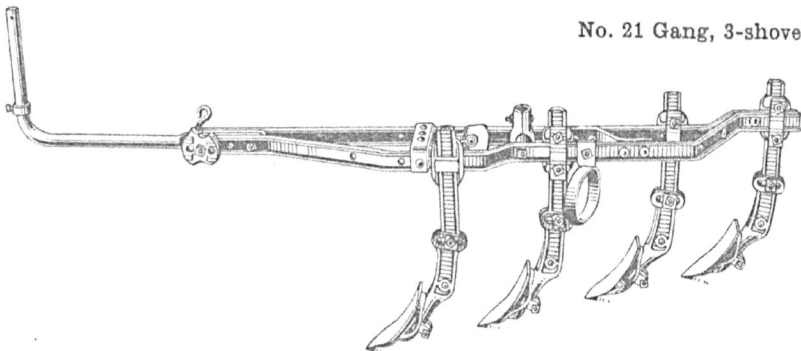

No. 41 Gang, 4-shovel, pin break—for "ZD" Cultivator

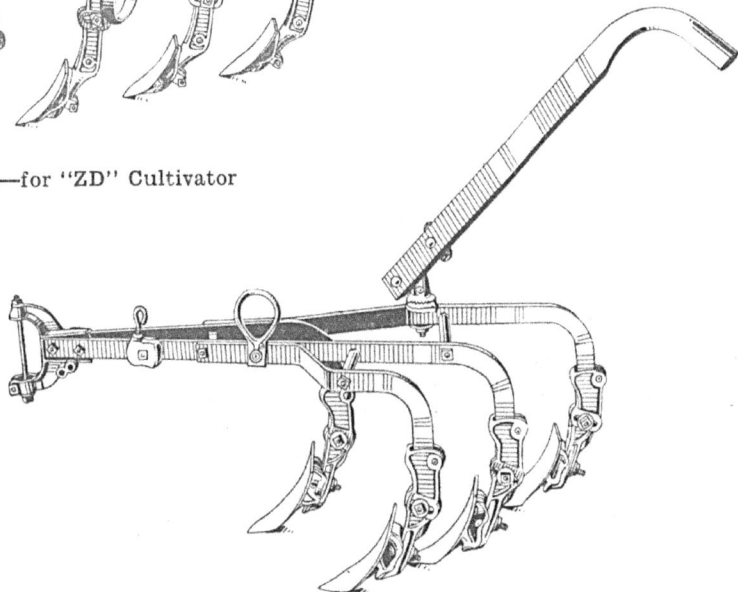

No. 241 Gang, 4-shovel, pin break—for "A" and "K" Cultivators

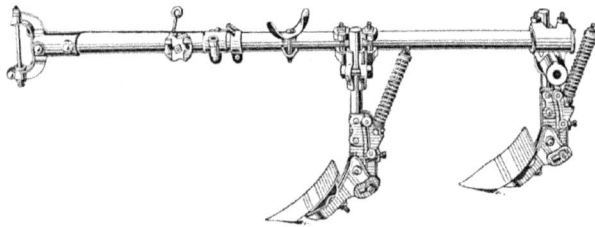

No. 220 Gang, 2-shovel, spring trip—for "TA" Cultivator

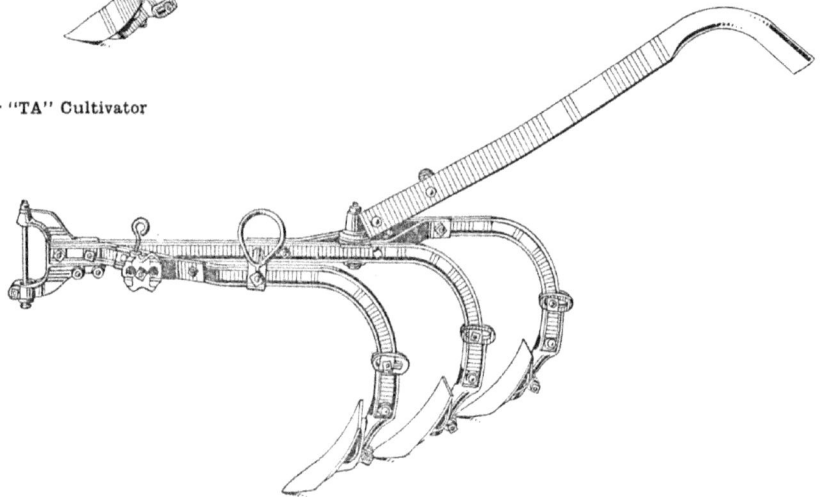

No. 21 Gang, 3-shovel, pin break—for "A" and "K" Cultivators

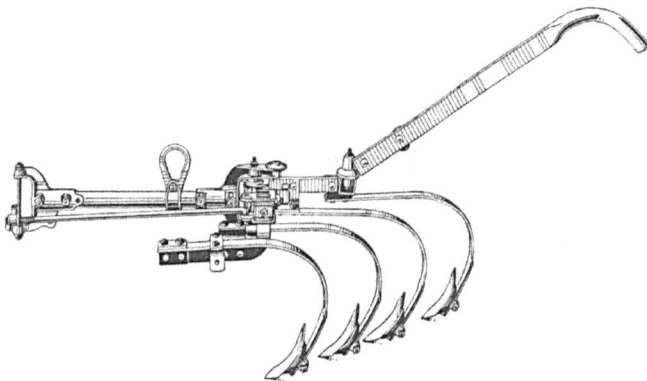

McCormick

No. 79 Gang, 4-shovel, spring tooth—for "A" and "K" Cultivators

No. 230 Gang, 2-shovel, spring trip—for "A" and]"K" Cultivators

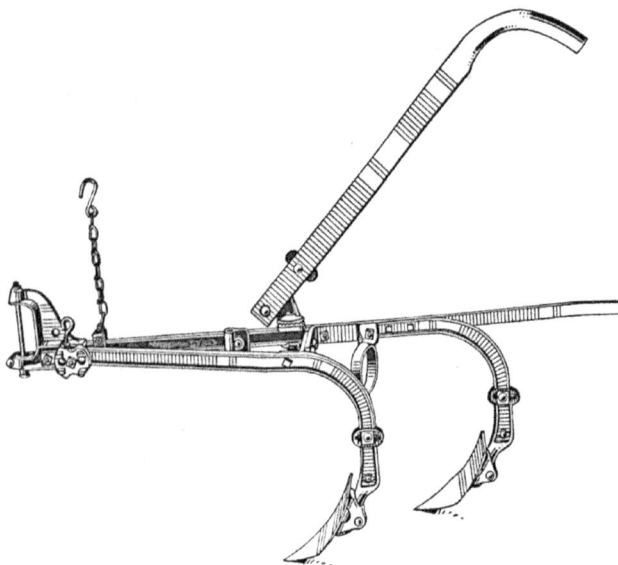

No 1 Gang, 2-shovel, pin break—for "SC" Cultivator

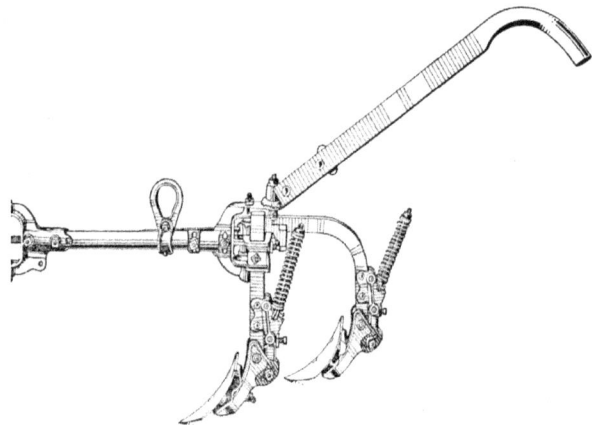

Rather than have you get bored with all these technical drawings I thought to throw in this picture from France in 1917 showing three women pulling a single section of spike tooth harrow. A hard pull. To equal the working ease of a single draft horse you'd have to have 6 women on this (or 8 men.)

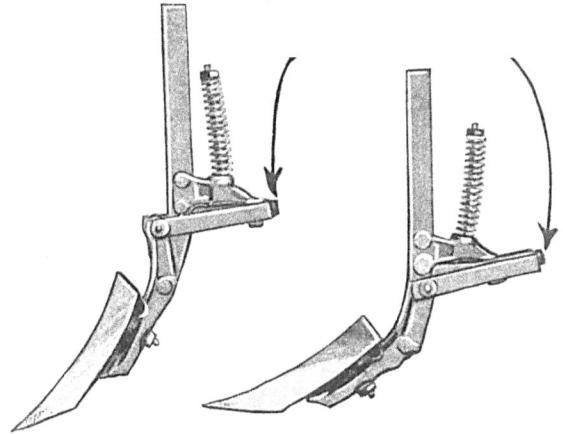

With Moline cultivators you can change the pitch of the shovel by turning the bolt for vertical or horizontal position

Moline

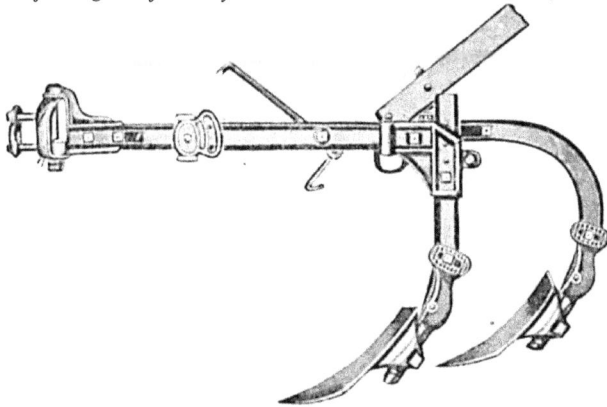

BF-4, 4-Shovel, Steel Beam, Adjustable Inside
Shank, Round Sleeve

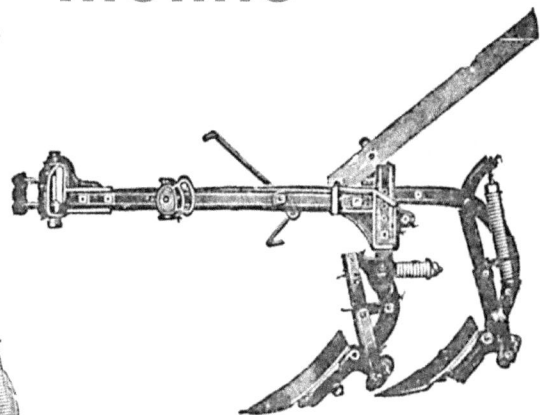

BF-8, 4-Shovel, Steel Beam, Spring Trip, Adjustable Shank, Open Sleeve

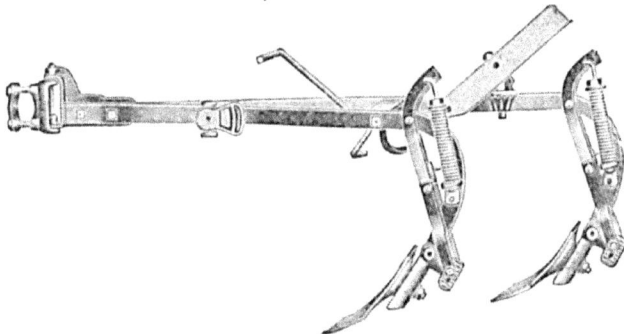

BF-6, 4-Shovel, Steel Beam, Spring Trip, Open
Sleeve

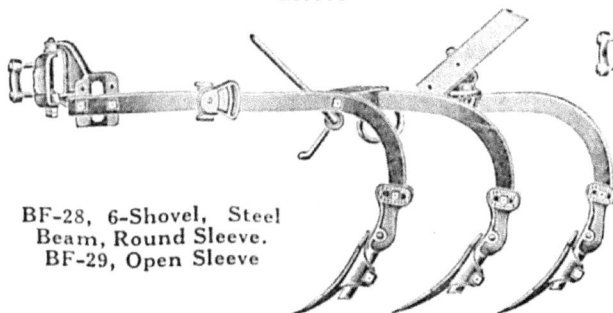

BF-28, 6-Shovel, Steel
Beam, Round Sleeve.
BF-29, Open Sleeve

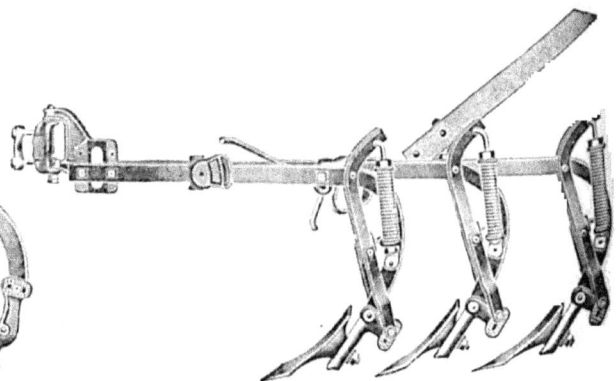

BF-30, 6-Shovel, Spring Trip

Shovels, Points, & Gangs

Moline Balance-Frame Narrow Tread Cultivator Gangs

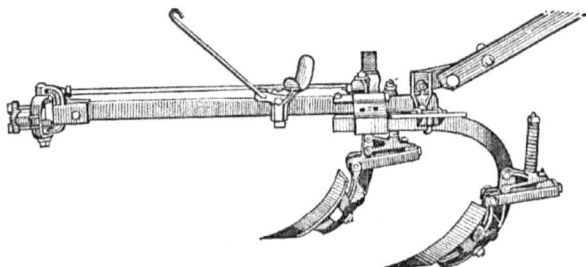

TD-21, 4-Shovel, Parallel, Spring Trip, Square
Pipe, Open Sleeve

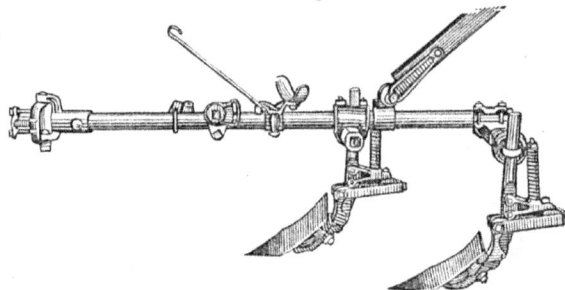

TD-24, 4-Shovel, Round Pipe, Spring Trip, Open
Sleeve, Round Shank

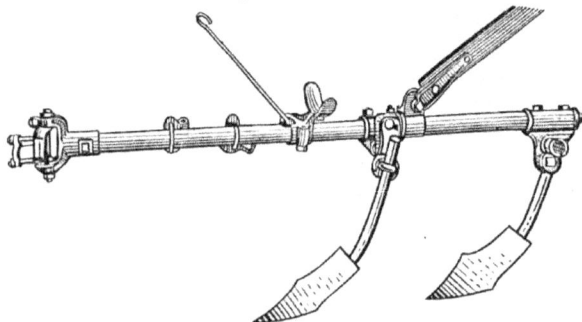

TD-59, 4-Shovel, Round Pipe, Round Shank,
Friction Break

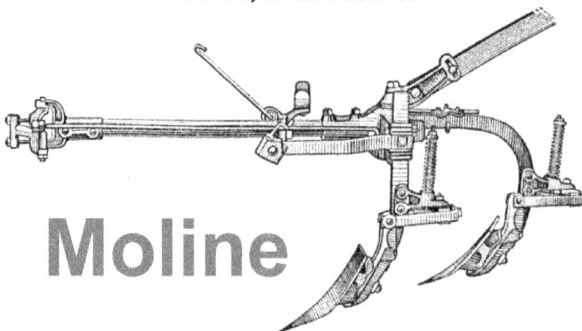

TD-644, 4-Shovel, Round Pipe, Parallel, Spring
Trip, Open Sleeve

MOLINE BALANCE-FRAME LOW-WHEEL CULTIVATOR GANGS

KA-2, 4-Shovel, Steel Beam, Pin Break, Round
Sleeve, Adjustable Shanks

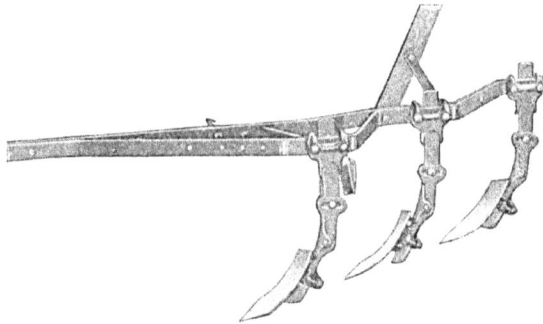

KA-28, 6-Shovel, Pin Break, Round Sleeve, Adjustable Shank.   KA-29, Open Sleeve

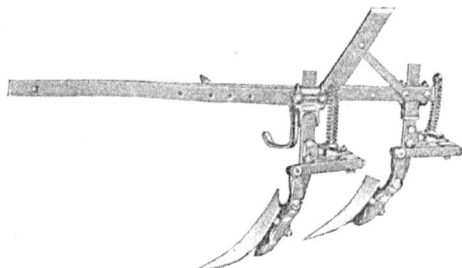

KA-8, 4-Shovel, Steel Beam, Spring Trip, Adjustable Shanks, Open Sleeve

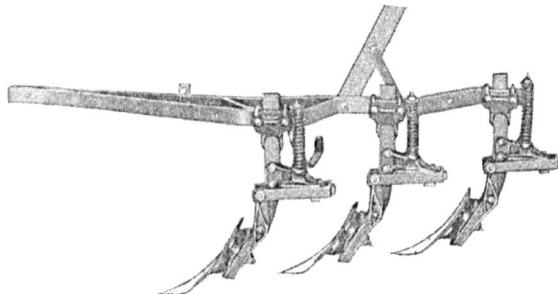

KA-30, 6-Shovel, Spring Trip, Open Sleeve

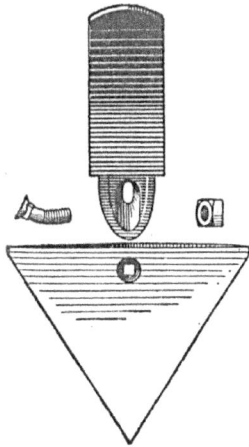

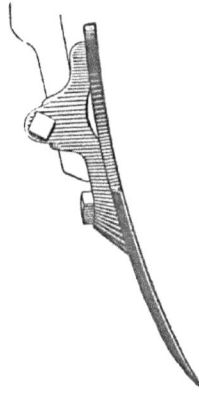

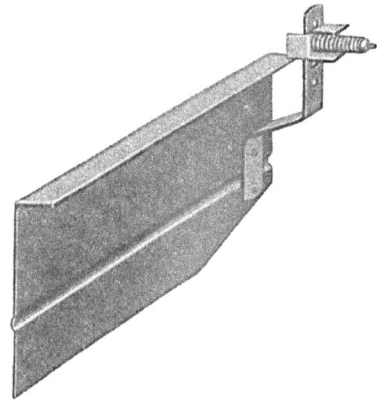

Moline Quick-Detachable Shovels

Moline Shield Held with
Spring Tension

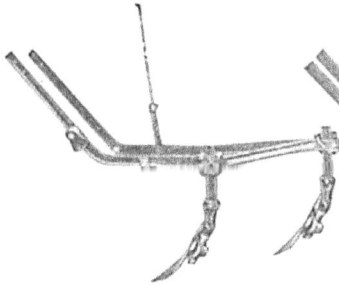

BH-1 and 2

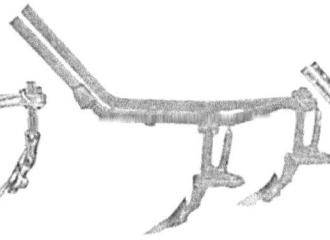

BH-7

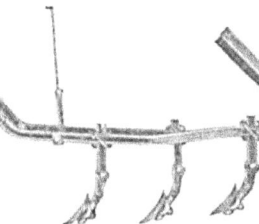

BH-28 and 29

BH-32

BH-63

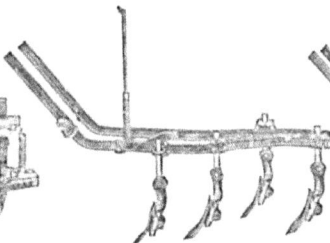

BH-76

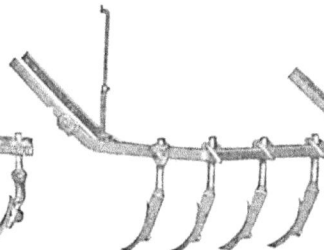

BH-106

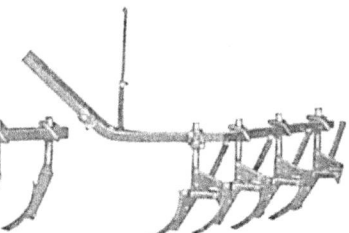

BH-107

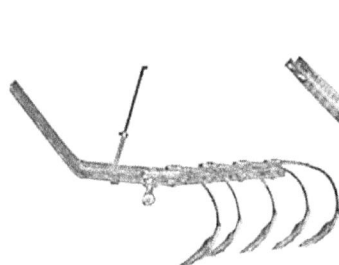

BH-110

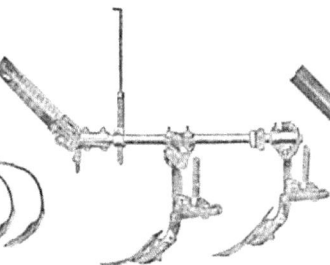

BH-126

BH-129

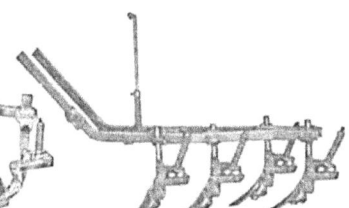

BH-131

Moline

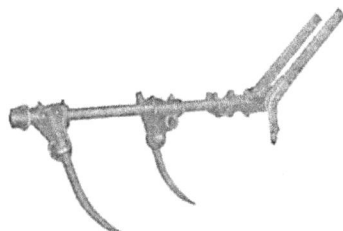

BH-59

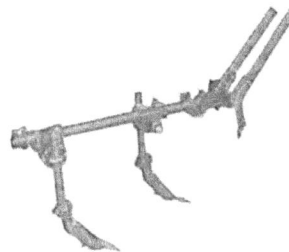

BH-127

Moline Seat-Guide Cultivator Gangs

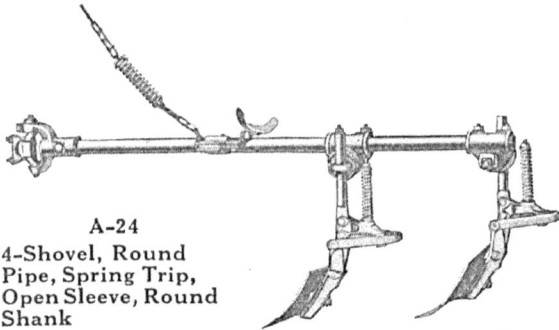

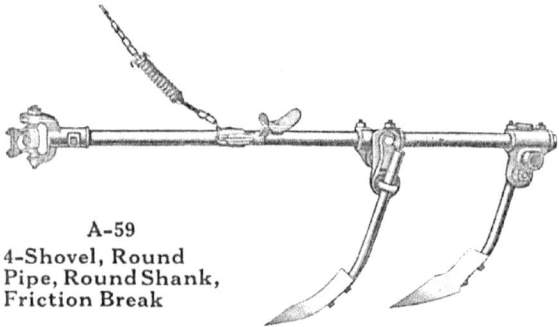

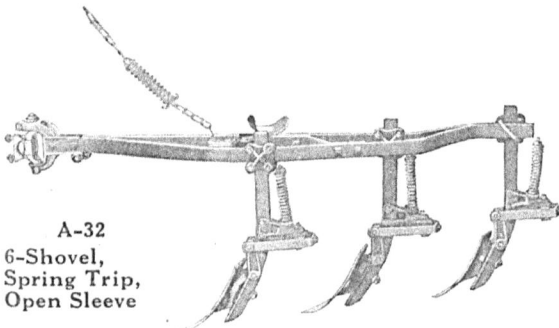

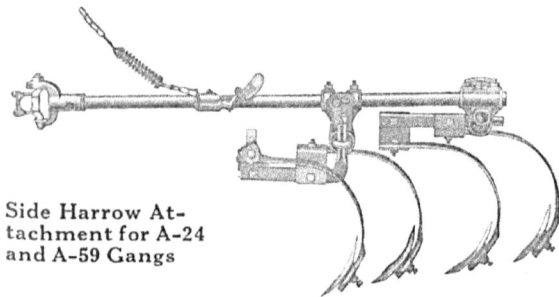

A-24
4-Shovel, Round
Pipe, Spring Trip,
Open Sleeve, Round
Shank

A-59
4-Shovel, Round
Pipe, Round Shank,
Friction Break

A-32
6-Shovel,
Spring Trip,
Open Sleeve

Side Harrow Attachment for A-24
and A-59 Gangs

Moline

No. 10—Wood Beam, Steel Shank, 4-Shovel
Used on AA10, DF10, E10, F10, S10, SP10, SA10, WA10, XA10,
YA10, Cultivators

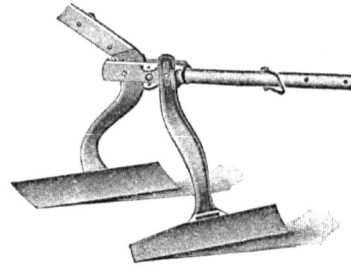

No. 12—Gopher or Surface Rig
Used on A12, W12, X12, Y12 Cultivators

John Deere

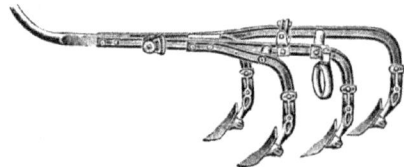

No. 14—Steel Beam, 8-Shovel, Three Aligned
Used on R14, RA14, T14, TA14 Cultivators

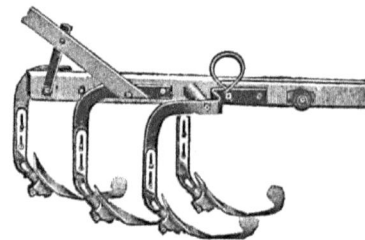

No. 15—Wood Beam, Eagle Claw, 8-Shovel
Used on AA15, E15, EA15 WA15, XA15, YA15 Cultivators

KO-143, 8-Shovel, Steel Beam, Friction Break, Round Sleeve.   KO-145, Open Sleeve

KO-48, 8-Shovel, Steel Beam, Spring Trip, Open Sleeve

KO-138, 8-Shovel, Round Pipe, Friction Break, Round Shank

KO-116, 8-Shovel, Round Pipe, Spring Trip, Open Sleeve

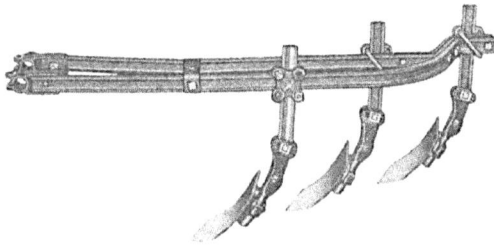

KO-144, 12-Shovel, Steel Beam, Friction Break, Round Sleeve.   KO-146, Open Sleeve

KO-49, 12-Shovel, Steel Beam, Spring Trip, Open Sleeve.   KO-78, Round Sleeve

Moline

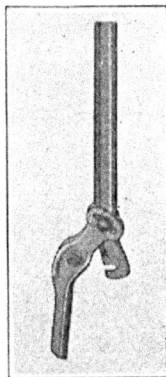

Moline Friction-Break Shank

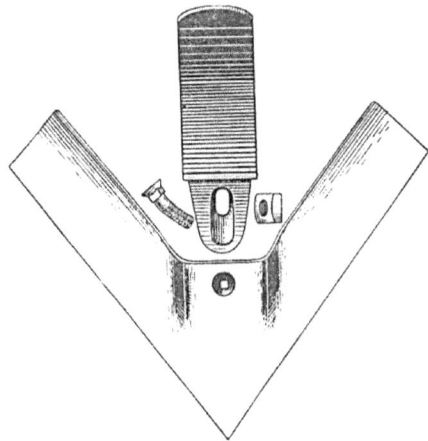

Detachable Sweep

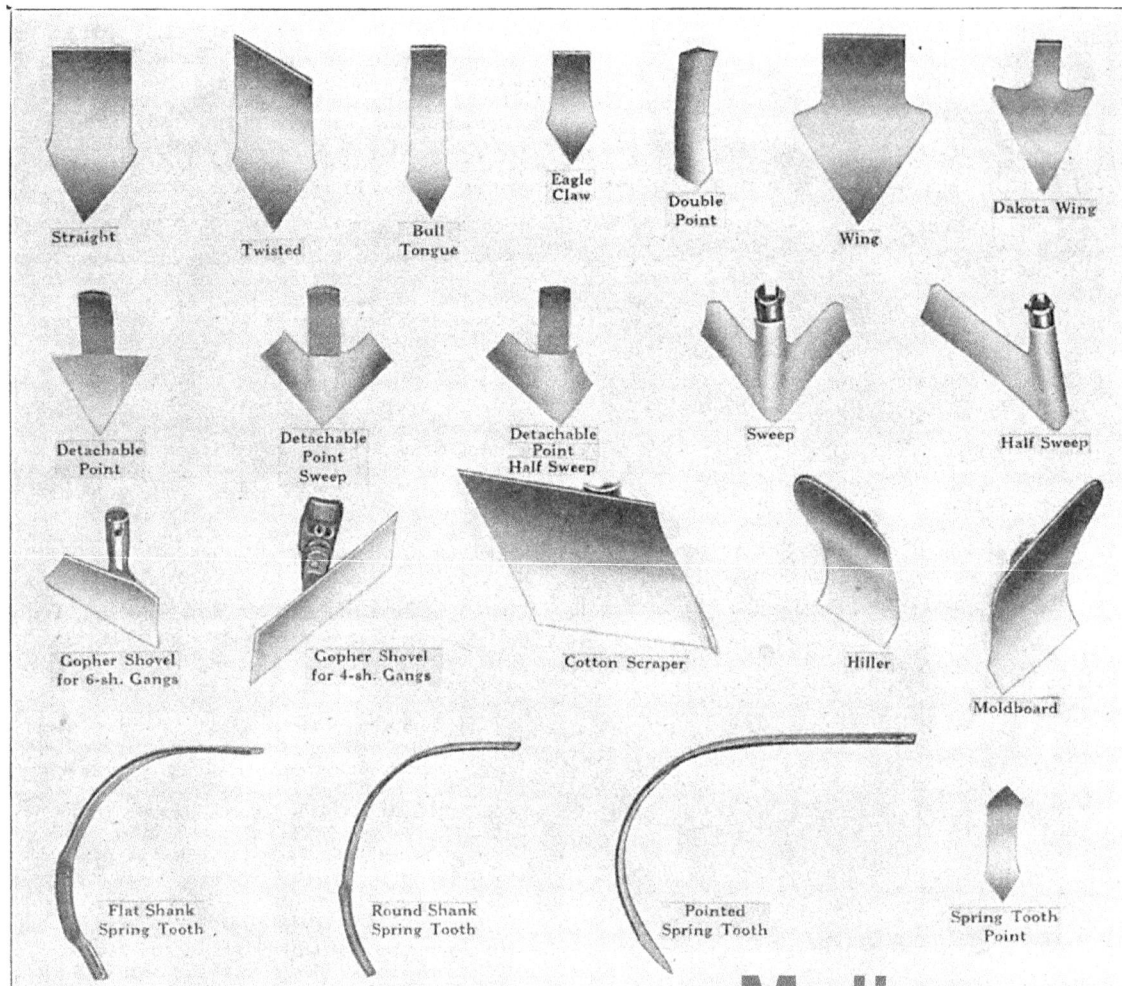

Straight

Twisted

Bull Tongue

Eagle Claw

Double Point

Wing

Dakota Wing

Detachable Point

Detachable Point Sweep

Detachable Point Half Sweep

Sweep

Half Sweep

Gopher Shovel for 6-sh. Gangs

Gopher Shovel for 4-sh. Gangs

Cotton Scraper

Hiller

Moldboard

Flat Shank Spring Tooth

Round Shank Spring Tooth

Pointed Spring Tooth

Spring Tooth Point

Moline

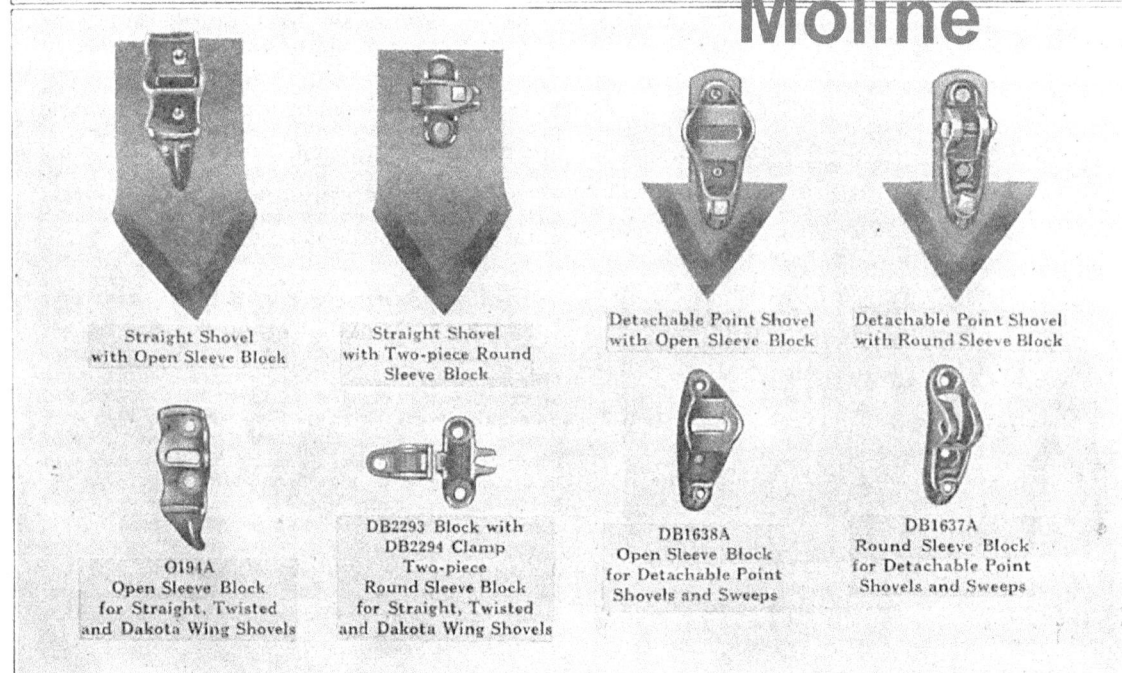

Straight Shovel with Open Sleeve Block

Straight Shovel with Two-piece Round Sleeve Block

Detachable Point Shovel with Open Sleeve Block

Detachable Point Shovel with Round Sleeve Block

O194A
Open Sleeve Block for Straight, Twisted and Dakota Wing Shovels

DB2293 Block with DB2294 Clamp Two-piece Round Sleeve Block for Straight, Twisted and Dakota Wing Shovels

DB1638A
Open Sleeve Block for Detachable Point Shovels and Sweeps

DB1637A
Round Sleeve Block for Detachable Point Shovels and Sweeps

Rotary Shield.

Moline Rotary Shields

Many farmers prefer this type of shield because it permits the pulverized soil to sift through the openings in the shield as it revolves and yet it protects the young plant from being covered with clods. It is hardly possible to secure such favorable results with ordinary shields.

The fine, pulverized soil thrown around the young corn plants forms a mulch to hold moisture. It also covers and destroys many small weeds close to the row, giving the corn every opportunity to make a healthy vigorous growth.

The Moline rotary shield is durably made of heavy wire loops fitted into an inner pressed steel rim. The wire loops, coming in contact with the ground, cause the shield to revolve. It will fit any Moline cultivator.

Adjustable Yoke.

Moline Disc Hillers

Moline disc hillers are especially designed for hilling crops and barring off cotton. A twelve inch blade, made of high carbon steel is used. It will retain a keen cutting edge. The bearing is dust-proof, allowing the disc to revolve freely at all times. It is securely supported by a heavy steel stem and clamp. The connecting yoke is adjustable for width of rows and substantially made.

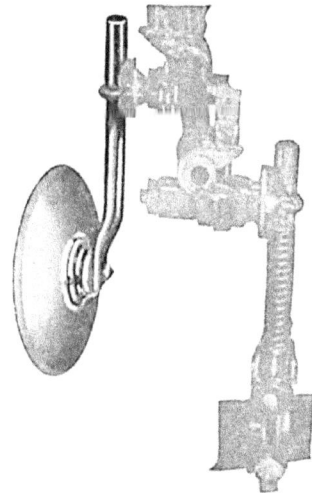

Moline

Side Harrow Attachments

A variety of side harrow attachments are furnished for gangs used in cultivation of cotton ridges. All types are furnished with either flat shank or round shank teeth.

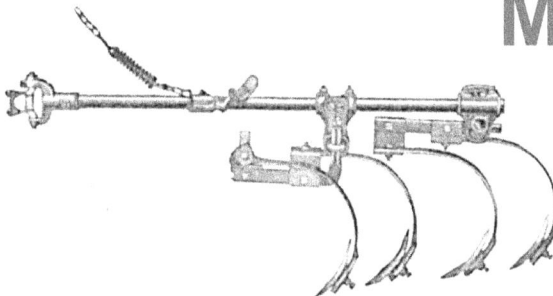

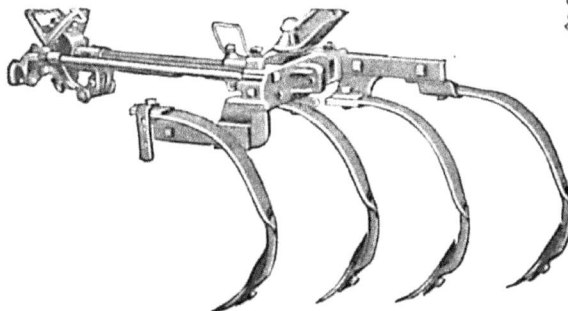

No. 4 R Spring-Tooth Side Harrow Attachment for No. 24 and No. 59 Gangs, with No. 132 Shovels. Also Furnished with No. 39 Shovels.

No. 5R Adjustable Spring-Tooth Side Harrow Attachment for No. 644 Gang with No. 132 Shovels.

Also Furnished with No. 39 Shovels.

No. 2 F Spring-Tooth Side Harrow Attachment for No. 644 Gang, with No. 39 Shovels. Also Furnished with No. 132 Shovels.

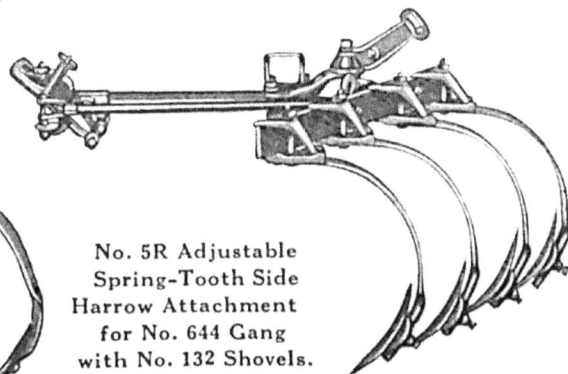

R-55

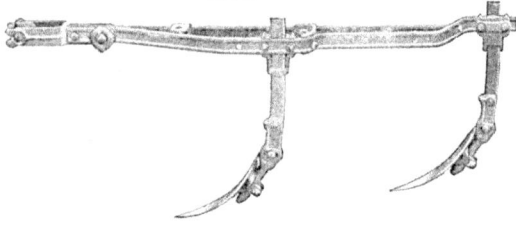

R-15

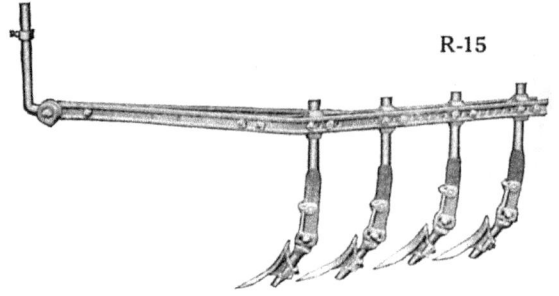

R-58X

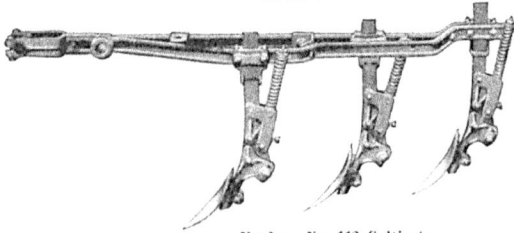

Used on No. 112 Cultivator

R-11X

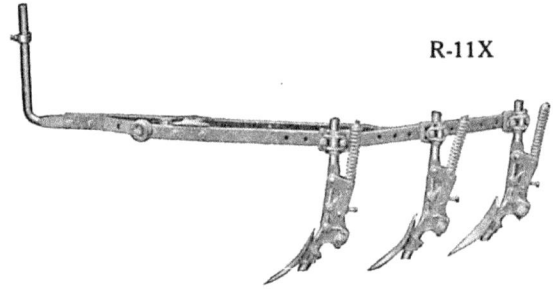

R-52

R-15X

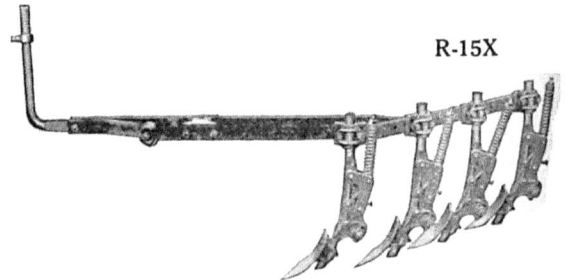

Rock
Island

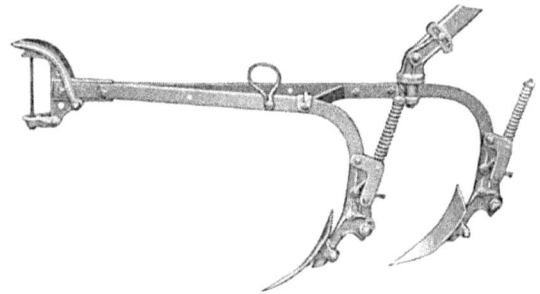

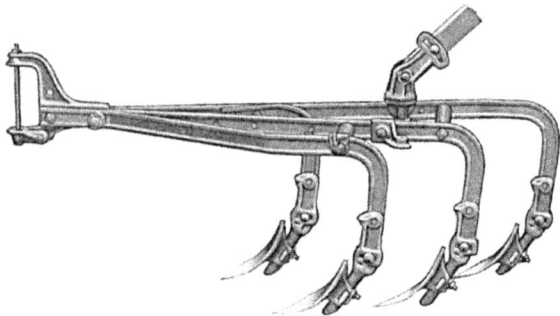

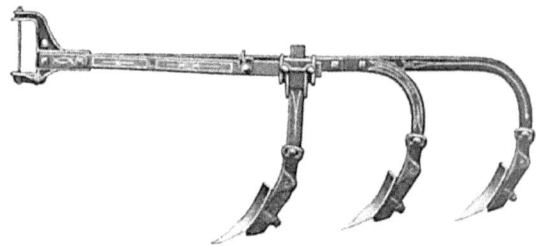

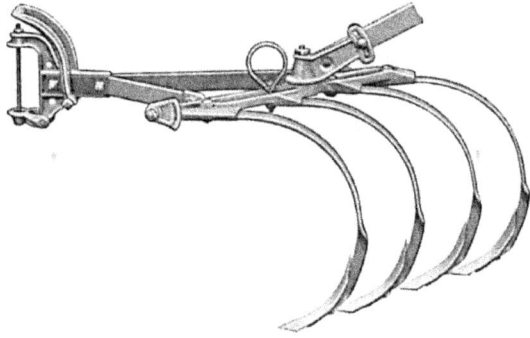

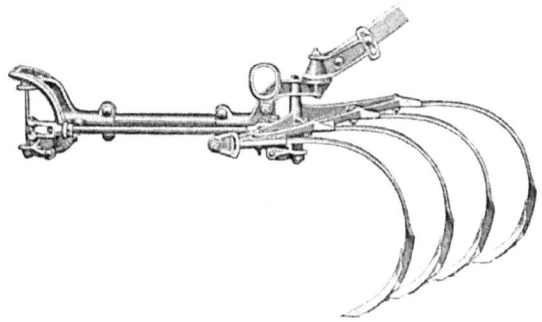

Rock Island

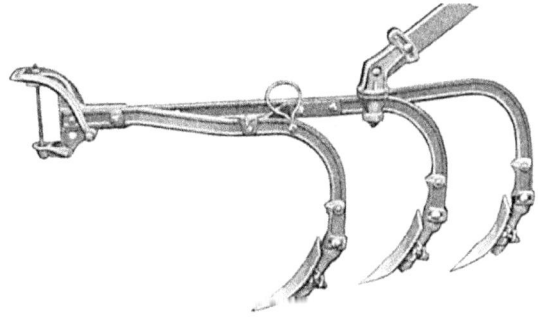

ROCK ISLAND CULTIVATOR GANGS

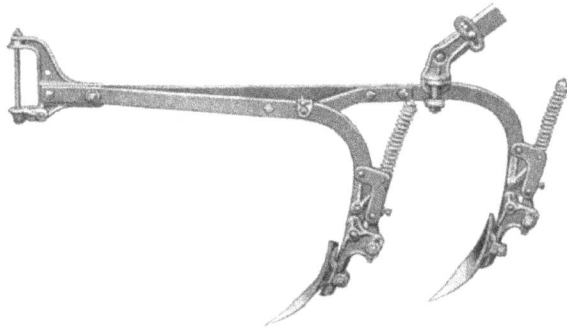

R-21X

R-5X

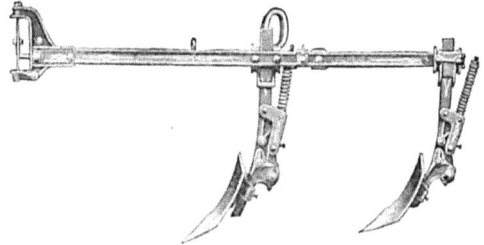

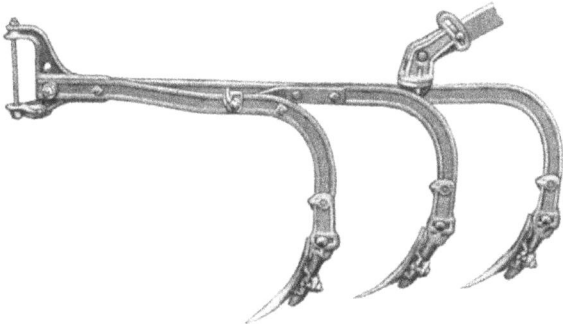

R-41

R-18X

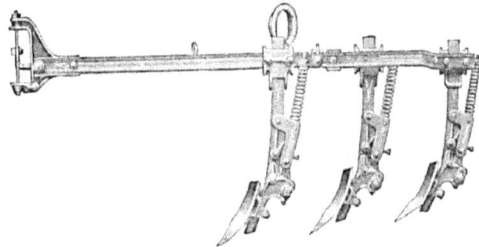

Used on No. 86
Cultivator

O-25X

O-23

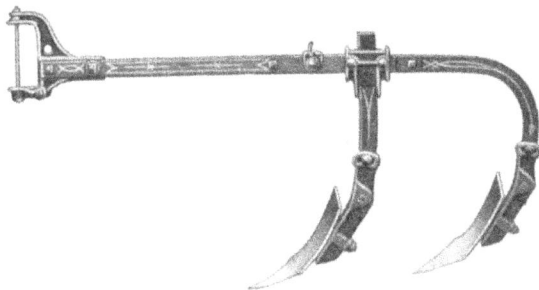

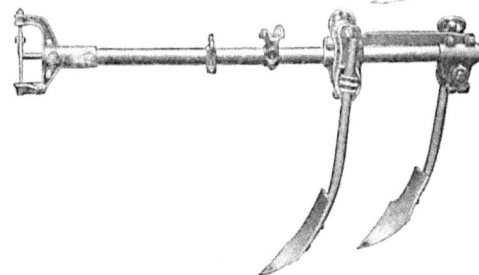

26

O-24X

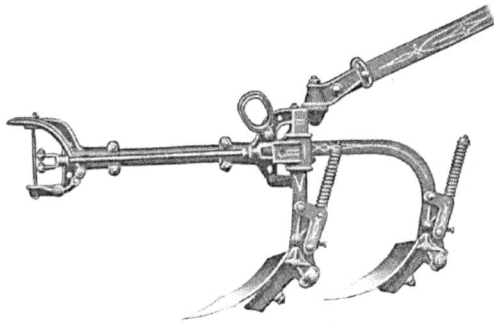

O-18X (Southern)

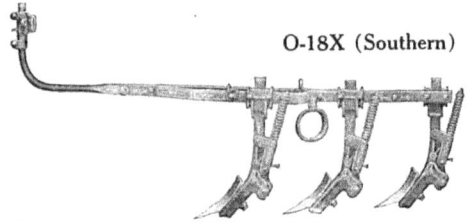

O-24X

O-25X

O-25X

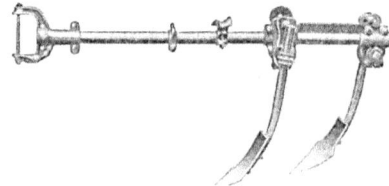

No. 26

O-32X

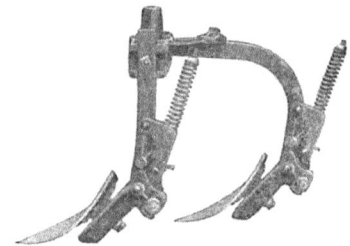

Rock
Island

R-18X

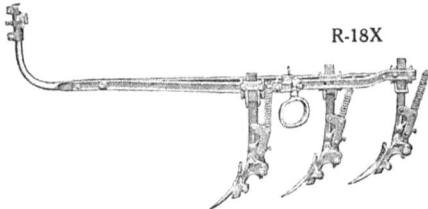

R-12

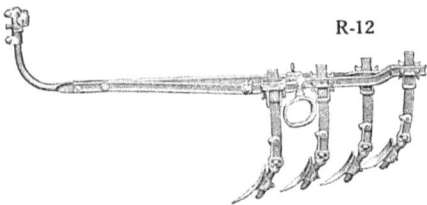

O-31O

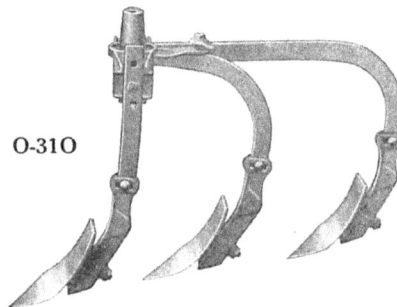

O-5 (Southern)

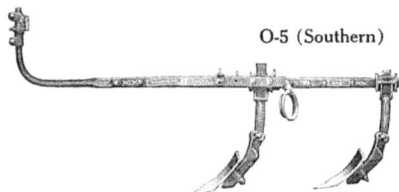

R-38S

ROCK ISLAND SPRING TRIP

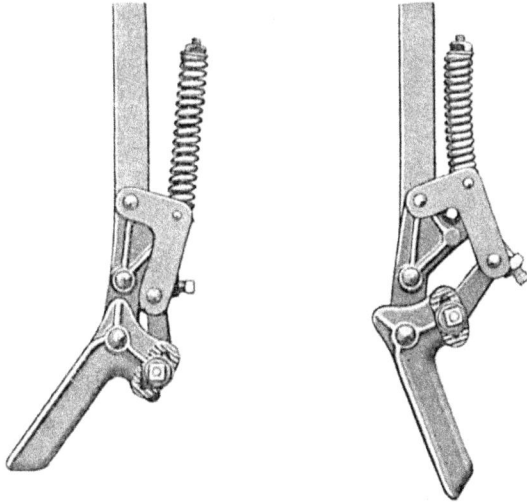

Fig. 1
Position of Shovel Sleeve before
Spring has been tripped

Fig. 2
Position of Shovel Sleeve with
Spring tripped

**Fits Any
Rock Island Pipe
Beam Cultivator**

**Double Pointed
Detachable Shovels**

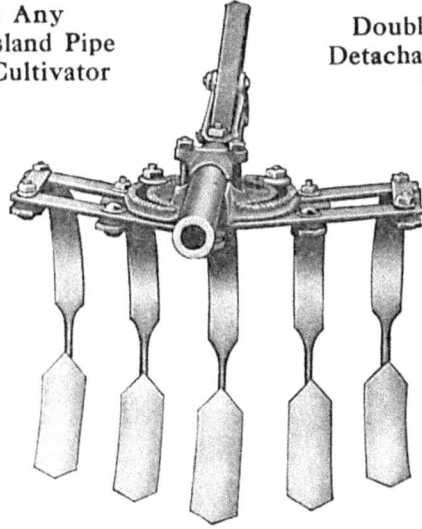

Rock Island

R-19

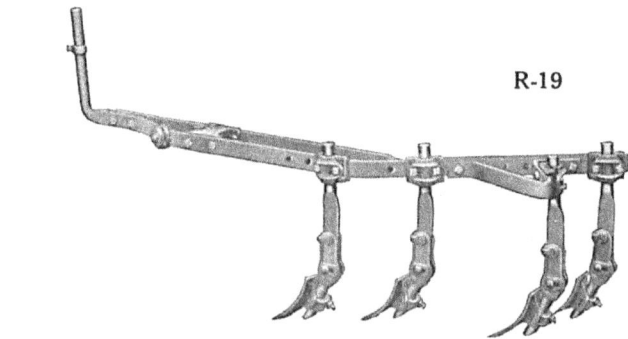

R-31

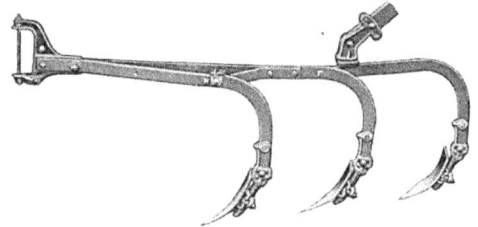

R-18X

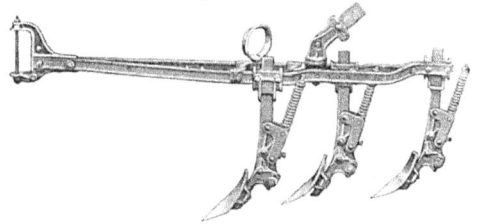

O-12S

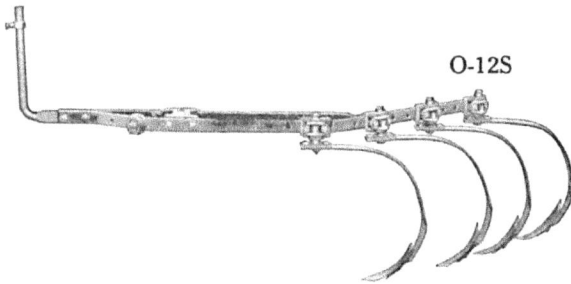

R-12

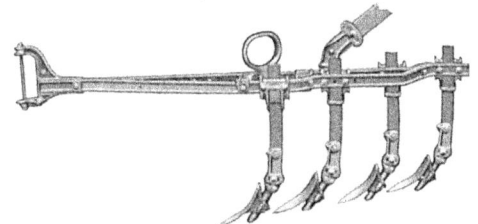

R-5

R-11X

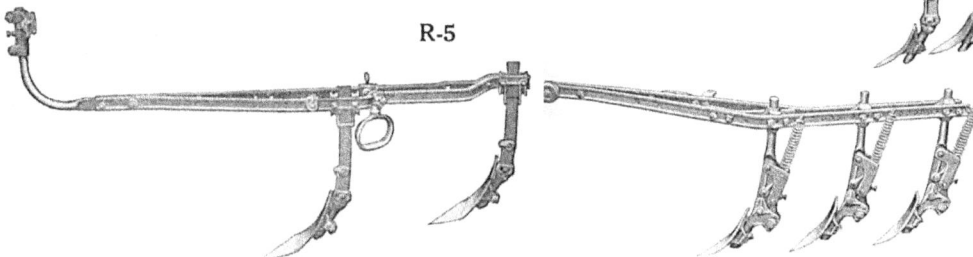

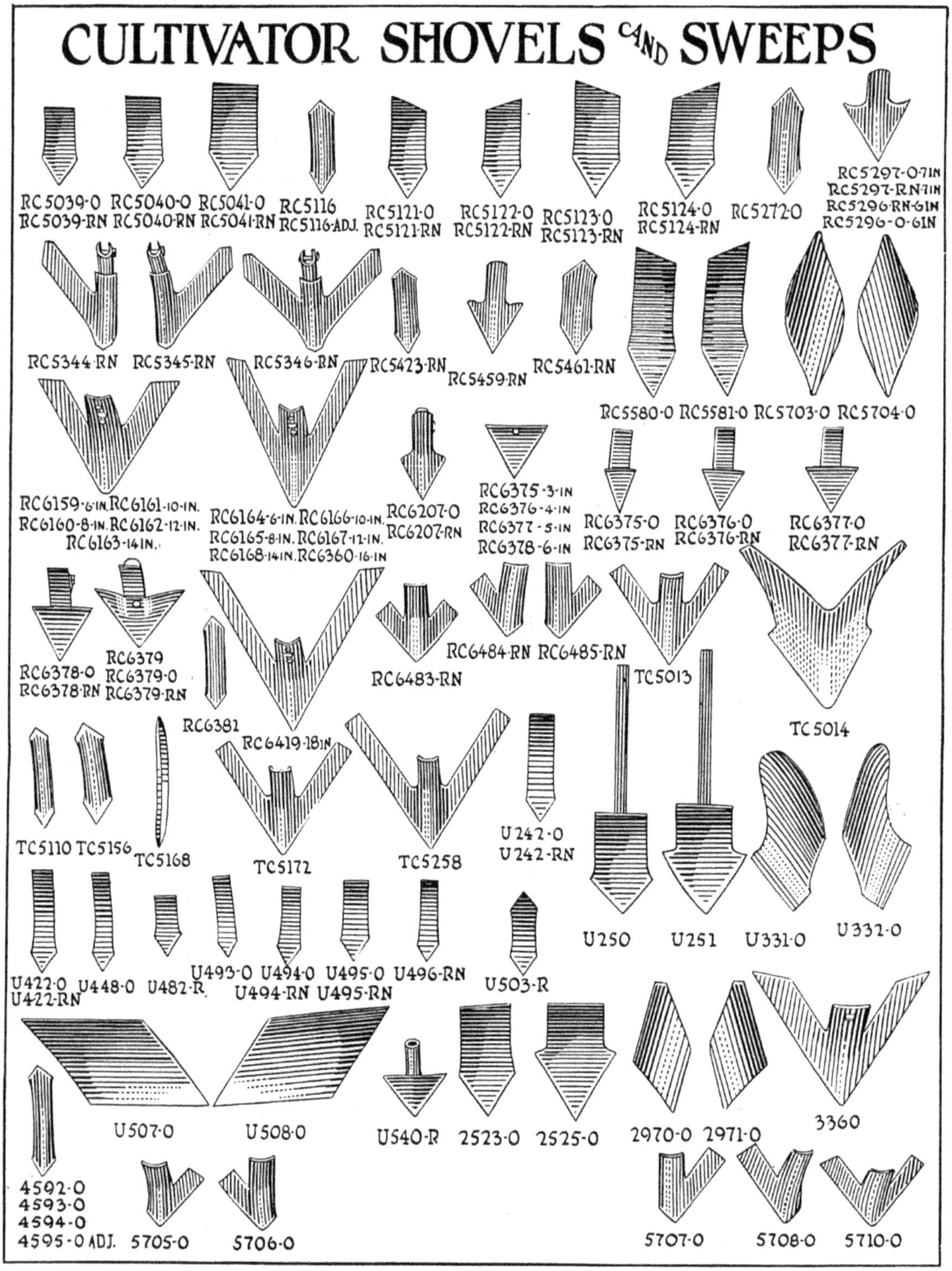

CULTIVATOR SHOVELS ᴬᴺᴰ SWEEPS

RC5039-O RC5040-O RC5041-O RC5116 RC5121-O RC5122-O RC5123-O RC5124-O RC5272-O
RC5039-RN RC5040-RN RC5041-RN RC5116-ADJ. RC5121-RN RC5122-RN RC5123-RN RC5124-RN

RC5297-O-7IN
RC5297-RN-7IN
RC5296-RN-6IN
RC5296-O-6IN

RC5344-RN RC5345-RN RC5346-RN RC5423-RN RC5461-RN
RC5459-RN

RC5580-O RC5581-O RC5703-O RC5704-O

RC6159-6-IN. RC6161-10-IN. RC6164-6-IN. RC6166-10-IN. RC6207-O RC6375-3-IN RC6375-O RC6376-O RC6377-O
RC6160-8-IN. RC6162-12-IN. RC6165-8-IN. RC6167-12-IN. RC6207-RN RC6376-4-IN RC6375-RN RC6376-RN RC6377-RN
RC6163-14IN. RC6168-14IN. RC6360-16-IN. RC6377-5-IN
RC6378-6-IN

RC6378-O RC6379 RC6483-RN RC6484-RN RC6485-RN TC5013 TC5014
RC6378-RN RC6379-O
RC6379-RN

RC6381
RC6419-18IN.

TC5110 TC5156 TC5168 TC5172 TC5258 U242-O U250 U251 U331-O U332-O
U242-RN

U422-O U448-O U482-R U493-O U494-O U495-O U496-RN U503-R
U422-RN U494-RN U495-RN

U507-O U508-O U540-R 2523-O 2525-O 2970-O 2971-O 3360

4592-O
4593-O
4594-O
4595-O ADJ. 5705-O 5706-O 5707-O 5708-O 5710-O

Oliver

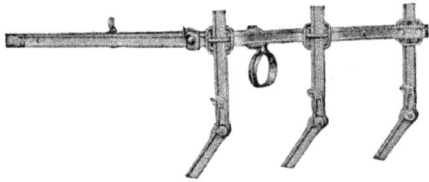

No. 1 Imp. CO 6-shovel, pin break, open sleeve, diagonal gangs.

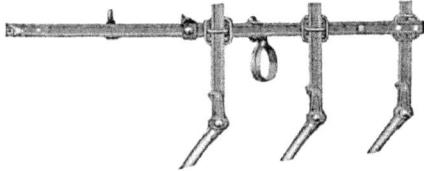

No. 1 Imp. CR 6-shovel, pin break, round sleeve, diagonal gangs.

No. 1 Imp. BSO 4-shovel, spring trip, open sleeve, diagonal gangs.

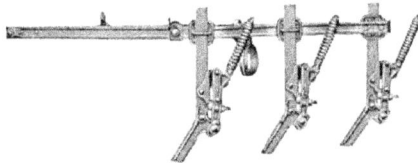

No. 1 Imp. DSO 6-shovel, spring trip, open sleeve, diagonal gangs.

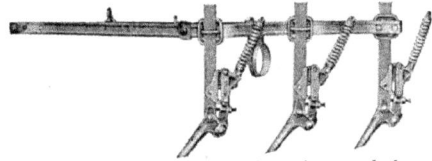

No. 1 Imp. DR 6-shovel, spring trip, round sleeve, diagonal gangs.

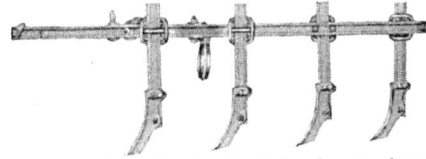

No. 1 Imp. VDPO 8-shovel, pin break, open sleeve, diagonal gangs.

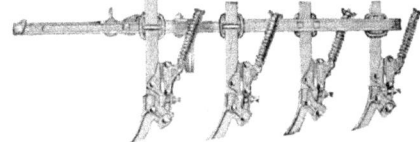

No. 1 Imp. VDSPO 8-shovel spring trip, open sleeve, diagonal gangs.

Oliver

Showing the No. 1 Imp. M 16" disc gang.

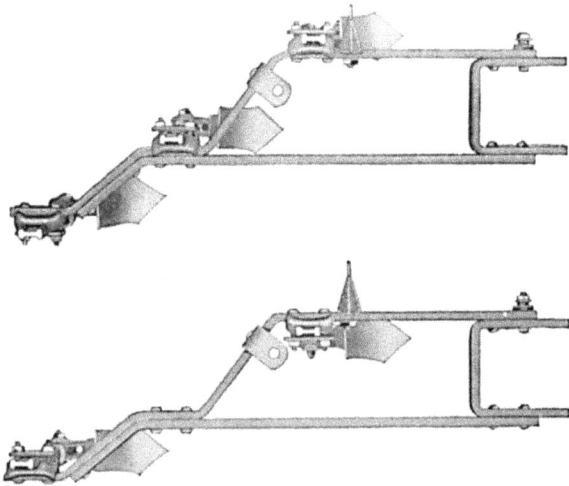

This illustration shows the Western type gangs and the different positions in which the shovels can be set. Either pin break or spring trip gangs are furnished.

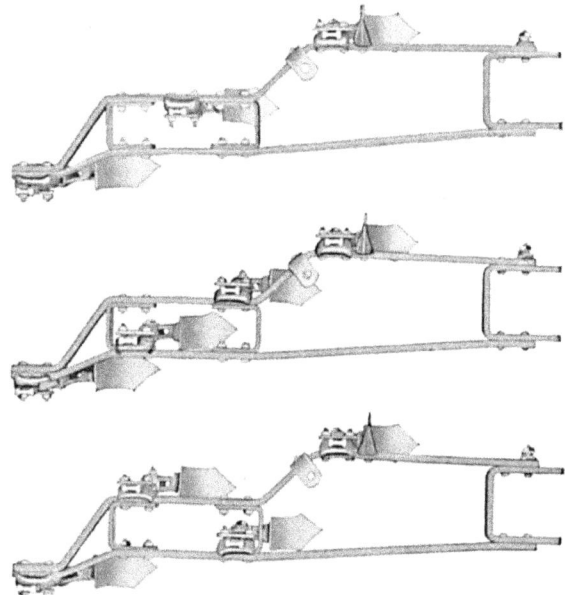

Above are illustrated the different positions for setting the shovels on the Eastern Type gangs. Either pin break or spring trip gangs are furnished.

Shovels, Points, & Gangs

Cultivator Gang Attachments

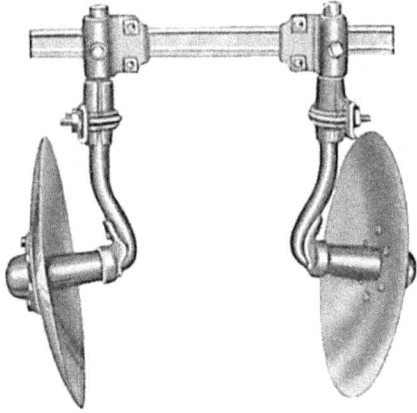

No. 1-A 12" and 14" Front disc attachments.

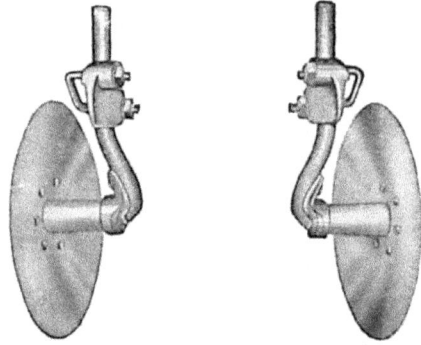

No. 22 12" Front disc attachment.

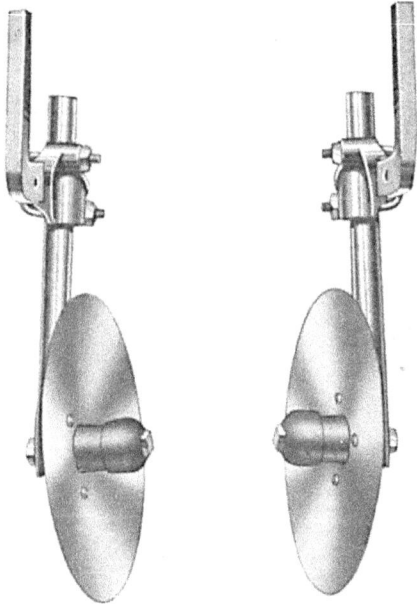

No. 2 Cultivator 12" Front disc attachment.

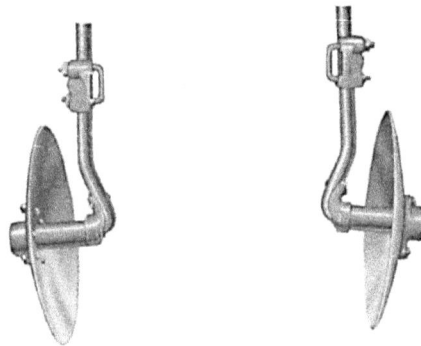

No. 35 12" and 14" Front disc attachments.

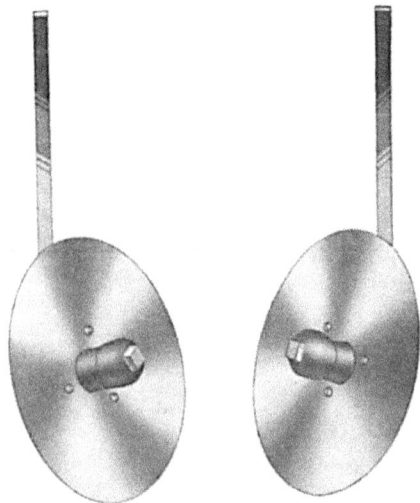

Oliver

No. 3 12" Front disc Attachment.

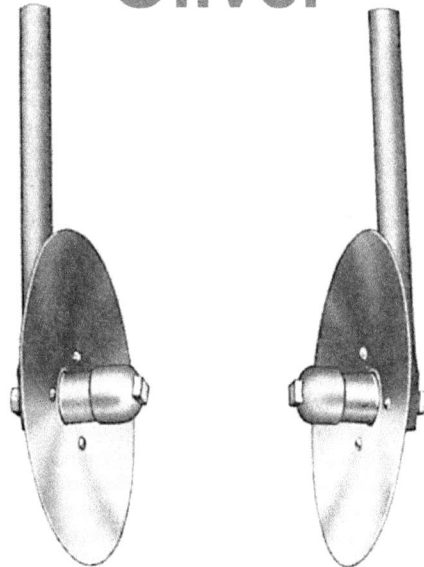

No. 350 12" and No. 351 14" Front disc attachments. The Nos. 352 12" and 353 14" are similar except they have clamps for use with channel beam gangs. The Nos. 354 12" and 355 14" are equipped with cross bar attachment for use on channel beam gangs.

SPRING TOOTH ATTACHMENTS FOR P. & O. CULTIVATORS

GOPHER ATTACHMENTS FOR P. & O. CULTIVATORS.

P & O

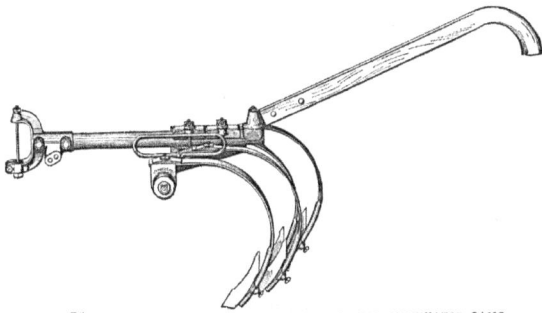

NO. 76 SPRING TOOTH ATTACHMENT ON WALKING CULTIVATOR GANG.

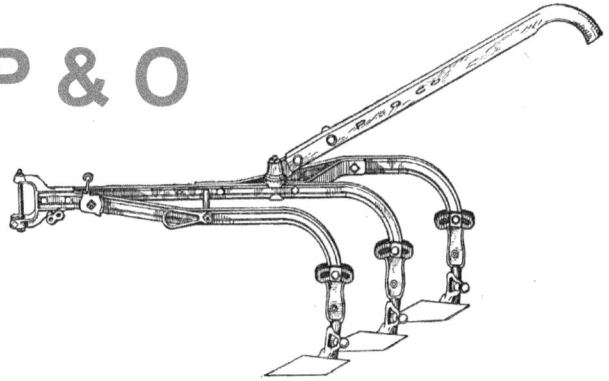

SIX-SHOVEL GOPHER ATTACHMENT ON NO. 21 WALKING GANG.

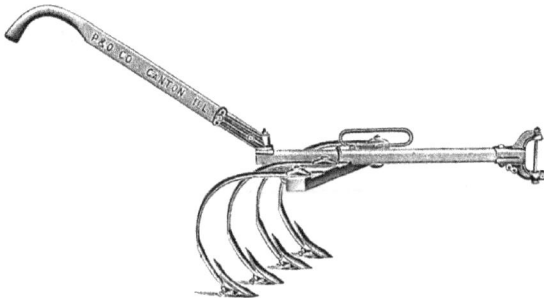

SPRING TOOTH ATTACHMENT ON WALKING CULTIVATOR GANG.

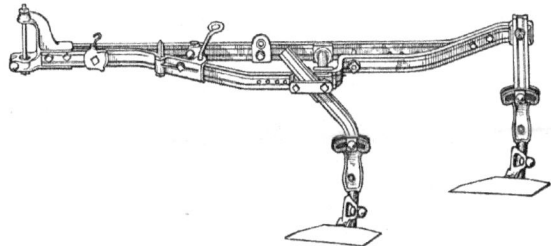

FOUR-SHOVEL GOPHER ATTACHMENT ON NO. 1 YBX GANG.

P. & O. CULTIVATOR GANGS.
Spring Trips.

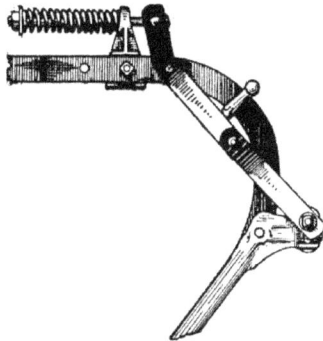

No. 1.

No. 2.

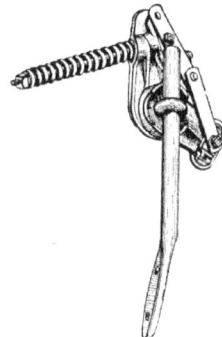

No. 3.

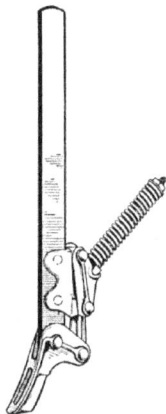

No. 4.

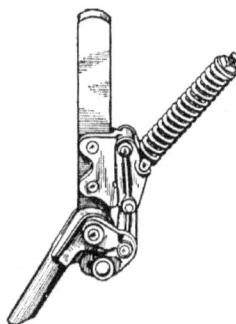

No. 5.

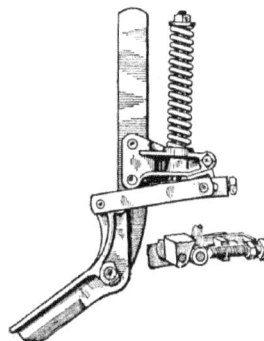

No. 6.

Shovels, Points, & Gangs

SPRING TOOTH ATTACHMENT FOR P. & O. CULTIVATORS.

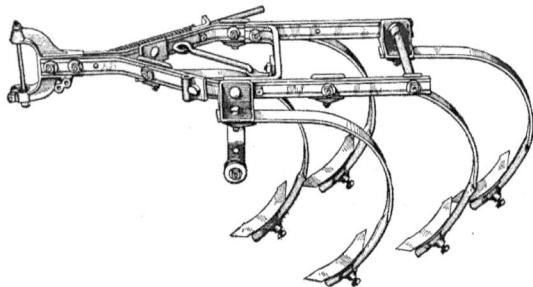

NO. 71 SPRING TOOTH ATTACHMENT FOR YBX CULTIVATOR.

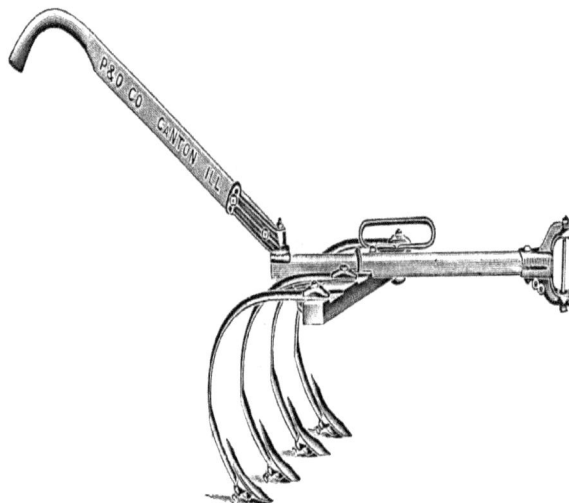

NO. 74 SPRING TOOTH ATTACHMENT ON WALKING CULTIVATOR GANG.

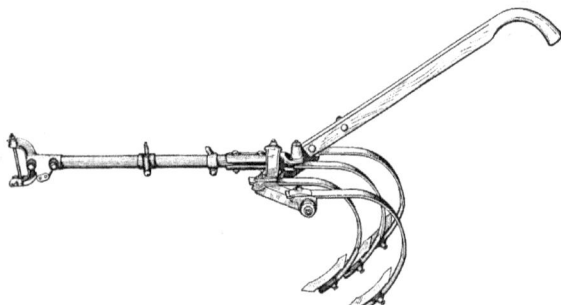

NO. 73 SPRING TOOTH ATTACHMENT ON WALKING CULTIVATOR GANG.

P & O

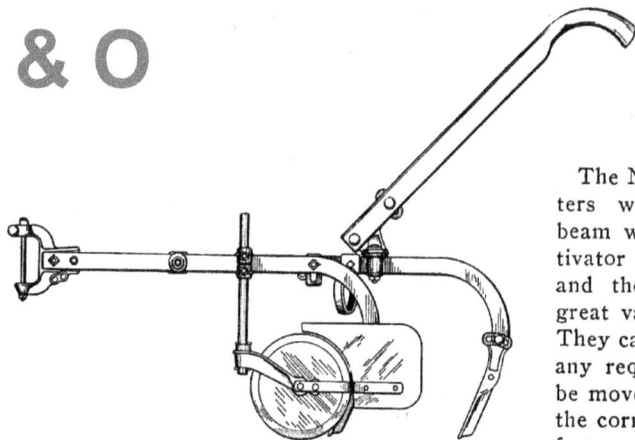

The No. 2 Swivel Vine Cutters work with any steel beam walking or riding cultivator gangs of any make and they are capable of a great variety of adjustments. They can be set vertically for any required depth, and can be moved away from or near the corn as desired. A strong feature is the fact that the hardened steel rolling cutter blade swivels with the gang, and the shield, being a part of the cutter, invariably remains directly in line with it. The shield arm is pivoted on the blade axle and can be raised or lowered, maintaining the same distance from the blade at all times. It is easy to cultivate close to the corn without danger. There is nothing clumsy or bunglesome to get out of order. Absolutely simple and as light as practical to make it.

The No. 1 Disc Vine Cutters are built for the Cantonian, Double Row, or any Pivot Wheel Cultivators, and are substituted for the front shovels. Being in advance of the shovels, the vines are cut so that the blades do not become clogged, requiring frequent stopping. The shields can be dispensed with.

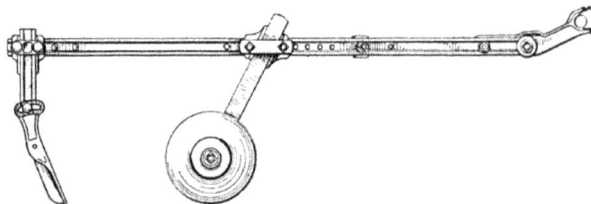

Chapter Seventeen

Oddities

Here's just a sampling, just a taste, of how various and sometimes odd the design course was for horsedrawn tillage tools. Some of these such as the harrow on the right, are pure elegance in their utility. Some, such as the Vineyard Hoe on page 351 and the Orchard Spading Disc below, deserve consideration to be reinstated within the lineup of tools for the future. And my favorite, the *Mumbler* (is it the name?), should be on every farmer's list of tools to make at home. (I know I'm going to try to build one.) We've added some advertisements from the turn of the century to showcase not only the implements but also the "language" of the industry and the times.

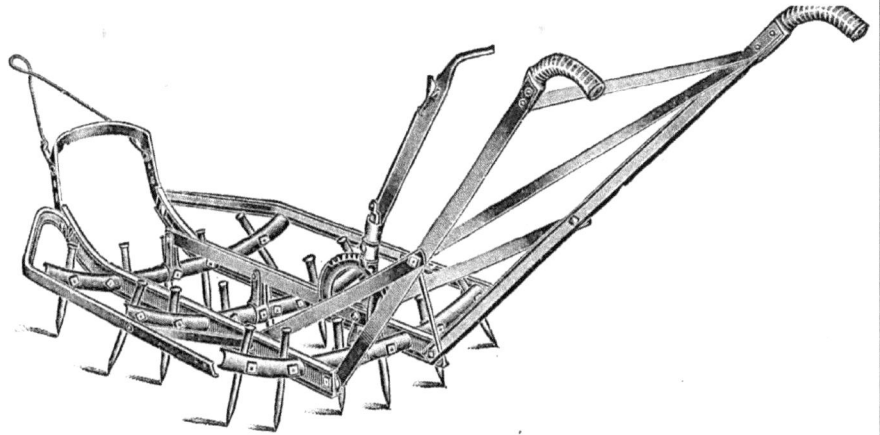

John Deere Mulch Cultivator
With 17 Teeth

A highly unusual Orchard Spade Disc. Fully adjustable, gangs will throw both ways.

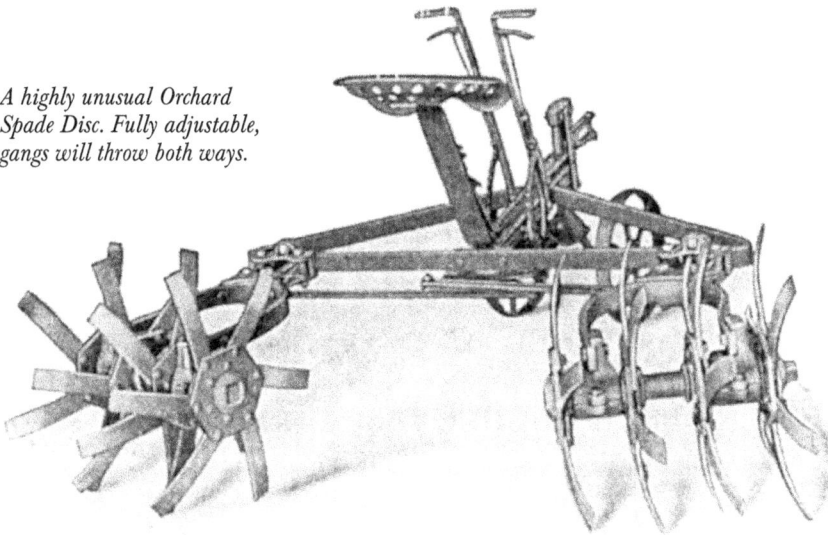

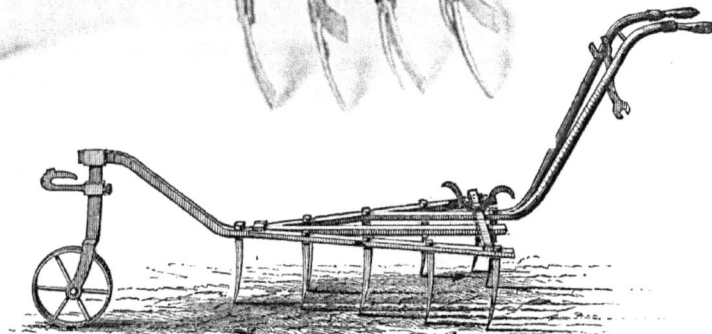

A rare Tooth Hiller

Vineyard harrow. Souchu-Pinet, Langeais, France.

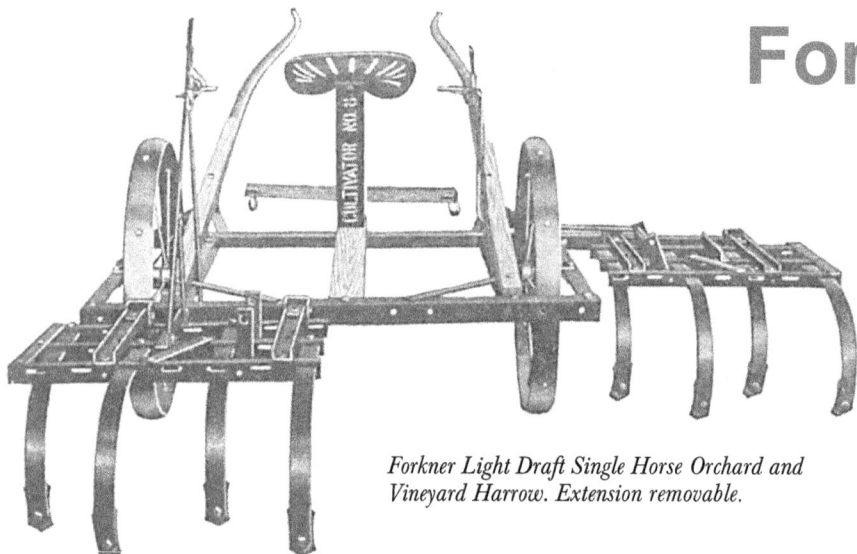

Forkner

Forkner Light Draft Single Horse Orchard and Vineyard Harrow. Extension removable.

Cultivator Attachments for Forkner "Light Draft" Harrow

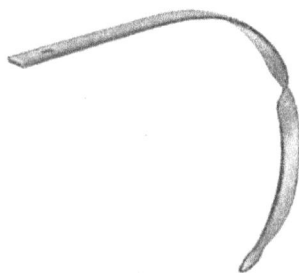

No. 5A Alfalfa Tooth. Wt. 3½ lbs. List . . $0.80

10-inch Sweep. Wt. 1¾ lbs. List $0.75

15-inch Sweep. Wt. 3 lbs. List, $0.95

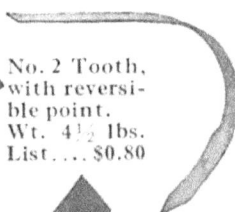

No. 2 Tooth, with reversible point. Wt. 4½ lbs. List $0.80

Reversible Point for No. 2 tooth. List . . $0.25.

10-inch Furrower. Wt. 5 lbs. List, $0.95

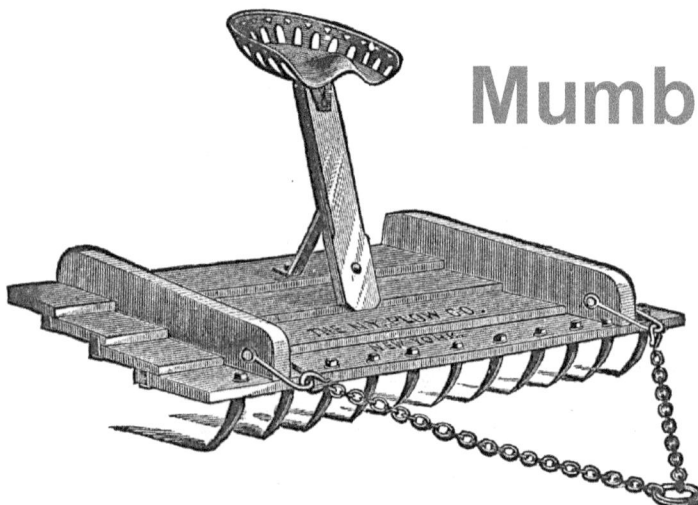

Mumbler

THE MUMBLER.

Kramer

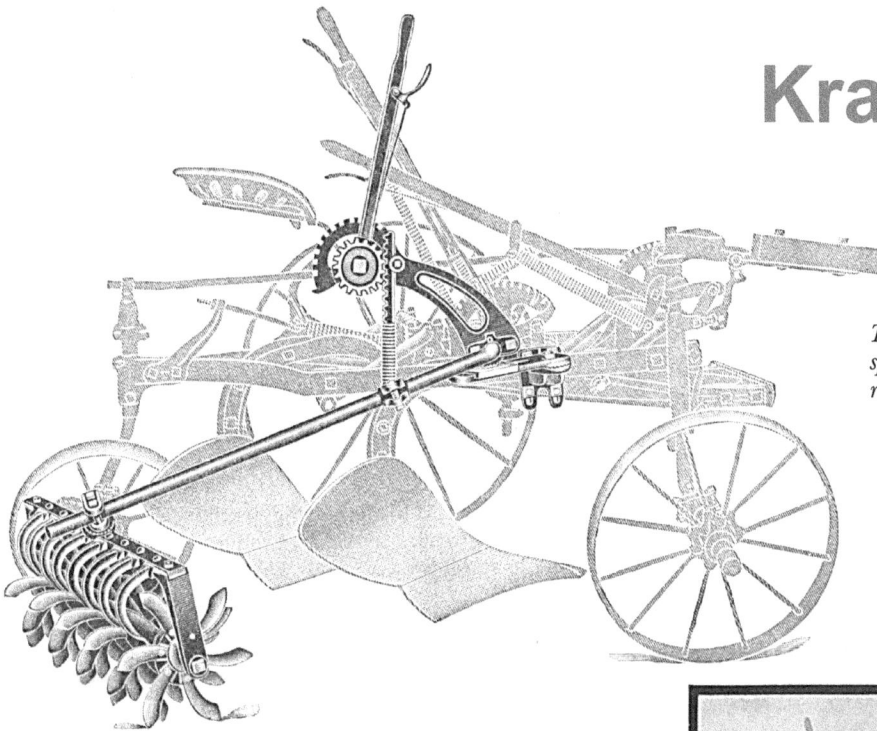

Two views of the Kramer built spading Harrow attachment for riding plows.

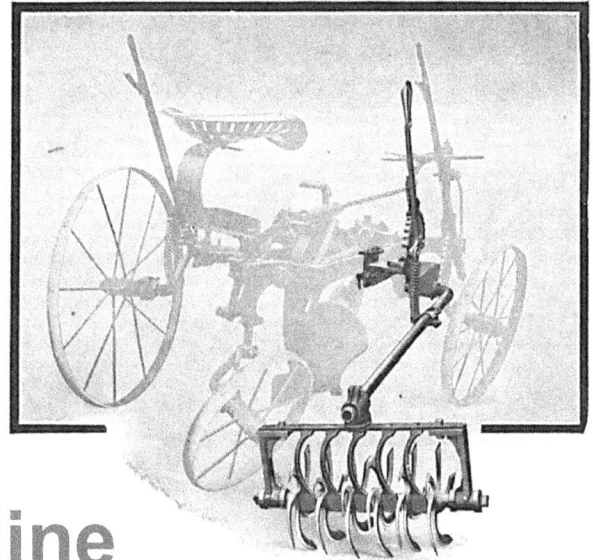

Moline

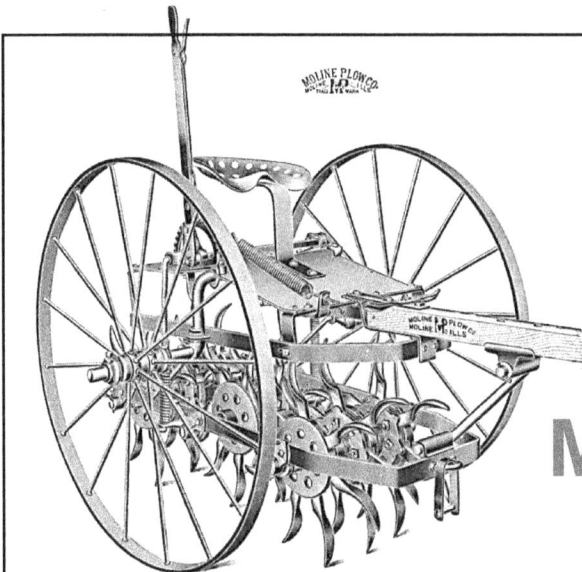

CAPTAIN LAFITTE TWO OR THREE HEAD STUBBLE DIGGERS

The Captain Lafitte Stubble Digger was designed solely for chopping sugar cane stubble.

The heads are of steel and the bearings are so designed that they will always remain free from dirt, and at all times leave the teeth free and flexible.

The two head digger, takes four heads in the rear and three in front, while the three head has three in the front and rear and two in the center.

The lower frame is held parallel with the ground by means of a heavy yoke at the front end, and two parallel yokes at the rear end, and is held down by means of spring pressure.

The spring pressure may be increased or decreased by the operating lever, while the set screw collar on the spring rods may be adjusted to fit different heights of cane drill. The lifting spring helps the operator to raise the lower frame.

The rear frame is entirely covered by a sheet iron top to ensure the safety of the operator.

The draft is applied directly to the lower frame.

Oddities

Deere

JOHN DEERE
MOLINE, ILL.
THE TRADE MARK OF QUALITY
MADE FAMOUS BY GOOD IMPLEMENTS

Unusual Oliver Orchard Harrow

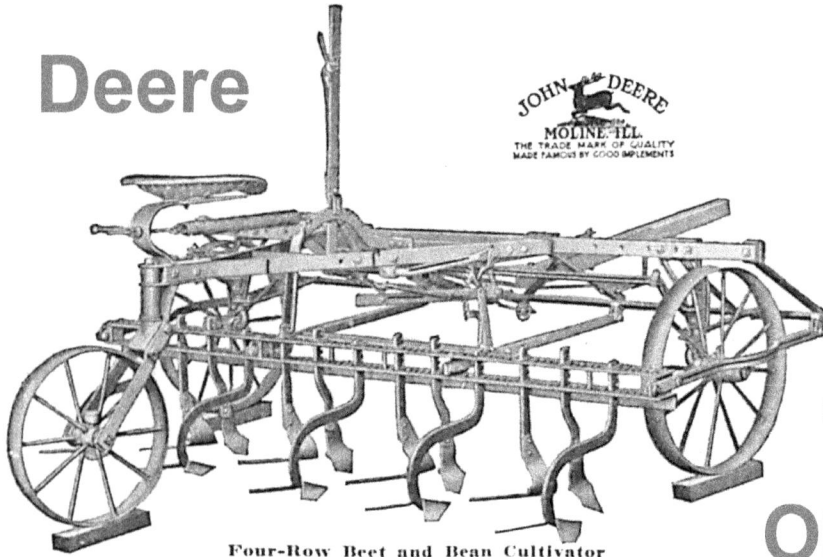

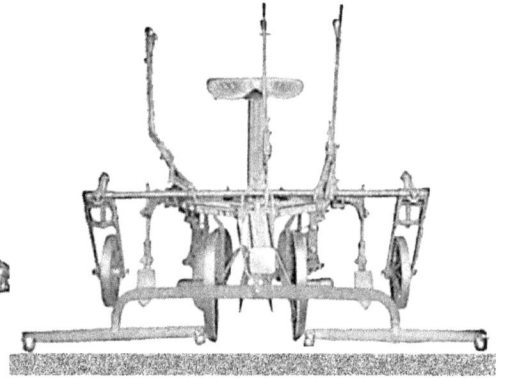

Four-Row Beet and Bean Cultivator

Oliver

Rear view of the No. 34 with discs angled by individual levers.

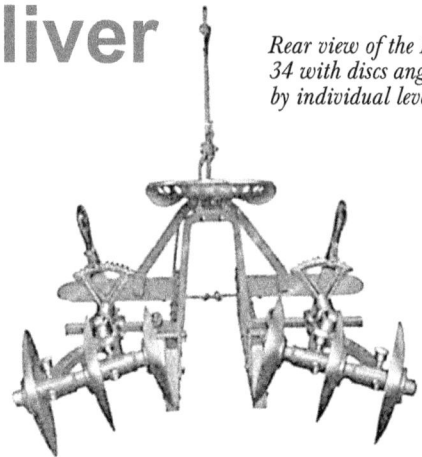

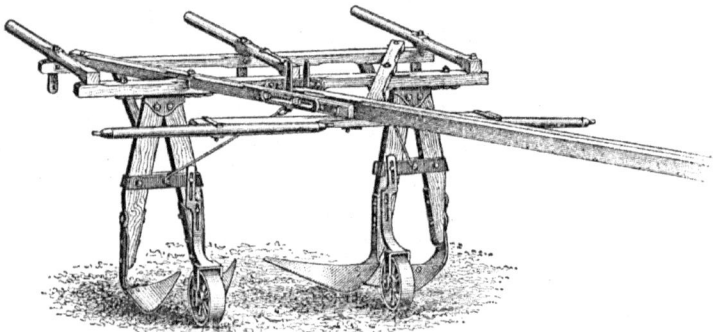

An old Tobacco Cultivator

P. & O. DIVERSE CULTIVATORS.

P & O

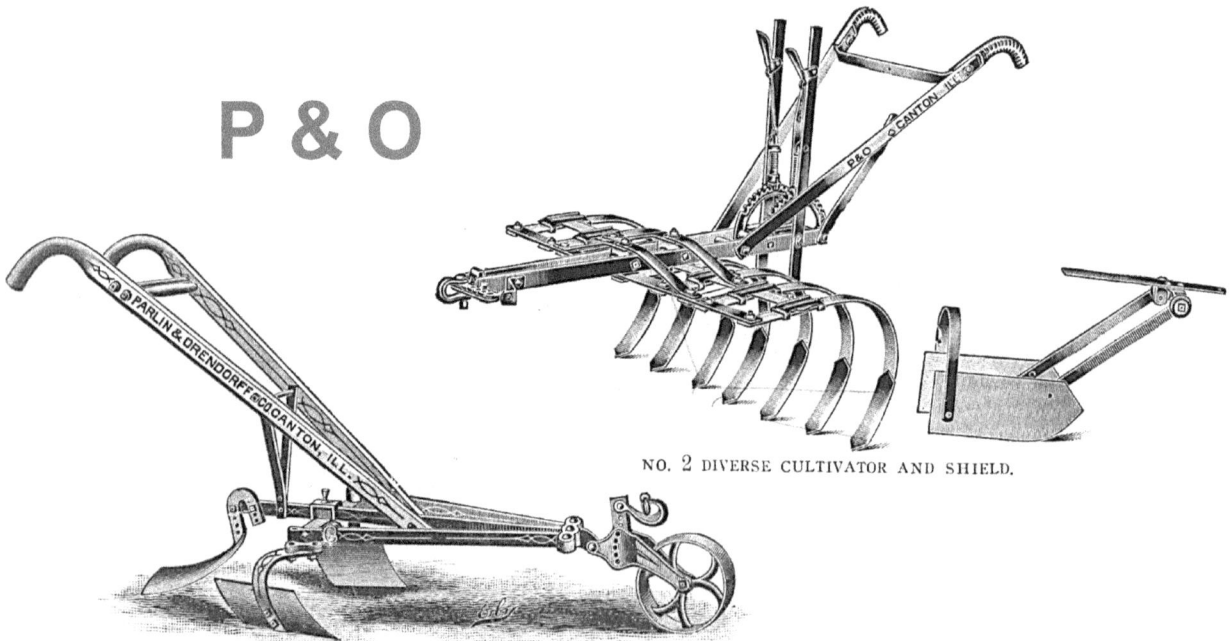

NO. 2 DIVERSE CULTIVATOR AND SHIELD.

STEEL FRAME FIVE TOOTH CULTIVATOR AS A COVERING IMPLEMENT

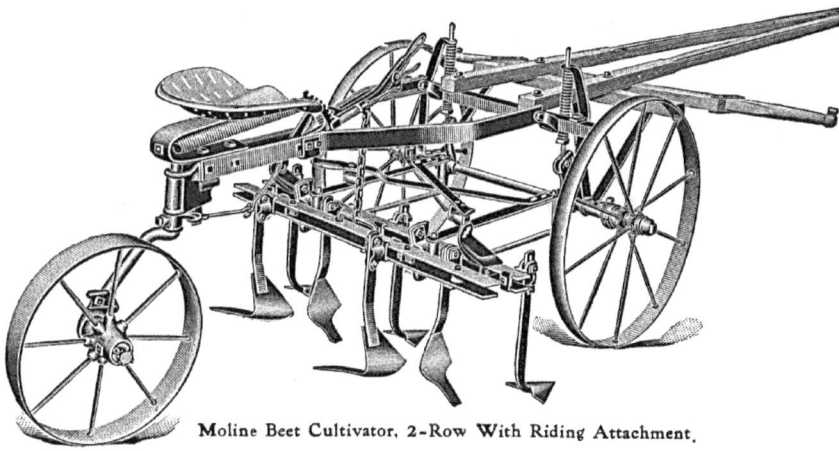

Moline Beet Cultivator, 2-Row With Riding Attachment.

Moline

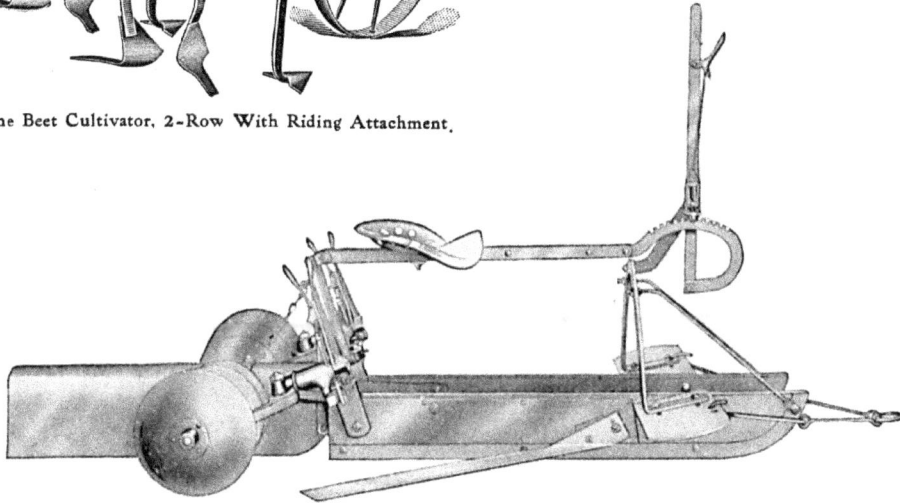

The No. 34 sled cultivator equipped with adjustable side knives which are special equipment. This is an efficient, inexpensive cultivator.

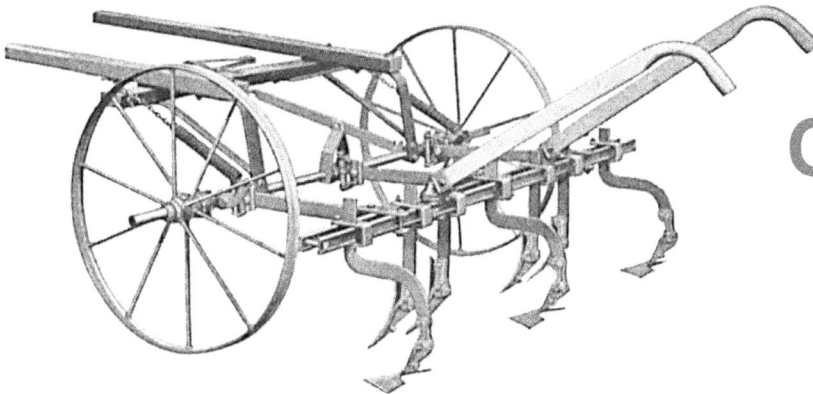

Oliver

Oliver One Horse Two Row Beet cultivator

ROCK ISLAND NO. 129 SLED
LISTER CULTIVATOR

Rock
Island

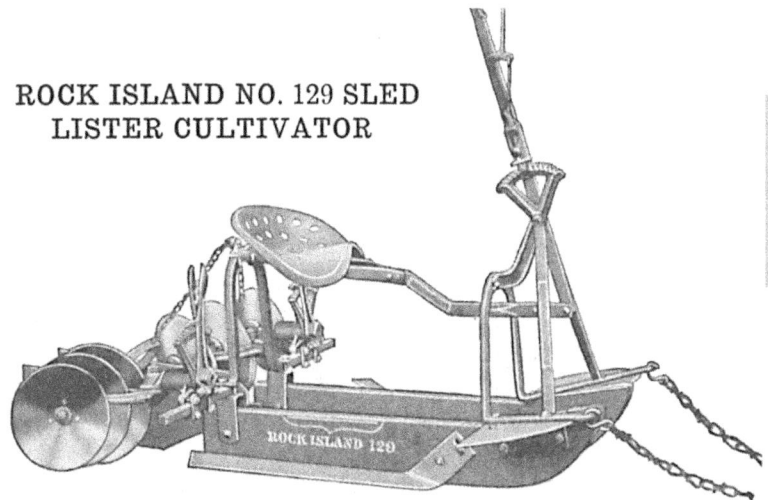

Odditles

Orchard Cultivators

P & O

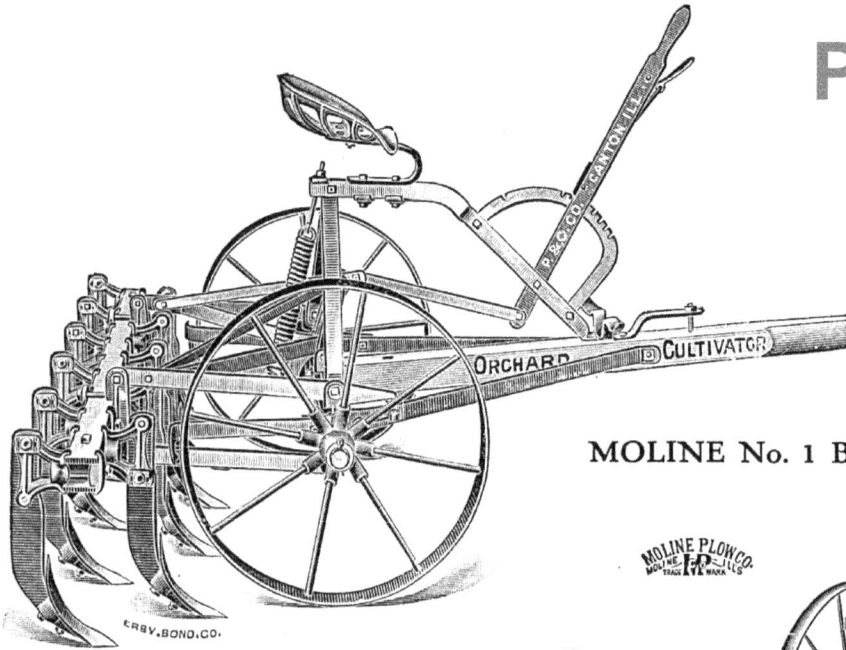

MOLINE No. 1 BEET CULTIVATORS

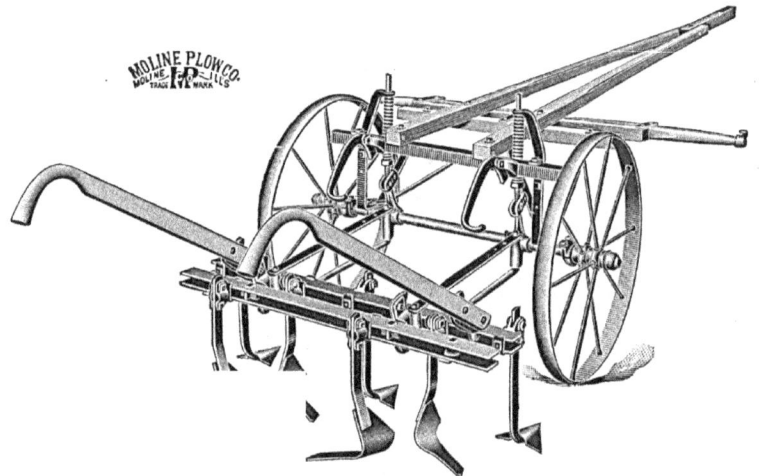

2-Row Without Riding Attachment.

Moline

P & O

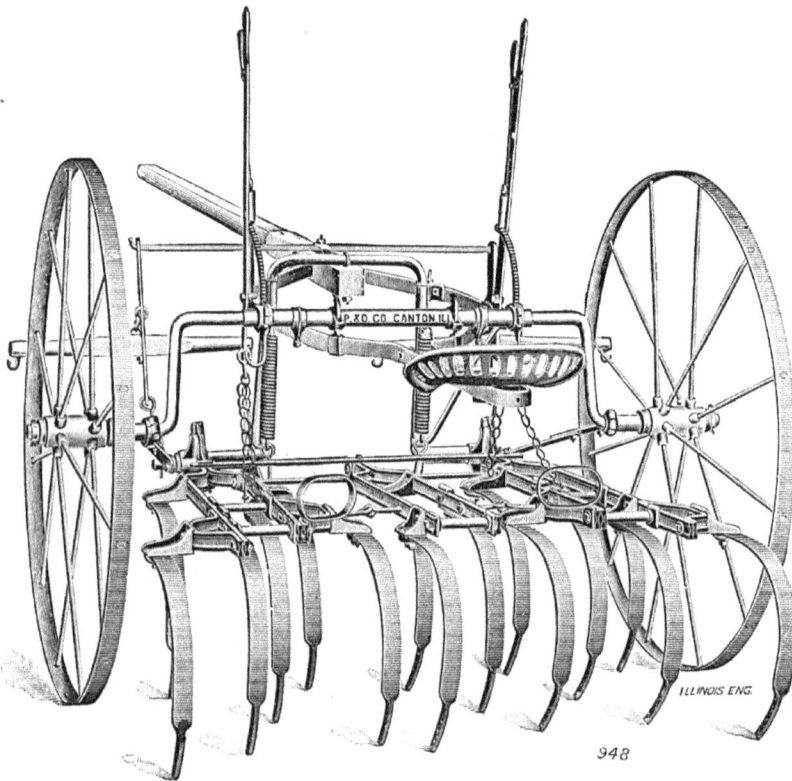

Victor Junior Cultivator, with Fifteen Tooth Gangs

MOLINE 2-ROW BEET CULTIVATOR No. 4

Moline

Moline Vineyard Hoe.
One horse was hitched to the right side of the pole. The right handle was hinged to move hoe closer or away from row. Left handle was hinged separate from the right and served to angle disc blade steering hoe.

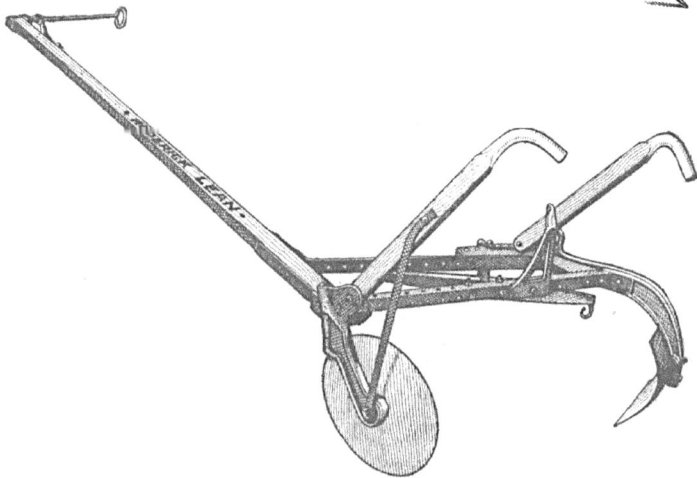

Roderick Lean

WHEEL TYPE

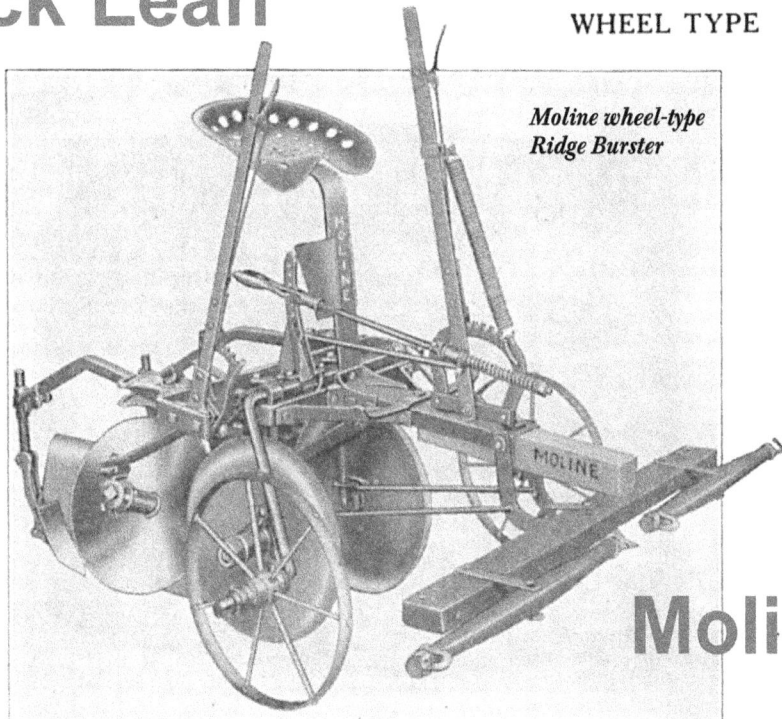

Moline wheel-type Ridge Burster

Moline

Oddities

The Improved Patent Tiger Protection Spring Tooth Harrow.

Steel Frame, Never Wears Out, Never Bends or Breaks.

EVERY HARROW A GUARDED HARROW.

The Steel "KING"

Spring Tooth Combined Riding and Walking Harrow, Corn Cultivator and Seeder.

The finest and most complete machine of its kind ever put on the market. All parts are MADE OF STEEL, excepting pole and axle. It has FIFTEEN teeth and cuts all ground, including wheel track, and weighs less than other harrows with twelve teeth only. It has a patent swing seat attachment which prevents jolting and jerking.

It is a pleasure to ride on it. Send for descriptive circular.

THE LAWRENCE & CHAPIN CO.,
Sole Makers. Kalamazoo, Michigan.

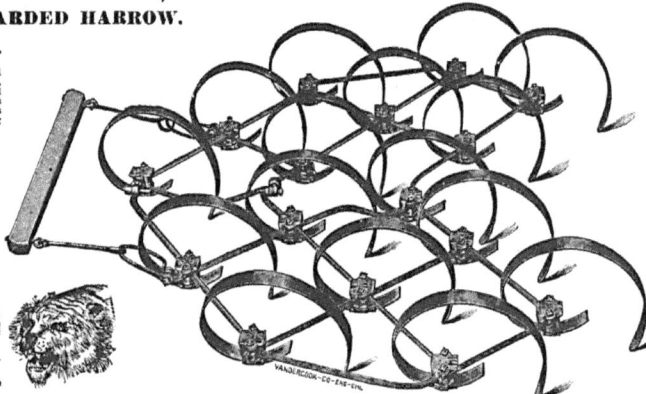

Licensed by G. B. Olin & Co. and D. C. & H. C. Reed & Co., under Patent Fastening Sustained by Judge Gresham, December 22, 1888.

More than 20,000 of these frames already in use. Frames made of same material as teeth. Every Tooth Holder of tempered steel, and forming a shoe, which keeps the frame up from the ground, thereby saving one horse in draft in most soils.

Sold as low as inferior harrows. Wood-frame Harrows made with same Teeth and Holders.

Send for circulars and prices.

Mention this paper.

The Lawrence & Chapin Co., Sole Makers, Kalamazoo, Mich.

OUR NEW STANDARD SPRING TOOTH HARROW {THOUSANDS SOLD in 1888

STANDARD SPRING TOOTH HARROW.

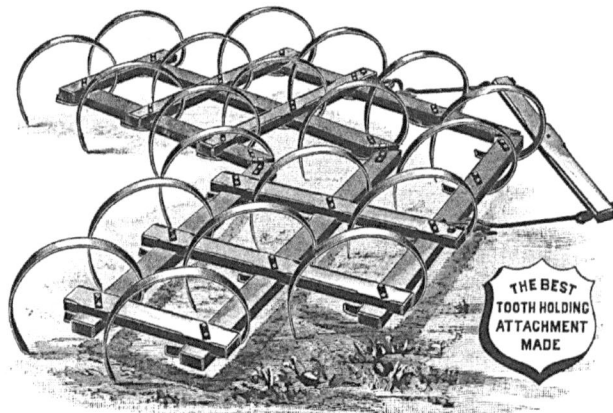

THE BEST TOOTH HOLDING ATTACHMENT MADE

Tooth or Tooth-Holder cannot slip or work sideways.
No Strain on Bolts.
Bolt Heads Fully Protected with Steel Clip.
Our Harrows are Well Made and First-Class in Every Respect.
Guarded Harrows for Stumpy Land.
Iron-Plated Harrows for Stony Land. (Plates extending full length of frame, fully protecting it.)
Agents Wanted in Every Town.
Write for Prices, Terms, Etc.
We also continue to manufacture our well-known Perfection Harrow, Tooth fastens on top of frame, and the New Clipper Spring Harrow.

LICENSED

To MANUFACTURE and SELL under the **REED** Patent,

By G. B. Olin & Co., of Canandaigua, N. Y., and D. C. & H. C. Reed & Co., of Kalamazoo, Mich., who are the exclusive owners of above patent for all of the United States. This patent does not expire until 1895, and has been sustained by several United States Circuit and District Judges. The last decision was rendered at Chicago, Dec. 22, 1888, by the Hon. Walter Q. Gresham, sustaining the patent, and an injunction was ordered to issue against a dealer in Agricultural Implements in Indiana who was selling Spring Tooth Harrows manufactured by Ira J. Hunt, of Kalamazoo, Mich. The license which we have taken not only protects ourselves but our customers as well. We have also settled all back royalties. We have succeeded to the business of Chase, Henry & Co., and F. J. Henry & Co., and all Harrows sold by them were included in our settlement with the owners of the above Patent. We want AGENTS in every town in the United States. Write for Price List, Terms, etc.

CHASE, TAYLOR & CO., Mfrs., - KALAMAZOO, MICH.

THE NEW REED IMPROVED TWO-HORSE CORN CULTIVATOR.

OUR IMPROVED No. 10 FOR 1889.

MANUFACTURED BY
D. C. & H. C. REED & CO., Kalamazoo, Mich.
ONE of the BEST SPRING-TOOTH CORN CULTIVATORS in the MARKET
With the Third Section a Complete Fallow Cultivator.
WE ALSO CLAIM FOR IT
A First-Class Bean Cultivator.
And Fourth, by attaching the Seed Box makes it
A PERFECT BROADCAST SEEDER.
Every Machine Guaranteed. Send for Circulars and Prices.

The Original Kalamazoo Spring-Tooth Harrows.
Bolt Heads fully protected by Channelled Clips. Its reputation recommends it to every one wanting a first-class Harrow.

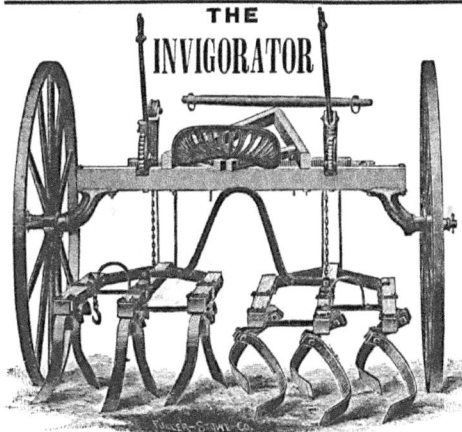

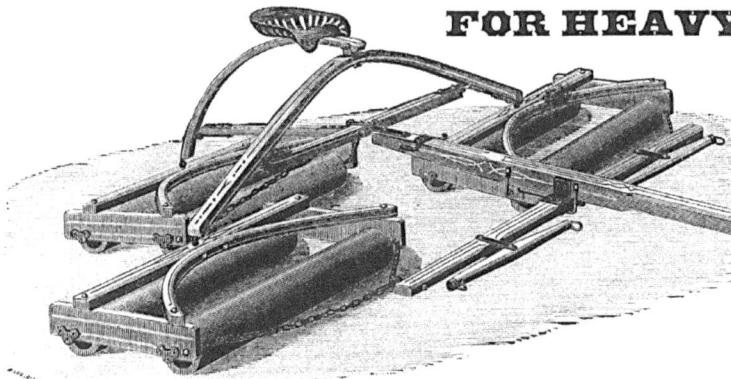

Oddities

In Closing

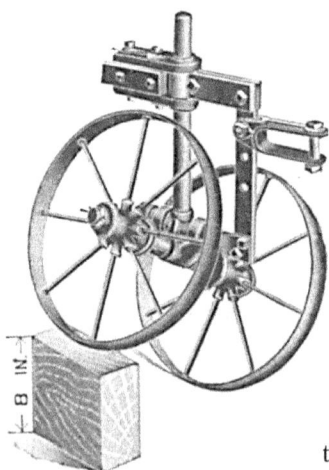

Looking back over this book, and over the time and experience of creating this book, I am struck by the dichotomy, the opposing halves, the warring sentiments, of assessed valuation. I think this volume is both important and valuable while at the same time being silly and vastly incomplete.

A few years back I had the grandiose notion of compiling, in one book, all the information I had gathered and would continue to gather, on horsedrawn farming implements from the recent (or accessible) past and present. I organized the idea into chapters which I felt followed a reasonable view of a season's work, progressing from plowing through tillage and seeding to harvest. My first mistake, or good piece of luck, was to announce to a few folks my intentions. What followed was a parade of stated urgencies; "I need the plowing chapter right away!" "When can you get to the haying bit, don't make me wait!" These admonitions - dovetailed into the discovery that an effort at a complete book, on plows alone, would run to well over a thousand pages - forced me to give up on the one big book and divide the project into sections. So the plow book and the haying book were born, and well received. This book is, was, not so urgently sought after, but it needed doing. If I continue with the series there "might" be another six to ten volumes. We shall see. As for this one I realize that it is not so much a book as it is a catalog of hardware images. It is our sincere wish that it serves a few of you in your interests and pursuits. We are confident that it serves one of our main concerns, that of preservation.

They may be seen today by many as 'relic' technologies but we, the many, forget them at our own peril because they represent a 'bridge' in the history of our

MAKING FARM LIFE EASIER

Rock Island Implements

working as farmers. A bridge we may have to return to, and perhaps even cross - repeatedly. When we can no longer afford the fancy complex computerized implements, when we can't find parts for them because of their planned obsolescence, when we can't afford the fuel, we'll appreciate having access to a simple and accessible set of tools for getting the farm work done.

We also know very well that, to a large group of people, the contents of this book already represents, **today,** 'appropriate' and/or 'transitional' technologies - stuff we can use, stuff we can learn from, stuff we still get excited about. We don't need to sell the notion about this way of working. Those inclined to it will find it on their own. And for those of you who sometimes feel awkward when the general populace makes fun of us for choosing to work horses I say let them laugh, it can't hurt us. Not while we're so busy enjoying our work and feeling the warmth of the bloom on our lives.

Glad to put this book on the shelf.

Be thinking about you all while I'm discing my field. LRM

Index

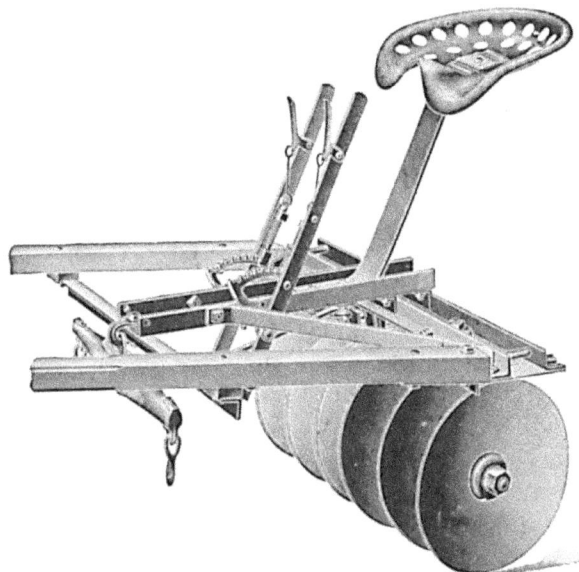

Bibliography

Bailey, L. H., The Principles of Agriculture, A Text-Book for Schools and Rural Societies, The MacMillan Company, New York, NY, 1912.

Burkett, Charles William, Agriculture For Beginners, Ginn & Company, Boston, MA, 1902.

Cook, G.C., Scranton, L.L., McColly, H.F., Farm Mechanics Text and Handbook, The Interstate Printers & Publishers, Danville, IL, 1951.

Corbett, Lee Cleveland, Garden Farming, Ginn and Company, Boston, MA, 1913.

Crozier, William and Peter Henderson, How the Farm Pays, Peter Henderson & Co., New York, NY, 1902.

Farm Practices, Plowing, Harrowing, Planting, Cultivating for Quality Crops & Larger Profits, Allis-Chalmers Manufacturing Co., Tractor Division, Milwaukee, WI.

I.C.S. Reference Library, Farm Crops, Tobacco, Soiling Crops, International Textbook Company, Scranton, PA.

International Harvester Company, TM-1 Parts Catalog for Tillage Machines, Chicago, IL.

John Deere, The Operation, Care and Repair of Farm Machinery Eighth Edition, Published by John Deere, Moline, IL.

John Deere-Van Brunt Repair Catalog No. 130-M, Van Brunt Mfg. Co., The, Horicon, WI, 1929.

John Deere-Van Brunt Repair Catalog No. 132-M, Van Brunt Mfg. Co., The, Horicon, WI, 1940.

McCormick-Deering Tillage Implements Repairs Catalog No. 25-TM, International Harvester Company of America Inc., Chicago, IL.

McCormick-Deering Tillage Implements Repairs Catalog No. 27-TM,

International Harvester Company of America Inc., Chicago, IL.

McCormick-Deering Tillage Implements Repairs Catalog No. 28-TM, International Harvester Company of America Inc., Chicago, IL.

McLennan, John, PH.M., A Manual of Practical Farming, The MacMillan Company, New York, NY, 1913.

Moline Flying Dutchman Farm Implements General Catalog No. 60, Moline Implement Company, Moline, IL.

Operation, Care, and Repair of Farm Machinery Tenth Edition, John Deere, Moline, IL.

Parlin & Orendorff Co., Catalogue No. 59, Dean & Co., Minneapolis, MN.

Parlin & Orendorff Co., Canton Plows, Inland Press, Chicago, IL.

R. Herschel Manufacturing Co. Catalogue No. 79, Peoria, IL, 1927.

Ramsower, Harry C., Equipment for the Farm and the Farmstead, Ginn and Company, Boston, MA, 1917.

Seymour, E.L.D., B.S.A., Farm Knowledge, Volume III: Farm Implements and Construction, Doubleday, Page & Company, Garden City, NY, 1919.

Shopwork On The Farm, McGraw-Hill Book Company, Inc., 1945.

Smith, Harris Pearson A.E., Farm Machinery and Equipment, McGraw-Hill Book Company, Inc., New York, 1937.

Taylor, Dr. W. E., Soil Culture and Modern Farm Methods, Deere & Webber Company, Minneapolis, MN.

United States Commissioners, Reports of the, to the Paris Universal Exposition 1878, Published under direction of the Secretary of State by Authority of Congress, Volume V., Washington: Government Printing Office, 1880.

Warren, G. F., Elements of Agriculture, The MacMillan Company, New York, NY, 1914.

Manufacturers

B. W. Macknair & Son
3055 US Highway 522 North
Lewiston, PA 17044
717-543-5136

Beiler's Machinery
609 Musser School Road
Leola, PA 17540
717-656-9733

I & J Manufacturing
10 South New Holland Road, Suite 2
Gordonville, PA 17529
717-442-9451

Pioneer Equipment Inc.
16875 Jericho Road
Dalton, OH 44618
330-857-6340

White Horse Machine
5566 Old Philadelphia Pike
Gap, PA 17527
717-768-8313

Lynn Miller, born in Kansas City in 1947, is the founder, editor, and publisher of the award winning Small Farmer's Journal (established in 1976). He has a couple of college degrees hanging on the wall. He is a farmer, painter and writer who has authored 20+ books (these include agricultural titles plus essays, poetry, and fiction). He has three surviving adult children and five grandchildren and currently lives and works with his wife on their remote central Oregon cattle and horse ranch.

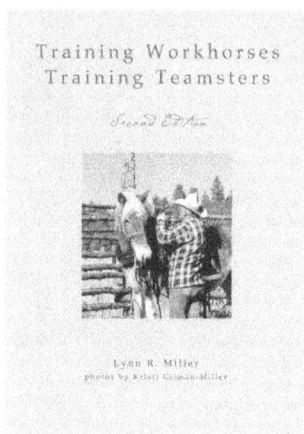

Training Workhorses / Training Teamsters. This book covers such subjects as understanding horses, training horses to work in harness - on the farm, in the woods, and on the road - correcting behavior problems with work horses, care & feeding of work horses, training tips for every age of horse, from imprinting newborn foals to starting older horse, harness and bits, harnessing, the dynamics of pulling, learning to drive horses and driving dynamics.

©2018. 352 pages, hundreds of color photos & illustrations.
$60 softcover

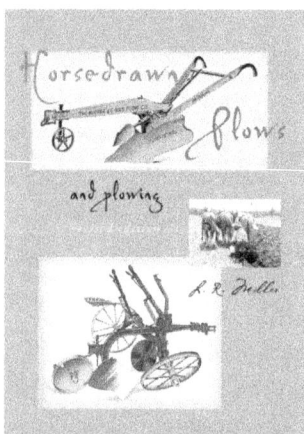

Horsedrawn Plows and plowing. With over 1,000 drawings and photos covering how to plow with horses using older equipment and new implements. Here you will find simple diagrams explaining tricky adjustments for both riding and walking plows. Detailed engineer's drawings of John Deere, Oliver, McCormick Deering, Parlin and Orendorff, Avery, and many other older manufacturers will be immensly helpful to folks restoring equipment. Also includes closeup photos and information on new makes of animal-drawn plows including Pioneer and White Horse.

©2018. 368 pages, photos & illustrations.
$45 softcover

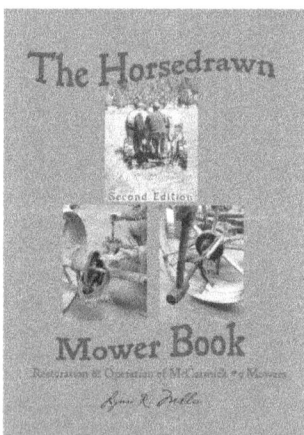

The Horsedrawn Mower Book. With hundreds of photos and drawings, the profusely illustrated text covers restoration, rebuilding, repair, and tune-up with a focus on the very popular McCormick Deering (International) No. 9. It also includes references to other makes and models as well as resource information for updating cutter bar assemblies to new materials and functions.

©2018. 352 pages, hundreds of photos & illustrations.
$45 softcover

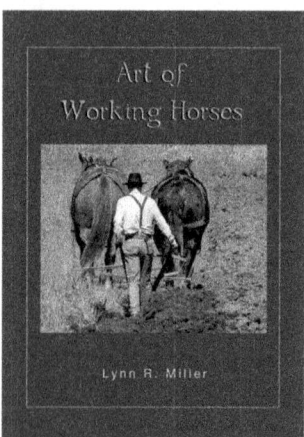

Art of Working Horses follows on the heels of his other seven Work Horse Library titles. This book tells the inside story of how people today find success working horses and mules in harness, whether it be on farm fields, in the woods, or on the road. Over 500 photos and illustrations accompany an anecdote-rich text which makes a case for the future of true horsepower.

©2016. 368 pages, over 500 photos & illustrations.
$65 hardcover, $45 softcover

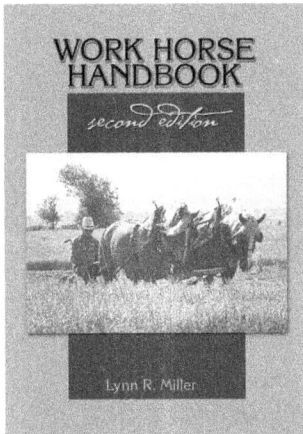

Work Horse Handbook. The original has become a classic and the standard reference on working horses. The second edition expands that reputation with hundreds of new drawings and photographs to illustrate the expanded text. From care and feeding thru hitching and driving: every aspect is covered. New subjects and material include; extensive equipment coverage including forecart advances, synthetic harnesses, new hitch gear apparatus, additional training and procedural information, care and feeding additions, additional breeds coverage, plus an expansive source directory.

©2015. 386 pages, hundreds of photos & illustrations.
$45 softcover

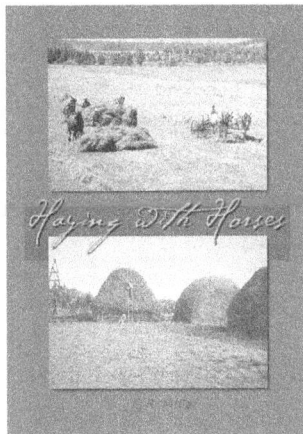

Haying With Horses. To our knowledge this is the first comprehensive text on the subject. A new practical reference text with 1,000 illustrations covering all aspects of Haymaking with Horses and Mules in harness. Offering in-depth information on Mowers, Rakes, Hayloaders, Buckrakes, Stackers, Tracks and Trollies for barns, Hay Fork systems, Balers, Wagons, Feed Sleds, and Forecart adaptations etc. Unloading systems, and feeding systems are also covered.

©2000. 368 pages, 1,000 illustrations.
$45 softcover

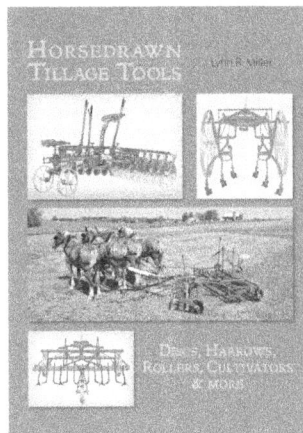

Horsedrawn Tillage Tools. This important book covers operation, care and repair of animal-powered cultivators, field cultivators, discs, harrows, ridge busters, listers and rollers as well as seedbed preparation and crop cultivation tools. It includes specifics on how animals are hitched and driven with these implements. Detailed engineer's drawings of John Deere, Oliver, McCormick-Deering, Parlin-Orendorff, Avery and many other older manufacturers will be immensely helpful to folks restoring equipment. This cornerstone of the Work Horse Library has over 1,000 photos and illustrations covering the new and old, the useful and the historic.

©2001. 368 pages, photos & illustrations.
$45 softcover

Also by Lynn R. Miller
Brown Dwarf
elastic signature: notes on painting
Talking Man
Old Man Farming
Why Farm
The Glass Horse
Farmer Pirates & Dancing Cows
thought small: poems & drawings

For many years Anne & Eric Nordell have contributed articles to *Small Farmer's Journal* about (among many other subjects) tillage and cultivation processes and equipment under the heading "Cultivating Questions." Please reference this partial list:

*out of print

To order back issues ($15 ea):
Small Farmer's Journal
PO Box 1627, Sisters, OR 97759
800-876-2893 • 541-549-2064
www.smallfarmersjournal.com

www.ingramcontent.com/pod-product-compliance
Lightning Source LLC
Chambersburg PA
CBHW081801200326
41597CB00023B/4108